Springer Water

The book series Springer Water comprises a broad portfolio of multi- and interdisciplinary scientific books, aiming at researchers, students, and everyone interested in water-related science. The series includes peer-reviewed monographs, edited volumes, textbooks, and conference proceedings. Its volumes combine all kinds of water-related research areas, such as: the movement, distribution and quality of freshwater; water resources; the quality and pollution of water and its influence on health; the water industry including drinking water, wastewater, and desalination services and technologies; water history; as well as water management and the governmental, political, developmental, and ethical aspects of water.

More information about this series at https://link.springer.com/bookseries/13419

Omid Bozorg-Haddad
Editor

Climate Change in Sustainable Water Resources Management

Springer

Editor
Omid Bozorg-Haddad
Irrigation and Reclamation Engineering
University of Tehran
Karaj, Iran

ISSN 2364-6934 ISSN 2364-8198 (electronic)
Springer Water
ISBN 978-981-19-1897-1 ISBN 978-981-19-1898-8 (eBook)
https://doi.org/10.1007/978-981-19-1898-8

This Springer imprint is published by the registered company Springer Nature Singapore Pte Ltd.
The registered company address is: 152 Beach Road, #21-01/04 Gateway East, Singapore 189721, Singapore

Preface

One of the most important and complex concerns of the present and future centuries is climate change. The phenomenon of climate change is not confined to our era, and evidence indicates that the Earth has faced climate change in different periods, although this issue in the present era attracts the attention of many of the world's scientific and political societies. Any change in the climate can be due to natural climatic change or the result of human activities. Such a change can occur in temperature, rainfall, humidity, weather patterns, wind, radiation, and other climatic variables. However, even a small change in the climate conditions can lead to tangible changes in hydrology and the sensible water resource systems. The impacts of climate change on water resources include the changes in rainfall and runoff patterns, sea level, land use, water demand, and many other aspects. The negative effects of this phenomenon on water resources can cause irreparable damages. Thus, identification of these effects and their reasons is essential. Investigation of such changes can facilitate the studies on a series of hydroclimatic, economic, and social problems such as droughts, floods, foods, and human migration.

In this book, to gain a comprehensive understanding of climate change, water experts can benefit the content of three separated sections according to their background. In the Part I, basic concepts of climate change, its natural and anthropogenic drivers, and its effects on water resources, from both quality and quantity aspects, are discussed. Results of the accredited researches show that greenhouse gases take the first place in the consistent growth of the radiative forcing and the Earth's energy level since the 1950s. Increasing temperature and decreasing precipitation have reduced the discharge of surface runoffs in different parts of the world, which leads to a decline in the level of groundwater aquifers. In Part II, climatic scenarios and IPCC reports are discussed. In order to analyze future climate conditions in water resource studies, we explain several categories of climate models, including Energy Balance Models (EBMs), radiative-convective models, and General Circulation Models (GCMs). Being most comprehensive, GCMs have been widely applied for future climate change projections using different scenarios of population growth, greenhouse gas emissions, and land-use changes. As the final step in climate change studies, downscaling method, including statistical and dynamic, is explained thoroughly. The last Part focuses on water resource modeling under clime change conditions. To broaden the viewpoint of experts on available mitigation and adaptation measures, numerous case studies of all continents on the earth are presented. Generally, this book is arranged in a manner to meet the need of readers with diverse background through choosing the most appropriate section.

Karaj, Iran Omid Bozorg-Haddad

Contents

Chapter 1
Overview of Climate Change in Water Resources Management Studies

Omid Bozorg-Haddad, Saba Jafari, and Xuefeng Chu

1.1 Introduction

In general, climate change takes place whenever the climate condition in a region changes compared to its long-term behavior (Karamouz and Araghinejad 2005). The rise of global temperatures and the changes in precipitation patterns are the effects of climate change, which lead to a decrease in streamflow. Global climate change in the future also is a threat to the world's water resources, which makes it difficult to access these resources.

Mirza et al. (2003) examined the impacts of global warming and climate change on the possibility of flooding of the Ganges–Brahmaputra–Meghna (GBM) river basin in Bangladesh using four General Circulation Models (GCMs). The output of the GCM models was used as an input of the Mike11-GIS hydrological model. Their results showed an increase in the average of maximum discharges in the GBM, which potentially led to flooding.

Rosenzweig et al. (2004) assessed the influences of the changes in agricultural water demand and the availability of water induced by climate change on irrigation reliability using a series of models, including the Water Balance (WATBAL) model for water supply, the Ceres-Maize Index Model, the Soygro model, and the Cropwat

O. Bozorg-Haddad (✉) · S. Jafari
Department of Irrigation and Reclamation Engineering, Faculty of Agricultural Engineering and Technology, College of Agriculture and Natural Resources, University of Tehran, 31587-77871 Karaj, Tehran, Iran
e-mail: OBHaddad@ut.ac.ir

S. Jafari
e-mail: Saba.Jafari@ut.ac.ir

X. Chu
Department of Civil and Environmental Engineering, Dept. 2470, North Dakota State University, Fargo, ND 58108-6050, USA
e-mail: Xuefeng.Chu@ndsu.edu

O. Bozorg-Haddad (ed.), *Climate Change in Sustainable Water Resources Management*, Springer Water, https://doi.org/10.1007/978-981-19-1898-8_1

model for product operation, and the Water Evaluation and Planning (WEAP) model for water demand forecasting, evaluation, and planning. These models were used in conjunction with different climate change scenarios to evaluate the adaptation strategies of water resources in major agricultural areas in Argentina, Brazil, China, Hungary, Romania, and the United States. The results showed that even in relatively water-rich areas, the effects of global warming on agricultural water demand and the increased water demand due to urbanization would require improved irrigation technology and water management.

Zhao et al. (2005) investigated the response of climatic variables to greenhouse gas emissions in South Africa by using the outputs of three GCMs and B2 scenarios from the set of Special Report on Emission Scenario (SRES). The simulations for most parts of South Africa indicated that by the end of the twenty-first century, rainfall would drop by 8.2%. Bates et al. (2008) found from their study on the Tsha Rolpa Glacier Lake in Nepal that due to rising temperatures in recent years and the melting of the lake's glaciers, the lake's water area increased from 0.23 km^2 in 1957 to 1.65 km^2 in 1997, which added about 100 million m^3 of water to the lake. Since the water in the lake was maintained only by its icy masses, the increase in this volume elevated the risk of devastating floods.

By analyzing the effects of climate change on river flow in Northern China, Zhenmei et al. (2008) also examined the trend of annual flow changes in the past 50 years using the Mann–Kendall test. They found that climate change reduced the average annual flow rate by 64% and decreased the amount of rainfall. Traynham et al. (2010) assessed the effects of climate change and population growth on the water resources system of the Puget region in the United States. The response of the water resources system to future water demands under the influence of climate change was examined for different cities in the Puget area. Three GCMs and two emission scenarios were used to assess the region's water supply over a 75-year period. The performance of the water supply system in each city was determined by the reliability and firm yield criteria. The results showed that climate change would reduce the future system's yield, and the existing operation policies needed to be changed to meet the future demand.

Tabari et al. (2011) studied the average, maximum, and minimum annual trends of rainfall and temperature series at 13 stations in western, southern, and southwestern Iran for the period of 1966–2005 to understand how global warming affected the regional climate. Their results showed that the average, maximum, and minimum annual temperatures increased 0.412, 0.452, and 0.493 °C per decade, respectively, while the rainfall time series exhibited different changing patterns (i.e., increasing–decreasing trends) throughout the region. Their study also indicated that there was a need for more research on human impacts on the environment as a factor of climate change.

Georgakakos et al. (2012) assessed the adaptive reservoir management strategies under climate change in the Central Valley of Northern California. The reservoir management assessment included the adaptive policy with the developed Inform Decision Support System (INFORM DSS) and the current policy obtained from the Department of Water Resources Planning Simulation (DWRPS). Two sets of hydrological data were compared for basic and future periods. The results showed that the adaptive management under both basic and future conditions was effective in mitigating the adverse effects of climate change and reducing the system's vulnerability.

Ashofteh et al. (2012) evaluated the effect of climate change on the reservoir inflow and the tail water demand by utilizing three scenarios in East Azerbaijan, Iran. The monthly series of temperature and rainfall were achieved from the third version of the Hadley Centre Coupled Model (HadCM3) under the A2 scenario (Mendoza-Ponce et al. 2021). The obtained data were input to the recognition of UHs[1] and IHACRES.[2] A monthly runoff series was simulated for the period of 2026–2039, and the required water volume was estimated. The simulation of reservoir operation with the WEAP showed that the reliability for the future period decreased by 4% compared to that for the base period (1987–2000) and that the vulnerability and flexibility increased by 38% and 4%, respectively. The optimal operation of the reservoir was also determined by LINGO, which showed that in the future, the reliability decreased by 21% and, in contrast, the vulnerability and flexibility increased by 31% and 14%, respectively.

Alvarez et al. (2014) examined the effect of climate change on the administration of some reservoirs in a basin in Quebec, Canada and the adaptation strategies. For this purpose, a method was developed, which combined the HESPs,[3] a SOM,[4] an ANN[5] model, and a WBM.[6] The RCM[7] was used for the base period (1961–2000) and the future period (2041–2070). Their results showed that providing adaptive solutions in the future would reduce the risk of flooding and the vulnerability of the system under climate change.

Zareian et al. (2014) examined the effects of climate change on the inflow of the Zayandeh-Rud reservoir in Iran by using different various weighting approaches. The AOGCM15 was selected to study the effects of climate change on temperature and precipitation of the upstream area in 2015–2074. The modelling results were compared with the observation data in 1971–2000, and three patterns of climate change (ideal, moderate, and critical) were determined. The meteorological data were downscaled for 2074–2015 with the Lars-WG model, and then the IHACRES hydrological model was used to predict the monthly inflows of the reservoir for different climate change patterns, which were further used for the optimization of the downstream water resources.

Ahmadi et al. (2015) studied the rules of adaptive operation of Karoon-4 Reservoir in Iran with climate change. The HadCM3 model under Scenario A2 was used to predict the future temperature and precipitation, and the IHACRES model was also

[1] Unit Hydrograph.

[2] Identification of Unit Hydrographs and Component Flows from Rainfall, Evaporation and Streamflow Data.

[3] Hydrological Ensemble Streamflow Prediction.

[4] Stochastic Optimization Model.

[5] Artificial Neural Network.

[6] Water Balance Model.

[7] Regional Climate Model.

used to simulate the incoming flow to the reservoir. The rules of operation were extracted by the Non-Dominated Sorting Genetic Algorithm (NSGA-II), and the reservoir was operated adaptively and non-adaptively to climate change. The results showed that adaptive operation with climate change increased the reliability and reduced the vulnerability related to hydropower generation.

Moursi et al. (2017) evaluated the effects of climate change on water deficit in the Sevier River basin in Utah using a decision-scaling framework. In their study, 31 general circulation models (GCMs) were used for predictions to assess the vulnerability of water deficit in 2000–2099, which was defined by using an index to measure the ratio of available water to the agricultural water demand predicted by the AquaCrop model. The results revealed that the GCMs showed an increase in available water for agriculture in the basin and also indicated a significant risk of agricultural water deficit in 2025–2049 with the RCP4.5 emission scenario, which emphasized the need for adaptive strategies.

Khare et al. (2017) evaluated the effect of climate change on soil erosion in the Mandakini Basin in India. The rise of rainfall intensity, apart from other factors such as land-use change, led to an increase in soil erosion. In their study, the future precipitation generated by downscaling the GCMs data was used to determine the impact on soil erosion, which was estimated by the Universal Soil Loss Equation (USLE). They found that soil erosion in the future exhibited an increasing trend due to an increase in precipitation.

1.2 Symbols and Definitions

ANN: Artificial Neural Network.
AOGCM: Atmosphere–Ocean General Circulation Model.
AR4: Fourth Assessment Report of IPCC.
AR5: Fifth Assessment Report of IPCC.
CMIP: Coupled Model Inter-Comparison Project.
CMWG: Climatic Models Working Groups.
CRU: Climatic Research Unit.
DSS: Decision Support System.
DWRPS: Department of Water Resources Planning Simulation.
FAR: First Assessment Report of IPCC.
Flexibility: A combination of other performance criteria [Reliability × Resiliency × (1-Vulnerability)].
GCMs: General Circulation Models.
GHG: Greenhouse Gases.
GHGES: Greenhouse Gas Emission Scenarios.
GIS: Geographic Information System.
HadCM3: Third Version of the Hadley Centre Coupled Model.
IHACRES: Identification of Unit Hydrographs and Component Flows from Rainfall, Evaporation and Stream flow Data

IPCC: Intergovernmental Panel on Climate Change.
LARS-WG: Long Ashton Research Station Weather Generator.
RCM: Regional Circulation Model.
RCPs: Representative Concentration Pathways.
Resiliency: How quickly the system recovers from failure.
RF: Radiative Forcing.
SAR: Second Assessment Report of IPCC.
SDM: System Dynamics Model.
SRES: Special Report on Emission Scenarios.
TAR: Third Assessment Report of IPCC.
UNEP: United Nations Environment Program.
USLE: Universal Soil Loss Equation.
Vulnerability: How severe the consequences of failure may be.
WATBAL: Water Balance Model.
WEAP: Water Evaluation and Planning System.
WGEN: Weather Generator.
WGMS: World Glacier Monitoring Service.
WMO: World Meteorological Organization.

1.3 Significance of the Study

Climate change is a serious threat to human societies, which is potentially irreversible. According to the Intergovernmental Panel on Climate Change (IPCC) (Mandal 2017), the Earth will warm and its surface temperature could rise 0.6 °C during the twentieth century, and based on the greenhouse gas emissions estimates, the temperature will increase 1.0–3.5 °C by 2100. By the end of the twenty-first century, global warming will be greater than it has occurred in the last 10,000 years (IPCC 2001). Undoubtedly, in the coming years, due to the increase in human activities, greenhouse gas emissions will increase, which will intensify the change in climatic variables. On the other hand, even if the emission of greenhouse gases is stopped now, due to the long shelf life of greenhouse gases that have already been released into the atmosphere, humans would face climate change in the twenty-first century. Among the 16 most threatening factors for humans in the twenty-first century (e.g., food shortages, poverty, drought, floods, and nuclear weapons), climate change ranks first (Martin 2007). Studies show that countries in low latitudes will be disturbed further from the negative consequences of climate change (Lane et al. 1999). Climate change has affected economic growth in such a way that it can bankrupt up to one-fifth of the economy unless an effective measure is taken (Peston 2006). Generally, climate change in the future could have dissimilar effects on water resources, environment, industry, health, agriculture, and all other systems that interact with the climate system (Ashofteh 2014). Therefore, it is necessary to identify the negative effects of this phenomenon on the intended systems, and adaptive strategies should be anticipated to deal with them. Since the phenomenon of climate change and its effects on water resources are one of the most important challenges for water resources managers (Mendoza-Ponce et al. 2021), a number of relevant studies have been conducted since the late twentieth century. Various climatic models have been developed and used to simulate the affecting processes and to predict the future climate conditions for a variety of possibilities scenarios (Nouri-Tirtashi et al. 2015).

According to IPCC Technical Paper VI, the frequency and severity of borderline events such as droughts and floods are increasing due to climate change (Bates et al. 2008). The regions with higher forecasted rainfall are at a greater risk of flooding (Solomon et al. 2007). The climate change effects are also reflected in the changes in surface runoff and groundwater levels. It is estimated that the average change in runoff due to climate change is more than the precipitation amount at most, which is higher in arid regions than in humid areas (Lettenmaier and Gan 1990). Therefore, the increased intensity of extreme events such as droughts and floods induced by climate change is undeniable. Thus, examining the influences of climate change on water resources (Mandal et al. 2019a, b) is one of the necessities for water resources planning and management (Karamouz and Araghizejad 2005). Evidence of the influences of climate change shows that in addition to analyzing the occurrence of this phenomenon, special attention should be paid to its impacts on the water resources planning and management at a basin scale. This can be achieved by changing the way for data acquisition and processing based on the variability of basin data, paying attention to long-term scenarios affecting the change of basin water resources, and paying attention to extreme hydrological values. It is aimed to optimize available water resources. In this regard, it is important for the managers of water systems to provide necessary operation policies by considering the needs of consumers in the best way based on the available water resources. Instead of considering current conditions only, optimal adaptive policies that take into account climate change conditions should be adopted and implemented for the future.

1.4 Basic Principles

Climate is the average of climate conditions in a place, including its constituent factors such as temperature, rainfall, and humidity. In other words, it is the long-term average climate of an area (Bradley et al. 1985). Climate change is the change in climate that lasts for a long period of time (e.g., decades or longer). This term can be defined as any change in climate over time (IPCC 2007a, b, c). This change can be in the averages of temperature, rainfall, humidity, wind, radiation, and other variables. Climate can get hotter or colder, and the average of each variable can increase or decrease over time. This complex long-term global atmospheric-oceanic phenomenon can be influenced by natural factors such as volcanoes, solar activity, oceans, and atmospheres that interact with each other or by human activities (Buchdahl 1999).

1.5 Climate Change and Dangers Ahead

In recent decades, greenhouse gases (GHG), especially carbon dioxide (CO2), have increased dramatically due to the growth of industries and factories since the beginning of the industrial revolution and the increase in fossil fuel consumption, deforestation, and land-use change (Carter et al. 2007). Continuation of this trend and the increase of greenhouse gases have increased the average temperature of the Earth, resulting in global warming and changes in other climatic variables such as rainfall. That is why the most important factor in climate change is the increase of greenhouse gases. By changing climatic variables, other systems affected by these variables, such as water resources, agriculture, environment, and economy, will also change (Mearns et al 2010). However, this study will focus on water resources. According to various reports of the World Glacier Monitoring Service (WGMS), mountain glaciers have been exposed to melting since 2005, which is more than three times faster than the one in the 1980s. As a result, the average diameter of 30 glaciers in the world has been reduced by 60–70 cm. Studies show that rising water levels, along with the increase of glaciers melting, will exacerbate flooding (IPCC 2007a, b, c). Satellite data show that since 1978, the average number of polar glaciers has dropped by 2.7% every ten years. In addition, Rothrock et al. (1999) determined that the sea ice thickness has considerably decreased. As shown in Fig. 1.1, six different regions experienced a decrease in the thickness of sea ice between 1993 and 1997 compared to the data in 1958–1976.

Many studies on assessment of the effects of climate change indicate that many years between 1995 and 2006 are the warmest years in the global surface temperature record (since 1850). The updated 100-year temperature linear trend (1906–2005) of

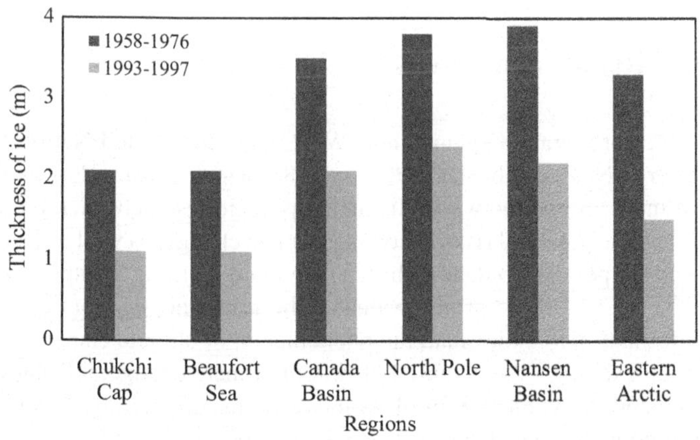

Fig. 1.1 Changes in thickness of the sea ice in six different regions (Based on Rothrock et al. 1999)

Fig. 1.2 Observed changes in global average temperatures 1850–2020 (Based on Jones and Palutikof 2006)

0.74 °C per year (0.56 to 0.92 °C) is greater than the corresponding trend for 1901–2000 provided in the TAR of 0.6 °C (0.4 to 0.8 °C) (IPCC, 2007a, b, c). According to the climatic research unit (CRU), Fig. 1.2 shows the observed changes in the global average temperatures (anomaly) in 1850–2020 (Jones and Palutikof 2006).

The effect of climate change on precipitation by examining the past statistics shows that precipitation increased in most areas in the world, especially in parts of South and North America, Northern Europe, and Central Asia from 1900 to 2005. Drought intensities also increased in parts of South Africa and parts of South Asia (IPCC 2007b).

1.6 IPCC Reports and Steps for the Forecast

The World Meteorological Organization (WMO) and the United Nations Environment Program (UNEP) established IPCC in 1988, which consisted of thousands of scientists from all around the world. Its main task is to research scientific and technical issues and the potential risks following climate change, as well as its impact on the world and set policies to deal with it. Three main working groups were formed by IPCC. The first working group evaluates the scientific, economic, social, and technical information to understand the phenomenon of climate change; the second working group examines the potential effects of climate change, adaptation to this phenomenon, and the vulnerability of various systems affected by it; and the third working group investigates the reduction of the effects of this phenomenon (IPCC 2001).

One of the main goals of this study is to predict the Earth's climate. Today, with the advancement of computers and the high speed and accuracy of their calculations, it is easier to make climatic predictions. In general, there are three significant steps to perform a climate forecast. However, it is worth noting that each of these steps has a very high diversity. The first step is to select a scenario for the Earth's forward state. In other words, at this stage, assumptions about the expected conditions for the Earth are considered, which take into account various topics such as the growth rate of the future population, the technological progress, and the emission of greenhouse gases. The second step, after selecting the scenario, is to simulate the Earth's climate. At this stage, the Earth's climatic behavior is defined in the form of a model and simulated according to the selected scenario in the first step as the initial conditions in the model. While the Earth's future climate can be predicted after these two steps, the results of these predictions, which are done at large scales in the atmospheric general circulation models, cannot be generalized to smaller scales (e.g., a basin scale). To solve this problem, it is necessary to take another step which is known as downscaling. In addition, it is essential to use precipitation-runoff models to estimate runoff in the study of climate change impacts on water resources. In the following section, after reviewing the IPCC reports, the scientific logic of each of these steps is examined, and some of the conventional methods and models for each step are described.

The IPCC has so far published five series of its reports: First Assessment Report (FAR), Second Assessment Report (SAR), Third Assessment Report (TAR), Fourth Report (AR4), and Fifth Report (AR5). The FAR was completed in 1990. One of the most important achievements of this report was its overall forecast for the average global temperature changing pattern up to 2005. In addition, the report emphasizes that greenhouse gas emissions are growing, and human activities increase the gases such as CO2, CH4, Nitrous Oxide (NO and NO2), and Chlorofluorocarbons (CFCs). This increase can cause the rise of temperature on a global scale. The SAR published in 1995 was a report on the socio-economic information at the time and its relationship to climate change. The report used a set of scenarios called IS92 for climate forecasting (IPCC 2013, 2014).

The final version of the TAR was prepared and published in 2001. It was commonly used as a reference to demonstrate a scientific agreement on global warming. The highlight of this report was the definitions of a total of 40 scenarios as a special report on SRES scenarios. This series of scenarios with slight changes have been used by climate researchers for nearly 13 years as the best and most complete scenarios.

The AR4 was a complete climatic report that had been published by that time. It has a scientific basis similar to that of the TAR. The biggest advance in the AR4 in comparison with the TAR and older versions is the better conception related to interpretations of past and present climate changes and consequently the better simulation of the future climate change relying more on the uncertainty in the models.

The decision to prepare the AR5 was made by IPCC members at its 28th meeting in Budapest, Hungary, on 9 and 10 April 2008. IPCC introduced a new procedure for preparing scenarios called Representative Concentration Pathways (RCPs), with the help of the scientific community in the climate change field. Thus, four concentration pathways of greenhouse gases (rather than emissions) were proposed. The difference between these four hypothetical pathways was the imaginal amount of the Radiative Forcing (RF) for them. A brief overview of the concept of RF is provided in the section regarding the concepts and theories used in the IPCC scenarios.

1.6.1 IPCC Scenarios (from SERS to RCPs)

1.6.1.1 SERS

Undoubtedly, in the coming years, humans will face more changes in climate variables due to the increase in greenhouse gas emissions. As noted, any changes in the concentration of greenhouse gases in the Earth's atmosphere will change the balance between the elements of the Earth's climate system (Mandal 2017). However, it is not clear how much gases will enter the Earth's atmosphere by human societies in the future and what will happen to the Earth's climate system. Therefore, they are presented in a completely uncertain way and under different emission scenarios (Ashofteh 2014). These scenarios are divided into Non-Climatic Scenarios and Climatic Scenarios, as detailed below.

- **Non-climatic Scenarios**

The information on greenhouse gas emissions, population growth, economic and social status, as well as technological and agricultural issues is the basis of non-climatic scenarios. A number of emission scenarios were introduced by IPCC in 1996 in the Special Report on Emission Scenario (SRES).

SRES defined four main titles, including A1, A2, B1, and B2, as families of emission scenarios to describe the relationship between greenhouse gas emissions and airborne particles and their outcomes in different parts of the world. Figure 1.3 shows these four titles (Nakicenovic et al. 2000).

Figure 1.3 reflects the emphasis of each scenario group on the global, regional, environmental, and economic perspectives. In summary, scenarios A1 and A2 consider more economic issues, while scenarios B1 and B2 focus on environmental issues. Scenarios A2 and B2 consider regional solutions, and scenarios A1 and B1 target global solutions (IPCC 2007c). Each family of the scenarios offers different conditions (e.g., environmental, social, and industrial conditions) and consists of different groups. Three different groups for the family of A1 are considered based on the types of industry used in the twenty-first century:

A1F1: Intensify the use of fossil fuels.
A1T: Use non-fossil energy sources.
A1B: Use fossil and non-fossil resources in a balanced way.

In total, these scenarios are divided into six groups: A1F1, A1T, A1B, A2, B1, and B2, each of which is further divided into several sub-branches. They form 40 different scenarios. It should be noted that based on the production of radiative forcing, the status of the scenarios is A1F1, A2, A1B, B2, A1T, and B1 (IPCC 2000).

- **Climatic Scenarios**

In general, climatic scenarios can be defined as the detection of the extent and manner of the change in climatic variables which will happen in the future periods and on a regional scale (Ashofteh 2014).

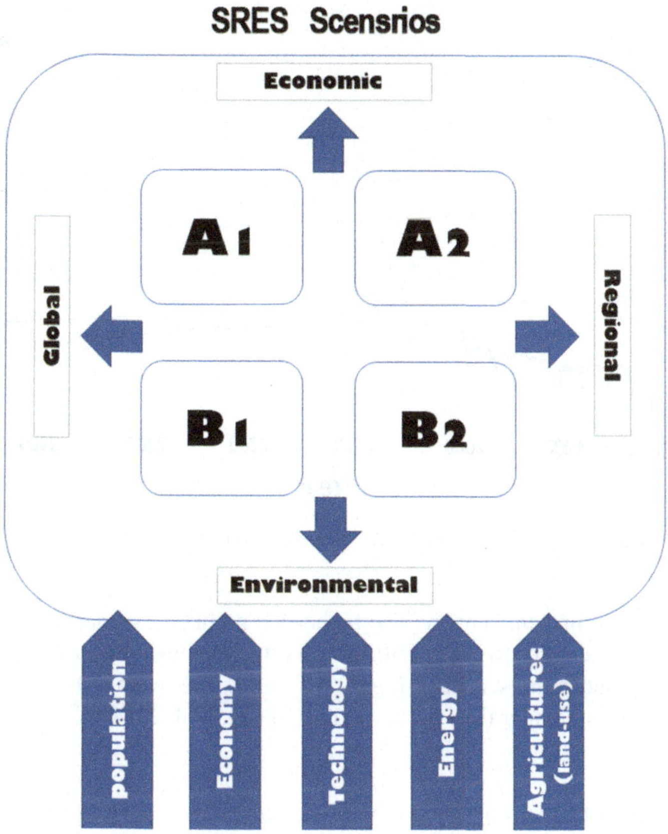

Fig. 1.3 Four families of emission scenarios and their triggers (Based on Nakicenovic et al. 2000)

1.6.1.2 RCP

In 2014, the scientific community decided to design a set of scenarios to consider not only the greenhouse gas concentration but also other parameters such as land use in these new scenarios. The goal is to use advanced models in parallel for various scenarios that could justify the set of climate stimuli, which help the development and definition of possible routes for the future of the Earth's climate. This is exactly the opposite of the process that was used to define the older IPCC climate scenarios previously (Zolghadr-asli 2017).

In these scenarios, a set of RF components are used and applied in chain processes in climate forecasting. In fact, the deliberate use of the term of "concentration", instead of "emission", suggests that emission is just one of the outputs of these scenarios. Finally, four general pathways are expressed for the rates of RF: 8.5, 6.0, 4.5, and 2.6 w/m^2, represented by RCP 8.5, RCP 6, RCP 4.5, and RCP 2.6, respectively (Mandal 2017). In fact, the simulation has been performed for each case

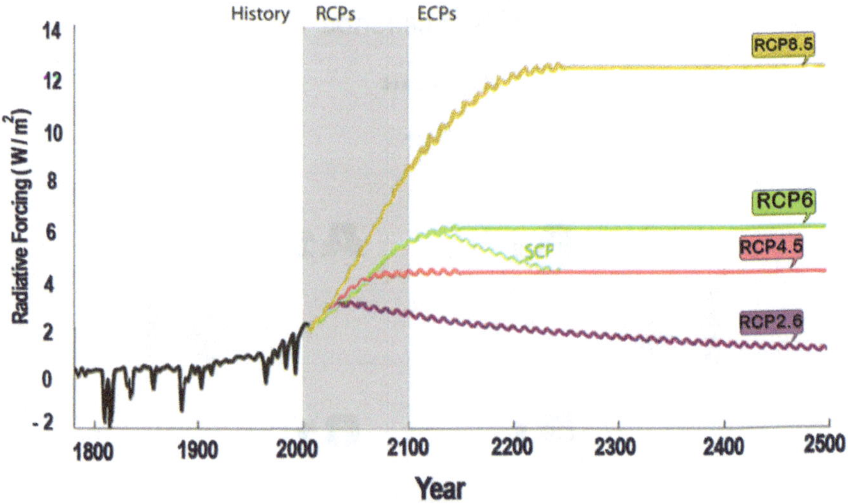

Fig. 1.4 Illustrated paths for RCP scenarios (Based on Clarke et al. 2014)

by the end of the present century. The higher RF number a scenario has, the more energy the Earth absorbs from the Sun, and therefore the higher average temperature the Earth has under the scenario. Figure 1.4 shows the average behaviour of the Earth's temperature under the four scenarios (Clarke et al. 2014).

- **RF**

The concept of RF, first proposed in the TAR, was calculated on a global scale with acceptable accuracy in AR4 and used in AR5 as a tool for the definition of climatic forecasting scenarios. In climatic sciences, radiative forcing or climate forcing is defined as the difference between insolation (sunlight) absorbed by the Earth and energy radiated back to space. Typically, the unit of RF in the Tropopause space is watts per square meter (w/m^2) which itself indicates the amount of energy absorbed per unit area of the planet. Positive values of RF mean that the system is warmer and negative values indicate a decrease in the system temperature. The reasons for the change in the amount of RF include the changes in the concentration of greenhouse gases and the amount of dust in the air. It should be noted that all greenhouse gases have a positive RF value, which means that by increasing their concentrations in the atmosphere, these gases will be able to increase the energy absorbed by the Earth.

Fig. 1.5 An overview of the General Atmospheric Circulation Model (Based on Viner and Hulme 1997)

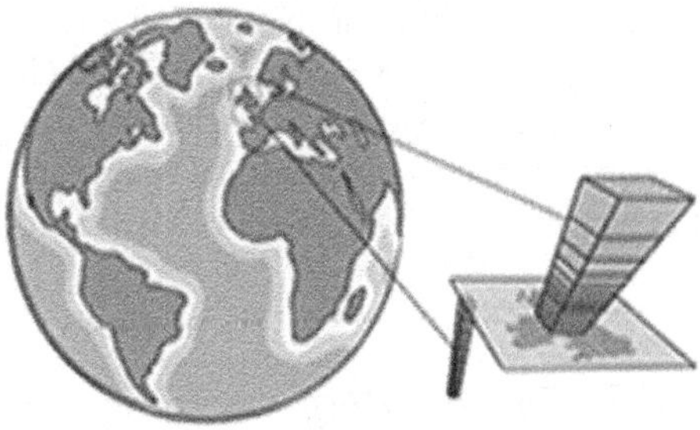

1.6.2 *Climatic Models*

After defining the behavioral pattern of climate stimuli, it is necessary to simulate how they affect the planet's climate. The solution lies in the AGCMs[8] that are designed to model the Earth's existing climate and can predict changes in the Earth's climate in the future (Xu 1999). Computer modelling based on the knowledge of the global atmospheric and oceanic thermodynamics is an acceptable method and has been widely used for estimating future climate change. IPCC has examined several GCMs (Mendoza-Ponce et al. 2021), which have been employed to simulate climatic variables such as temperature and precipitation during a period (IPCC 2011). These models are developed based on the physics rules described by mathematical equations that are solved on a three-dimensional network on Earth. To simulate the Earth's climate, the key climatic processes are simulated in separate sub-models, and then all sub-models related to the atmosphere and ocean are integrated to form the AOGCMs—the most reputable tools for predicting climatic variables (Lane et al. 1999; Mitchell 2003; Wilby and Harris 2006).

In general, spatial GCM models typically simulate the atmospheric space in the form of a vertical column and a grid of 5 to 20 unequal layers. To improve the simulation performance of these models, the layers located near the Earth's surface are closer to each other. Obviously, a larger number of layers used in the model lead to a smaller size of the defined cells and a shorter time interval in the model. As a result, the modelling of higher accuracy requires more computing resources and a longer running time (Fig. 1.5).

In the third assessment report, IPCC cited seven AOGCMs including HadCM3 from the Hadley Climate Research and Forecasting Center in the United Kingdom, CGCM2 from the Canadian Climate Modeling Center, CCSR-NIES from the Japan Research Center, ECHAM4 from the German Research Center, GFDL-R30 from the AGFDL,[9] CSIRO-MK2 from the Australian Scientific and Industrial Center, and the NCAR-DOE PCM from the American Atmospheric Research Center (Ashofteh 2014).

[8] Atmospheric General Circulation Model.

[9] American Geophysical Fluid Dynamics Laboratory.

In 1995, in order to integrate the results of the models, the WMO created the Climatic Models Working Groups (WGCM) responsible for creating an overall modelling framework of GCMs. The relevant result was presented in the form of a project called Coupled Model Inter-comparison Project (CMIP). Finally, in 2014, the fifth phase of this project (CMIP5) began to work on the results of AR5.

1.6.3 Downscaling

One of the main limitations in the use of GCMs is that their spatial and temporal resolutions do not meet the required accuracy of regional hydrological models. The spatial resolution of the GCMs is about 200 km, which is not particularly suitable for the study of mountainous areas and climatic variables such as precipitation and temperature (Wilby and Dettinger 2000). The output of these models can be converted from large-scale to local-scale variables over the study basin using certain downscaling methods. Commonly used downscaling methods include Proportional Downscaling, Dynamical Downscaling, and Statistical Downscaling (Wilby and Harris 2006).

1.6.3.1 Proportional Downscaling

The Proportional Downscaling method is one of the simplest methods for retrieving large-scale data, in which climate variables simulated by AOGCM are derived from the information about the cells of the study area. In this method, the difference between the basic and future AOGCM data is added or multiplied to the observed values by the Change Factor method (Hay et al 2000; Diaz-Nieto and Wibly 2005; Ashofteh et al. 2013a).

In this method, it is necessary to model the climatic conditions in the intended base period by GCMs. The model then produces temperature and rainfall climate change scenarios. Finally, it is necessary to calculate the average changes observed each month for the selected variables between the simulations for the basic and future periods, as follows (Ashofteh 2014):

$$\Delta T_i = \overline{T}_{GCM,fut,i} - \overline{T}_{GCM,base,i} \quad (1.1)$$

$$\Delta P_i = \frac{\overline{P}_{GCM,fut,i}}{\overline{P}_{GCM,base,i}} \quad (1.2)$$

in which i ($1 \leq i < 12$) = month i; $\overline{T}_{GCM,fut,i}$ and $\overline{P}_{GCM,fut,i}$ = average temperature and precipitation simulated by GCMs in the future period for month i; and $\overline{T}_{GCM,base,i}$ and $\overline{P}_{GCM,base,i}$ = average temperature and precipitation simulated by AOGCMs in the baseline period for month i.

$$T_i = T_{obs,i} + \Delta T_i \tag{1.3}$$

$$P_i = P_{obs,i} \times \Delta P_i \tag{1.4}$$

in which $T_{obs,i}$ and $P_{obs,i}$ = time series of temperature and precipitation observed in month i of the baseline period, respectively; and T_i and P_i = time series of predicted temperature and precipitation for month i, respectively (Ashofteh 2015). Although this method presents acceptable results in some cases, especially in short-term climatic forecasts, in many cases, it may not be able to provide the best solution for climate forecasting because it does not account for a variety of variables in the climatic prediction.

1.6.3.2 Statistical Downscaling

Various statistical downscaling methods have been developed to convert the large-scale GCM output to smaller scale data. The basis of statistical methods is to establish a favorable relationship, import large-scale climatic variables from AOGCM as the input and obtain the regional-scale climatic variables (Qian et al. 2005; Fowler et al. 2007). In statistical methods, the production of weather conditions is mainly based on the modification of the patterns provided by conventional meteorological data generators, such as WGEN, LARS-WG, and EARWIG. These methods enable users to use multiple chains of scenarios, which are ideal for risk analysis, instead of using just one time series. On the other hand, the inability of these methods to investigate decadal climate variability, which stems mainly from the weakness of the inter-annual data series produced by these models, in some cases, may lead to irrational responses.

1.6.3.3 Dynamical Downscaling

In the dynamical downscaling method, a high-resolution Regional Circulation Model (RCM) is commonly used. In such a way, the GCM outputs are used as the boundary conditions of the RCM. Due to the time and cost constraints of these models, the physical scales of these methods range from 20 to 50 km^2. The greatest advantage of these methods is their accuracy. Since the RCM is coupled with a GCM, the overall quality of dynamically downscaled RCM output is tied to the accuracy of the large-scale forcing of the GCM and its biases (Seaby et al. 2013). The fact that these models depend on the output from GCMs has led to some limitations for users and climate researchers, such as the size of the domain covered by them, the number of experiments, and ultimately the duration of the simulation.

1.6.4 Precipitation-Runoff Models

Runoff estimation has always been a challenge. This is important because it determines how much water from a rainfall event can be expected, as well as to what extent a flood event may have a destructive risk. The first attempt associated with this matter was made about a century and a half ago, when an Irish engineer, Thomas James Mulvaney (1822–1892), tried to estimate the peak discharge of a flood in 1851. This way of thinking has shown remarkable progress over time, but it is still one of the most important issues in water science (Beven 2011). Finally, according to the methodology section, Fig. 1.6 shows the flowchart related to the basic principles of studying and examining the effects of climate change on water resources planning and management at the basin level.

1.7 Practical Examples

The influences of climate change on various parts have been studied by many researchers in the water resources field. In this section, climate change studies are illustrated with three examples in the context of water resources development and management.

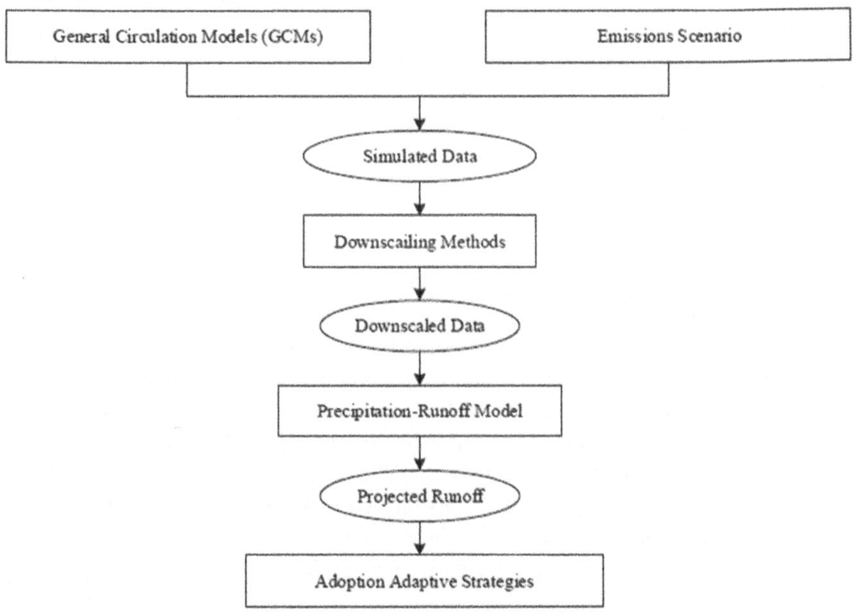

Fig. 1.6 Flowchart of assessment of climate change effect on water resources

Table 1.1 Computed Nonparametric Test Statistics (Confidence Level of 99%) of Hydroclimatic Variables (Ashofteh et al. 2017; Mandal et al. 2019a, b)

	Mann–Kendall trend tests	Spearman trend tests
Variable	(Z_M)	(Z_S)
Temperature	0.25	0.26
Rainfall	−0.20	−0.23
Runoff	−0.07	−0.10

1.7.1 Example I

Influences of "Climate Change on the Conflict between Water Resources and Agricultural Water Use" (Mandal et al. 2019a, b; Ashofteh et al. 2017).

Increasing CO_2 emissions will affect the climate and lead to adverse human and environmental effects. Water resources can be particularly vulnerable to this phenomenon. Since agricultural water use accounts for about 70% of the total, it is necessary to study the impacts of climate change on water resources and analyze the future of agriculture.

This sample is related to the research by Ashofteh et al. (2017). The Aidoghmoush irrigation network in Iran was assessed as a case study of the conflict between water resources and water consumption under climate change conditions" (Mandal et al. 2019a, b). In the study, the trend of climatic variables was determined by the Mann–Kendall (Z_M) and Spearman (Z_S) tests (Table 1.1). The statistics in Table 1.1 indicate that temperature increased while rainfall and runoff decreased due to the climate change in the observation period.

Seven AOGCMs (including HadCM3, CCSR-NIES, CSIRO MK2, CGCM2, GFDL R30, NCAR DOE PCM, and ECHAM4) (Mandal et al. 2019a, b) were used to examine the future changes under emission scenario A2. Emission scenario A2 was selected since it produced a high amount of CO2 (Mandal et al. 2019a, b). The input data of GCMs (i.e., monthly climatic data) were obtained from the IPCC's site (IPCC-DDC 1988) for the baseline period (1971–2000), first (2010–2039), second (2040–2069), and third (2070–2099) future periods (Mandal et al. 2019a, b). In addition, the technique of proportional downscaling with the GCM-Retrieve Data Program was used (Mandal et al. 2019a, b). The differences of the long-term AMT[10] and rainfall in the future and in the baseline period were calculated with the seven AOGCMs (Mandal et al. 2019a, b).

Results showed that the temperature in the future third period was greater than those in the first and second future periods (Mandal et al. 2019a, b). The changes in rainfall in winter, spring, summer, and autumn ranged from −27% to 27%, from −52% to 37%, from −94% to 140%, and from −19% to 119%, respectively, in the first future period; from −41% to 51%, from −81% to 20%, from −59% to 108%,

[10] average monthly temperature.

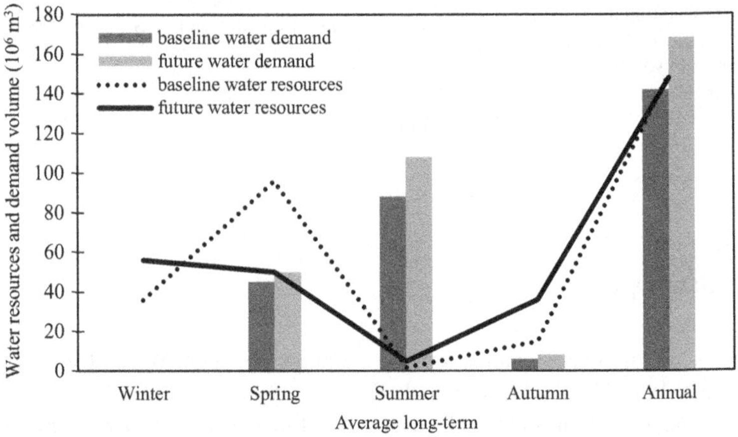

Fig. 1.7 Comparison of the long-term average water resources and water demands (Based on Ashofteh et al. 2017)

and from −35% to 94%, respectively, in the second future period; and from −51% to 38%, from − 74% to 118%, from −98% to 464%, and from −35% to 254%, respectively, in the future third period (Mandal et al. 2019a, b).

The irrigation requirement in the coming periods under changing climatic conditions will put more pressure on the agricultural water supply system and worsen food security (Mandal et al. 2019a, b). The annual increase in the irrigation required for walnuts, alfalfa, and potatoes will be approximately 20%, 16%, and 19%, respectively, in the first future period compared to the baseline period (Mandal et al. 2019a, b). All in all, water availability for agriculture would decline in the future. Figure 1.7 shows the emerging conflict from the effect of climate change on the temporal distribution of water resources and water consumption.

Figure 1.7 indicates that the future water resources in spring (wet season) will be drastically reduced (around 46%) due to a significant decrease in rainfall compared to the baseline rainfall (about 25%) in comparison with other seasons (Mandal et al. 2019a, b). However, there would be an increase of about 53 and 35% in the volume of water resources in the future periods in autumn and winter, respectively, due to the lower water demands in these seasons (Mandal et al. 2019a, b). Finally, the summer water demand is high, while the water availability is limited. Therefore, summer will be the most important season of the water crisis in this case study.

In conclusion, there would be a growing discrepancy between available water resources and agricultural water demand under the future climate change (Mandal et al. 2019a, b). So, awareness of this crisis can encourage decision-makers to adopt various strategies such as improving the irrigation efficiency (Mandal et al. 2019a, b), delaying planting dates, changing the system management methods, and/or implementing other adaptive methods.

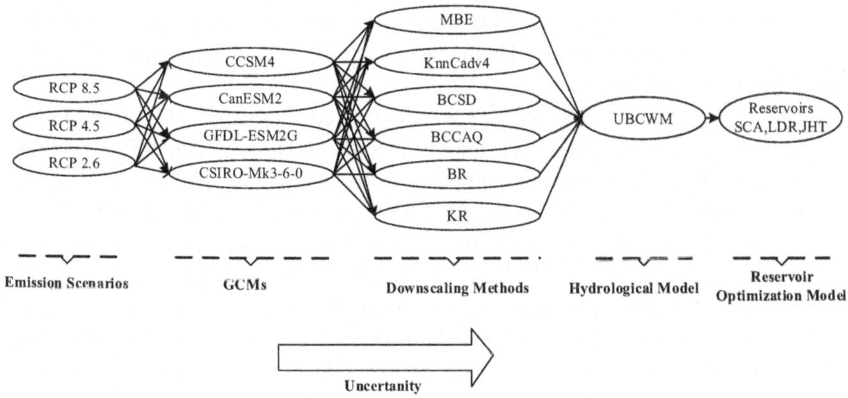

Fig. 1.8 Framework for assessing climate change impacts on reservoir operations (Based on Mandal et al. 2019a, b)

1.7.2 Example II

"Reservoir Operations under Changing Climate Conditions: Hydropower-Production Perspective" (Mandal et al. 2019a, b).

The second sample is associated with the study by Mandal et al. (2019a, b). The purpose of this study was to examine the impacts of climate change on reservoir system operations in the Campbell River basin in British Columbia, Canada that consisted of three reservoirs: Strathcona, Ladore, and John Hart (Mandal 2017). Three GHGES,[11] four GCMs, six downscaling methods, one hydrologic model, and an SDM[12] were integrated and used for the assessment. The modelling was performed for a future time period (2036–2065) and compared with the historical time period (1984–2013) (Mandal et al. 2019a, b). The models for the historical and future time periods were respectively used for validation and planning purposes (Mandal et al. 2019a, b).

The uncertainty analysis of the climate change process and the modelling involved: (I) different processes in GCMs, (II) choice of GCMs, (III) selection of emission scenarios, (IV) selection of downscaling models, (V) uncertainty analysis of hydrologic model parameters, and (VI) different hydrologic model structures. As shown in Fig. 1.8, 72 different scenarios, which consisted of three RCPs, four GCMs, and six downscaling models, were analyzed.

Additionally, the GCMs used in this study are from the CMIP5. The six downscaling models were BCSD, bias-corrected constructed analogues with quantile mapping reordering (BCCAQ), delta change method coupled with a nonparametric k-nearest neighbor weather generator, delta change method coupled with a maximum

[11] Greenhouse Gas Emission Scenarios.

[12] System Dynamics Model.

entropy-based weather generator, nonparametric statistical downscaling model based on the kernel regression (KR), and beta regression (BR) based statistical downscaling model (Mandal et al. 2019a, b).

At the first stage, climate variables were extracted for GCMs and different emission scenarios (RCPs). These variables included maximum temperature (Tmax), minimum temperature (Tmin), precipitation (Pr), mean sea level pressure (mslp), specific humidity (hus) at 500 hPa, zonal wind (u-wind), and meridional wind (v-wind) (Mandal et al. 2019a, b). Then, the climate variables were spatially interpolated at 10 downscaling locations in the basin and used as the input of the four downscaling models (MEB, KR, BR, and KnnCAD V4) (Mandal et al. 2019a, b). The downscaled climate variables (Tmax, Tmin, and Pr) (Mandal et al. 2019a, b) were directly derived from the IPCC databases (BCCAQ and BCSD) (IPCC 2014). At the last step, the downscaled climate variables were inputted to the UBC Watershed Model to generate daily runoff. BC Hydro was used for calibrating and validating the UBCWM. Note that an SDM was used to simulate the operation of multiple reservoirs in the Campbell River basin. Using the streamflow data from the UBCWM, the SDM simulated water inflow, storage, and release of all reservoirs. The power generated by this system was estimated by using the storage and release data.

The modelling results indicated that power generation was reduced due to the decrease of inflow of all three reservoirs over summer and autumn. The results also showed that the highest level of uncertainty was associated with the downscaling models, and the hydropower reliability declined more than 50% for all three reservoirs under climate change conditions (Mandal et al. 2019a, b). Table 1.2, 1.3and 1.4 show the comparisons of the predicted and historical hydropower generated by the Campbell River system for varying RCPs, GCMs, and downscaling methods.

1.7.3 Example III

"Reservoir water-quality projections under climate-change conditions" (Azadi et al. 2019).

Azadi et al. (2019) assessed water quality variations in the Aidoghmoush reservoir located in East Azerbaijan in Iran under climate change conditions in 2026–2039 (Azadi et al. 2018). The HadCM3 model was applied to calculate the temperature and rainfall under emission scenario A2 in the baseline period (1987–2000), and these two climate variables were then predicted over a future study period (2026–2039). In addition, the IHACRES model was used to simulate the average annual runoff, and the CE-QUAL-W2 model was used to simulate the reservoir water quality under climate change conditions. The streamflow velocity decreased as entering the reservoir, which led to thermal stratification because of uneven mixing and the induced changes in the density of water with depth (Azadi et al. 2018). The water quality of incoming streamflow was affected by reservoirs. Such changes can be harmful to the downstream ecosystem. Thus, it is necessary to predict water quality variations under climate change and also determine the related water withdrawal

Table 1.2 Comparison of the historical and future mean seasonal power production (MW) for different emission scenarios for Strathcona, Ladore, and John Hart reservoirs in the Campbell River System, British Columbia, Canada (Mandal et al. 2019a, b)

2036–2065								
			RCP 2.6		RCP 4.5		RCP 8.5	
Reservoir	Season	Historical (1984–20,013)	Mean	Change in mean value (%)	Mean	Change in mean value (%)	Mean	Change in mean value (%)
Strathcona	Winter	29.4	26.8	−9	27.3	−7	27.0	−8
	Spring	24.5	22.4	−8	22.5	−8	22.1	−9
	Summer	22.6	8.5	−61	7.8	−65	7.1	−67
	Fall	25.9	18.2	−27	18.3	−27	17.7	−30
Ladore	Winter	31.5	20.9	−33	21.4	−32	21.3	−32
	Spring	27.0	19.7	−26	19.9	−26	19.6	−27
	Summer	20.7	7.0	−65	6.5	−68	5.9	−70
	Fall	25.1	15.2	−38	15.3	−37	14.9	−40
John Hart	Winter	106.4	87.8	−17	88.6	−16	87.9	−17
	Spring	92.0	80.9	−12	81.5	−11	80.5	−12
	Summer	68.7	30.5	−54	28.3	−58	25.9	−60
	Fall	84.5	59.3	−28	59.5	−28	57.4	−31

policies. Figure 1.9 shows the modelling components and framework for simulation of climatic variables, estimation of reservoir inflows, and assessment of reservoir water quality (Azadi et al. 2018).

TDS was selected as the water quality indicator in this area associated with an arid climate and relatively high suspended solids. The raw data were obtained from the meteorological administrations in Iran. The climate data such as temperature and rainfall were obtained from the IPCC website by applying the HadCM3 model under the scenario of A2 for both base time period and climate change period. Using the downscaled climatic data, the IHACRES model was used to simulate future runoff. Then, reservoir inflow and water demand for the future period were calculated. Finally, the water quality of the reservoir was simulated by CE-QUAL-W2. The input data of CE-QUAL-W2 included geometric and meteorological data, as well as initial and boundary conditions such as initial water depths and temperatures, inlet and outlet data, and inlet water temperatures (Azadi et al. 2018).

The results illustrated that the future mean annual runoff would decline by around 1%, while the agricultural water demand would rise by 16%. The surface air temperature would increase by 1.3 °C in comparison with the base time period. The bottom and surface water temperatures of the reservoir increased by 1.19 °C and 1.24 °C, respectively. Moreover, climate change would also affect the total dissolved solids of the reservoir, especially for the options with irrigation. Compared to the TDS in the base time period, the average TDS near the reservoir water surface for the climate

Table 1.3 Comparison of the historical and future mean seasonal power production (MW) for different GCMs for Strathcona, Ladore, and John Hart reservoirs in the Campbell River System, British Columbia, Canada (Mandal et al. 2019a, b)

2036–2065			CanESM2		CCSM4		CSIRO-MK3-6–0		GFDL-ESM2G	
Reservoir	Season	Historical (1984–20,013)	Mean	Change in mean value (%)	Mean	Change in mean value (%)	Mean	Change in mean value (%)	Mean	Change in mean value (%)
Strathcona	Winter	29.4	24.1	−18	23.0	−22	24.1	−18	23.0	−22
	Spring	24.5	17.4	−29	16.6	−32	16.5	−33	16.1	−34
	Summer	22.6	4.7	−79	4.7	−79	5.1	−78	5.3	−77
	Fall	25.9	12.4	−52	10.8	−58	12.4	−52	11.4	−56
Ladore	Winter	31.5	18.4	−42	18.3	−42	18.4	−42	18.2	−42
	Spring	27.0	15.9	−41	15.3	−43	15.1	−44	14.8	−45
	Summer	20.7	5.1	−75	4.8	−77	5.5	−74	4.3	−79
	Fall	25.1	11.8	−53	10.6	−58	11.5	−54	11.2	−55
John Hart	Winter	106.4	81.5	−23	77.7	−27	81.1	−24	78.8	−26
	Spring	92.0	68.7	−25	65.8	−28	64.9	−29	63.2	−31
	Summer	68.7	10.1	−85	9.5	−86	12.5	−82	13.6	−80
	Fall	84.5	41.4	−51	36.4	−57	41.1	−51	39.4	−53

Table 1.4 "Comparison of the historical and future mean seasonal power production (MW) for different downscaling models for Strathcona, Ladore, and John Hart reservoirs in the Campbell River System, British Columbia, Canada" (Mandal et al. 2019a, b)

2036–2065														
			BCCAQ		BCSD		KF		BR		MBE		KnnCAD V4	
Reservoir	Season	Historical (1984–20,013)	Mean	Change in mean value (%)	Mean	Change in mean value (%)	Mean	Change in mean value (%)	Mean	Change in mean value (%)	Mean	Change in mean value (%)	Mean	Change in mean value (%)
Strathcona	Winter	29.4	27.5	−7	26.9	−8	34.[illegible]	18	19.0	−35	23.0	−22	28.3	−4
	Spring	24.5	22.6	−8	23.5	−4	30.[illegible]	23	18.6	−24	16.1	−34	21.7	−11
	Summer	22.6	15.6	−31	17.6	−22	29.[illegible]	20	13.3	−41	9.0	−60	14.2	−37
	Fall	25.9	19.3	−26	18.0	−30	30.[illegible]	18	14.9	−42	11.5	−56	17.8	−31
Ladore	Winter	31.5	19.9	−37	19.8	−37	32.1	2	17.4	−45	18.2	−42	20.6	−35
	Spring	27.0	19.4	−28	19.6	−27	27.[illegible]	1	16.6	−38	14.8	−45	18.8	−30
	Summer	20.7	15.1	−27	16.4	−21	23.9	15	13.7	−34	9.9	−52	14.0	−32
	Fall	25.1	15.0	−40	15.0	−40	24.7	−2	13.1	−48	11.2	−55	14.9	−41
John Hart	Winter	106.4	92.0	−14	88.8	−16	94.4	−11	65.4	−38	78.8	−26	93.0	−13
	Spring	92.0	87.0	−5	88.9	−3	89.1	−3	72.1	−22	63.2	−31	84.9	−8
	Summer	68.7	67.0	−2	73.2	−7	97.1	4	58.7	−15	41.2	−40	61.8	−10
	Fall	84.5	63.0	−25	58.1	−31	91.9	9	53.6	−37	39.4	−53	58.8	−30

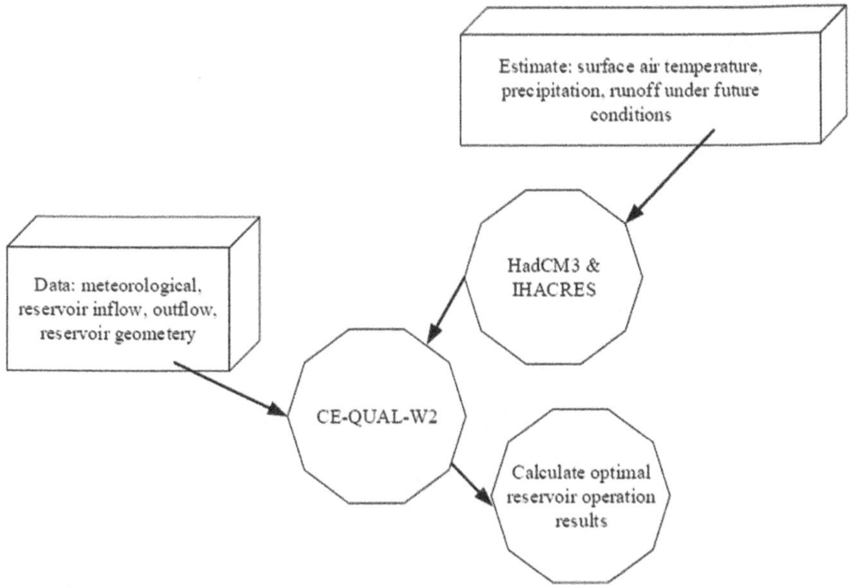

Fig. 1.9 Modelling components and framework

change period would increase by 4.3%. The simulation indicated that January 2038 would reach the maximum TDS 1,980 (g/m^3) at the water surface of the reservoir in the climate change period. For the base time period, it occurred on February 15th, 2000 (1,616 g/m^3). It was demonstrated that the average TDS in the future under climate change conditions would be higher than that in the base time period" (Azadi et al. 2018).

In summary, it was concluded from the study that climate change had significant effects and the modelling methods helped identify the optimal strategies of reservoir operations to deal with the changes in thermal stratification and TDS (Azadi et al. 2018). That is, the modelling identified the reservoir water levels where water can be extracted with the various water quality characteristics.

1.8 Summary

The results of the aforementioned studies have repeatedly proven that climate change alters the current climate of the Earth fundamentally. Therefore, management and planning for the uncertain future require climate analysis and forecasting. For this purpose, it is necessary to define and explain the possible perspectives and to simulate the behavioral pattern of the Earth for different scenarios. AOGCMs can be used for such assessments. However, these models are often implemented on a large scale. To analyze climate change at a local scale, it is important to downscale the information obtained from the previous stage. A variety of downscaling methods have been developed, and they have their own advantages and limitations. Thus, it is necessary to select the appropriate method, depending on the conditions of the problem. Such modelling studies provide the planners with a new perspective and valuable data concerning the regional climate

References

Ahmadi M, Bozorg-Haddad O, Loáiciga HA (2015) Adaptive reservoir operation rules under climatic change. Water Resour Manage 29(4):1247–1266

Alvarez UFH, Trudel M, Leconte R (2014) Impacts and adaptation to climate change using a reservoir management tool to a northern watershed: application to Lièvre river watershed, Quebec, Canada. Water Resour Manage 28(11):3667–3680

Ashofteh P-S (2014) Climate change and water: tools and approaches. Javedankherad, No. 1, Tehran, Iran

Ashofteh P-S, Bozorg-Haddad O, Loáiciga HA (2015) Evaluation of climatic-change impacts on multiobjective reservoir operation with multiobjective genetic programming. J Water Resour Plan Manag 141(11):04015030

Ashofteh P-S, Bozorg-Haddad O, Loáiciga HA (2017) Impacts of climate change on the conflict between water resources and agricultural water use. J Irrig Drain Eng 143(4):02516002

Ashofteh P-S, Bozorg-Haddad O, Mariño MA (2012) Climate change impact on reservoir performance indexes in agricultural water supply. J Irrig Drain Eng 139(2):85–97

Ashofteh PS, Bozorg-Haddad O, Mariño MA (2013) Climate change impact on reservoir performance indexes in agricultural water supply. J Irrig Drain Eng 139(2):85–97

Azadi F, Ashofteh P-S, Loáiciga HA (2018) Reservoir water-quality projections under climate-change conditions. Water Resour Manag 3(1 33):401–421

Azadi F, Ashofteh P-S, Loáiciga HA (2019) Reservoir water-quality projections under climate-change conditions. Water Resour Manage 33(1):401–421

Bates B, kundzewicz Z W, Wu S, Palutikof J (2008) Climate change and water technical paper of the intergovernmental panel on climate change. IPCC Secretariat Geneva 210

Beven KJ (2011) Rainfall-runoff modelling: the primer. John Wiley and Sons Publication, West Sussex, UK

Bradley RS, Kelly PM, Jones PD, Goodess CM, Diaz HF (1985) A climatic data bank for northern hemisphere land areas, 1851–1980. Technical Report TRO17, U.S. Dept. of Energy, Carbon Dioxide Research Division, pp 335

Buchdahl JM (1999) Global climate change student information guide. Manchester Metropolitan University, Atmospheric Research and Information Center, p 98

Carter TAK, Alfsen E, Barrow B, Bass X, Dai P, Desanker SR, Gaffin F, Giorgi M, Hulme M, Lal LJ, Mata LO, Mearns JFB, Mitchell T, Morita R, Moss D, Murdiyarso JD, Pabon-Caicedo J, Palutikof ML, Parry C, Rosenweig B, Seguin RJ, Scholes, Whetton PH (2007) Guidelines on the use of scenario data for climate impact and adaptation assessment. Version 2. Intergovernmental Panel on Climate Change, Task Group on Scenarios for Climate Impact Assessment, p 71

Clarke L, Jiang K, Akimoto K, Babiker M, Blanford G, Fisher-Vanden K, Hourcade JC, Krey V, Kriegler E, Löschel A, McCollum D (2014) Assessing transformation pathways

Diaz-Nieto J, Wibly RL (2005) A comparison of statistical downscaling and climate change factor methods: impacts on low flows in the River Thames United Kingdom. Climatic Change 63(2–3):245–268

Fowler HJ, Blenkinsop S, Tebaldi C (2007) Linking climate change modelling to impacts studies: recent advances in downscaling techniques for hydrological modeling. Int J Climatol 27(12):1547–1578

Georgakakos AP, Yao H, Kistenmacher M, Georgakakos KP, Graham NE, Cheng FY, Spencer C, Shamir E (2012) Value of adaptive water resources management in Northern California under climatic variability and change: reservoir management. J Hydrol 412:34–46

Hay LE, Wibly RL, Leavesley GH (2000) A comparison of delta change and downscaled GCM scenarios for three mountainous basins in the United States. J Am Water Resour Assoc 36(2):387–397

IPCC-DDC (1988) Data distribution centre. http://ipcc-ddc.cru.uea.ac.uk/ (Oct. 1, 2016)

IPCC (2000) Intergovernmental panel on climate change. In: Nakicenovic N, Swart R (eds) Special report on emissions scenarios. Cambridge University Press, UK. p 570

IPCC (2001) Climate change, the scientific basis. Contribution of working group I to the third assessment report of the intergovernmental panel on climate change. Cambridge University Press, Cambridge, U.K

IPCC (2007) Summary for policy makers, Climate change 2007a: The physical science basis. Contribution of the working Group I to the fourth assessment report of the intergovernmental panel on climate change. Cambridge University Press, Cambridge, UK

IPCC (2007a) Climate change: the physical science basis. Contribution of working group I to the fourth assessment report of the intergovernmental panel on climate change. Cambridge University Press, Cambridge, pp 1–18

IPCC (2007b) Climate change 2007b: the scientific basis. Houghton JT, Ding YDJG, Griggs DJ, Noguer M, van der Linden PJ, Dai X, Maskell K, Johnson CA (eds) Contribution of working groups I to the fourth assessment report of the intergovernmental panel on climate change. Cambridge, United Kingdom and New York, NY, US

IPCC (2007c) Climate change 2007c: mitigation. Davidson O, Swart R (eds) Contribution of working groups I to the fourth assessment report of the intergovernmental panel on climate change. Cambridge, United Kingdom and New York, NY, USA, 881

IPCC (2011) IPCC intergovernmental panel on climate change web site: organization page. http://www.ipcc.ch/organization/organization.shtml

IPCC (2013) Climate change 2013: the physical science basis. Contribution of working Group I to the fifth assessment report of the intergovernmental panel on climate change. Cambridge University Press, Cambridge, United Kingdom and New York, NY

IPCC (2014) Climate change 2014: impacts, adaptation, and vulnerability. Contribution of working Group II to the fifth assessment report of the intergovernmental panel on climate change. Cambridge University Press, Cambridge, United Kingdom and New York, NY

Jones P, Palutikof P (2006) Global temperature record. Climate research unit, University of East Anglia. http://www.cru.uea.ac.uk/. Accessed 22 October 2020

Karamouz M, Araghinejad S (2005) Advanced hydrology. Industrial University of Amir Kabir (Poly Technics), No. 1, Tehran, Iran

Khare D, Mondal A, Kundu S, Mishra PK (2017) Climate change impact on soil erosion in the Mandakini River Basin, North India. Appl Water Sci 7(5):2373–2383

Lane ME, Kirshen PH, Vogel RM (1999) Indicators of impacts of global climate change on U.S. water resources. J Water Resour Plan Manag 125(4):194–204

Lettenmaier DP, Gan TY (1990) Hydrologic sensitivities of the Sacramento-San Joaquin River basin, California, to global warming. Water Resour Res 26(1):69–86

Mandal S, Arunkumar R, Breach PA, Simonovic SP (2019a) Reservoir operations under changing climate conditions: hydropower-production perspective. J Water Resour Plan Manag 145(5):04019016

Martin J (2007) The meaning of the 21th century: A vital blueprint for ensuring our future. Penguin Group publication, New York, U.S.

Mearns LO, Giorgi F, Whetton P, Pabon D, Hulme M, Lal M (2010) Guidelines for use of climate scenarios developed from regional climate model experiments, DDC of IPCC TGCIA, final version, p 38

Mandal S (2017) Uncertainty modeling in the assessment of climate change impacts on water resources management

Mandal S, Arunkumar R, Breach PA, Simonovic SP (2019b) Reservoir operations under changing climate conditions: hydropower-production perspective. J Water Resour Plan Manag 145:04019016

Mendoza-Ponce A, Corona-Núñez RO, Nava LF, Estrada F, Calderón-Bustamante O, Martínez-Meyer E, Carabias J, Larralde-Corona AH, Barrios M, Pardo-Villegas PD (2021) Impacts of land management and climate change in a developing and socioenvironmental challenging transboundary region. J Environ Manag 300, 113748

Mirza MMQ, Warrick RA, Ericksen NJ (2003) The implications of climate change on floods of the Ganges, Brahmaputra and Meghna rivers in Bangladesh. Clim Change 57(3):287–318

Mitchell TD (2003) Pattern scaling: an examination of the accuracy of the technique for describing future climates. Clim Change 60(3):217–242

Moursi H, Kim D, Kaluarachchi JJ (2017) A probabilistic assessment of agricultural water scarcity in a semi-arid and snowmelt-dominated river basin under climate change. Agric Water Manag 193:142–152

Nakicenovic N, Alcamo J, Grubler A, Riahi K, Roehrl RA, Rogner HH, Victor N (2000) Special report on emissions scenarios (SRES), a special report of Working Group III of the intergovernmental panel on climate change. Cambridge University Press

Nouri-Tirtashi M, Sharifi M-B, Zarghami M (2015) The effect of climate change on the inflow to the reservoirs of dams in conditions of uncertainty (case study: Bustan and Golestan dams in the Wood catchment). Irrigation Drain Iran 9:367–380 (In Persian)

Peston R (2006) Report's stark warning on climate. BBC News, 29

Qian BD, Hayhoe H, Gameda S (2005) Evaluation of the stochastic weather generators LARS-WG and AAFC-WG for climate change impact studies. Climate Res 29(1):3–21

Rosenzweig C, Strzepek KM, Major DC, Iglesias A, Yates DN, McCluskey A, Hillel D (2004) Water resources for agriculture in a changing climate: international case studies. Glob Environ Chang 14(4):345–360

Rothrock DA, Yu Y, Maykut GA (1999) Thinning of the Arctic sea-ice cover. Geophys Res Lett 26(23):3469–3472

Seaby LP, Refsgaard JC, Sonnenborg TO, Stisen S, Christensen JH, Jensen KH (2013) Assessment of robustness and significance of climate change signals for an ensemble of distribution-based scaled climate projections. J Hydrol 486:479–493

Solomon S, Manning M, Marquis M, Qin D (2007) Climate change 2007-the physical science basis: Working group I contribution to the fourth assessment report of the IPCC. (4). Cambridge university press

Tabari H, Somee BS, Zadeh MR (2011) Testing for long-term trends in climatic variables in Iran. Atmos Res 100(1):132–140

Traynham L, Palmer R, Polebitski A (2010) Impacts of future climate conditions and forecasted population growth on water supply system in the Puget Sound region. J Water Resour Plan Manag 137(4):318–326

Viner D, Hulme M (1997) The climate impacts LINK project: applying results from the hadley centre's climate change experiments for climate change impacts assessment. Climatic Research Unit, Norwich, UK, pp 17

Wilby RL, Dettinger MD (2000) Streamflow changes in the Sierra Nevada, California, simulated using a statistically downscaled general circulation model scenario of climate change. Link Clim Change Land Surface Change 6(12):99–121

Wilby RL, Harris I (2006) A framework for assessing uncertainties in climate change impacts: low-flow scenarios for the River Thames, UK. Water Resour Res 42(2)

Xu CY (1999) From GCMs to river flow: a review of downscaling methods and hydrologic modeling approaches. Prog Phys Geogr 23(2):229–249

Zareian MJ, Eslamian S, Hosseinpour EZ (2014) Climate change impacts on reservoir inflow using various weighted approaches. World Environmental and Water Resources Congress, Portland, Oregon, USA, June 1–5, pp 2136–2145

Zhao Y, Camberlin P, Richard Y (2005) Validation of a coupled GCM and projection of summer rainfall change over South Africa, using a statistical downscaling method. Climate Res 28(2):109–122

Zhenmei M, Kang S, Zhang L, Tong L, Su X (2008) Analysis of impact of climate variability and human activity on streamflow for a river basin in arid region of northwest China. J Hydrol 3523–4:239–249

Zolghadr-Asli B (2017) Discussion of "Multiscale assessment of the impacts of climate change on water resources in Tanzania" by Adhikari U, Nejadhashemi AP, Herman MR, Messina JP. J Hydrol Eng. In Press

Part I
Introduction to Climate Change with Focus on Water Resources

Introduction

The change is an inseparable part of nature. Although the intensity and the rate of change vay, their effects on terrestrial life are arguably irrefutable. Our planet has experienced several changes that impacted the distribution and abundance of biodiversity and species richness (Thomas 2008). The changes in climatic variables, such as temperature and precipitation, have always been regarded as one of the main causes of consequent changes on the Earth. Assessing the historical trajectory of changes of the climatic variables can provide an opportunity to understand the past, the present, and the future climate conditions of the Earth, and clarify how these conditions impact the Earth system. In addition to historical assessment, presenting the evidence of climate change phenomena can reduce debates on the importance of studying climate change, including the reasons, the effects, and mitigation/adaptation measures. In the second chapter, the causes of climate change, particularly sources of greenhouse gas emission are expressed in detail. Since the share of natural sources is negligible in comparison to the anthropogenic sources, recognizing the roles of humans can be worthwhile in either controlling or reducing the emission. Ultimately, for more awareness of climate change impact on water systems, water resources and water demand are also assessed. Analyzing the changes in water resources and water demand can be a worthwhile guidance for the policymakers to satisfy the growing water demand in a sustainable way. All these steps are covered in Sect. 1 as the foundation of the next two sections.

Reference

Thomas CD, Cameron A, Green RE, Bakkenes M, Beaumont LJ, Collingham, YC, Erasmus BFN., Siqueria MF, Grainger A, Hannah L, Hughes L, Huntley B, Jaarsveld AS, Midgley GF, Ortega-Huetra MA, Townsend Peterson A, Phillips OL, Williams SE (2008) Extiction risk from climate change. Nature 427:145–148

Chapter 2
Basic Concepts

Arezoo Boroomandnia, Omid Bozorg-Haddad, Scott Baum, Christopher Ndehedehe, Kefeng Zhang, and Veljko Prodanovic

2.1 The Earth System

The Earth system plays an active role in determining climatic conditions. Therefore, understanding Earth systems is crucial to developing an understanding of climate change. The Earth system is divided into two parts: the Geosphere and the Biosphere. The Geosphere is a collective term for the Lithosphere, Atmosphere, Hydrosphere, and Cryosphere; it contains the contents of the Earth's interior, mantle, rocks and

A. Boroomandnia
Department of Irrigation & Reclamation Engineering, Faculty of Agricultural Engineering & Technology, College of Agriculture & Natural Resources, University of Tehran, 31587-77871 Karaj, Tehran, Iran
e-mail: aboroomandnia@ut.ac.ir

O. Bozorg-Haddad (✉)
Department of Irrigation & Reclamation Engineering, Faculty of Agricultural Engineering & Technology, College of Agriculture & Natural Resources, University of Tehran, Karaj, Tehran, Iran
e-mail: OBHaddad@ut.ac.ir

S. Baum
School Engineering and Built Environ–Arch and Design, Nathan Campus, Griffith University, N13 1.32, Brisbane, QLD, Australia
e-mail: s.baum@griffith.edu.au

C. Ndehedehe
School of Environment and Science-Marine, Nathan Campus, Griffith University, N78 4.12, Brisbane, QLD, Australia
e-mail: c.ndehedehe@griffith.edu.au

K. Zhang · V. Prodanovic
Water Research Centre, School of Civil and Environmental Engineering, Vallentine Annex, University of New South Wales, Building H22, Sydney, NSW 2052, Australia
e-mail: kefeng.zhang@unsw.edu.au

O. Bozorg-Haddad (ed.), *Climate Change in Sustainable Water Resources Management*, Springer Water, https://doi.org/10.1007/978-981-19-1898-8_2

minerals, oceans, surrounded gases. Biosphere, is defined as the ecological system that integrates all living things and their interactions.

2.1.1 *Lithosphere*

The lithosphere is the solid outer-most part of the earth and includes all parts of the crust, mantle, and core of the Earth, all of which contain stones, mountains, minerals, lava, molten materials, and soil particles. The Earth's crust involves floating oceanic and continental tectonic plates on porous viscous materials known as the Earth mantle. While the thickness of oceanic plates is lower than continental plates, their density is higher. The movement of tectonic plates, caused by thermal energy from the mantle part of the lithosphere, is known as tectonic activity. When tectonic plates move and collide, significant geological events occur including earthquakes, orogeny (mountain building) and the formation of deep ocean trenches.

2.1.1.1 Atmosphere

The envelope of gases surrounding the Earth is known as the atmosphere. This layer contains various gaseous combinations, which are vital for life on the Earth. These gases include, Nitrogen (78%), Oxygen (21%), Argon (0.93%), Carbon dioxide (0.04%), and small amount of other gases like Hydrogen and steam. Not only are these components of atmosphere vital for the maintenance of life on earth, but they also provide protection against solar ultraviolet radiation, solar wind and cosmic rays.

Based on changes in temperature, pressure, and composition of the element, the atmosphere is divided into six layers (Fig. 2.1). The troposphere is the closest to the Earth's surface with varied depth from 17 km at the equator to 7 km at the poles. This layer contains around 75% of the atmosphere's total mass and it accounts for most of our weather including clouds, snow and rain. The temperature in the troposphere gets colder as the distance from the Earth's surface increases. The next layer is stratosphere with the depth of 50 km above the troposphere. The stratosphere contains the ozone layer and protects the Earth against ultraviolet radiation. The stratosphere's temperature increases with height due to the absorption of ultraviolet radiation from the sun. The temperatures are highest over the summer pole and lowest over the winter pole. The next region in the atmosphere is the mesosphere which extends to about 85 kms above the stratosphere and reaches a minimum temperature of around −90 °C or 180 °k. The thermosphere (or ionshpere) lies above the mesosphere above the 80 km mark. The temperature in the thermosphere increases with height due to the absorption of energetic ultraviolet and X-Ray radiation from the sun. The absolute temperature varies between day and night and between the different seasons. The region above 500 kms is called the exosphere. It comprises mostly oxygen and

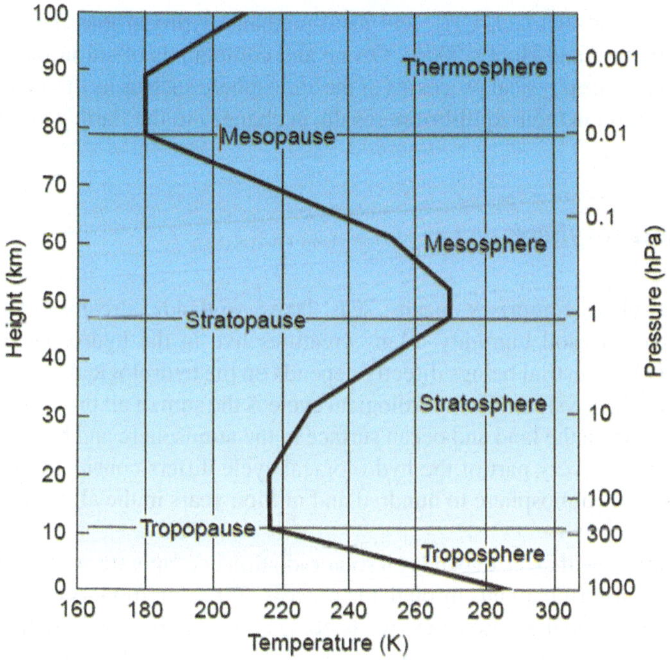

Fig. 2.1 The vertical temperature and pressure profile, represented by the U.S. Standard Atmosphere (Wallace and Hobbs 2006)

hydrogen atoms. The pressure and temperature fluctuation of the different spheres by height is illustrated in Fig. 2.1.

The Earth's terrestrial weather is developed within the troposphere. The composition of the long-lived gases, including Nitrogen, Oxygen, Argon, and Carbon Dioxide is almost uniform and irrelevant to altitude. However, chemical reactions control the concentration of elements and their transport in the atmosphere. Cyclopentene (C5H8), Propene (C_3H_6), Nitrogen Oxides (NO_x), Hydrogen Peroxide (H_2O_2), Sulfur Dioxide (SO_2), Water (H_2O), Ozone (O_3), and Carbon Monoxide (CO) are categorized as moderately long-lived elements whose residence time varies from 1 h to 1 year respectively. Short-lived elements, Hydroxide (OH), Nitrate (NO_3), CH_3O_2, remain in the troposphere for less than an hour. The concentration of short-lived elements is higher close to the emission source, while long-lasting elements have more uniform concentration over the whole troposphere. Several sources of atmospheric gases exist, and they include biogenic, lithosphere, the oceans, human activities, and in situ formation. For instance, oxygen formation dates back to 3.8 billion years ago and was emitted from biological activities. Organic matter degeneration and volcanoes also lead to the Methane (CH_4) emission. Conversion of N_2 into NH_3, N_2O, and NO (but also degradation of nitrogen into N_2 gas) is often carried out by microorganisms living in soils. The combustion of fossil fuels is another source of NO_x, SO_x, and VOC emission. Tropospheric ozone is created via photochemical reactions among OH, CO, CH_4, and nano-methane hydrocarbons in the presence of NO_x (Wallace and Hobbs 2006). Ozone and compounds of sulfur, nitrogen, and carbon control the physical processes of the atmosphere such as its thermal structure. The interruption in their equilibrium results in changes to the Earth's temperature.

2.1.2 Hydrosphere

The Hydrosphere comprises oceans, seas, lakes, wetlands, streams, groundwater, water vapor, and soil humidity. Many creatures live in the hydrosphere, and the survival of all terrestrial beings directly depends on the hydrological cycle within all parts of the Earth system. The hydrological cycle is the sum of all processes in which water moves from the land and ocean surface to the atmosphere and back. The dwell time of water in every part of the hydrological cycle differs considerably. It ranges from days in the atmosphere to hundred and million years in the alpine glaciers and oceans.

The hydrological cycle depends on solar radiation to evaporate water from oceans and seas and also from plants through the process of transpiration. As water converts to steam it rises into the atmosphere while being cooled and condensed, before being returned to the land and oceans as rain. As rain falls on the earth's surface, it's runoff (both surface and subsurface) results in the creation of lakes and rivers. A part of the water also penetrates the ground moving deeper through the soils to forms and recharge underground aquifers. Finally, a part of the surface and underground water leads to seas and oceans.

This exchange of moisture between the land surface and the atmosphere can be affected by anthropogenic actions and oceanic perturbations, resulting in complex hydrological processes and the redistribution of continental water storage. The transferred volume between water cycle component is illustrated in Fig. 2.2.

The oceans contain the largest volume of water in the hydrosphere. Since oceans are able to dissolve gases, they can be considered as both a sink and a source of soluble gases. Water bodies are mainly a source of sulfur compounds and a sink of carbon gaseous compounds. Water reservoirs also act as heat reservoirs, affecting the solar reflection to the atmosphere via changes in albedo. These interactions with the atmosphere play a key role in climate conditions.

2.1.3 Cryosphere

The Earth's cryosphere consists of all the solid forms of water on the planet and is the largest single source of freshwater on the planet. It includes sea ice, lake ice and river-ice, snow cover, glaciers, ice caps and ice sheets, and frozen ground. The thermal inertia of the climate system is related to the cryosphere due to its influence on the Earth's albedo and oceanic thermohaline circulation. Two major

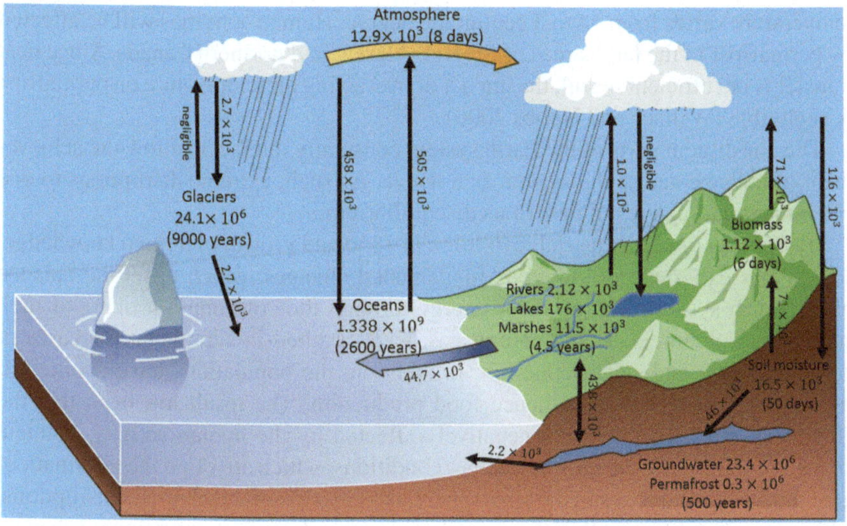

Fig. 2.2 Global hydrologic cycle, storage volume (km3), fluxes (km3/year), and residence time (Henshaw et al. 2000)

parts of the cryosphere are found in ice sheets located in Antarctica and Greenland. Ice sheets are formed by the accumulation of snow layers over tens of thousands of years. As snow falls to the surface, the pressure on the older snow layers increases, transforming them into ice. Analysis of ice cores indicate the age of ice sheets are approximately 100,000 years for Greenland and 500,000 years for Antarctica. Ice sheets (like glaciers) are not stagnant and generally flow slowly towards their margins. However, in some regions (called ice streams), the flow is much faster than in other parts, sometimes reaching several kilometres per year. Mountain glaciers are similar to ice sheets in many aspects, but are smaller in scale. As such, they are more rapidly affected by ambient atmospheric / weather conditions.

In contrast to ice sheets which are formed by snow fall, sea ice is formed when sea water freezes. Sea ice covers a similar area in both the northern and southern hemispheres. Its seasonal cycle is larger in the Southern Ocean where the majority of the ice cover is referred to as first-year sea ice (i.e., sea ice that has not survived one summer). Because of the large thermal inertia of the ocean, the minimum and maximum sea ice extent differ to that of snow cover on the land by around two months. Sea ice is at a maximum around March and a minimum around September in both hemispheres. The sea ice is thinner in the Southern Hemisphere, with a mean thickness of less than 1 m, while the mean ice thickness in the central Arctic is around 3 m. Ice pieces can float on the sea even for years, while they move clockwise toward the north of Alaska and experience transpolar drift from Siberia to Greenland. Land snow covers an area greater than sea ice, but the total area covered diminishes when the temperature of the land surface increases at the beginning of spring. The frozen underlying land known as permafrost is another component of cryosphere. Its temperature varies from –4 to 3 centigrade degree. Human activities will be affected by permafrost in the large areas of Siberia, Alaska, and northern Canada. Since heat transfer is not efficient in soil, the impact of overlaying air temperature on permafrost is negligible (Wallace and Hobbs 2006).

The constituent parts of the Earth system constantly interact to allow the achievement of balance across the system as a whole. As such, extreme disruptions to one constituent part can result in serious disequilibrium.

By the start of the industrial revolution in 1830 and growth of human knowledge, society has experienced numerous fundamental changes, which upset the balance of the Earth system. The need for energy through the consumption of fossil fuel, such as petroleum, coal, and natural gas, have resulted in significant increases in carbon dioxide and steam emission. In addition, the populations have resulted in rising demand for food and hence food production. The result has been that the climate of the Earth has been negatively affected by the human activity. Climate change or long-term changes in weather conditions, which are often the aftermath of greenhouse gas increases in the atmosphere, is a clear example of human disruptions. Although various factors have been attributed to the phenomenon of climate change, excessive emissions of greenhouse gases in the atmosphere have been cited as the main cause of this phenomenon (Burlando and Rosso 2002; Wilby and Harris 2006; Schewe et al. 2014; Bär et al. 2015; Sarzaeim et al. 2017). Climate change causes many problems, such as gradual warming, rising sea levels, torrential rains and severe droughts in different parts of the world (extreme events). Thus, in recent decades, concerns about the increase in the average global temperature and its devastating effects on the world has attracted the attention of both scientists and politicians.

2.2 Nature of Climate Change

The climate change is a complex atmospheric-oceanic phenomenon on a global and long-term scale that is caused by various natural factors such as solar activity, volcanic activity and abnormal factors, including greenhouse gases' increase in the Earth's atmosphere and their interaction (Taghdisian and Minapour 2003). These factors lead to changes in climate, changes in temporal and spatial distribution of precipitation and its type (solid or liquid) and are followed by changes in the pattern of climate-related natural disasters such as floods, changes in the characteristics of surface streams, evaporation and transpiration, groundwater recharge and water quality. Generally, the term climate refers to the long-term average of climatic data, and any change in it is called "climate change" (Ashofteh 2014). In other words, climate change can be interpreted as statistically significant changes in the average climate behavior of a region. These changes can be observed in the amount of received solar radiation, the average temperature, wind speed and direction, the amount and distribution of precipitation, ocean circulation, and other climatic and hydrological variables. When the climate of a region becomes either warmer or cooler, the amount of precipitation can increase or decrease, but the severity of the impact of climate change on the temporal and spatial dimensions can be highly variable (Henshaw et al. 2000).

In order to get a proper understanding of climate change, we must first get acquainted with the mechanism of the atmospheric system and heat absorption on the planet. As shown in Fig. 2.3 solar radiation reaches the Earth's atmosphere after passing through space. Most of it passes through the atmosphere and is absorbed by the Earth. This process will warm the Earth and provide energy for survival. Part of the solar radiation is also reflected and released from the Earth's atmosphere without being absorbed. As is described in key processes in Physics, objects that obtain energy reflect part of it. As such the Earth reflects part of its absorbed energy, which has a longer wavelength and less energy than the original received radiation. Research has shown that the Earth emits infrared radiation after warming (Ben Slama 2016). Some of these infrared rays pass through the atmosphere, while others are absorbed by greenhouse gases in the atmosphere and are reflected back to the Earth. In fact, greenhouse gases act as semi-transparent object. They do not pass all the infrared radiation emitted from the Earth's surface and return some of them back to the Earth's surface, which causes the Earth to receive more heat energy.

Greenhouse gases involved in this process are water vapor (H_2O), Carbon dioxide (CO_2), Nitrous oxides (N_2O), Methane (CH_4), Ozone (O_3), Chlorofluorocarbons (CFCs), Hydrofluorocarbons (HFCs), and Perfluorocarbons (PFCs). These gases

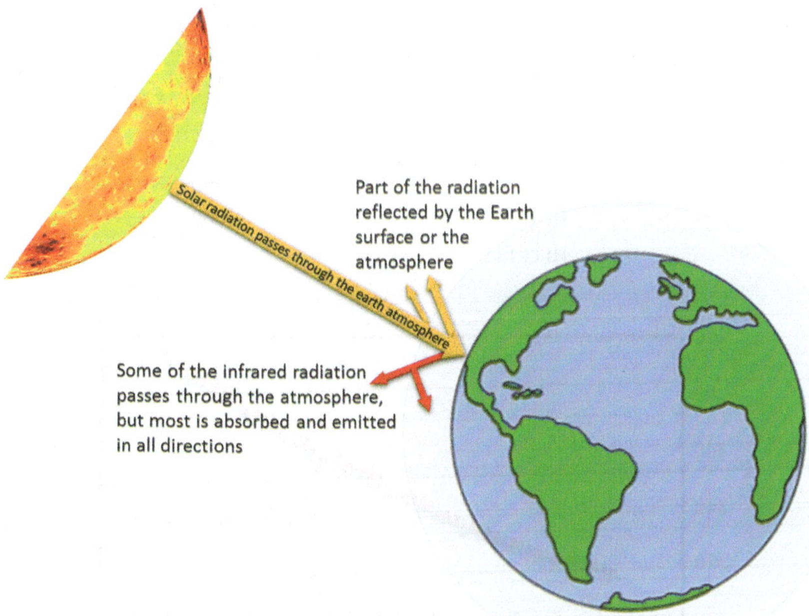

Fig. 2.3 The natural process of energy absorption by the Earth

are considered as direct radiative forces which participate in radiation absorption. However, there are also indirect greenhouse gases, which, after chemical transformation, result in the formation of direct forces and global warming. Indirect greenhouse gas are as follows: non-methane Volatile Organic Compounds (NMVOCs), Nitrogen oxides (NOx) comprised of Nitrogen monoxide (NO) and Nitrogen dioxide (NO_2), Sulfur dioxide (SO_2), and Carbon monoxide (CO).

To compare the effectiveness of greenhouse gases on the surface-troposphere system with each other, the concept of Global Warming Potential (GWP) has been introduced (Table 2.1). It shows the conceivable warming effect resulting from the emission of each gas relative to CO_2 for the contemporary atmosphere. Yet, future changes in chemical composition of the atmosphere would change the GWP since the relation between the radiative forcing and CO_2 atmospheric concentrations is non-linear.

Amongst the greenhouse gases water vapor, carbon dioxide and methane have a greater effect on heat retention due to the relatively high concentration of gasses in the atmosphere. Concentration of greenhouse gases has dramatically risen over the last century, with the changes in the CO2 concentration is shown in Fig. 2.4.

As previously mentioned, the exchange of energy naturally stabilizes the Earth's temperature to a suitable level, so that in the absence of greenhouse effects in the

Table 2.1 Direct GWP for 100-year time horizon (FAR-1992)

Gas	GWP
Carbon dioxide	1
Methane	11
Nitrous oxide	270
CFC-11	3400
CFC-12	7100
HCFC-22	1600
HFC-134	1200

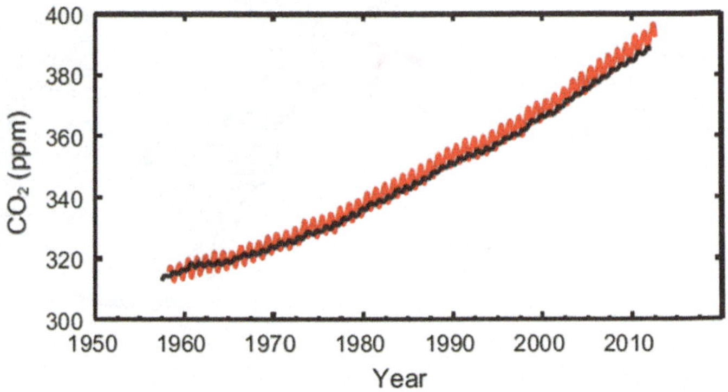

Fig. 2.4 Changes in CO_2 concentration over the last 6 decades (IPCC 2014)

atmosphere, the Earth's temperature would be on average be 15.5 degrees Celsius lower than what it is now. Such low temperature would make the planet uninhabitable (IPCC 2014). However, if the concentration of greenhouse gases inside the atmosphere surpasses the normal level, the Earth's energy balance will be disturbed, causing more heat energy to remain inside the Earth's atmosphere, leading to an abnormal warming of the Earth. The phenomenon of rising average temperature in the atmosphere and oceans, which is mainly due to increased concentrations of greenhouse gases and heat energy trapped in the atmosphere, is called global warming. This phenomenon, which began in the late nineteenth century and is expected to continue in the coming years will lead to increasing global climate change.

Figure 2.5 shows the proportion of different greenhouse gases that constitute the greenhouse gas effect. It should be noted that water vapor is usually calculated separately, because its amount is much higher than other greenhouse gases and the humidity of different areas varies depending on different temperature and weather conditions.

In Fig. 2.5, the proportion of carbon dioxide from all sources (emitted from fossil fuels combustion, industrial activities and deforestation) is 76%, illustrating the significant impact that the release of CO_2 has on the greenhouse effect. Methane accounts for the second-highest greenhouse gas in the atmosphere that causes global warming. Other greenhouse gases have less effect on global warming than these two gases.

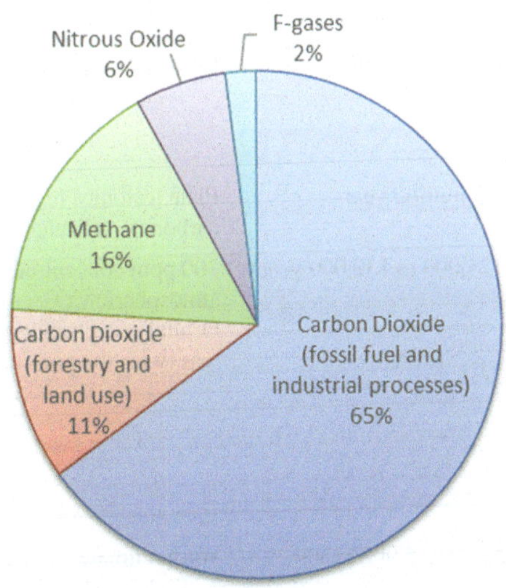

Fig. 2.5 The proportion of different greenhouse gases in greenhouse effect (IPCC 2014)

2.3 History of Climatic Change

Despite contemporary concerns regarding climate change, it has been the case that the Earth has experienced vast ongoing changes to climate over its life time (Table 2.2). Scientists estimate that the formation of solar system begun around 4.5 billion years ago. During planetesimals, gravity combined dust in space into solid chunks which formed Earth. Scientists believe the sun's luminosity has risen about 30% over the life time of the solar system. However, the geosphere has not been frozen when the sun was rather faint, which might be due to the presence of high concentration of greenhouse gases at that time. Before oxygen accumulation in the atmosphere, methane with two or three-fold of today's concentration was the dominant greenhouse

Table 2.2 Historic millstone of the Earth climate variation (Wallace and Hobbs 2006; FAR 1992)

Years before present	Incidence	Results
4.5 billion years	continual bombardment by smaller planetesimals	Liberated water vapor and other volatile substances, forming a primordial atmosphere and the oceans
3.8 billion years	Stable condition	Microbial life form
2.4–1.9 billion years	Photosynthesis emergence	Oxygen accumulation in the atmosphere Ozone layer formation First major glaciation
65 million years	Cretaceous epoch	High concentration of CO_2 in comparison with today High air temperature particularly at high latitude
2.5 million years	Plate tectonics regulate carbonate–silicate cycle	Start point of declining CO_2 concentrations
125,000 to 130,000 years	300 ppm concentration of atmospheric CO_2 Orbital eccentricity of the Earth was twice the present	Some of the Pleistocene interglacial were noticeably warmer (1 to 2 °C) than present Higher precipitation in majority parts of the Northern Hemisphere (30–50% higher than today) Maximum global mean sea level (5–10 m higher than present)
5,000 to 6,000 years	Warm climate	The Early and Middle Holocene Summer temperature was 3–4 °C higher than today in high latitudes Increased precipitation in high latitude and subtropical regions Lakes in mid-latitude had lower level than today

gas. The earth experienced major glaciations three times: 2.2–2.4 million years ago, 600–750 million years ago, and 280 million years ago. Fluctuation in CO_2 concentration plays a key role in glaciation and ice-free state, which are known as Ice-house and Hot-house. Volcanic activities and metamorphism emissions on one side, and plate limestone formation in the oceans on the other side led to severe increase and decrease of CO_2 levels, respectively. The Earth's climate has recorded Ice-house state over the last 3 million years; however, anthropogenic greenhouse gas emissions drive the planet toward the Warm-house and Hot-house status after 34 million years.

Over the 10,000 to 15,000 years before present (BP), the last ice age occurred. In this glacial era, North America and Scandinavia were covered by ice sheets and sea level around the world was around 120 m lower than present value. The periodic change in orbital pattern of the earth around the sun (the Earth distance from the sun and the angle of the Earths' axis) as a reason for these regular variations is approved for 15, 40, and 100 thousands years period and known as Milankovitch Theory (Berger 1980; Cuffey and Brook 2000). In addition, over the Quaternary period variations in CO2 and CH4 concentrations are significant factors as a relationship between their concentration and the Earth's temperature is observed (Fig. 2.6). In fact, there is an assumption that these two greenhouse gases modify or intensify the other driving impacts.

Current interglacial epoch has begun since 10 thousand years ago. In this period, although global temperature fluctuation has been greatly smaller, some of the fluctuations, such as Little Ice Age, continued several centuries (Fig. 2.7). Changes in the atmosphere composition or the Earth orbital alternation cannot be the reason for climate change over the centuries. Although there is much debate on the reason of the century-scale fluctuations, changes in sea surface temperature have been attributed to these fluctuations. The Younger Dryas cold episode around 10,500 BP, is another

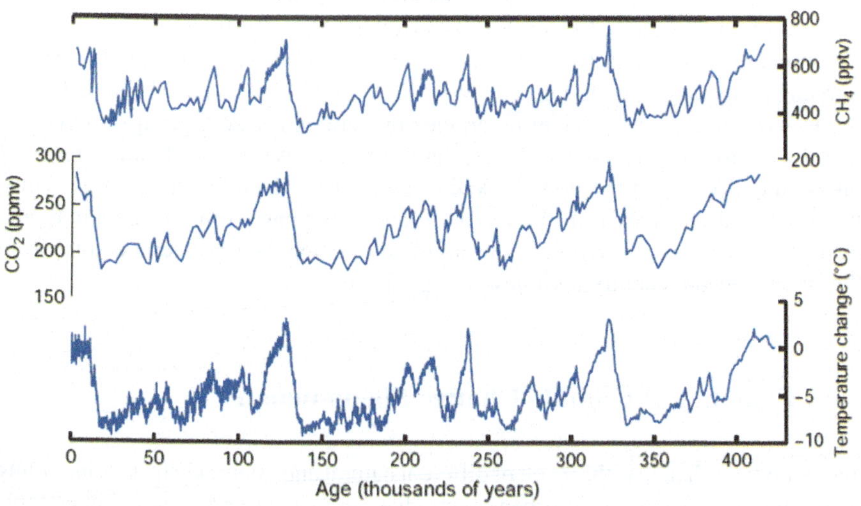

Fig. 2.6 Concentration of methane and carbon dioxide, and projected temperature from the Vostok ice core, Antarctica, over the last 440 thousand years (Petit et al. 1999)

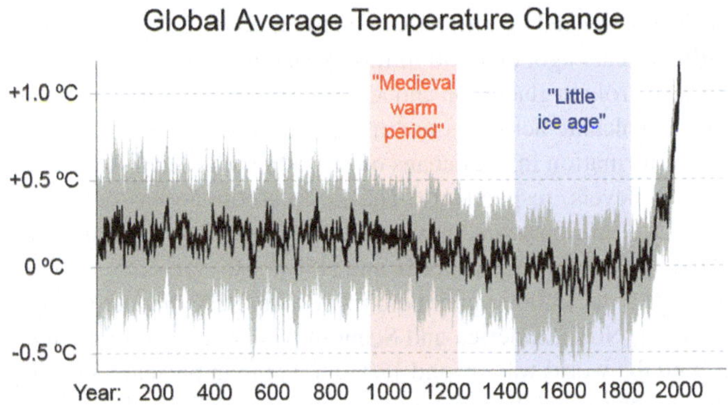

Fig. 2.7 Global average temperature over the last two thousand years (Hawkins, 2020)

example of continental glaciation recorded in New Zealand (Salinger 1989). The assumption is that large scale melting of the Laurentide Ice sheet occurred to offset the decrease in deep water production in the North Atlantic. So massive amounts of freshwater with low density flowed into the North Atlantic Ocean which was followed by alterations in the global oceanic circulation (Broecker et al. 1985). A reoccurrence of the Younger Dryas is highly probable. Indeed, changes in the thermohaline circulation (less than in the Younger Dryas) as a result of global warming may increase the precipitation rate over the extratropical North Atlantic.

The global average temperature following the last glaciation has had minor changes (less than 2 °C). But changes in regional scale have been significant. For instance, the humidity in the Sahara has increased significantly over time (Hulme, 1992 #158), and where as once (12,000–4000 years BP), Lake Chad occupied an area as large as the Caspian Sea, following 4000 BP climate conditions changed and the lake's basin began to dry out (Street-Perrot and Harnson 1985).

Observations by satellite images show a decrease in the volume of polar ice due to rising temperature and rising sea levels. These observations have been published in the form of accredited scientific articles by various research groups around the world with different access levels to technology, databases and different sources of uncertainty (Turner and Overland 2009; Cogley 2009; Marzeion et al. 2012). Thus, it can be said that based on the research, and according to the observations and measurements made, the unusual upward trend of rising the Earth temperature from the late nineteenth century until now, is significant.

2.4 Evidence of Climate Change Phenomenon

Today, climate change is the result of different natural and manmade factors including biotic changes, changes in absorbed solar radiations by the Earth, changes in tectonic plates, volcanic eruptions, and increases in greenhouse gas concentration. Despite

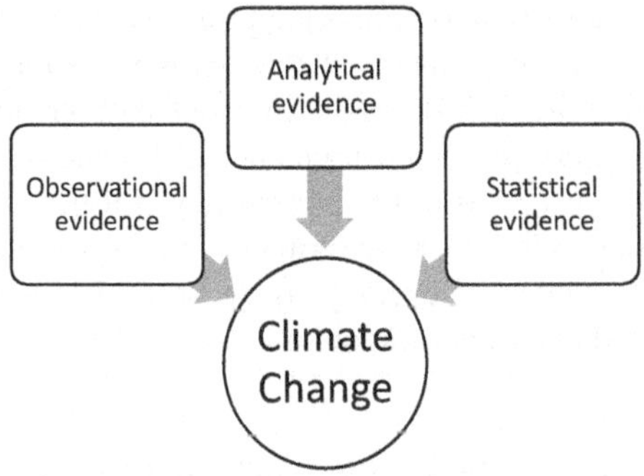

Fig. 2.8 Axes of climate change phenomenon proof

significant agreement by scientists on the existence of climate change there are still many scientific discussions regarding the proportion, intensity and factors influencing the change. The evidence of climate change can be proven via three main assessment category which are shown in Fig. 2.8, and explained separately in this section.

2.4.1 Observational Evidence

Climate change has not only been recorded via measurement tools, but it has been accompanied with many vivid effects on hydrologic cycle and biologic system. Tree rings, marine sedimentary cores, and ice cores provide quantitative information about past regional and global climates and changes in atmospheric composition. Paying attention to these changes can provide important information about how the Earth system behaves in the face of its internal and external changes on regional and even global scale. Therefore, such an approach leads to understanding the climate mechanism in the past and predicting its behavior and changes in the future (Zhang 2015; Yang et al. 2017). Information from gases trapped in glaciers confirm an increase in the concentration of greenhouse gases, including carbon dioxide and methane. It is proven that increase in the average temperature of the Earth is directly related to the concentration of greenhouse gases (IPCC 2014).

2.4.2 Statistical Evidence

Observations in climate systems are made through direct physical and biogeochemical measurements as well as remote sensing by ground stations and satellites, which provide long-term information on the conditions of the climate system. These measurements have been done recently, through the advancement in technology, but statistical analysis allows for data to be extrapolated to look further into the past. For example, advanced statistical methods using satellite imagery like the Gravity

Recovery and Climate Experiment have been used to quantify the unabated mass loss in the Patagonian ice-field caused by warming of the climate system (Ndehedehe and Ferreira, 2020). Large-scale global observations began in the middle of the nineteen century and led to the creation of climate system databases from hundreds to millions of years ago to the present day. All this information allows for a comprehensive observation of long-term changes in the lithosphere, atmosphere, hydrosphere, and cryosphere. Generally, the recorded data are divided to two class: climatic variables and hydrologic variables.

2.4.2.1 Climatic Variables and Historic Trends

Climatic parameters allow comparison between various geographical and temporal scales (Table 2.3). The long-term records of the indicators provide useful data for analyzing the changes of climate condition. These data are collected by instruments on land synoptic stations, satellite, buoys, and other platforms.

Table 2.3 Applied climate variables

Indicator name	Definitions
Frost days	Annual count when TN(daily minimum) <0 °C
Summer days	Annual count when TX(daily maximum) >25 °C
Ice days	Annual count when TX <0 °C
Tropical nights	Annual count when TN >20 °C
Growing season length	Annual (1st Jan to 31st Dec in NH, 1st July to 30th June in SH) count between first span of at least 6 days with TG >5 °C and first span after July 1 (January 1 in SH) of 6 days with TG <5 °C
Max Tmax	Monthly maximum value of daily maximum temp
Max Tmin	Monthly maximum value of daily minimum temp
Min Tmax	Monthly minimum value of daily maximum temp
Min Tmin	Monthly minimum value of daily minimum temp
Cool nights	Percentage of days when TN < 10th percentile
Cool days	Percentage of days when TX < 10th percentile
Warm nights	Percentage of days when TN > 90th percentile
Warm days	Percentage of days when TX > 90th percentile
Warm spell duration indicator	Annual count of days with at least 6 consecutive days when TX > 90th percentile
Cold spell duration indicator	Annual count of days with at least 6 consecutive days when TN < 10th percentile
Diurnal temperature range	Monthly mean difference between TX and TN

TN: Minimum air Temperature; TX: Maximum air Temperature; TG: Temperature Gradient; SH: South Hemisphere; NH: North Hemisphere

Generally, these indicators show variations in temperature. These parameters affect the hydrologic cycle directly and biologic system indirectly. As can be seen in Fig. 2.9a, until 1900 there was no significant increase in global average temperature. But since 1900, there has been a significant upward trend in both the Earth's average temperature and the ocean's average temperature, with the last three decades consistently warmer than the previous decades, while the 2000s finally being recognized as the warmest decade (IPCC 2014).

Between 1983 and 2012, the Northern Hemisphere recorded the warmest period over the last 1400 years. The linear trend of the global average temperature (combined land and ocean surface) indicates 0.85 °C (0.65–1.06) increase from 1880 to 2012. Almost all parts of the world have experience higher temperature 1–2.5 °C, except northern part of the north Atlantic Ocean.

Since short term data are sensitive to beginning and end dates, they are not suitable for reliably illustrating long-term trends. For instance, data assessment shows 0.05 °C per decade increase in temperature for 1998–2012 period, while the value for 1951–2012 is 0.12. The second value is more compatible with occurrence of severe El Nino. Reconstruction of data shows the earth surface temperature during 950–1250 epoch (the Medieval Climate Anomaly) was as high as the last 30 years, however, it was not as extended as observed in recent times. In addition to the Earth surface, observations show an increase in the troposphere temperature since the mid-twentieth century particularly in the extratropical Northern Hemisphere. The number of cold days and nights has fallen, while the number of warm days and nights has risen. Also more heat waves have been recorded in Asia, Australia, and majority of Europe (IPCC 2014).

According to the recorded data, the average heat content of oceans around the globe has increased. The upper layer of oceans with 700 m depth has warmed from 1971 to 2010 since it stores more than 60% of the net energy rise in the climate system. In this period, temperature of the upper 75-m layer has increased at 0.11 °C. Temperature of the lower depth also has changed as the remained 30% of the increased energy is stored in it. The depth of 700–2000 m became warmer from 1957 to 2009, while there were no major changes in the depth of 2000–3000 m in the 1992–2005 period. The depth below 3000 m had been warmer in this period and the Southern Ocean recorded the highest warming rate (IPCC 2014; Fig. 2.10).

Other important ocean changes relate to the level of salinity. In areas with higher initial salinity and higher levels of evaporation, the overall level of salinity has increased. In contrast, in areas with lower initial levels of salinity and higher levels of precipitation salinity has decreased.

Another important indicator of increased temperature is ice sheets' shrinkage. As shown in Fig. 2.11, ice extent has decreased since 1980s, and consequently sea level has risen in comparison to the average of 1986–2005. From 1971 to 2009, the average loss in ice was 226 Gt per year, with significant increased losses since 1993. Rate of ice sheets' melting in Greenland rose from 34 to 215 Gt/yr over the 1992–2001 period in comparison to the 2002–2011 period. The northern Antarctic Peninsula and the Amundsen Sea sector of West Antarctica have had the highest proportion of ice mass loss. The loss rate of the Arctic sea ice was 3.5–4.1% per decade from 1979

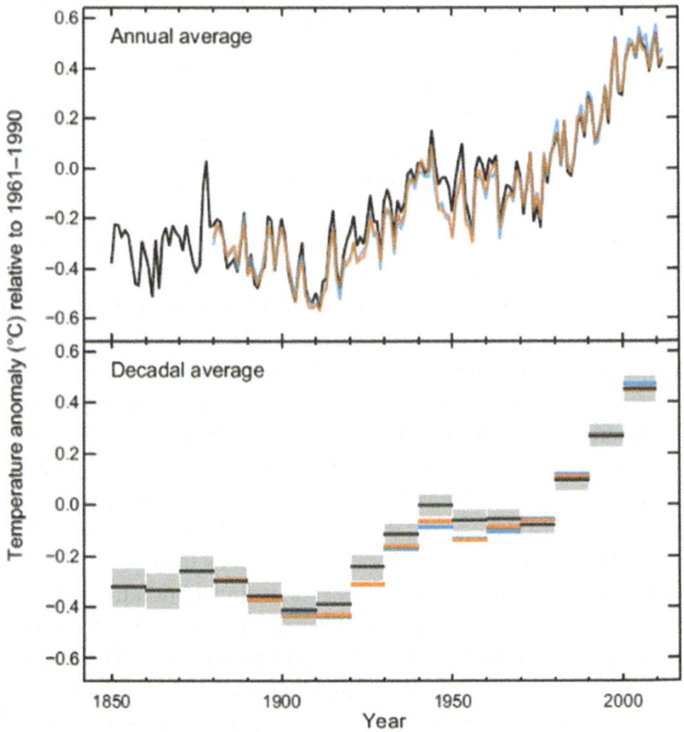

a)

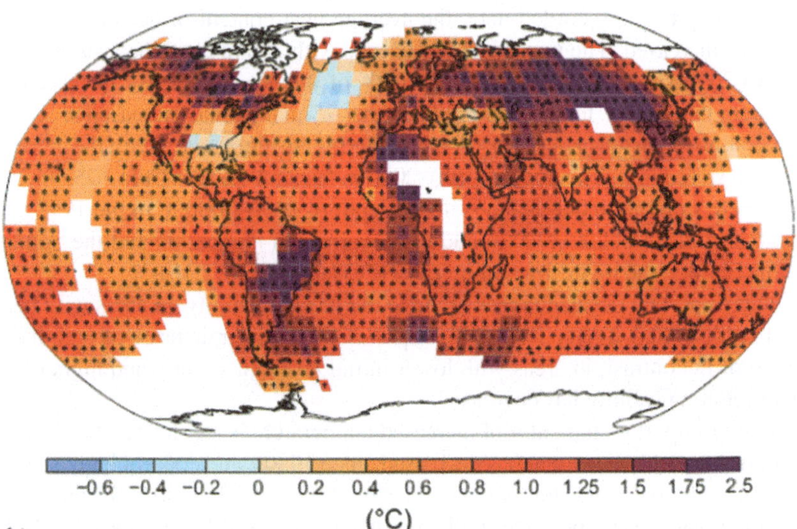

b)

Fig. 2.9 Changes in global average temperature relative to 1961–1990 (a) average of the land and ocean temperature in annual and decade scale, (b) map of the changes in the Earth's surface temperature (IPCC 2014)

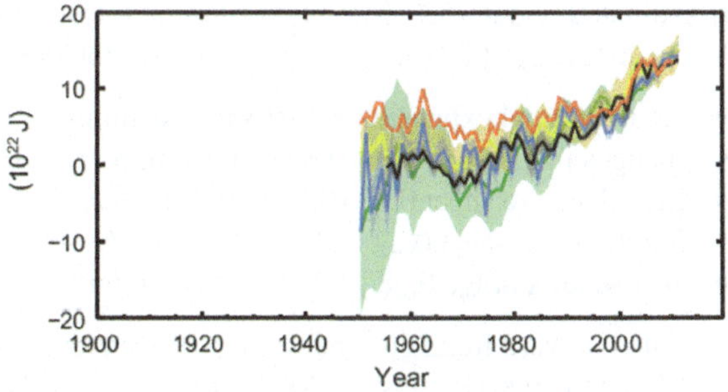

Fig. 2.10 Changes in global average heat content of upper 700-m in ocean (IPCC 2014)

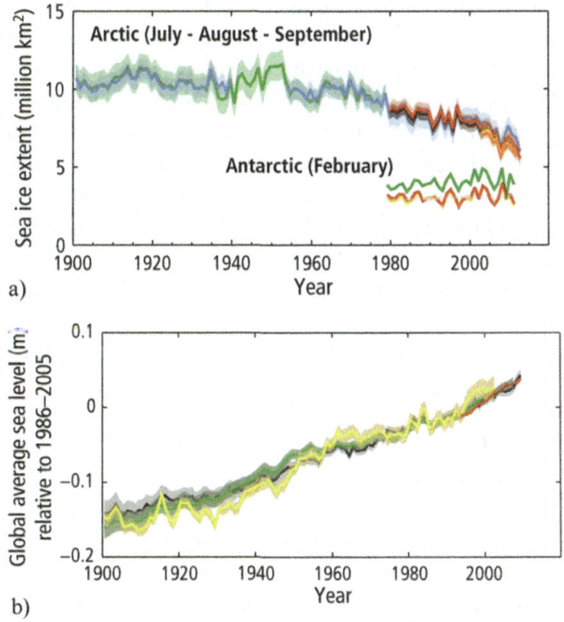

Fig. 2.11 **a** Arctic (July to September average) and Antarctic (February) sea ice extent, **b** Global mean sea level relative to the 1986–2005

to 2012 with fluctuations during different seasons. Reduction in the volume of polar glaciers is reported in many scientific articles (Turner and Overland, 2009; Cogley, 2009; Marzeion et al. 2012). In contrast, the Antarctic sea ice extent grew 1.2–1.8% on average per decade between 1979 and 2012.

Furthermore, global warming of 1.5 to 4.5 °C can raise sea levels by 20 to 140 cm, respectively (Taghdisian and Minapour 2003). Over the past 100 years, sea levels have risen by an average of 17–21 cm resulted from an increase in lower atmospheric temperature only 0.3 °C from 1860. It was higher than the mean rate of the previous two millennia. Sea level increased at the beginning of the twentieth century and has continued with an upward trend. The annual average sea level rise was 1.7 mm between 1901 and 2010, with much higher rises in the period 1993–2010. The role of glacier mass loss as well as ocean thermal expansion since 1970s in sea level rising is estimated about 75%. The proportion of various resources in increasing the sea level equal to 2.8 (2.3–3.4) mm per year over the period 1993–2010 is:

- ocean thermal expansion 1.1 (0.8 to 1.4) mm/yr
- changes in glaciers 0.76 (0.39 to 1.13) mm/yr
- Greenland ice sheet 0.33 (0.25 to 0.41) mm/yr
- Antarctic ice sheet 0.27 (0.16 to 0.38) mm/yr
- land water storage 0.38 (0.26 to 0.49) mm/yr.

If this upward trend continues, in 2100, sea levels will rise to about 0.5 to 1.4 m compared to 1990 (Rahmstorf 2007). Rising sea level is a devastating issue. Not only will it cause land areas to submerge, it will lead to infiltration of saline water from the seas and oceans into the freshwater resources, making them unusable. Additionally, in the present century, it is predicted that the lives of millions of people in different parts of the world will be threatened by rising sea levels. Economic and environmental damages are also important effects of these changes (Dasgupta et al. 2008).

In terms of the permafrost temperatures, since the early 1980s it has risen in almost all regions. Parts of Northern Alaska became warmer by up to 3 °C for the 1980–2000 period and this has increased in parts of the Russian European North up to 2 °C during 1971 to 2010, which was accompanied by a decline in permafrost thickness.

Finally, it is noteworthy that uncertainty is an inseparable part of climate variables assessment, especially when long-term data are not accessible. In Table 2.4, the reliability of climate phenomenon assessment by IPCC is summarized.

Table 2.4 Global-scale assessment of recent observed changes in extreme climate events

Phenomenon and direction of trend	Assessment that changes occurred (typically since 1950 unless otherwise indicated)
Warmer and/or fewer cold days and nights over most land areas	90–100% probability
Warmer and/or more frequent hot days and nights over most land areas	90–100% probability
Warm spells/heat waves. Frequency and/or duration increases over most land areas	Medium confidence on a global scale 66–100% probability in large parts of Europe, Asia and Australia
Heavy precipitation events. Increase in the frequency, intensity, and/or amount of heavy precipitation	66–100% probability
Increases in intensity and/or duration of drought	Low confidence on a global scale 66–100% probability changes in some regions
Increases in intense tropical cyclone activity	Low confidence in long term (centennial) changes Virtually certain in North Atlantic since 1970
Increased incidence and/or magnitude of extreme high sea level	66–100% probability (since 1970)

2.4.2.2 Hydrologic Variables and Historic Trends

Precipitation is the most important hydrologic variable for describing the effect of climate change on the hydrologic cycle. In Table 2.5, the hydrologic variable applied to illustrate climate change are listed.

According to several databases, precipitation has changed considerably all over the world during the last 30 years. Globally, areas in the mid-latitudes of the Northern Hemisphere have experienced more precipitation since 1901. This is due to observed increase in frequency of heavy rainfalls on different temporal and spatial scales. Local observations show an increase in the number of droughts on the one hand, and a growth in the occurrence of extreme rainfall on the other hand since 1950 and particularly between 1990 and 2010. Figure 2.12 highlights changes in precipitation around the world. More extreme rainfall and severe droughts have been observed since the second half of the last century. The changes are much more severe today than in the early 1950s (IPCC 2014). Although changes in snowfall in comparison to changes in rainfall are uncertain, global warming leads to the shorter snowfall season in most parts of the northern hemisphere, and earlier beginning of the cold season (Takala et al. 2009). The snow cover in the Northern Hemisphere has reduced since the 1950s, with the average reduction from 1967 to 2012, of 1.6 (0.8 to 2.4) % for March and April, and 11.7 (8.8–14.6) % for June.

In Norway, for example, an increase in average temperature has reduced snow water equivalent (Wong et al. 2011). Additionally, continuous increase in local and global evapotranspiration since 1960s, have been attributed to changes in precipitation, daily temperature, aerosols' density, vapor pressure deficiency and wind speed (McVicar et al. 2010; Miralles et al. 2010, Sushanta Sarkar and Sarkar 2018). Other

Table 2.5 Applied hydrologic variables

Indicator name	Definitions
Max 1-day precipitation amount	Monthly maximum 1-day precipitation
Max 5-day precipitation amount	Monthly maximum consecutive 5-day precipitation
Simple daily intensity index	Annual total precipitation divided by the number of wet days (defined as PRCP > = 1.0 mm) in the year
Number of heavy precipitation days	Annual count of days when PRCP > = 10 mm
Number of very heavy precipitation days	Annual count of days when PRCP > = 20 mm
Number of days above nn mm	Annual count of days when PRCP > = nn mm, nn is user defined threshold
Consecutive dry days	Maximum number of consecutive days with RR < 1 mm
Consecutive wet days	Maximum number of consecutive days with RR > = 1 mm
Very wet days	Annual total PRCP when RR > 95th percentile
Extremely wet days	Annual total PRCP when RR > 99th percentile
Annual total wet-day precipitation	Annual total PRCP in wet days (RR > = 1 mm)

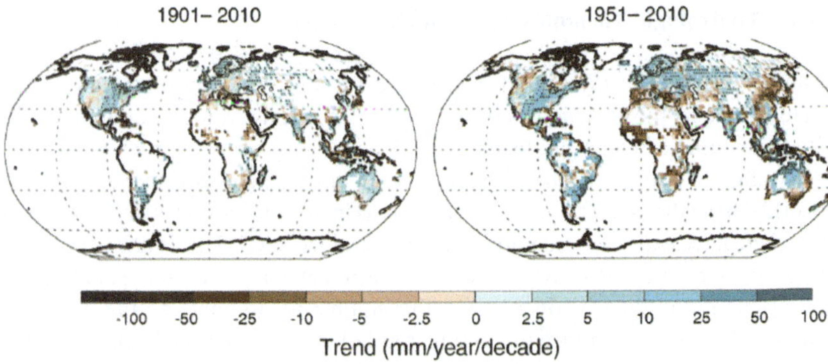

Fig. 2.12 Map of observed changes in precipitation over land; (IPCC 2014)

research in the arid region of Northwest China indicates increase in temperature during winter and rise in precipitation during summer since 1961. The headwater of some rivers in the region has fallen due to decline in precipitation. Some rivers have recorded higher flow rate on account of rapid glacier shrinkage and snowmelt and more precipitation in mountainous area (Wang and Qin 2017).

The observed trends in stream flow values are associated with local changes in rainfall and temperature values since 1650s. In Europe, the amount of stream flow in the southern and eastern regions shows a significant decrease between 1962 and 2004, while in other parts of Europe, especially in the northern altitude, the amount of stream flow is increasing (Stahl-2010). In North America, stream flow in the Mississippi basin increased between 1951 and 2002. On the other hand, stream flow has declined in the Northwest Pacific as well as in the South Atlantic (Kalra-2008). In China, the Yellow River's flow has declined from 1960 to 2000 due to the 12% reduction in summer and autumn rainfall, while observations made on the Yangtze River show a slight increase in the annual flow due to the increase in monsoon rains (Piao-2010). However, such changes in stream flow values must be carefully and thoroughly considered because various factors in addition to climate change, land use change, irrigation systems, and urban planning can affect the amount of river flow.

Global analysis and simulation of river flow rates show a dramatic increase in stream flow for about one-third of the world's 200 major rivers, including the Congo, Mississippi, and Yenisei rivers, and a decrease in 45 of these rivers. The flow reduction in low and medium altitudes is due to the drought and heat periods in West Africa, southern Europe, southern and eastern Asia, eastern Australia, western Canada and the America, and northern parts of South America. While pointing out that climate change has far-reaching effects on the hydrological cycle, Berlando and Rousseau (Burlando and Rosso 2002) examined the effects of this phenomenon on runoff and rainfall-runoff in the Arno River Basin in central Italy, indicating a reduction in available water. But this reduction does not necessarily mean a reduction in the amount of water entering the river, it might be changes in temporal and spatial distribution of rainfalls. The results showed that changes in rainfall can lead to significant changes in runoff. Jamali et al. (2012) stated that Middle Eastern countries are vulnerable to climate change. Examining the effects of climate change on the hydrology of the Karkheh basin showed that in the short run, the average annual temperature and stream flow increase by 0.9 °C and decrease by 10–15%, respectively. The average annual rainfall in Karkheh basin does not change significantly, while the time of occurrence changes. In the long run, the average annual temperature increases by 2–4 °C and the average annual rainfall and runoff decrease by 15–17 and 25–32%, respectively. Such changes in runoff have a major impact on water resources and related management and planning.

Other important variation is soil moisture. Local fluctuations in soil moisture levels also show a significant decrease on average. For example, the trend in soil moisture values for China indicate an increase in soil drought, a longer drying time, and an increase in drought intensity to more than 37% of the land surface for the 1950–2006 (Wang et al. 2011). This is due to the increase in the number of dry days and the prolongation of dry periods (Gemmer et al. 2011; Fischer et al. 2013). It can be attributed to the increase in hot days and warm periods. Other regions in the tropics are also showing increased acceleration of the hydrological cycles. The predominance of extreme droughts in Brazil, for instance shows a rise in drought affected areas from 25% in 2012 to 70% in 2015 (Ndehedehe et al. 2020).

Although increased evapotranspiration, prolonged hot and dry periods, reduced glaciers, rising sea levels, and obvious changes in hydrological variables especially precipitation are the results of climate change phenomenon and global warming. But how this phenomenon affects the hydrological cycle is very controversial. In other words, since the hydrological cycle is a complex chain of phenomena and is affected by various factors, the effect of climate change on regional scale cannot be generalized. Many parts of the world, including medium altitude areas experience significant rainfall reductions, while higher altitudes show more tangible changes in rainfall patterns.

2.4.2.3 Analytical Evidence

Processes affecting climatic conditions can represent significant natural diversity. Even in the absence of effective external forces, different periodic changes are observed at large spatial and temporal scales. Many of these changes are represented by simple statistical distributions. But at present, many climatic components represent several states, for example, icy and intra-glacial cycles, as well as certain states of internal change in El-Nino oscillations. Moving between different states can be caused by natural changes or in response to external forces. The relationship between these changes, external forces and reactions to them reflects the complexity of the dynamics in the climate system. Climate change, whether due to natural changes or due to the use of external forces, can lead to changes in the likelihood or severity of certain climatic events (IPCC 2014).

Conceptual and numerical models of the Earth's climate system are used to confirm the existence of climate change. These models are made using the knowledge of the Earth's climate system and determine the degree of climate change and their effects on the planet. Conceptual models are based on the physics, chemistry, biology approaches, as well as complex climatic processes such as cloud formation and precipitation. Since the models are the product of current human knowledge and technology, they may be defective. However, the most important tool for analyzing uncertainty is to test various hypotheses to find cause-and-effect relationships, as well as to predict possible future changes in the Earth's climate, which show significant changes over time.

2.5 Importance of Studying Climate Change

The consequences of climate change during the twenty-first century have proved to be devastating and is considered to be among one of the most significant global threats of our times alongside poverty, nuclear weapons, famine, flood and drought (Martin 2012). Climate change would widely affect not only the human well-being, but also other living things. Therefore, assessing the reason of these changes and their consequent effects on national, regional, and international level is an essential and urgent task. These kinds of assessments lead to identifying the triggering factors, and facilitating the prediction and projection of probable consequences in the future. Acquiring knowledge in terms of the severity and occurrence of changes are important in developing successful mitigation and adaptation plans. Information gathering and integrated investigation improve the quality and safety of these plans.

To mitigate the destructive effect of climate change or to take appropriate adaptation measures relies on methods and measures introduced by national and international organizations. The Intergovernmental Panel on Climate Change (IPCC), in collaboration with the World Meteorological Organization and the United Nations Environment Program, identified climate change phenomenon, examined its effects, and addressed adaptation strategies. The IPCC has published five assessment reports so far in 1990, 1995, 2001, 2007 and 2014, and has planned to publish the sixth report by April 2021. The fifth and final evaluation report, published in 2013 and 2014, includes four separate reports:

1. The report of the first working group consisting of 258 experts with the aim of focusing on the physical sciences of the phenomenon of climate change;
2. The report of the second working group consisting of 302 experts with the aim of evaluating the effects of climate change and adaptation strategies;
3. The report of the third working group consisting of 271 experts with the aim of focusing on strategies to prevent the climate change and assessing the framework of risk and its uncertainty and;
4. Consolidated report as a summary and review of results.

Previous assessments and research undertaken by the IPCC show sufficient evidence to demonstrate the significant role of human activities in climate change around the globe (IPCC 2014). The most important and convincing evidence is based on observations of the properties of the lithosphere, atmosphere, hydrosphere, and cryosphere. The incontrovertible evidence of observations and statistics of ice cores in the past and present shows a significant increase in the important greenhouse gases of carbon dioxide, methane and nitrous oxide over the past few decades, all of which stemmed from human activities. Therefore, addressing climate change under an organizational framework is an urgent task. The establishment of organizations such as the IPCC has improved public awareness about climate change and attracted the attention of authorities to limit greenhouse gases emissions via legislation. Annual conferences held by the IPCC put more emphasize on the threats of climate change and effects of human activities on its deterioration.

Finally, to vividly demonstrate the importance of studying climate change, the destructive effects of climate change on the earth and terrestrial life is summarized in Table 2.6. Identifying these effects all around the world and taking appropriate measure to reduce their impacts are explained in the last three chapters of this book.

Table 2.6 Summary of the negative effects of climate change

Affected sector	Damage
Flora and fauna	Increased extinction of species
	Changes in phenomena such as the plants' blooming
	Changes in phenomena such as the plants' blooming
	Migration of insects, birds and mammals
	Hurt to the plant communities due to changes in river and lake areas, and water quality
	Overfeeding and consequent changes in species composition or flora communities' replacement
	Negative impact on the population of birds depending on the network of wetlands and lakes due to the changes in rainfall and flood patterns in large areas located in arid regions
	Decreased in the flora and fauna reproduction rate on account of the temperature stress (higher maximum temperature, more hot days, and heat waves)
	Higher minimum temperature, reduction of the number of frosty and cold days and reduction of cold waves will cause distribution and activity expansion of some insects and pathogens
	increased fire hazards and disease outbreaks, displacement of species and habitats due to drought and desertification

(continued)

Table 2.6 (continued)

Affected sector	Damage
Water bodies	Increasing lakes' temperature will lead to changes in the temperature cycle, solubility of oxygen and performance of the ecosystem
	Changes in the frequency and intensity of rainfall increase soil erosion and the amount of sediment in rivers
	Sedimentation along with the increased atmospheric nitrogen deposition affect the river's chemical status
	Increased evaporation due to the increase in average temperature can have a significant effect on the water balance of wetlands
	Increased the rate of evaporation rises the concentration of water solutes and consequent cost of water treatment
	Reduction in available water resources (2–6 °C increase in the air temperature rises the annual evaporation rate by 6–12%)
Soil	Increased soil erosion due to heavy and torrential rains
	Changes on the amount of different gases in the soil and the amount of vegetation
	decreases in the level of soils' organic matter as a result of increasing soil temperature
Air	Increase in the frequency of dust storms
	Increase formation of Tropospheric ozone
	Increase the air temperature
Health	spread of various infectious diseases such as malaria, snow fever and yellow fever
	Threaten people with heart disease in hot weather since the circulatory system is more active to keep the body temperature cool
	Problematic for people with asthma and lung patients due to the increase of tropospheric ozone concentration in hot weather
	Increase the mortality rate among children and the elderly during the hot days
	Damage human mental health as the disastrous phenomena (floods, fires) increase
Agriculture	Reduce crop production because of the occurrence of droughts frost, hail, flood, erosion, change in the rainy season, reduced irrigation efficiency, the reduction of water available for irrigation, and increasing crop water demand due to higher temperature
	Reducing soil moisture and increasing the rate of decomposition of organic matter in the soil
	Increasing drought stress and early maturity of plants
	Increasing new diseases and pests adapted to warm climates
	Increasing agricultural products' prices

(continued)

Table 2.6 (continued)

Affected sector	Damage
Energy	Reduced the power generation capacity of power plants because of increasing the water temperature reduces the efficiency of condensation process in steam power plants
	Reduced the power generation in gas power plants as the ambient temperature is more than the designed temperature
	Require more treatment and disinfection systems to monitor the corrosion of power plant equipment due to the increase in the salt concentration of water bodies resulted of rainfall reduction
	Limited production capacity of hydropower plants resulted of water shortage
	Rising electricity demand as the need for cooling systems increase
Society	Increase migration to urban area because of long drought in rural area
	Lose the cultural place, cemetery, and social relevancy due to rising sea level particularly in poor countries and islands
	Raise of poverty in society with agriculture dependent economy

References

Ashofteh P (2014) Climate change: tools and approaches. Kherad Publisher, Mashhad Iran

Bär R, Rouholahnejad E, Rahman K, Abbaspour KC, Lehmann A (2015) Climat change and agricultural water resources: a vulnerability assessment of Black Sea catchment. Environ Sci Policy 46:57–69

Ben Slama R (2016) Green house effect vs. infrared radiation emissions. J Climatol Weather Forecast 4(161):2

Berger A (1980) The Milankovitch astronomical theory of paleoclimates: a modern review. Vistas Astron 24:103–122

Broecker WS, Peteet DM, Rind D (1985) Does the ocean-atmosphere system have more than one stable mode of operation? Nature 315:21–26

Burlando P, Rosso R (2002) Effects of transient climate change on basin hydrology. 2. Impacts on runoff variability in the Arno River, central Italy. Hydrol Process 16(6):1177–1199

Cogley JG (2009) A more complete version of the World Glacier Inventory. Ann Glaciol 50:32–38

Cuffey KM, Brook EJ (2000) 18–Ice sheets and the ice-core record of climate change. In: Jacobson MC, Charlson RJ, Rodhe H, Orians GH (eds) International geophysics, vol 72. Academic Press, pp 459–497

Dasgupta S, Laplante B, Meisner C, Wheeler D, Yan J (2008) The impact of sea level rise on developing countries: a comparative analysis. Clim Change 93:379–388

FAR (first assessment report IPCC) (1992) Climate change: the supplementary report to the IPCC scientific assessment [Houghton JT, Callander BA, Varney SK (eds)]. Cambridge University Press, Cambridge, United Kingdom and New York, NY, USA, 116 p

Fischer EM, Knutti R (2013) Robust projections of combined humidity and temperature extremes. Nat Clim Chang 3:126–130

Gemmer M, Fischer T, Jiang T, Su B, Liu LL (2011) Trends in precipitation extremes in the Zhujiang River Basin, South China. J Clim 24:750–761

Henshaw PC, Charlson RJ, Burges SJ (2000) 6–Water and the hydrosphere. In: Jacobson MC, Charlson RJ, Rodhe H, Orians GH (eds) International geophysics, vol 72. Academic Press, pp 109–131

IPCC (2014) Climate change 2014: impacts, adaptation and vulnerability-regional aspects. Cambridge University Press, Cambridge, United Kingdom and New York, NY
Jamali S, Abrishamchi A, Marino M (2012) Climate change impact assessment on hydrology of Karkheh Basin. Water Manag 166(2):93–104
Martin J (2012) The meaning of the 21st century: a vital blueprint for ensuring our future. Random House
Marzeion B, Jarosch AH, Hofer M (2012) Past and future sea-level change from the surface mass balance of glaciers. Cryosphere 6:1295–1322
McVicar TR, Donohue RJ, O'Grady AP, Li L (2010) The effects of climatic changes on plant physiological andcatchment ecohydrological processes in the high-rainfall catch-ments of the Murray-Darling Basin. Commonwealth Scientific and Industrial Research Organiza-tion (CSIRO) Water for a Healthy Country National Research, Canberra, ACT, Australia
Miralles DG, Gash JH, Holmes TRH, De Jeu RAM, Dolman AJ (2010) Globalcanopy interception from satellite observations. J Geophys Res-Atmos 115
Ndehedehe CE, Agutu N, Ferreira VG, Getirana A (2020) Evolutionary drought patterns over the Sahel and their teleconnections with low frequency climate oscillations. Atmos Res 233:104700
Petit J-R, Jouzel J, Raynaud D, Barkov NI, Barnola J-M, Basile I, Bender M, Chappellaz J, Davis M, Delaygue G (1999) Climate and atmospheric history of the past 420,000 years from the Vostok ice core, Antarctica. Nature 399(6735):429
Rahmstorf S (2007) A semi-empirical approach to projecting future sea-level rise. Science 315(5810):368–370
Salinger MJ (1989) CO_2 and climate change: impacts on New Zealand agriculture. Pers Commun 5p
Sarkar S, Sarkar S (2018) A review on impact of climate change on evapotranspiration. Pharma Innov J 7(11):387–390
Sarzaeim P, Bozorg-Haddad O, Fallah-Mehdipour E, Loáiciga HA (2017) Environmental water demand assessment under climate change conditions. Environ Monit Assess 189(7):1–18
Schewe J, Heinke J, Gerten D, Haddeland I, Arnell NW, Clark DB, Dankers R, Eisner S, Fekete BM, Colón-González FJ (2014) Multimodel assessment of water scarcity under climate change. Proc Natl Acad Sci 111(9):3245–3250
Street-Perrot FA, Harrison SP (1985) Lakelevels and climate reconstruction. In: Hecht AD (ed) Paleoclimate analysis and modelling. John Wiley, New York, pp 291–340
Taghdisian H, Minapour S (2003) Climate change. Environmental Protection Agency, Tahran, Iran p 1
Takala OM, Pulliainen J, Metsämäki S, Koskinen J (2009) Detection of snowmeltusing spaceborne microwave radiometer data in Eurasia From 1979 to 2007. Ieee Trans Geosci Remote Sensing 47:2996–3007
Turner J, Overland JE (2009) Contrasting climate change in the two polar regions. Polar Res 28:146–164
Wallace JM, Hobbs PV (2006) Atmospheric science, 2nd edn. Academic Press, San Diego
Wang Y-J, Qin D-H (2017) Influence of climate change and human activity on water resources in arid region of Northwest China: an overview. Adv Clim Chang Res 8(4):268–278
Wang X et al (2011) Trends and low-frequency variability of storminess over Western Europe, 1878–2007. Clim Dyn 37:2355–2371
Wilby RL, Harris I (2006) A framework for assessing uncertainties in climate change impacts: low-flow scenarios for the River Thames, UK. Water Resources Res 42(2)
Wong WK, Beldring S, Engen-Skaugen T, Haddeland I, Hisdal H (2011) Climate change effects on spatiotemporal patterns of hydroclimatological summer droughts in Norway. J Hydrometeorol 12(6):1205–1220

Yang B, He M, Shishov V, Tychkov I, Vaganov E, Rossi S, Ljungqvist FC, Bräuning A, Grießinger J (2017) New perspective on spring vegetation phenology and global climate change based on Tibetan Plateau tree-ring data. Proc Natl Acad Sci 114(27):6966–6971

Zhang Z (2015) Tree-rings, a key ecological indicator of environment and climate change. Ecol Ind 51:107–116

Chapter 3
Climate Change Drivers

Hossein Ahmadi, Omid Bozorg-Haddad, Steven Lucas, Veljko Prodanovic, and Kefeng Zhang

3.1 Climate Change Origins

Changes in climatic variables and hydrological variables are confirmed by observations and long-term data analysis at ground and space stations. The changes in the climatic behavior of an area compared to the behavior expected on account of the long term recorded data in that area are called climate change and accompanied with devastating effects (Karamooz and Araghinejad 2005). Studies show that the Earth's surface and air temperature in the presence of greenhouse gases (GHGs) is 33 °C warmer than in the absence of these gases (Earth's current temperature is 15 °C, which in the absence of GHGs will drop to −18 °C). Accordingly, if the

H. Ahmadi
Department of Irrigation and Reclamation Engineering, Faculty of Agricultural Engineering and Technology, College of Agriculture and Natural Resources, University of Tehran, 31587-77871 Karaj, Tehran, Iran
e-mail: hahmadi134@ut.ac.ir

O. Bozorg-Haddad (✉)
Department of Irrigation and Reclamation Engineering, Faculty of Agricultural Engineering and Technology, College of Agriculture and Natural Resources, University of Tehran, Karaj, Tehran, Iran
e-mail: obhaddad@ut.ac.ir

S. Lucas
The Tom Farrell Institute for the Environment, University of Newcastle, Callaghan, NSW 2308, Australia
e-mail: steven.lucas@newcastle.edu.au

V. Prodanovic · K. Zhang
Water Research Centre, School of Civil and Environmental Engineering, University of New South Wales, Vallentine Annex, Building H22, Sydney, NSW 2052, Australia
e-mail: v.prodanovic@unsw.edu.au

K. Zhang
e-mail: kefeng.zhang@unsw.edu.au

O. Bozorg-Haddad (ed.), *Climate Change in Sustainable Water Resources Management*, Springer Water, https://doi.org/10.1007/978-981-19-1898-8_3

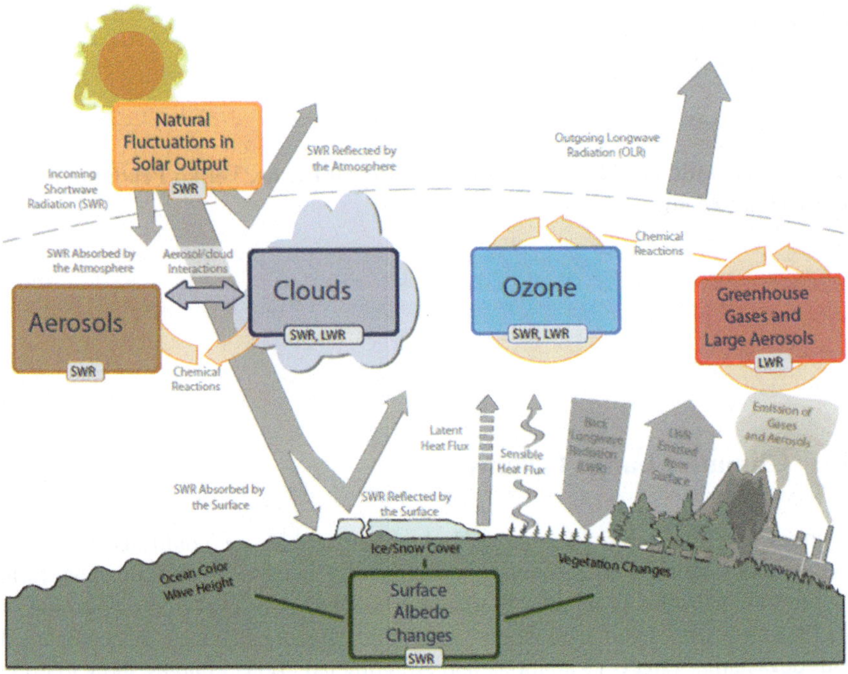

Fig. 3.1 Drivers of the climate change phenomenon (Solomon et al. 2007)

amount of GHGs in the Earth's atmosphere increases, more of the solar energy will be stored in the Earth, and this will further increase the temperature. According to the available evidence, the earth's temperature has increased by 0.3–0.6 °C over the past hundred years, which is proportional to the increase in GHGs in the Earth's atmosphere. However, climatologists believe that other factors also contribute to global warming and believe that the effect of greenhouse gas alone is not the cause of global warming. Generally, recognition of the origins of the climate change is the first step in tackling with this phenomenon. In this chapter, while examining the changes in the amount of GHGs in the Earth's atmosphere, other drivers of climate change either natural or anthropogenic are assessed. The factors which interrupt the radiative balance between received solar shortwave radiation (SWR) and outgoing longwave radiation (OLR) and consequently change the Earth climate condition are briefly shown in Fig. 3.1.

3.2 Greenhouse Gases

From about 1800, at the same time as the Industrial Revolution, there was a sudden increase in GHG emissions, which is still continuously increasing. Concentration of trapped GHGs in the ice cores have reached its highest value over the past 800 thousand years (Solomon et al. 2007). Over the past 200 years, and especially in the last 50 years, the emission of these gases have increased dramatically. Today, GHG emissions play a major role in the world economy, and modern life is unimaginable without them. However, effect of various GHGs is not equal due to the difference in their lifetime and radiative efficiency. To compare the role of various GHGs in climate change these two features are presented in Table 3.1.

Presenting the changes in climate system, radiative forcing which expresses the changes in net irradiance of the Earth system (i.e., the difference between solar irradiance absorbed by the Earth and energy radiated back to space, W/m^2) is defined. Numerous agents affect the radiative forcing which are discussed in this chapter, however, in terms of GHGs radiative efficiency (W/m^2.ppb) is the proper technical term for presenting the effect of GHGs on climate system. Molecular structure of the gases determines their radiative efficiency. Since changes in the concentration of the tropospheric gases is expressed in ppb, the radiative efficiency indicates the radiative forcing of a greenhouse gas for one ppb concentration. Therefore, multiplying radiative efficiency by the gas concentration, its radiative forcing can be calculated. For instance, in spite of the trifle amount of CO_2 radiative efficiency, its high tropospheric concentration poses it in the first place regarding radiative forcing. Life time is also another significant factor in analyzing climate change agents. In addition to

Table 3.1 Important features of dominant GHGs (Solomon et al. 2015; Panagopoulos et al. 2019)

Gas	Lifetime	Radiative efficiency (W/m^2.ppb)
CO_2	5–2000	1.37×10^{-5}
CH_4	9.1	3.63×10^{-4}
N_2O	131	3.03×10^{-3}
SF_6	3200	0.575
CF_4	50,000	0.1
C_2F_6	10,000	0.26
HFC-125	28.2	0.219
HFC-134	13.4	0.159
HFC-143	47.1	0.159
HFC-152	1.5	0.094
HFC-23	222	0.176
CFC-11	45	0.263
CFC-12	100	0.32
CFC-113	85	0.3
HCFC-22	11.9	0.2
HCFC-141	9.2	0.152
HCFC-142	17.2	0.186
CCl_4	26	0.175
CH_3CCl_3	5	0.069

ppb: part per billion

the mentioned GHGs in Table 3.1, water vapor is considered as the most effective gas with greenhouse effect in many scientific sources. In this section the role of the GHGs is explained separately.

3.2.1 Water Vapor

The largest proportion of GHG in the Earth's climate is water vapor. Air temperature controls evaporation rates and consequently the water vapor concentration in the atmosphere. Some experts classify this as a feedback mechanism instead of a forcing factor (IPCC 2014). Although the effect of water vapor in intensifying climate change is significant, it is not under scientists' attention as high as CO_2 due to the following reasons. Firstly, the lifetime of water vapor in the atmosphere is short (10 days). In high humidity and cool condition, the water vapor will condense and fall down to the land in the form of rain or snow. Secondly, share of its natural sources (evaporation from water bodies) is radically greater than the share of produced water vapor directly by anthropogenic sources (evaporation of agriculture sector, cooling system in power plants and industries, fossil fuels' combustion). When the air temperature increases by one degree Celsius, the atmosphere ability in containing water vapor grow at 7%. Therefore, it is essential to be aware of the forcing factors, particularly manmade agents, which leads to the increase in water vapor concentration by increasing the Earth temperature.

The stratospheric water vapor portion is affected by other GHGs, including CH_4 and CO_2. Based on the analyzed historical data, the increase in CO_2 concentration led to the rise of the Earth temperature and water evaporation. Indeed, the sustain level of the atmospheric water vapor owe to the presence of CO_2, otherwise the earth surface would be frozen. The oxidation of CH_4, emitted from human activates, in the atmosphere leads to the formation of water vapor. The amount of water vapor in stratosphere rose 1 ± 0.2 ppm between 1980 and 2010, in which around 25% is caused by CH_4 oxidation. Increase in the stratospheric water vapor leads to the fall in stratosphere temperature and rise in the troposphere temperature. It also participates in reduction of stratospheric ozone (Stenke and Grewe 2005). Changes in the H_2O vapor content in the stratosphere is illustrated in Fig. 3.2. H_2O vapor concentration, shown in Fig. 3.2, has been measured by satellite for 60° S to 60° N, and by balloon at 40° N (green point) and averaged over 30° N and 50° N (black line).

Concentration of water vapor in the stratosphere has experienced numerous fluctuations in long-term. While water vapor has had an upward trend since the second half of the twentieth century, no increase has been recorded since 1996. There is a close connection between the observed variations in stratospheric water vapor and changes in tropopause temperature in tropical regions (Fueglistaler and Haynes 2005). However, CH_4 oxidation and changes in tropical tropopause temperature and no other reason cannot provide convincing explanation for water vapor fluctuations from 1980 to 2010.

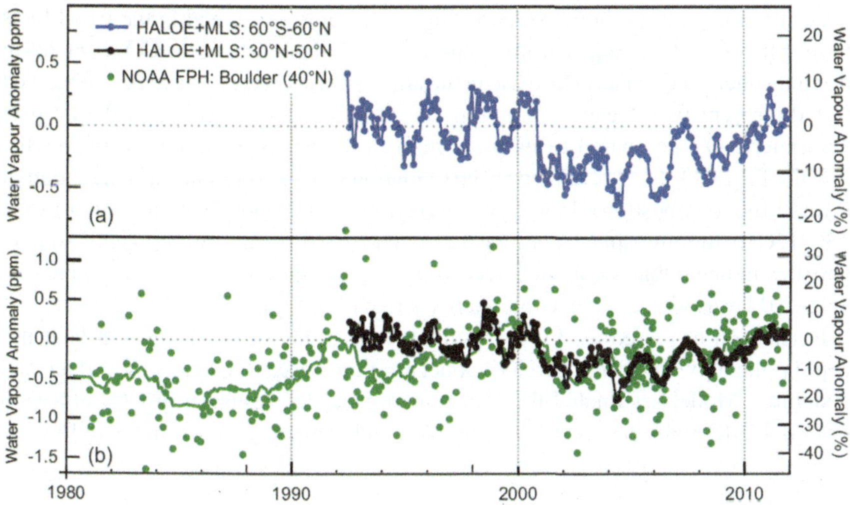

Fig. 3.2 Water vapor irregularities in the lower stratosphere at monthly scale measured by **a** HALOE and MLS satellite sensors, **b** balloon-borne instruments (Solomon et al. 2007)

In summary, even though the water vapor is not a forcing agent, it is a robust feedback intensify the primary forcing. It is noteworthy that the amplifying effect of water vapor is considered in all climate models.

3.2.2 Carbon Dioxide (CO_2)

Examining carbon dioxide emissions over a 160,000-year period, it is projected that as the concentration of CO_2 increases, so does the Earth's temperature, and vice versa. In other words, there is a direct relationship between the concentration of CO_2 and the earth's temperature with high confidence (Taghdisian and Minapour 2003). Important peak of CO_2 concentration and the Earth's temperature is listed in Table 3.2.

The main reasons of increase in CO_2 emissions between 1750 and 2011 are fossil fuel combustion and cement production. One other factor reflecting the carbon

Table 3.2 Significant rise of atmospheric CO_2 concentration

Period	Atmospheric CO_2 concentration	Global mean surface temperature
The mid-Pliocene	350–450 ppm	1.9–3.6 °C warmer than pre-industrial climate
The Early Eocene	Exceeded 1000 ppm	9–14 °C warmer than pre-industrial climate
1750–2011	345–405 Gigatonnes	–

concentration is atmospheric oxygen. Since oxygen is consumed in the fossil fuels burning process, the changes in atmospheric oxygen demonstrate the CO_2 contents in the atmosphere. The steady decrease in the atmospheric O_2 levels have been observed over the recent two decades (Manning and Keeling 2006). Although the drop in atmospheric O_2, compared to its proportion in the air, is negligible, it proves the growth of the CO_2 level is related to the oxidation process (organic carbon oxidation or fossil fuel combustion). In addition, reduction in the stable isotopic ratio of CO_2 (13C/12C) in the atmosphere approve the increase of CO_2 with anthropogenic source. Statistics indicate that the annual average CO_2 concentration was 8.3 Gigatonne in 2002–2011 period with an average increase rate of 3.2% per year (Fig. 3.3).

In addition to the emitted CO_2 resulted from combustion, changes in land use amplify the upward trend of CO_2 concentration based on land cover data and modelling. Models estimated the share of tropical deforestation in CO_2 emission about 0.9 Billion ton per year between 2002 and 2011 (Fig. 3.4). Active deforesta-

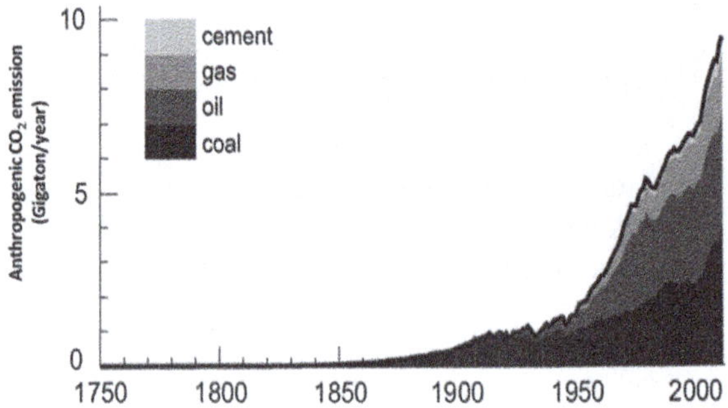

Fig. 3.3 CO_2 concentration change emitted by fossil fuel combustion and Cement production since 1750 (Solomon et al. 2007)

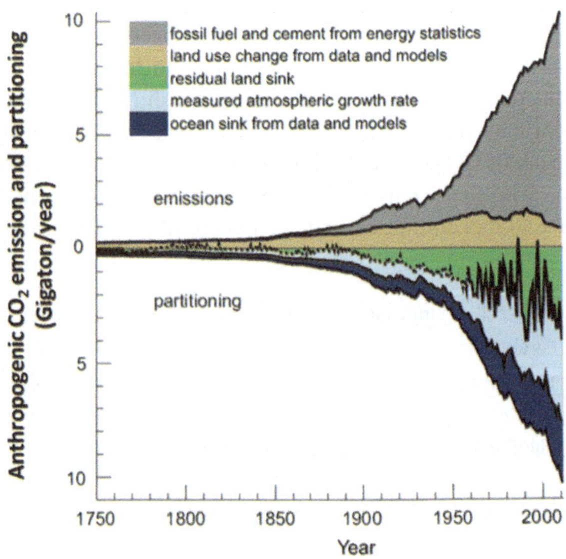

Fig. 3.4 CO_2 concentration change divided to sources and sinks since 1750 (Solomon et al. 2007)

tion for farming purposes causes three phenomena: (1) The soil composition changes rapidly and the organic matter in the soil is oxidized and enters the atmosphere as CO_2. (2) Harvested wood from forests either directly (consumed as fuel) or indirectly (such as use in chemical processes) can be a source of CO_2 emissions. (3) By deforestation, the possibility of photosynthesis process, in which CO_2 is consumed, is eliminated and more CO_2 will remain in the atmosphere.

As it is illustrated in Fig. 3.4, the ocean and terrestrial ecosystems absorb part of the extra carbon released to the atmosphere. Due to the result of many investigations, carried out based on the measurements of the atmospheric oxygen to nitrogen ratio or from CO_2 partial pressure data, 1.0–3.2 Gigatonnes of carbon have been absorbed by the oceans. In general, CO_2 is slightly soluble in water, but due to the vastness of the oceans, the amount of CO_2 dissolved in water is extremely significant. The carbon absorption by the oceans have led to the increase in pH (0.1 since the industrial epoch commencement) and have negatively affected the marine ecology. Forests also need and consume CO_2 through the photosynthesis process. Higher photosynthesis results from higher atmospheric CO_2 levels, and nitrogen deposition. Longer growing season is recognized as the role of the terrestrial ecosystems in carbon capture. The amount of carbon uptake by the forests is about 4 billion tonnes per year (de Moraes Sá et al. 2017). Absorbed CO_2 by land ecosystem is stored in leaves, roots, stems, and soil. However, the effect of these two natural carbon sinks is not compatible with (further less than) the emitted CO_2, as a constant increase in atmospheric CO_2 level is observed. Scientists revealed that from the emitted amount of CO_2, 15 to 40% of emissions will remain in the atmosphere for up to two millenniums. If the concentration of CO_2 continues to rise at the same rate during the current century, the concentration of this gas will reach a level that is equivalent to 50 million years of normal human activity. As the lifetime and huge volume of CO_2 emission make it the most important GHG, the turnover time for carbon transfers among its main reservoirs in global scales is depicted in Fig. 3.5.

On account of the long-time remaining of CO_2 in the ocean and soil, scientist make endeavor to remove atmospheric CO_2 and store it in the ground or ocean (Janzen 2004; Lal 2008). Summary of their effective measures, known as "Carbon Dioxide Removal" (CDR) is listed in Table 3.3. These measures are can be permanent and non-permanent. The permanent measures refer to the long-term (for over 1000 years) removal of CO_2 which affect the radiative forcing. The non-permanent measures, however, delay the transfer of CO_2 to the atmosphere for several years. The estimated effect of CDR measures on the future concentration of CO_2 is depicted in Fig. 3.6. As it expressed former, the non-permanent methods (green line) can only delay the carbon emission to the atmosphere.

It is noteworthy that most of the carbon dioxide removal techniques, implemented on land can change the surface albedo. For example, expanding the greenspace and plant cover, affects the evapotranspiration and hydrological cycle at the regional scale. The ocean-based measures also disrupt the marine ecosystems in long-term. Generally, by taking CDR actions, increased variability in the pH of soil and water bodies (streams, lakes, and oceans) is inevitable.

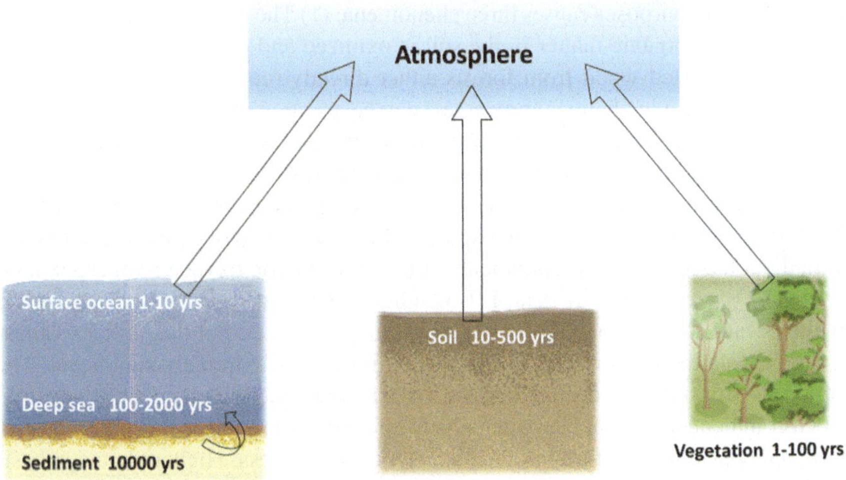

Fig. 3.5 The turnover time for carbon transfers among its main reservoirs

Distribution pattern of CO_2 concentration in the Earth demonstrates higher concentration in north of the equator in comparison to the south. Higher reduction rate of atmospheric oxygen in the north hemisphere also confirmed the role of industrialized countries in the northern parts in CO_2 concentration increase. The proportion of CO_2 emitted by its main producers is illustrated in Fig. 3.7.

As it is shown in Fig. 3.7, approximately half of the world CO_2 emission is produced by China and the United States. Power plants and steel industries in China are the pioneer in CO_2 emission. These two are followed by India, Russia, Japan, Germany, Iran, South Korea, and Saudi Arabia, which released 25% of the total CO_2 to the atmosphere.

How CO_2 emissions have changed over time by its largest producers from 1960 to 2018 can also contain significant points (Fig. 3.8). Based on this information published by the Global Carbon Atlas (GCA) in 2020, China has a noteworthy upward trend in CO_2 production, especially since 2001. Although the production of CO_2 in the United States had a growth between 1990 and 2007, its emission intensity decreased from 2007 to 2014 and had a decreasing trend. European countries and Russia have also reduced their CO_2 emissions.

3.2.3 Methane (CH_4)

Methane is a durable greenhouse gas. From a molecular point of view, CH_4 absorbs much more infrared radiation than CO_2 (28 times more) but its half-life is much shorter (IPCC 2014). Its higher radiative efficiency in Table 3.1 also approve this statement. Considering both factors, it has been found that the effect of one molecule

Table 3.3 Carbon dioxide removal measures

CDR Method Name	Storage Location	Some carbon cycle and climate implications
Afforestation/reforestation	Land (biomass, soils)	–
Improved forest management	Land (biomass, soils)	–
Sequestration of wood in buildings	–	–
Biomass burial	Land/ocean floor	Permanent if buried on the ocean floor
No till agriculture Conservation agriculture	Land (soils)	–
Biochar	Land (soils)	–
Fertilization of land plants	Land (biomass, soils)	–
Creation of wetlands	Land (wetland soils)	–
Biomass Energy with Carbon Capture and Storage (BECCS)	Ocean/geological formations	Permanent if stored in geological reservoir
Ocean iron fertilization	Ocean	–
Algae farming and burial	Ocean	–
Blue carbon (mangrove, kelp farming)	Ocean	–
Modifying ocean upwelling to bring nutrients from deep ocean to surface ocean	Ocean	Permanent removal
Enhanced weathering over land	Land/ocean floor	–
Enhanced weathering over ocean	Ocean	–
Direct-air capture with storage	Ocean/geological formations	Permanent removal if stored in geological reservoirs

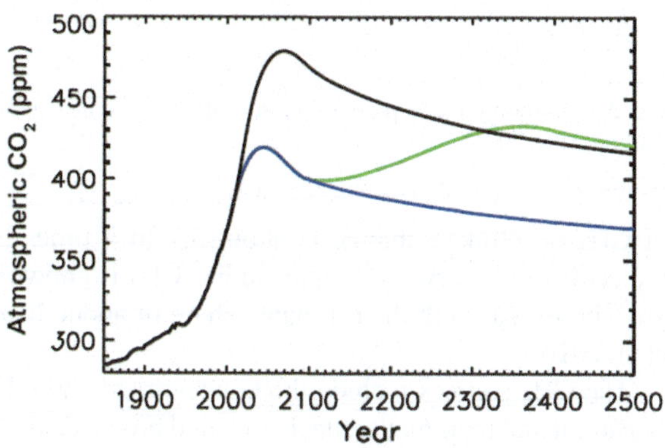

Fig. 3.6 Simulation the effects of CDR methods, black line: no removal, blue line: permanent confiscation, green line: non-permanent carbon confiscation. (Solomon et al. 2007)

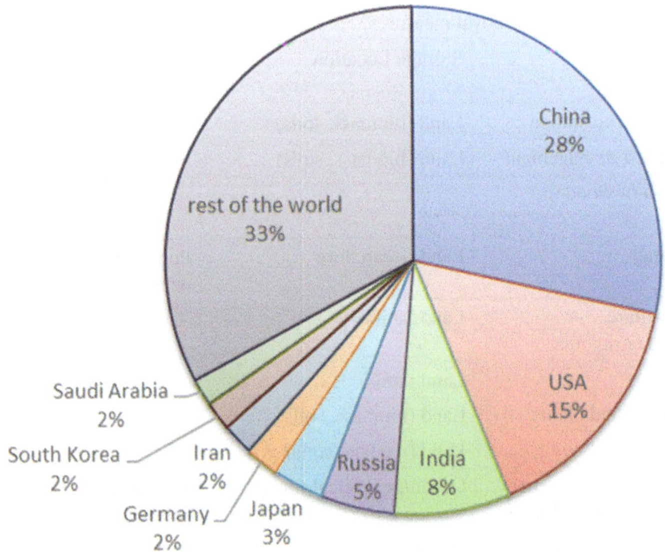

Fig. 3.7 Globally emitted CO_2 by fossil fuels and industrial process in 2018 (Global Carbon Atlas 2021)

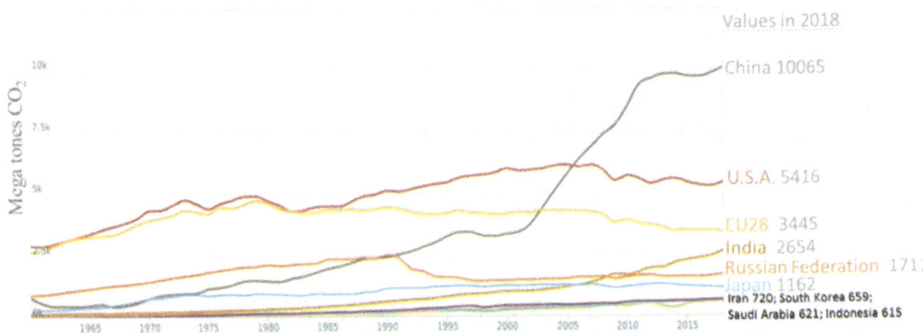

Fig. 3.8 Carbon dioxide production in millions of tonnes in large producers since 1960 (Global Carbon Atlas 2021)

of CH_4 on climate change is about 3.7 to 4 times higher than the effect of one molecule of CO_2. According to studies CH_4 is known as the second most important greenhouse gas with the estimated share of about 20% in global warming (Saunois et al. 2016).

The CH_4 sources include three categories: natural, anthropogenic, combination of natural and man-made (black, red, and brown arrows respectively in Fig. 3.9). The natural resources of CH_4 emission are: wetlands, lakes, plains and swamps, ocean, geothermal vents and mud volcanoes. The anthropogenic CH_4 sources, increasing its concentration in the Earth's atmosphere, include rice paddies, biomass fuels, dams

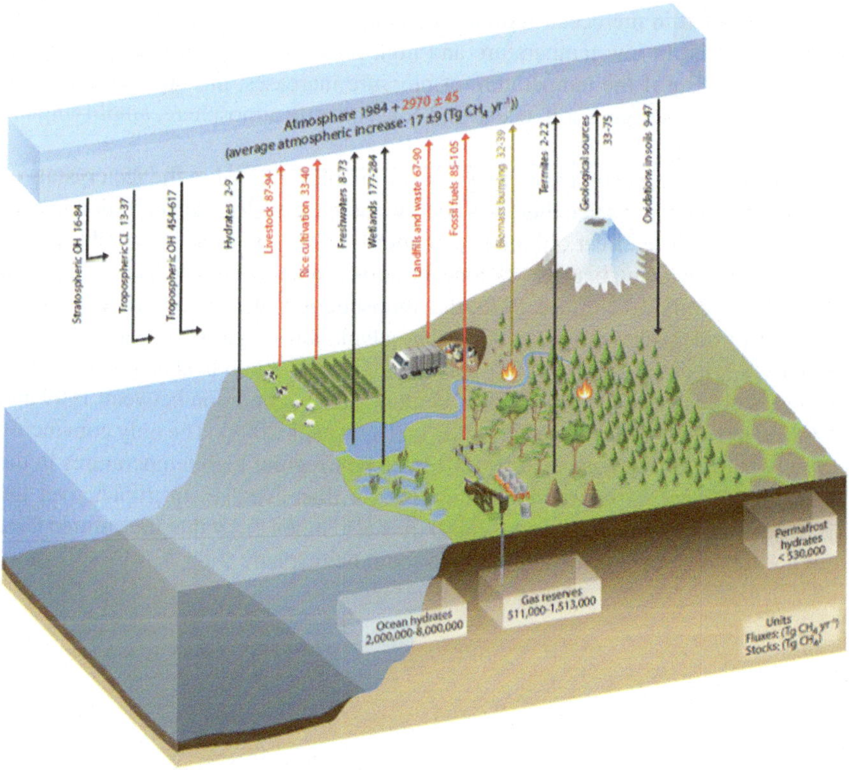

Fig. 3.9 Methane global cycle in 2000–2009 in million tonnes[1] (Global Carbon Atlas 2021)

and artificial lakes, ruminant livestock, waste treatment, natural gas leakage in the production and transfer process, processing and transfer oil and coal. According to research by the Brazilian National Space Research Institute (INPE), organic materials' degradation in dams' reservoir are the largest source of CH_4 emissions, releasing about a quarter of all CH_4 emitted in the world (Lima et al. 2008). Incomplete burning of plant biomass caused by natural disasters or human activities are also the combination of natural and anthropogenic source.

Fermentation processes of methanogenic microbes under low oxygen conditions leads to biogenic CH_4 emission (Conrad 1996). In contrast the chemical reactions in the atmosphere as the sink mechanism include oxidation by radical hydroxide molecules, reaction with oxygen and chlorine atoms, reaction with radical chlorine from sea salts in the layers near the sea level, and uptake by soil (Allan and Soden 2007). The contribution of each of the CH_4 adsorption sources is shown in Fig. 3.9. As can be seen, oxidation reactions with hydroxide molecules in the troposphere has the largest proportion in CH_4 adsorption (Kirschke 2013). A large amount of geological

[1] 1 Tg = 1 million ton.

CH_4 is deposited in the ocean sediments (shallow depth), continental shelves, and permafrost soils. In low temperature and high-pressure conditions, the geological CH_4 is stable. But if the temperature or pressure increases, this stored CH_4 might be release to the overlying soil, ocean and consequently atmosphere, amplifying the global warming trend.

Measuring the concentration of CH_4 in air bubbles trapped in the ice crystals of the Polar Regions shows that there has been an increase in CH_4 concentration in over time (Fig. 3.10). The effect of human activities on CH_4 emission growth during the industrial era is evident, because its concentration dramatically rose from 722 ppb in 1750–1803 ppb in 2011. Since 1978, the atmospheric CH_4 concentration has been measured in adequate station, making the calculation of global average feasible. From 1978, global average of CH_4 has experienced numerous changes, including downward growth rate by the end of 20 century, steady condition between 1999 and 2006, and upward trend from 2007 to 2011 (Rigby et al. 2008). The only convincing reason for the CH_4 increase in 2007–2011 can be irregular high temperatures in the Arctic in 2007 and extreme precipitation (larger than average) in tropics over the 2007–2008 period (Dlugokencky et al. 2009). It is also noteworthy that the emitted CH_4 from north hemisphere is greater than the south hemisphere (Fig. 3.11).

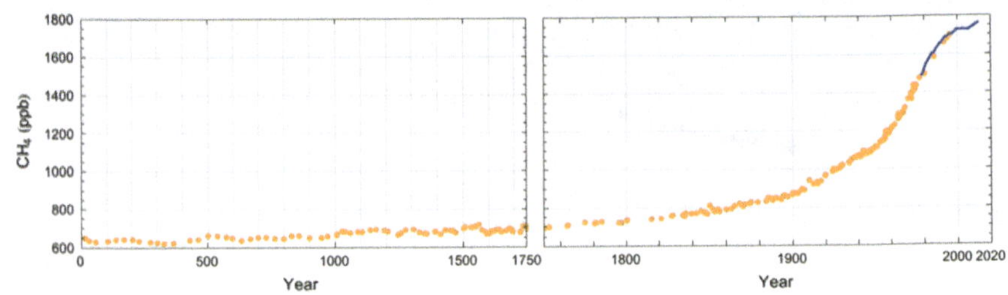

Fig. 3.10 Changes in concentration of the atmospheric CH_4 over the historical and industrial era, color symbols: measured by analyzing trapped air in ice cores, blue lines: measurements from the Cape Grim observatory (MacFarling-Meure et al. 2006)

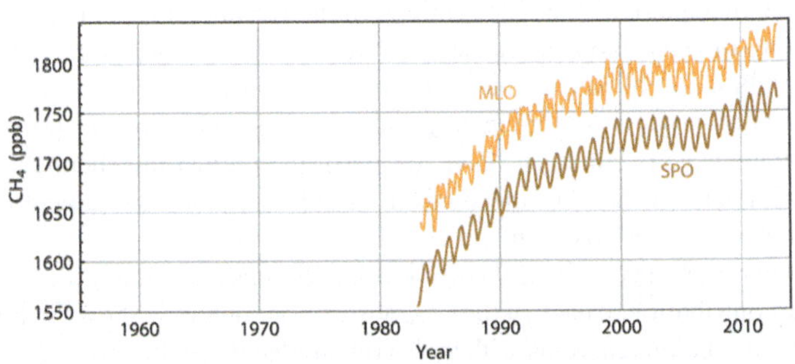

Fig. 3.11 Atmospheric concentration of CH_4 recorded from Mauna Loa (MLO) and South Pole (SPO) atmospheric stations in Northern and Southern Hemispheres (Solomon et al. 2007)

3.2.4 *Nitrous Oxide (N_2O)*

The third most influential GHG is the nitrous oxide (replace with CFC-12 due to the significant decrease in the CFCs concentration). Before industrial revolution, all the nitrogen species were produced naturally from non-reactive atmospheric N_2 via lightning and biological nitrogen fixation (BNF). In the biological process, considered as the main natural source, chemical reactions by microbes convert N_2 to ammonia. Before the industrial era, there was a balance between deposited nitrogen in the ground and water bodies and its output to the atmosphere via denitrification (Ayres et al. 1994). The role of natural factors in emission and sorption of different forms of nitrogenous gases is summarized in Table 3.4.

The main anthropogenic sources of N_2O emission are the industrial ammonia production process (Haber–Bosch), fossil fuels and biomass burning, and legumes' cultivation which increases BNF. While a great amount of reactive nitrogen is released into the atmosphere through human activates (Fig. 3.12), only about 30 to 60% of the total emission convert back to non-reactive N_2 (Bouwman et al. 2013).

Contributing effect of the reactive-Nitrogen (Nr) on terrestrial and atmospheric chemical reactions is called nitrogen cascade, in which Nr is a serious threat until it's converted back into the N_2 (Fig. 3.13). The NH_3 and NO_x are either deposited over the land or carried by winds and dissolved in the oceans, and will change into N_2O. The main Nr sinks are the processes of photolysis in the upper atmosphere and the reaction with atomic oxygen, which leads to the formation of nitrogen monoxide. N_2O is a strong GHG which is also produced during fertilizer production, fuel combustion, or microbial alteration of nitrogen-contained substrates.

In addition to N_2O, playing active role in global warming, NO_x can also impact climate change. The warming effect of NO_x is resulted from its role in tropospheric

Table 3.4 The global nitrogen budget (million ton/year)

Source/sink		NO_x[a]	NH_3[a]	N_2O[b]
Anthropogenic	Fossil fuel combustion and industrial processes	28.3	0.5	0.7
	Agriculture	3.7	30.4	4.1
	Biomass and biofuel burning	5.5	9.2	0.7
	Human excreta	–	–	0.2
Natural	Soil with natural vegetation cover	7.3	2.4	6.6
	Oceans	–	8.2	3.8
	Lightning	4	–	–
	Atmospheric chemistry	–	–	0.6
	Stratospheric sink	–	–	−14.3
Both	Atmospheric deposition on continent	– 27.1	−36.1	−1.1
	Atmospheric deposition on ocean	– 19.8	−17	−0.2

[a]Values were recorded in 2005
[b]Values were recorded in 2006

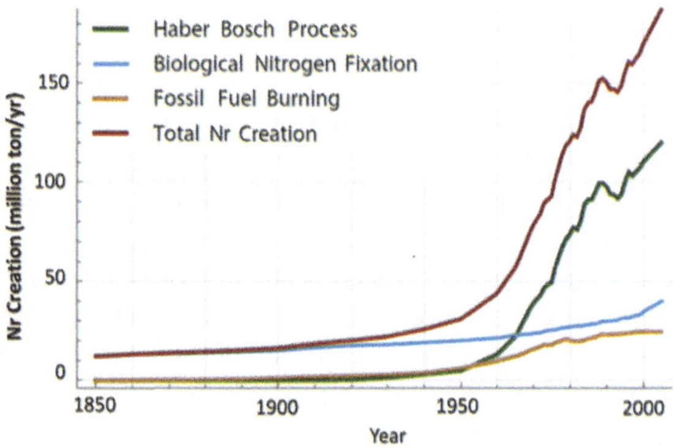

Fig. 3.12 Emitted reactive Nitrogen through main natural and anthropogenic sources (Solomon et al. 2007)

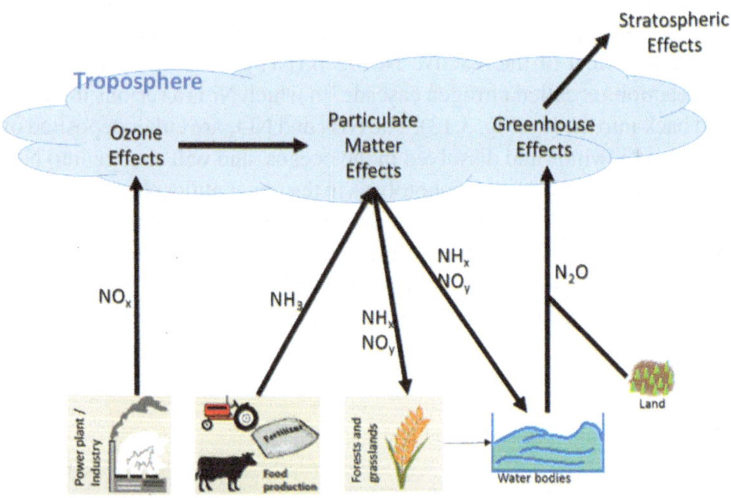

Fig. 3.13 Nitrogen cascade

ozone formation. In contrast, NO_x has cooling effect through decreasing CH_4 (due to the formation of OH radical) and creating nitrate aerosols. It is noteworthy that NH_3 also participate in the formation of nitrate aerosols. Nitrate aerosols, in addition to direct cooling, indirectly cool the atmosphere by improving formation of cloud core, providing the net cooling effect.

Analyzing the concentration of NO_x in the atmosphere is sophisticated in connection to their short lifecycle (around hours). However, satellite observations confirm 50% growth in NO_2 emission in the industrial zones of China between 1996 and 2004. In contrast, the figure declined by 30–50% in Europe, Japan, and the USA from 1996 to 2010 (Hilboll et al. 2013).

The connection between Nr and CO_2 is another important indirect impact of the nitrogen emission on climate change. Organisms require nitrogen and carbon and have established a tight connection. The biological productivity, availability, and requirement of organisms to these two elements is related to each other (Gruber and Galloway 2008), which will consequently affect the ecosystems' function. By increasing the concentration of Nr, the absorption capacity of biosphere as the main sink of CO_2 (absorb half of the emitted carbon) is accelerated. Indeed, increase in the Nr deposition improves the productivity in marine and terrestrial systems. On the other hand, formation of tropospheric ozone, resulting from NO_x and volatile organic compound emissions, reduce the sorption of CO_2 in the atmosphere.

In addition to the potential impact of the reactive-Nitrogen on climate, it participates in various chemical reactions that may lead to smog formation, soil and water acidification, eutrophication and interruption in biodiversity, and increasing nitrate concentrations in drinking water sourced from rivers, lakes, and groundwater (Davidson et al. 2011).

Scientists estimate that the increasing rate of the N_2O concentration in the atmosphere rise from 0.1% at the beginning of the last century to 0.3% per year in 1990. More details of anthropogenic sources of NO_x, NH_x, and N_2O is summarized in Table 3.4.

The global average of N_2O reached 324.2 ppb in 2011, which was 20% more than its estimated value in 1750 (Prather et al. 2012). Utilization of nitrogen soil fertilizer is considered as the dominant reason of this increase particularly from 1950s (Rockmann and Levin 2005). Based on the regular measurements at monthly scale by AGAGE[2] and NOAA/ESRL/GMD[3] since the late 1970s, the annual growth rate of N_2O concentration has been around 0.75 ppb (Fig. 3.14). The maximum amount of N_2O concentration is observed in the northern subtropics (Huang et al. 2008), which point to the application of fertilizers in the northern tropical to mid-latitudes regions. Also seasonal changes in concentration is evident which is resulted from air transfer between the troposphere and stratosphere (where photochemical processes abolish N_2O), thermal out-gassing of N_2O from the oceans, and ventilation (Jiang et al. 2007).

Generally, the net impact of Nr on global climate change is cooling. Indeed, Erisman et al. (2011) projected the Nr contributions (directly and indirectly) on the radiative balance equal to -0.24 W/m^2 (its uncertainty ranges from -0.5 to $+$ 0.2 W/m^2).

[2] Advanced Global Atmospheric Gases Experiment.

[3] National Oceanic and Atmospheric Administration, Earth System Research Laboratory, Global Monitoring Division.

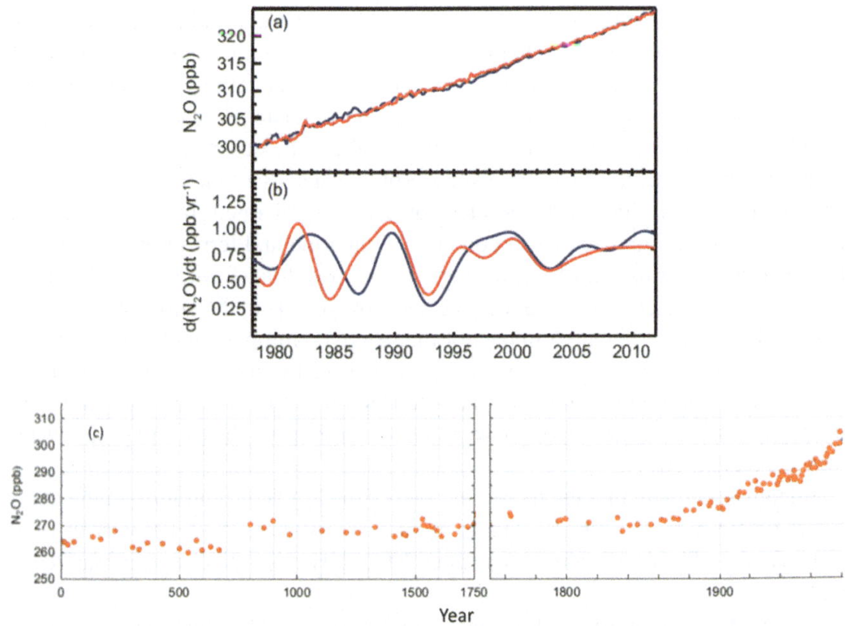

Fig. 3.14 **a** Globally average concentration of N_2O from AGAGE (red) and NOAA/ESRL/GMD (blue). **b** Atmospheric N_2O growth rates **c** Historical changes in N_2O concentration (Solomon et al. 2007)

3.2.5 Halogen-Containing Gases

Halogen-containing gases, such as Chlorofluorocarbons (CFC), Hydrofluorocarbons (HFC), Perfluorocarbons (PFC), Sulphur Hexafluoride (SF_6), and Nitrogen Trifluoride (NF_3) are chemical compounds containing Chlorine, Fluorine, Carbon, Sulphur and Nitrogen. These are most commonly used in everyday human life through applications such as foaming agents, solvents, ventilation, cooling, and insulation systems. The halogen-containing gases can remain in the atmosphere from 1.5 years to 50 thousand years (Table 3.1). The long life-span of some halogen-containing gases put priority on controlling their emission. According to research, these compounds can climb into the stratosphere, and if remained unchanged after several decades in the lower stratosphere, migrate to upper stratosphere where they chemically decompose. Materials released from the decomposition of chlorofluorocarbons eventually destroy ozone in a series of complex chemical reactions. In addition, some types of chlorofluorocarbons absorb infrared radiation in the lower atmosphere and act as a GHG.

Due to the low concentration of HFCs, their impact on radiative forcing is smaller than the CFCs and HCFC, which production is limited based on the Montreal Protocol. The emission of HFCs also should be controlled, otherwise it will gradually grow. In Fig. 3.15 and Table 3.5 changes of the global average concentration

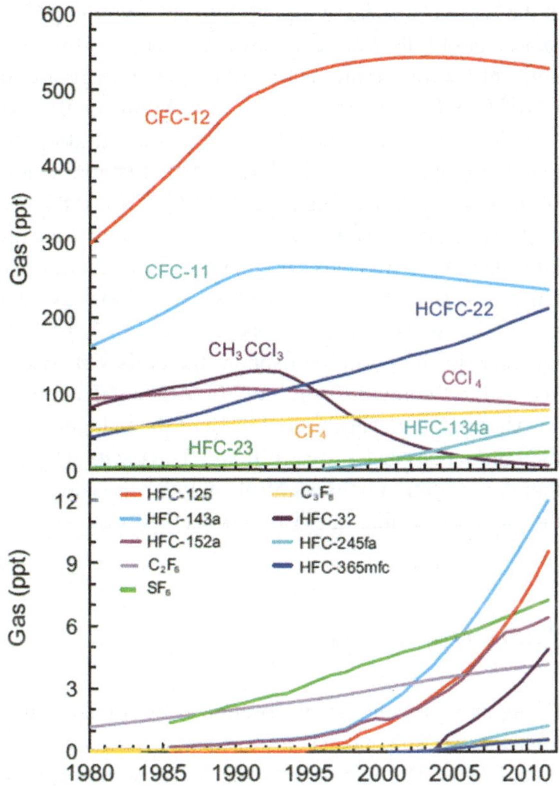

Fig. 3.15 Global average concentration of major halogen-containing GHGs (Solomon et al. 2007)

Table 3.5 Features of some halogen-containing gases

Halogen-containing gas	Explanation
HFC-134a	Replacement for CFC-12
	Largest producers are North America, Europe and East Asia
	Applied in automobile air conditioners and foam blowing applications
HFC-23	By-product of HCFC-22 Main producers are developing countries and East Asia
HFC-152a	Applied as foam blowing agent and aerosol spray propellant
HFC-143a HFC-125 HFC-32	Used as refrigerant blends
CF4 and C2F6	By-products of aluminum production and utilized in plasma etching of electronics
SF6	Emitted through electricity distribution systems, magnesium production, and semi-conductor manufacturing
HCFC-22	Declined emission by developed country while grow in south and Southeast Asia

at the Earth's surface of major halogen-containing GHGs and some important information about them are expressed. It is noteworthy that the estimations derived from atmospheric observations for some species is greater than the values reported to the UNFCCC[4] by countries' authorities (Muhle et al. 2010).

CFC-12 which has had the highest concentration in comparison to other halogen-containing gases, takes the largest radiative efficiency too. The downward trend of some halogen-containing gases (like CFCs) is on account of the Kyoto and Montreal protocols, which lead to the increase of their transitional substrate gases (such as HFC-134a). There is also no reason for the changing behavior of some other gases. On top of that, Fig. 3.15 illustrate the global average concentration, whereas the spatial distribution of the emissions is not same as their behavior in global scale. For instance, in the north hemisphere the emissions decrease in regions north of 30°N and add to the regions south of 30°N (Montzka et al. 2009).

Although concentration of most of the halogen-containing gases continue to rise, their influences on RF are less than 1% of the total by well-mixed GHGs. It is expected that the limitation policies for the CFCs' emissions confine their future emissions to the gases remaining in existing tools or stores.

3.2.6 *Ozone (O_3)*

Ozone is a greenhouse gas produced by highly complex chemical reactions in the atmosphere. This gas has a short half-life (a few hours to a few days) compared to other GHGs. The abundance of ozone in the atmosphere is due to the balance between the reactions of production and degradation of this gas. Due to its small amounts in the atmosphere, the concentration of ozone is expressed in parts per trillion (ppt). But this small amount also plays a very important role in adjusting the amount of radiation received to the Earth's surface. Most atmospheric ozone (about 90%) exists in the stratosphere, with a maximum concentration of 25 km at the equator and 15 km near the poles. Stratospheric ozone severely absorbs the ultra violet (UV) rays and protects the earth surface against this harmful radiation. Nowadays it is known that the stratospheric ozone layer is rapidly degraded. The highest degradation rate of ozone in stratosphere layer occurs over Antarctica during the spring of the Southern Hemisphere, while its depletion rate in the Arctic is lower and occurs from January to February. Since changes of stratospheric ozone concentration is uneven across latitudes, the temporal changes of ozone is summarized in Table 3.6. Furthermore, the degradation rate of ozone in not constant at different heights above the Earth's surface. The highest decline is observed in the upper stratosphere (35–45 km) at around 10% in the mid-1990s, whereas the reduction rate decrease to 7–8% at the height of 20–25 km. Overall, the rate of ozone depletion in the world is approximately 2.3% every 10 years and recently has remained constant at 3.5% lower than the average of 1964–1980 (Solomon et al. 2007).

---

[4] United Nations Framework Convention on Climate Change.

Table 3.6 Degradation rate of stratospheric ozone in various temporal and spatial scale

Assessed period	Changing rate in comparison to the 1964–1980 average (%)	Geographical scale
1980s–early 1990s	3.5	The entire globe
	2.5	60°S to 60°N
current	6	Extratropical 30°S to 60°S
	3.5	NH extratropical

It is approved that compounds such as nitric oxide (NO), hydroxide radicals (OH), chlorine (Cl) and bromine (Br) atoms all act as catalysts that break down ozone gas in various stages. Therefore, if the generating compounds of these catalysts increase in the atmosphere, the rate of ozone depletion will be higher than its production. Since there is a dynamic equilibrium between ozone production and degradation, increasing the catalyst concentration increases the rate of degradation until reach a new equilibrium with less ozone concentration.

The troposphere contains the residual ozone of the atmosphere (less than 10%) (Weber et al. 2012). The half-life of the tropospheric ozone is just a few weeks. It is formed in the troposphere through transfer from the stratosphere or chemical reaction between other anthropogenic emissions at presence of sunlight (Monks et al. 2009). Disregarding the source of tropospheric ozone, its radiative forcing of $0.4 \pm 0.2\ W/m^2$ shows its importance as a GHG. However, it is noteworthy to mention that the value radiative forcing has high uncertainty as most of the models are unable to simulate the concentration of ozone over the nineteenth century. Results of the analysis satellite-based data by Ziemke et al. (2005) revealed upward trend in tropospheric ozone growing rate in the mid-latitude (north and south hemisphere) from 5 to 9% every 10 years between 1979 and 2003. Beig and Singh (2007) reported this figure for the tropical South Atlantic, India, southern China, Southeast Asia, Indonesia and the tropical regions downwind of China about 2–9% for the same period. Finding the long-term behavior of tropospheric ozone depend on the situ measurements, done in nineteen rural station over the last 50 years (two measurements date back to 1950s) (Oltmans et al. 2013). Analysis results of the data collected by these stations can be seen in Table 3.7. Measurements in the Europe depict more than two-fold increase in tropospheric ozone concentration before begging of the new century.

3.2.7 *Other GHGs*

Some GHGs with anthropogenic source, affecting the climate change indirectly are non-methane volatile organic compounds (VOCs) and carbon monoxide (CO). These gases will participate in aerosol and tropospheric ozone formation as well as changing the CH_4 lifecycle by changing the OH concentration. Aliphatic, aromatic and oxygenated hydrocarbons are examples of VOCs, remaining in the atmosphere

Table 3.7 Summary of the long-term behavior of tropospheric ozone based on measurements in nineteen rural stations over the last 50 years

Location	Number of sites	Status	
		Before 1990	Since 1990
NH	13	• 11 sites show increase of 1 to 5 ppb per decade • 2 have no significant trend	• Rise in East Asia • level off and decline in the eastern USA and Western Europe
SH	6	• 3 sites show increase of 2 ppb per decade • 3 have no significant trend	• Increase over mid-latitude districts of the South Pacific Ocean

from several hours to months. Although the VOC measurement network is not well developed around the world, the measured values in North America, Europe, and Japan illustrate an upward trend until 1980 and then a falling trend up to 10% annually (Aydin et al. 2011; Worton et al. 2012). In contrast, formaldehyde loads in China and India increased 4% and 1.6% annually, between 1997 and 2007, respectively.

Carbon monoxide is another GHG which is produced through imperfect combustion of fuels and oxidation of hydrocarbons. Satellite measurements revealed the annual reduction of CO column (1%) in the polluted regions of north hemisphere between 2002 and 2010 (Worden et al. 2013). This decrease resulted from reduction in CO emitted from anthropogenic sources at the same period.

3.3 Aerosols

Aerosols are tiny particles with natural and anthropogenic sources, which includes a core of sulphate, nitrogen-containing elements, equivalent black carbon, elemental carbon, or other natural sources. As it is shown in Fig. 3.16, inorganic species have the highest proportion of total emitted aerosols. From the formation viewpoint, aerosols are divided to two class: primary and secondary particles. Primary particles considered as the aerosols once they are generated, while secondary particles will form aerosols after condensation and coagulation. In Fig. 3.16, the size of circles depicts each type of aerosols' share schematically.

In scientific research, aerosols are measured separately based on their formation core. Non-methane volatile organic compounds are the dominant core of the total emitted aerosols (more than 50% on average) in majority of the worlds. As it is depicted in Fig. 3.17, the share of anthropogenic sources in aerosols creation is radically higher than natural sources, which their proportion is just considerable in Africa and Oceania.

As far as areoles' size is concerned, $PM_{2.5}$ and PM_{10} are two types of prevalent aerosols in urban areas, which represent particulate matters with the approximate diameter of 2.5 and 10 μm, respectively. $PM_{2.5}$ and PM_{10} are generated from ash,

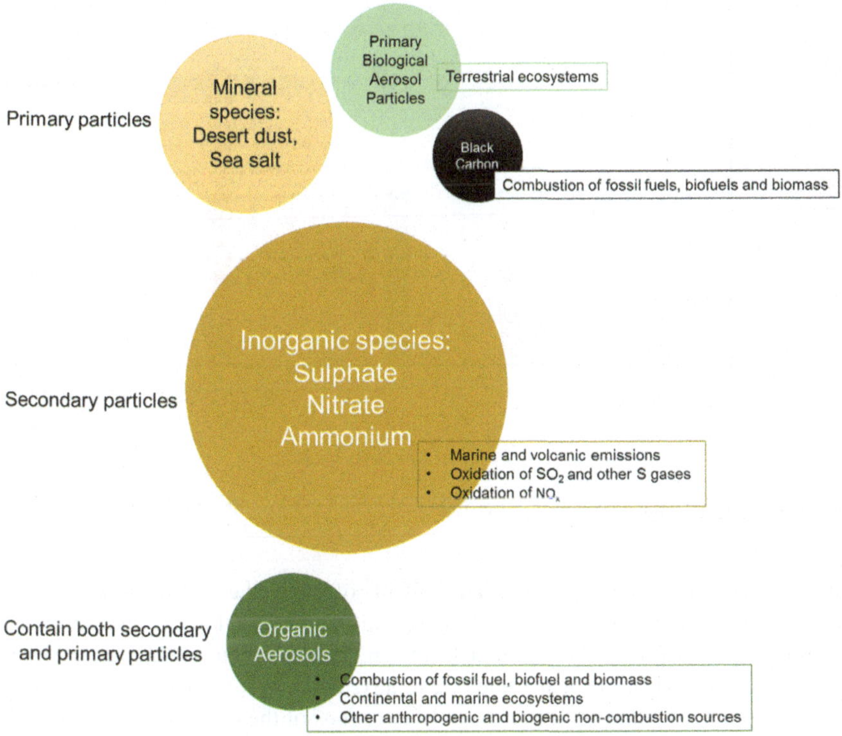

Fig. 3.16 All types of aerosols emitted from anthropogenic and natural sources

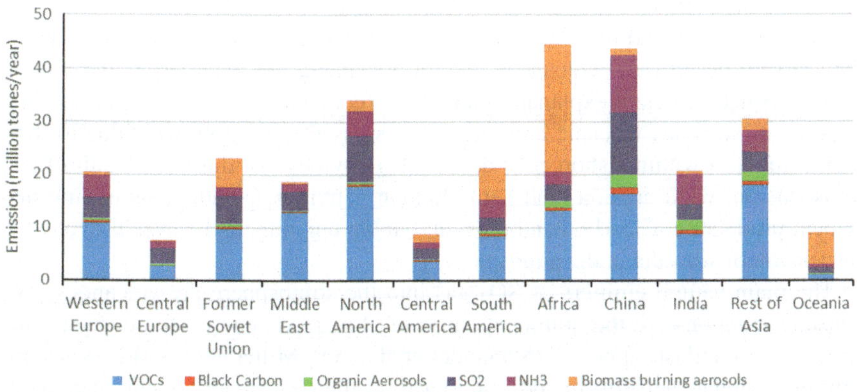

Fig. 3.17 Quantity of various aerosols emitted in different parts of the world in 2000

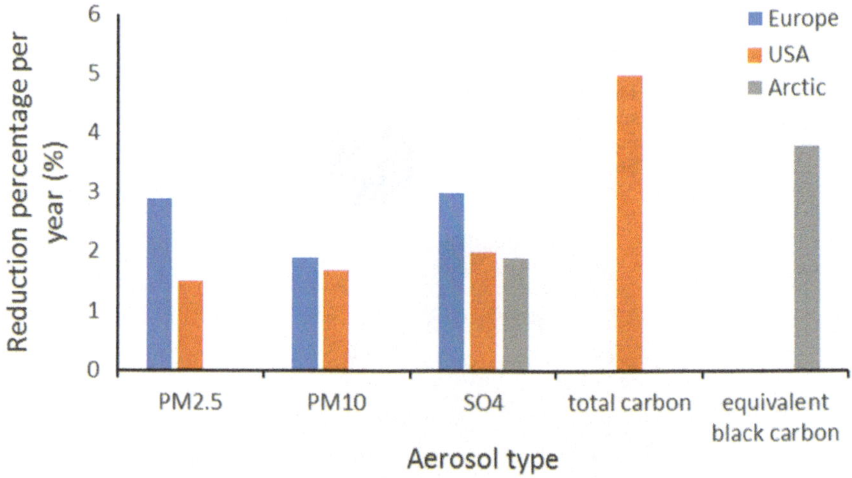

Fig. 3.18 Average annual reduction of aerosols in different parts of the world in 1990–2009

chemicals, metal, or diesel exhaust. The half-life of tropospheric aerosols varies from several days to weeks, leading to their significant impacts on the regional scale. The harmful impacts of aerosols on human health and being the cause of acid rains should also be considered on a local scale. Generally, majority of aerosols with anthropogenic sources are emitted from the polluted zones of the north hemisphere, while the natural sources participate in aerosols formation in both hemispheres (Carslaw et al. 2010). Reduction rate of aerosols, including all types, in Europe, USA, and Arctic over the 1990s–2000s is depicted in Fig. 3.18.

Burning biofuel and fossil fuels are the sources of anthropogenic aerosols, while the sea salt, desert dust, volcanoes and the biosphere are regarded as the natural sources. Since sufficient explanation about the fuel combustion has been presented in previous sections, volcanic emissions are assessed in the following due to their radical impacts within a short period. It is also obvious that dusts are result of soil resuspension, wind erosion, and some human activities including ploughing and cement production. Clearly, wind erosion, bubble bursting, and waves provoke the suspension of sea salts in the atmosphere.

The main natural emitters of SO_2 gas into the stratosphere are volcanoes. The volcanic eruptions are the dominant cause of climate change on the decadal time scales for pre-industrial period (Schneider et al. 2009; Miller et al. 2012). Volcanos emit the Sulphate aerosol (in addition to ash) to the stratosphere, where it remains for long time (one year in comparison to one week for tropospheric Sulphur) and effectively scatters the sunlight. As it is shown in Fig. 3.19 the last major volcanic eruption happened in 1991 (Mt Pinatubo), while several smaller eruptions led to the two-fold increase of RF for the 2008–2011 period in comparison to the 1999–2002 period (from –0.06 to –0.11 W/m^2) (Solomon et al. 2011).

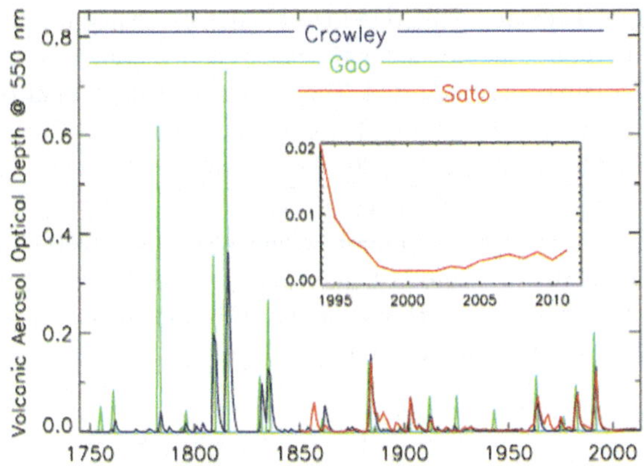

Fig. 3.19 Volcanic reconstructions of global average aerosol optical depth (at 550 nm) based on ice core data, surface and satellite observations carried out by Gao et al. (2008), Crowley and Unterman (2013), Sato et al. (1993)

In Fig. 3.19 the Mt Pinatubo eruption in 1991 is evident, however, there is no sign of the effect of the Eyjafjallajökull volcano eruption in Iceland in 2010. It is due to the volume of released SO_2 and its emission location. Since less than 50 kilo tones (10,000 times fewer than that of Mt Pinatubo) of SO_2 emitted in the troposphere (50 times weaker than stratospheric SO_2), its climatic effects are not obvious. It is also noteworthy that the role of volcanoes in CO_2 emission is not considerable in comparison to the anthropogenic emissions (100 times smaller). In terms of the volcanic effects on the climate change in the future, scientist expect numerous eruptions over the next two centuries, while short and deficient recorded data prohibit them from predicting the exact time of eruptions.

The impact of aerosols on climate is associated to two types of interactions: with radiation (ARI[5]) and with cloud (ACI[6]). ARI and ACI indicate the radiative flux changes resulted from aerosols' absorption or scattering, and affecting precipitating and non- precipitating clouds, respectively. The following features play active role in the ARI and ACI impacts: size diversity, chemical composition, hygroscopicity, mixing condition, cloud nucleation properties, and their spreading pattern in the atmosphere. To assess aerosol–radiation interactions, scientists introduced Aerosol Optical Depth (AOD) as a parameter measuring the columnar aerosols load. The long-term assessment is accompanied with high level of uncertainty due to their great spatial and temporal inconsistency and lack of related parameters. However, the data derived from surface-based remote sensing and satellite-based measurements, show AOD decline in the Europe and the eastern USA since the mid-1990s. This figure rose over eastern and southern Asia due to the dust storms in Arabian Peninsula since 2000s. The AOD changes in the other parts of the world do not show an obvious trend on account of the inter-annual variability. In global scale, effective radiative forcing of ARI is estimated −0.45 [−0.95 to +0.05] W/m^2 (Lohmann et al. 2010). As far as aerosol–cloud interactions are concerned, the relationship between individual storms' characteristic and cloud interactions is proved. Aerosols can change the

[5] Aerosols radiation interaction.

[6] Aerosols cloud interaction.

cloud albedo by altering their condensation and ice nuclei (Twomey impact). Cloud lifetime is another factor affected by aerosols and consequently participate in effective radiative forcing of ACI. The estimated changes of effective radiative forcing by ACI is −1.5 to −0.4 W/m^2. The combined effect of ACI and ARI for all types of aerosols is illustrated in Fig. 3.20. As it is shown all types of aerosols except black carbon have a cooling effect and decrease radiative forcing.

Considering both cooling and warming effects of aerosols, aerosols will cause a negative forcing which slightly moderates the warming effect of other GHGs (Fig. 3.21). As it is shown, aerosols can reduce the effective radiative forcing up to 2 W/m^2 with 90% certainty.

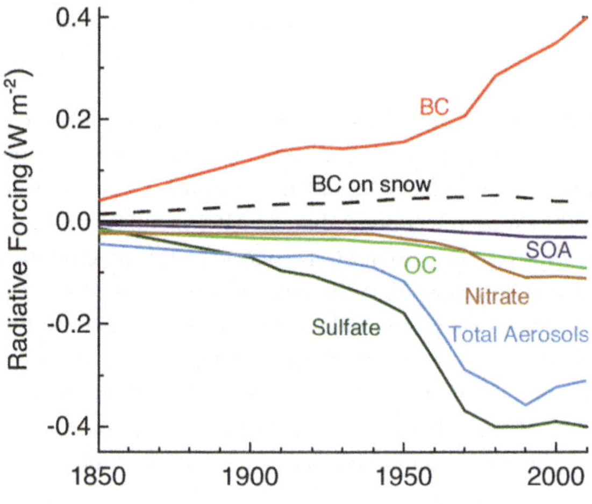

Fig. 3.20 Radiative forcing changes for different aerosols since 1850 (Solomon et al. 2007)

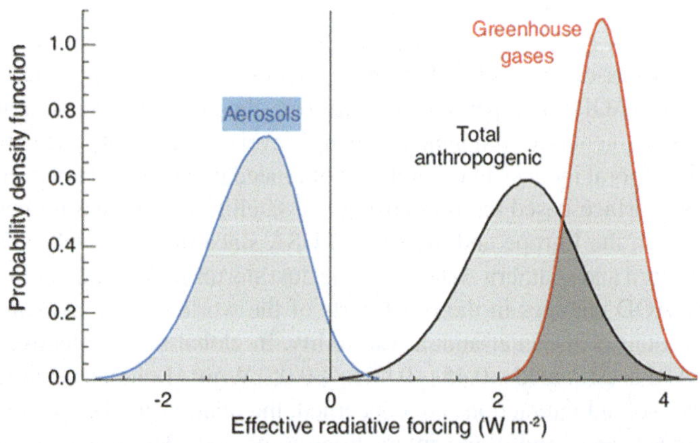

Fig. 3.21 Average annual reduction of aerosols in different parts (Solomon et al. 2007)

3.4 Clouds

The significance of clouds on climate system has been under scientists' attention since the 1970s. There is a sophisticated relationship between clouds and the climate system, known as cloud–climate feedback. The general effects of clouds on the earth climate include:

- Warming the atmosphere when water vapor is condensed
- Warming the Earth by affecting the sunlight flow
- Cooling the Earth by affecting the infrared light.

Cloud density (thickness), coverage (% of sky covered), and height in the atmosphere, directly affect the Earth's albedo. One of the natural elements that influence cloud features is cosmic ray flux. Based on a hypothesis (from laboratory, field and modelling studies) cosmic ray flux stimulates atmospheric ions creation and subsequently aerosol nucleation in the free troposphere and cloud formations (Dickinson 1975; Kirkby 2007). Reduction in the cosmic ray flux will be accompanied with less clouds which intensify the warming effect caused by solar activity. Nevertheless, the effect of cosmic ray–ionization mechanism on the cloud density and coverage at the global scale is minor, with high uncertainty (Snow-Kropla et al. 2011).

Cloud density (thickness), coverage (% of sky covered), and height in the atmosphere also drive reflectivity and are effective parameters in evaluating climate variables. In terms of cloud thickness, thicker/denser clouds reflect greater amount of infrared rays. Interaction of clouds in high altitudes with infrared light intensify the global warming. The impact of high clouds on the Earth surface temperature is greater than clouds found at low altitudes. When high clouds move upward, the emitted infrared light by land, oceans, and atmosphere decline, while changes of the reflected sunlight becomes negligible. Therefore, climate warming is expected due to the lower amount of infrared light released by the atmosphere and surface. Although clouds at low and mid-level altitudes strongly reflect the sunlight toward the space, their effect on emitting the infrared light is weak. Hence, their net contribution is climate cooling. Since the results of the most global climate models emphasize the reduction of low and mid-level cloud amount due to the rise of GHGs in future, growth in the sunlight absorption and consequent warming is expected.

Two other factors affecting the clouds in probable warmer climate condition are wind patterns and storm tracks. There is a connection between these two factors and the regional and seasonal trend in precipitation and cloudiness. Some researchers have determined the correlation of the poleward movement of the clouds with mid-latitude storm tracks (IPCC 2014). Since the received sunlight in the poles is minimum, the reflection of clouds will be futile in cooling the earth in comparison to the clouds locating in regions with high sunlight. In addition, the sunlight reflection by clouds containing liquid drops is higher than a cloud made of the same volume of water but in the ice crystals form. Changes in clouds also is accompanied with some regional effects, including increase in melting rate of the sea ice and decrease in evapotranspiration rates.

Table 3.8 Characteristics of longwave and shortwave clouds' radiative effect

Clouds' effect	Characteristics
LWCRE	• Pattern dominated by high clouds • Sensitive to optically thick clouds at all altitudes
SWCRE	• Rely on the available sunlight • Sensitive to the diurnal and seasonal cycles of cloudiness

Whereas there is not a widely acceptable interpretation of global cloud feedbacks based on the long and short-term observations, all current models regard net cloud feedbacks. Indeed, models simulate the clouds' function in the atmosphere and assess their impacts on the energy flows and conversions in the climate system. There is no doubt that the current climate and hydrological cycle are tightly connected to the clouds, however, there is uncertainty with their influence pattern on the future climate. Scientists benefit from global climate models to predict the cloudiness changes for a warmer climate in the future. The present evidence estimate that the global warming will be intensified by net cloud–climate response.

Finally, clouds directly affect the radiation budget which is measurable by satellite (Ramanathan et al. 1989). The effect of cloudy conditions on radiative forcing is both positive and negative through participating to the greenhouse effect and improving albedo, respectively. Clouds increase longwave radiative effect[7] about 30 W/m^2, while decrease the shortwave radiative effect[8] around 50 W/m^2 (Loeb et al. 2009). Therefore, the net effect of cloud on radiative forcing in global scale is cooling with the amount of 20 W/m^2. These large values prove the noteworthy potential of clouds on determining the climate condition (Table 3.8).

3.5 Solar Energy Density

Astronomers believe that magnetic phenomena at the solar surface (magnetic network, faculae, sunspots) represent changes in the surface of the sun at centennial and millennial scales, while the internal energy has not changed. The large number of spots indicates more activity of the sun as well as more intense solar wind, so by decreasing in the number of these spots, the sun and solar wind become calmer.

Total solar irradiance (TSI) over an 11-year solar cycle (1750-present) has fluctuated 0.1%. The most reliable estimate for the fluctuation range of radiative forcing resulted of TSI changes is 0.0–0.10 W/m^2. The interaction among dark sunspots, bright faculae and bright network elements has led to this variation (Foukal and Lean 1988). The measured TSI by various instruments over the recent decades is shown

[7] LWCRE.

[8] SWCRE.

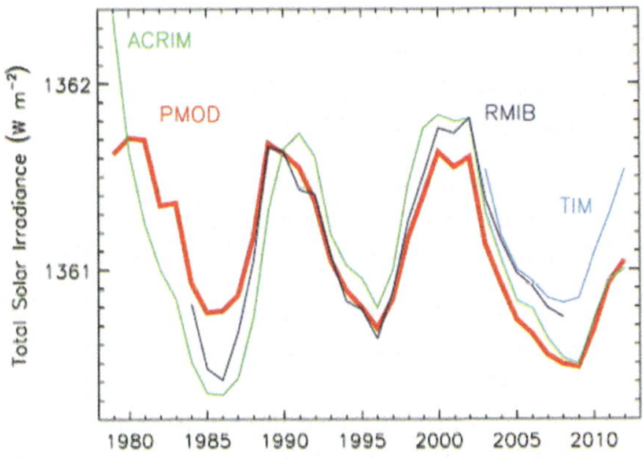

Fig. 3.22 Annual average composites of measured total solar irradiance (Solomon et al. 2007)

in Fig. 3.22. The TIM[9] is the most accurate as it is calibrated based on national standards. Among the satellite measurements, including ACRIM,[10] PMOD,[11] RMIB,[12] the more accurate one is PMOD.

The downward trend is anticipated for the TSI compared to its value over the past 30 years (Abreu et al. 2008). It can be due to the probable reduction in the mean magnetic field in sunspots and disappearance of sunspot activity (Penn and Livingston 2006). It is noteworthy that reduction in the solar activity is estimated with low certainty, while scientists are sure about the negligible effect of changes in solar radiation in comparison with the growing impact of greenhouse gases.

The ultraviolet, part of the spectral solar irradiance (SSI), has also changed several percent, which is successfully modeled between 1978 and 2003 (Crouch et al. 2008; Krivova et al. 2011). The Earth's surface is affected by the changes in the total solar irradiance, but UV alternation affects the stratosphere and subsequently the tropospheric circulation is influence via dynamical coupling (Haigh 1996). In fact, the changes of solar UV irradiance directly affect the ozone production rate in the stratosphere. In addition, the UV changes might impact indirectly on the tropospheric circulation and leads to transport-induced ozone production, variation in stratospheric temperature, and zonal winds (Frame and Gray 2010; Shindell et al. 2006). Effect of the SSI changes is considerably lower than the effect of TSI changes on the climate change (Gray et al. 2009). According to the space-based measurements over the recent 30 years, around 30% of the SC TSI variations belongs to the UV (Rottman 2006), however, it shares based on the models ranges from 30 to 90% (with high probable value of ~60%) (Ermolli et al. 2013). The results of the preindustrial UV reconstruction demonstrate gradual changes with larger trends at shorter wavelengths since 1750 as the following:

---

[9] Total Irradiance Monitor.

[10] Active Cavity Radiometer Irradiance Monitor.

[11] Physikalisch-Meteorologisches Observatorium Davos.

[12] Royal Meteorological Institute of Belgium.

- Near 25% at about 120 nm
- Around 8% at 130–175 nm
- Approximately 4% at 175–200 nm
- About 0.5% at 200–350 nm.

In connection with the Earth orbit around the Sun, three factors cause periodic perturbations, each of which changes the amount of radiation and the quality of solar energy emission across the Earth. These three factors are:

1. The orbit of the Earth around the Sun is the elliptical shape and the degree of ellipse or the degree of distance from its center changes slowly over 100,000 years.
2. The axial inclination of the Earth, or its curvature relative to the plane of its annual rotation around the Sun, is equal to 23.5° at present and deviates about one degree over 40,000 years.
3. The Earth deviates from its orbit and this deviation occurs periodically every 20,000 years.

Although these factors operate over a long period of time, they effect the distribution of solar energy. Scientists believe that these factors create ice ages. Also, nevertheless the density of solar energy plays a large role in the global climate, its changes are very small and must be developed on a large scale to affect the climate. Oceans are one of the elements that can greatly expand the effects of changes in the density of solar radiation. Changes in ocean water temperature, particularly surface water temperature, can cause changes in atmospheric pressure and humidity, and this phenomenon will also cause changes in the atmosphere (Solomon et al. 2007).

3.6 Changes in Land Use

The strong connection between land use and the surface albedo is confirmed scientifically. Since the Earth radiation budget is related to the albedo, any changes made on land cover will affect the radiation balance and consequently the climate system. In addition to the direct impact of land use on albedo, changes in the hydrological cycle (particularly river runoff) resulted from land cover changes including deforestation has impact on the climate system. Moreover, power plants and industrial sectors not only have direct warming effect on the climate via heat pollution, but also change the climate system through interruption in water cycle by increasing the water consumption. The mentioned impacts of land use on climate condition are explained in detail in the following.

Human activities such as the forests' exploitation and deforestation, the conversion of meadows into farms, the transformation of villages into cities, the creation of new artificial lakes and even the widespread use of solar collectors have changed the Earth's surface and increased the Earth's reflectivity. Land use changes from 1700 to the beginning of current century, is estimated about 42–68%, majority occurred in the temperate and tropical zones of the north hemisphere (Hurtt et al. 2006). The changes in radiative forcing and flux resulted from pervasive cultivation 1750, all around the world is illustrated in Fig. 3.23. Obviously, Europe and South Asia has been pioneered in agriculture, while expansion of farms in north and South America, and Australia leads to stronger reduction in radiative forcing. There is an assumption that the 'Little Ice Age' had been a result of this radical negative radiative forcing in combination with volcanic activity and solar activity changes (Betts et al. 2007). The shortwave flux and RF experience changes at -0.2 W/m^2 and -0.17 W/m^2 on average, respectively.

Satellite images also indicate the considerable growth of deforestation from 1980 to 2000, while its rate decrease over the past 20 years (FAO 2012). It is noteworthy that deforestation speeds up global warming due to the direct connection between forests and the atmospheric CO_2 concentration. To compensate for the devastating effect of land use changes, reforestation effort has been taken in Western Europe, North America and China. Although afforestation provides a balance between the effects of albedo and GHGs, its potential to alleviate climate change is restricted due to

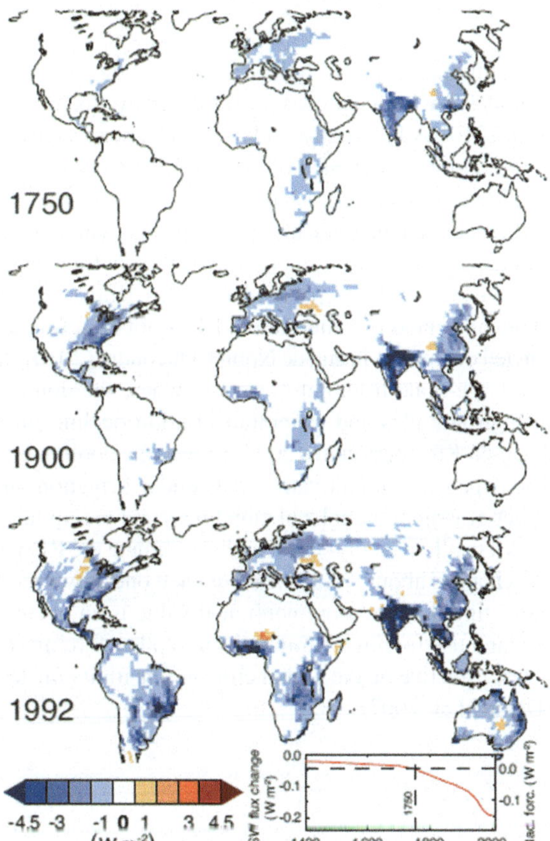

Fig. 3.23 Changes in radiative forcing and radiative flux in 1750, 1900, and 1992 (Betts et al. 2007; Pongratz et al. 2009)

Table 3.9 Effect of land cover on albedo and RF

Land use changes type	Explanation	Albedo	Radiative forcing
Deforestation	• Darkness level of forests is greater than grasses, croplands, barren land and desert, respectively • Long-lasting snow cover on land with low vegetation cover	↑	↓
Cultivation expansion	• Absorb more radiation in comparison to bare land	↓	↑
	• Increase dust generation	↑↓	–
Burn scars	• Reflect more radiation and easily covered by snow at high latitude, while the surface blackening lifetime is short	↑	↓
Urban areas expansion	• High probable to rise albedo via white roof covering used to moderate the heat island effect • The small global scale effect but large in local scale	↑	↓

the latitudinal contrast. Indeed, deforestation amplifies the greenhouse effect in low-latitudes, but high latitudes are impacted by albedo and evapotranspiration (Bathiany et al. 2010). Additionally, undesired and unexpected changes in precipitation patterns is the regional drawback of afforestation (Swann et al. 2012). Table 3.9 represents the effects of change in land cover on reflected radiation.

Land use changes are also accompanied with non-radiative forcing impacts on the climate system, such as modifying the evapotranspiration and surface roughness. To analyze the non-RF impact of land use, scientists assert that due to the heterogeneous incident of land use alternation, RF is not a useful parameter anymore. An analysis carried out by Davin and de Noblet-Ducoudre (2010) showed the influence on climate at a large-scale from deforestation, where the domination of albedo cooling effect in high latitudes and evapotranspiration decline in the tropical zones. In another research Kueppers et al. (2007) proved the cooling effect of irrigation in local scale. The expansion of cultivated areas under irrigation, highlight the importance of the effect of irrigation on local cloudiness and precipitation will rise.

All in all, most of the researches estimate the RF and non-radiative impact of land use changes about $-0.2\ W/m^2$ for each one, however, there are convincing evidence indicating that the aforementioned value is an overestimate. It is worth noting that the land use forcing in comparison to other forcings explained in previous sections, is less effective on the global climate sensitivity on account of its spatial distribution (Davin et al. 2007).

3.7 Summary

In this chapter, the balance of the Earth's thermal energy between short-wavelength radiant energy coming in from the Sun and long-wavelength radiant energy coming out of the earth is affected by a variety of factors. These factors stemmed of anthropogenic and natural sources, in which the natural effects are considerably lower than manmade. To make a clear understanding and comparison between factors, scientists introduce the parameter known as radiative forcing. As it is illustrated in Figs. 3.24 and 3.25, GHGs with high confidence takes the first place in increasing the Earth received energy. However, some of the GHGs participate in cooling effect through aerosols' creation. The CO_2 is highlighted as the most prominent GHG, amplifying the global warming. Among the natural factors, the solar energy fluctuations and volcanic activities will change the Earth's energy balance by warming and cooling the surface, respectively. The cumulative changes in the radiative forcing and the Earth's energy have grown consistently, but in various intensity since the 1950s. The upward trend has led to the increase in the global average temperature, an increase in the number of hot days and nights, a decrease in the volume of polar glaciers, and

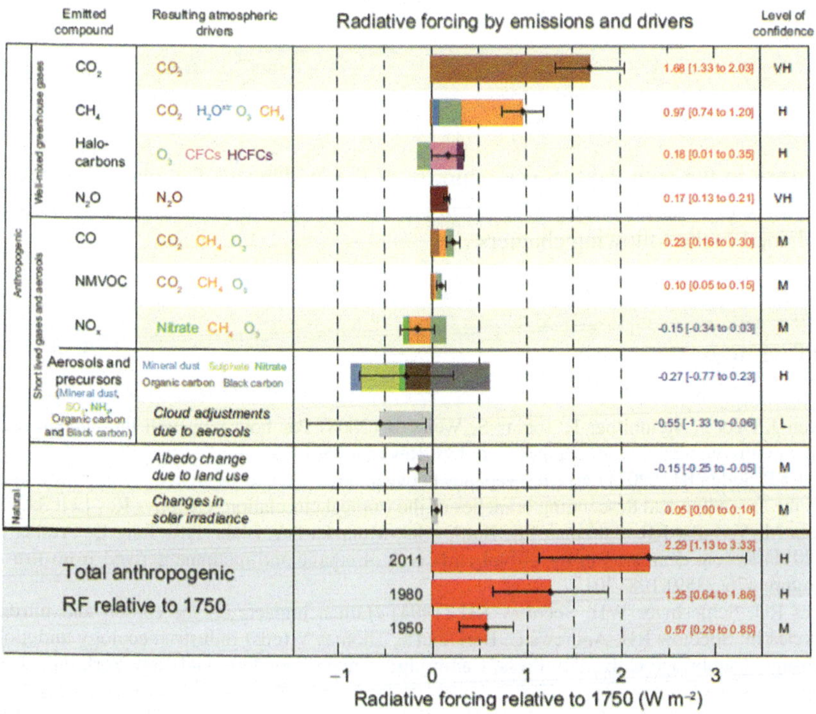

Fig. 3.24 Changes in radiative forcing for all affecting factors on climate change

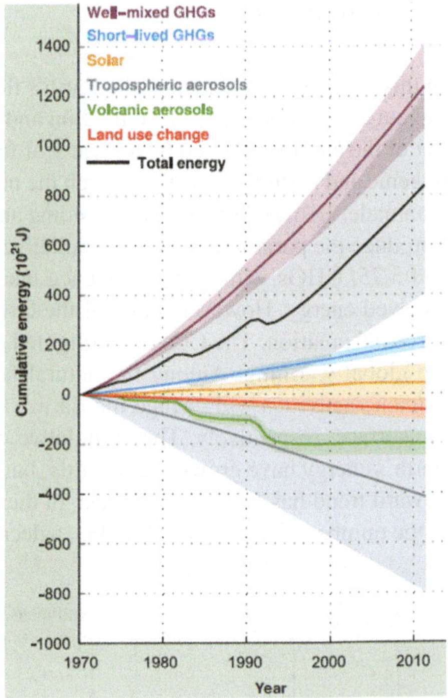

Fig. 3.25 Affecting factor's role on the earth cumulative energy over the last half century (Solomon et al. 2007)

decrease in the cold days. Since majority of the evidences of climate changes are considered as threat, taking adaptation and mitigation measures is vital, which are explained in the following chapters.

References

Abreu J, Beer J, Steinhilber F, Tobias S, Weiss N (2008) For how long will the current grand maximum of solar activity persists Geophys. Res Lett 35(20)

Allan RP, Soden BJ (2007) Large discrepancy between observed and simulated precipitation trends in the ascending and descending branches of the tropical circulation. Geophys Res Lett 34(18)

Aydin M, Verhulst KR, Saltzman ES, Battle MO, Montzka SA, Blake DR, Tang Q, Prather MJ (2011) Recent decreases in fossil-fuel emissions of ethane and methane derived from firn air. Nature 476(7359):198–201

Ayres RU, Schlesinger WH, Socolow RH (1994) Human impacts on the carbon and nitrogen cycles.In: Socolow RH, Andrews C, Berkhout F, Thomas V (eds) Industrial ecology and global change. Cambridge University Press, Cambridge, United Kingdom, and New York, NY, USA, pp 121–155

Bathiany S, Claussen M, Brovkin V, Raddatz T, Gayler V (2010) Combined biogeophysical and biogeochemical effects of large-scale forest cover changes in the MPI earth system model. Biogeosciences 7(5):1383–1399

Beig G, Singh V (2007) Trends in tropical tropospheric column ozone from satellite data and MOZART model. Geophys Res Lett 34(17)

Betts RA, Falloon PD, Goldewijk KK, Ramankutty N (2007) Biogeophysical effects of land use on climate: model simulations of radiative forcing and largescale temperature change. Agr Forest Meteorol 142(2–4):216–233

Bouwman AF, Beusen AHW, Griffioen J, Van Groenigen JW, Hefting MM, Oenema O, Van Puijenbroek PJTM, Seitzinger S, Slomp CP, Stehfest E (2013) Global trends and uncertainties in terrestrial denitrification and N_2O emissions. Philos Trans R Soc London Ser B 368(1621):20130112

Carslaw KS, Boucher O, Spracklen DV, Mann GW, Rae JGL, Woodward S, Kulmala M (2010) A review of natural aerosol interactions and feedbacks within the Earth system. Atmos Chem Phys 10(4):1701–1737

Conrad R (1996) Soil microorganisms as controllers of atmospheric trace gases (H_2, CO, CH_4, OCS, N_2O, and NO). Microbiol Rev 60(4):609–640

Crouch AD, Charbonneau P, Beaubien G, Paquin-Ricard D (2008) A model for the total solar irradiance based on active region decay. Astrophys J 677(1):723

Crowley TJ, Unterman MB (2013) Technical details concerning development of a 1200 yr proxy index for global volcanism. Earth Syst Sci Data 5(1):187–197

Davidson EA, David MB, Galloway JN, Goodale CL, Haeuber R, Harrison JA, Howarth RW, Jaynes DB, Lowrance RR, Thomas NB, Peel JL (2011) Excess nitrogen in the U.S. environnement: trends, risks, and solutions. Issues of Ecology, Report number 15. Ecological Society of America, Washington, DC.

Davin EL, de Noblet-Ducoudre N (2010) Climatic impact of global-scale deforestation: radiative versus nonradiative orocesses. J Clim 23(1):97–112

Davin E, de Noblet-Ducoudre N, Friedlingstein P (2007) Impact of land cover change on surface climate: Relevance of the radiative forcing concept. Geophys Res Lett 34(13)

de Moraes Sá JC, Lal R, Cerri CC, Lorenz K, Hungria M, de Faccio Carvalho PC (2017) Low-carbon agriculture in South America to mitigate global climate change and advance food security. Environ Int 98:102–112

Dickinson R (1975) Solar variability and lower atmosphere. Bull Am Meteorol Soc 56:1240–1248

Dlugokencky EJ et al (2009) Observational constraints on recent increases in the atmospheric CH_4 Burden. Geophys Res Lett 36:L18803

Erisman JW, Galloway J, Seitzinger S, Bleeker A, Butterbach-Bahl K (2011) Reactive nitrogen in the environment and its effect on climate change. Curr Opin Environ Sustain 3:281–290

Ermolli I, Matthes K, Dudok de Wit T, Krivova NA, Tourpali K, Weber M, Unruh YC, Gray L, Langematz U, Pilewskie P, Rozanov E, Schmutz W, Shapiro S, Solanki SK, Woods TN (2013) Recent variability of the solar spectral irradiance and its impact on climate modelling. Atmos Chem Phys 13:3945–3977

FAO (2012) State of the world's forests. Food and Agriculture Organization of the United Nations, Rome, Italy, 60p

Foukal P, Lean J (1988) Magnetic modulation of solar luminosity by phtospheric activity. Astrophys J 328:347–357

Frame T, Gray L (2010) The 11-yr solar cycle in ERA-40 data: an update to 2008 J Clim 23:2213–2222

Fueglistaler S, Haynes PH (2005) Control of interannual and longer-term variability of stratospheric water vapor. J Geophys Res Atmos 110:D24108

Gao CC, Robock A, Ammann C (2008) Volcanic forcing of climate over the past 1500 years: an improved ice core-based index for climate models. J Geophys Res Atmos 113:D23111. https://doi.org/10.1029/2008JD010239

Global Carbon Atlas (2021, January 12) Carbon emission share by country. http://www.globalcarbonatlas.org/en/CO2-emissions

Gray L, Rumbold S, Shine K (2009) Stratospheric temperature and radiative forcing response to 11-year solar cycle changes in irradiance and ozone. J Atmos Sci 66:2402–2417

Gruber N, Galloway JN (2008) An earth-system perspective of the global nitrogen cycle. Nature 451:293–296
Haigh JD (1996) The impact of solar variability on climate. Science 272: 981–984
Hilboll A, Richter A, Burrows JP (2013) Long-term changes of tropospheric NO_2 over megacities derived from multiple satellite instruments. Atmos Chem Phys 13:4145–4169
Huang J et al (2008) Estimation of regional emissions of nitrous oxide from 1997 to 2005 using multinetwork measurements: a chemical transport model, and an inverse method. J Geophys Res. 113:D17313
Hurtt GC et al (2006) The underpinnings of land-use history: three centuries of global gridded land-use transitions, wood-harvest activity, and resulting secondary lands. Global Chang Biol 12:1208–1229
IPCC Climate change (2014) Impacts, adaptation and vulnerability-regional aspects. Cambridge, United Kingdom and New York, NY, Cambridge University Press
Janzen HH (2004) Carbon cycling in earth systems—a soil science perspective. Agr Ecosyst Environ 104(3):399–417
Jiang X, Eichelberger SJ, Hartmann DL, Shia R, Yung YL (2007) Influence of doubled CO2 on ozone via changes in the Brewer-Dobson circulation. J Atmos Sci 64:2751–2755
Karamooz A, Araghinejad SH (2005) Advanced hydrology, Industrial University of Amir Kabir (Poly Technics). Iran, Publication Centre of Amir Kabir University, Tehran
Kirkby J (2007) Cosmic rays and climate. Surv Geophys 28:333–375
Kirschke S, Bousquet P, Ciais P, Saunois M, Canadell JG, Dlugokencky EJ, Bergamachi P, Bermann D, Blake D, Bruhwiler L, Cameron-Smith P, Castaldi S, Chevanllier F, Feng L, Fraser A, Heimann M, Hodson EL, Houweling S, Josse B, Fraser PJ, Krummel PB, Lamarque J-F, Langenfelds RL, Quéré CL, Naik V, O'Doherty S, Palmer PL, Pison I, Plummer D, Poulter B, Prinn RG, Rigby M, Ringeval B, Santini M, Schmidt M, Shindell DT, Simpson IJ, Spahni R, Paul Steele L, Strode SA, Sudo K, Szopa S, Werf GRVD, Voulgaraskis A, Welle MV, Weiss RF, Williams JE, Zeng G (2013) Three decades of global methan sources. Nat Geosci 6(10):813–823
Krivova N, Solanki S, Unruh Y (2011) Towards a long-term record of solar total and spectral irradiance. J Atmos Solar-Terres Phys 73:223–234
Kueppers LM, Snyder MA, Sloan LC (2007) Irrigation cooling effect: regional climate forcing by land-use change. Geophys Res Lett 34:L03703
Lal R (2008) Carbon sequestration. Philos Trans Royal Soc B Biol Sci 363(1492):815–830
Lima IB, Ramos FM, Bambace LA, Rosa RR (2008) Methane emissions from large dams as renewable energy resources: a developing nation perspective. Mitig Adapt Strat Glob Change 13(2):193–206
Loeb NG et al (2009) Toward optimal closure of the Earth's top-of-atmosphere radiation budget. J Clim 22:748–766
Lohmann U et al (2010) Total aerosol effect: radiative forcing or radiative flux perturbation? Atmos Chem Phys 10:3235–3246
MacFarling-Meure C et al (2006) Law Dome CO_2, CH_4 and N_2O ice core records extended to 2000 years BP. Geophys Res Lett 33:L14810
Manning AC, Keeling RF (2006) Global oceanic and land biotic carbon sinks from the Scripps atmospheric oxygen flask sampling network. Tellus B 58:95–116
Miller GH et al (2012) Abrupt onset of the Little Ice Age triggered by volcanism and sustained by sea-ice/ocean feedbacks. Geophys Res Lett 39:L02708
Monks PS et al (2009) Atmospheric composition change—global and regional air quality. Atmos Environ 43:5268–5350
Montzka S, Hall B, Elkins J (2009) Accelerated increases observed for hydrochlorofluorocarbons since 2004 in the global atmosphere. Geophys Res Lett 36:L03804
Muhle J et al (2010) Perfluorocarbons in the global atmosphere: tetrafluoromethane, hexafluoroethane, and octafluoropropane. Atmos Chem Phys 10:5145–5164
Oltmans SJ et al (2013) Recent tropospheric ozone changes—a pattern dominated by slow or no growth. Atmos Environ 67:331–351

Panagopoulos A, Haralambous KJ, Loizidou M (2019) Desalination brine disposal methods and treatment technologies-a review. Sci Total Env 693:133545
Penn M, Livingston W (2006) Temporal changes in sunspot umbral magnetic fields and temperatures. Astrophys J 649:L45–L48
Pongratz J, Raddatz T, Reick CH, Esch M, Claussen M (2009) Radiative forcing from anthropogenic land cover change since AD 800. Geophys Res Lett 36:L02709
Prather MJ, Holmes CD, Hsu J (2012) Reactive greenhouse gas scenarios: systematic exploration of uncertainties and the role of atmospheric chemistry. Geophys Res Lett 39:L09803
Ramanathan VL, Cess RD, Harrison EF, Minnis P, Barkstrom BR, Ahmad E, Hartmann D (1989) Cloud-radiative forcing and climate: results from the Earth Radiation Budget Experiment. Science 243:57–63
Rigby M et al (2008) Renewed growth of atmospheric methane. Geophys Res Lett 35:L22805
Rottman G (2006) Measurement of total and spectral solar irradiance. Space Sci Rev 125:39–51
Sato M, Hansen JE, McCormick MP, Pollack JB (1993) Stratospheric aerosol optical depth, 1850–1990. J Geophys Res Atmos 98:22987–22994
Saunois M, Bousquet P, Poulter B, Peregon A, Ciais P, Canadell JG, Dlugokencky EJ, Etiope G, Bastviken D, Houweling S, Janssens-Maenhout G (2016) The global methane budget 2000–2012. Earth Syst Sci Data 8(2):697–751
Schneider DP, Ammann CM, Otto-Bliesner BL, Kaufman DS (2009) Climate response to large, high-latitude and low-latitude volcanic eruptions in the Community Climate System Model. J Geophys Res 114:D15101
Shindell D, Faluvegi G, Lacis A, Hansen J, Ruedy R, Aguilar E (2006) Role of tropospheric ozone increases in 20th-century climate change. J Geophys Res Atmos 111: D08302
Snow-Kropla EJ, Pierce JR, Westervelt DM, Trivitayanurak W (2011) Cosmic rays, aerosol formation and cloud-condensation nuclei: sensitivities to model uncertainties. Atmos Chem Phys 11:4001–4013
Solomon S, Daniel JS, Neely RR, Vernier JP, Dutton EG, Thomason LW (2011) The persistently variable “background” stratospheric aerosol layer and global climate change. Science 333:866–870
Solomon S, Qin D, Manning M, Chen Z, Marquis M, Averyt KB, Tignor M, Miller HL (2007) Summary for policymakers. climate change 2007: the physical science basis. Contribution of working group i to the fourth assessment report of the intergovernmental panel on climate change. Cambridge University Press, pp 1–18
Stenke A, Grewe V (2005) Simulation of stratospheric water vapor trends: impact on stratospheric ozone chemistry. Atmos Chem Phys 5:1257–1272
Swann ALS, Fung IY, Chiang JCH (2012) Mid-latitude afforestation shifts general circulation and tropical precipitation. Proc Natl Acad Sci USA 109:712–716
Taghdisian H, Minapour S (2003) Climate Change. Tahran Iran Env Prot Agency 1
Weber M, Steinbrecht W, Long C, Fioletov VE, Frith SH, Stolarski R, Newman PA (2012) Stratospheric ozone, in ‘State of the Climate in 2011.’ Bull Am Meteorol Soc 93:S46–S49
Worden HM, Deeter MN, Frankenberg C, George M, Nichitiu F, Worden J, Aben I, Bowman KW, Clerbaux C, Coheur PF, De Laat ATJ (2013) Decadal record of satellite carbon monoxide observations. Atmos Chem Phys 13(2):837–850
Worton DR, Sturges WT, Reeves CE, Newland MJ, Penkett SA, Atlas E, Stroud V, Johnson K, Schmidbauer N, Solberg S, Schwander J (2012) Evidence from firn air for recent decreases in non-methane hydrocarbons and a 20th century increase in nitrogen oxides in the northern hemisphere. Atmos Environ 54:592–602
Ziemke JR, Chandra S, Bhartia PK (2005) A 25-year data record of atmospheric ozone in the Pacific from Total Ozone Mapping Spectrometer (TOMS) cloud slicing: implications for ozone trends in the stratosphere and troposphere. J Geophys Res 110 (D15)

Chapter 4
The Effect of Climate Change on Water Resources

Arman Oliazadeh, Omid Bozorg-Haddad, Hugo A. Loáiciga, Sajjad Ahmad, and Vijay P. Singh

4.1 Introduction

This chapter reviews the various effects of climate change on water resources. The discussion is divided into two main topics: water resources quality and quantity, and water use patterns. The following topics are covered under water resources quality and quantity: (i) surface water, including urban catchment and river basins; (ii) groundwater; (iii) seas and oceans; and (iv) wetlands and lakes. Similarly, agricultural demand, domestic use, energy production, and industry are covered in the topic of water use patterns. This chapter relies on current knowledge about the effects of climate change on water resources and the hydrologic cycle derived from measurements, and from projections of the future climate made with state-of-art general circulation models. The reader must be aware that such climate projections involve

A. Oliazadeh · O. Bozorg-Haddad (✉)
Faculty of Agricultural Engineering and Technology, Department of Irrigation and Reclamation Engineering, College of Agriculture and Natural Resources, University of Tehran, Tehran, Iran
e-mail: OBHaddad@ut.ac.ir

A. Oliazadeh
e-mail: Arman.Oliazadeh@ut.ac.ir

H. A. Loáiciga
Department of Geography, University of California, Santa Barbara, CA, USA
e-mail: Hugo.Loaiciga@ucsb.edu

S. Ahmad
Department of Civil and Environmental Engineering and Construction, University of Nevada, Las Vegas, NV, USA
e-mail: sajjad.ahmad@unlv.edu

V. P. Singh
Department of Civil and Environmental Engineering, Texas A&M University, College Station, TX, USA
e-mail: VSingh@tamu.edu

O. Bozorg-Haddad (ed.), *Climate Change in Sustainable Water Resources Management*, Springer Water, https://doi.org/10.1007/978-981-19-1898-8_4

probable ranges of impacts on the hydrologic cycle and water resources, which are inherently uncertain; yet, they provide useful scenarios with which to prepare for adaptation to and mitigation of climate-change effects.

4.2 Water Resources Quality and Quantity

Water is the most important component of the hydrological cycle, influenced by climate change and multiple processes. Climate change has a major influence on ecosystems and human life (Collins et al. 2013; Padrón et al. 2020). Changes in the water cycle components have affected existing water resources and water availability on land (Greve et al. 2018), there are also some human-induced changes in drought severity (Marvel et al. 2019), trends in soil moisture in the root zone (Gu et al. 2019), and precipitation (Zhang et al. 2007; Marvel and Bonfils 2013). However, a key concern is that due to the growing population and increasing demand for food, it is uncertain how well prepared human societies (especially in less-developed countries) are to withstand the new conditions imposed by a warming climate. This is a vital issue due to rising agricultural food prices and food insecurity threatening millions of people based on climate change studies (Nelson et al. 2014; IFPRI 2019; Janssens et al. 2020). It is time to examine the effects of climate change on water resources and how to manage them.

Recent research suggests that the increase in temperature observed since the middle of the twentieth century is related to the increase in greenhouse gases from human activities (IPCC 2007). In all continents except Antarctica the role of human activity in climate change is evident. Global warming brings about lasting changes in many components of the hydrological cycle, such as changes in precipitation patterns (Carrier et al. 2016; Tamaddun et al. 2019), intensification of extreme events, increased melting of snow and ice (Tamaddun et al. 2017), increased atmospheric water vapor, and changes in soil moisture (Puri et al. 2011) and runoff (Carrier et al. 2013). As surface air temperature rises governments, scientists, engineers, and water resource managers face daunting challenges, some of which are as follows:

- Increases in the concentration of pollutants and decreases in the level of dissolved oxygen in the water, which reduces water quality (Venkatesan et al. 2011)
- Increases in the growth rate of algae and microbes, which affects the quality of drinking water (Nazari-Sharabian et al. 2018)
- Necessity for review of pollutants control programs, and pollutants' releases to rivers and the water cycle (Sattari et al. 2020)
- Consideration of new financial resources and solutions to manage the increase in water consumption to reduce the pressure on existing water-supply facilities (Ahmad and Prashar 2010)
- Necessity to find suitable solutions for accurate estimation of evaporation, and compensation for lost water from wetlands and lakes (Saher et al. 2021).

Current global water resource management practices are not adequate to overcome the effects of climate change on water supply, flood management, water quality, agricultural production, energy use, or household consumption. Therefore, a better understanding of the climate change impacts on hydrological characteristics is necessary for achieving effective management solutions to respond to future climatic conditions. Reducing the effects of climate change on water resources requires comprehensive strategies for water demand and supply management, and the use of economic incentives to develop water markets, and attention to virtual water trading are essential. Furthermore, the improvement of water resource management must account for regional water supply and demand to plan water transfers between basins and be able to cope with altered conditions under climate change.

It is fair to state that the problems of water resource management have not been sufficiently addressed in the context of climate change. These problems are more pronounced in low-income, developing countries, especially in Asia and Africa. It is clear that the impact of climate change will be different in different regions, and countries in each region must devise suitable responses to these changes. For example, drought, desertification, and water resource shortages are the hallmarks of most Middle Eastern countries. Their rapid population growth and development contribute to the deterioration of water quality, while raising living standards increase the demand for water in the region.

The growth rate of populations differs among regions, and the world's average growth rate has decreased (WPP 2019). However, global water consumption is likely to increase due to the economic growth of developing countries. There is a lack of reliable information on future increases in water consumption in developing countries. Many countries and regions including the European Union, the United States, Canada and Australia have developed and implemented specific standards for wastewater treatment reuse (Dow et al. 2019). This technology can somewhat mitigate the effects of climate change on water resources.

Important factors in predicting the status of water resources, taking into account the effects of climate change, are: changes in water storage infrastructure; water transmission systems; wastewater treatment; reuse of treated water; desalination of saltwater; and control of pollutant emissions. Many dams are expected to be built in developing countries for hydropower generation and water supply. Many dams are obsolete or have caused irreversible environmental harm, and changes in runoff due to climate change may increase or reduce water storage in reservoirs. The need to maintain environmental flows is essential in reservoir operations, and highlights the need for the proper use of water resources in the future. In places where runoff might increase due to climate change there would be larger water storage, which would improve water supply.

Modification and review of reservoirs' operating rules must be considered taking into account the needs of the environment (Ahmad and Simonovic 2000), and the gradual improvement in water treatment is expected in developed countries. Pollution by untreated wastewater use in irrigation and discharged to streams, lakes, and the ocean will remain serious unless proper wastewater treatment is implemented in developing countries.

Water, as a critical and key factor to socioeconomic development (Chen et al. 2017), could imperil food security and the protection of ecosystems and the environment under climate change in many regions, especially in arid and semiarid regions (Cheng et al. 2007; Notter et al. 2012). This chapter reviews water resources' quality and quantity, and their consumption patterns under climate change.

4.2.1 Surface Water

The understanding of the hydrological cycle, water resources, land use, and human life under climate change remains daunting for scientists, engineers, and governments due to the complexity of the climate system (Jeuland and Whittington 2014; Zhang et al. 2015). The changing frequency and intensity of extreme weather under climate change, and the effects on global water resources and their status in arid and semiarid regions, has been investigated (Kundzewicz et al. 2008; Qiao et al. 2014; Ravazzani et al. 2014; Zuo et al. 2015). Precipitation intensity and pattern changes from increasing temperature in warming earth and these changes' effects on the hydrological cycle have also been studied (Hansen et al. 2006; Piao et al. 2007; Wang et al. 2012), see Figs. 4.1, 4.2 and 4.3.

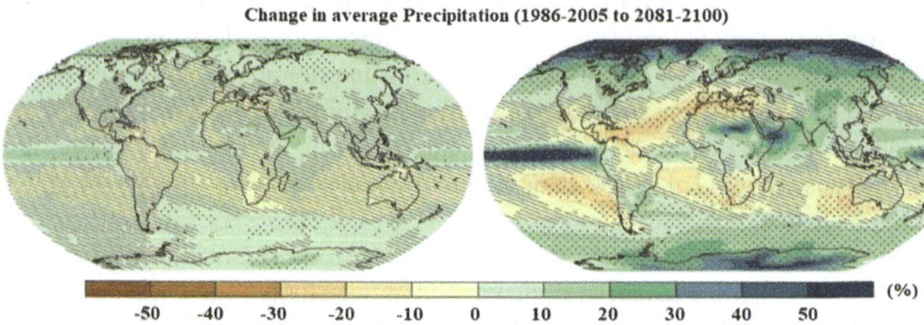

Fig. 4.1 Change in average precipitation (Left 1986–2005, Right 2081–2100, Source IPCC)

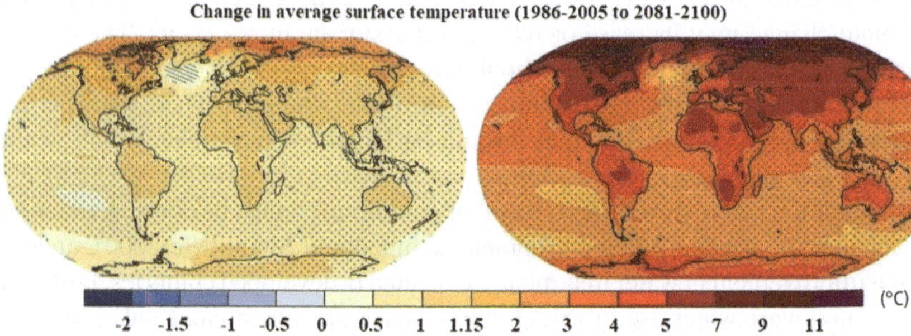

Fig. 4.2 Changes in surface temperature (Left 1986–2005, Right 2081–2100, Source IPCC)

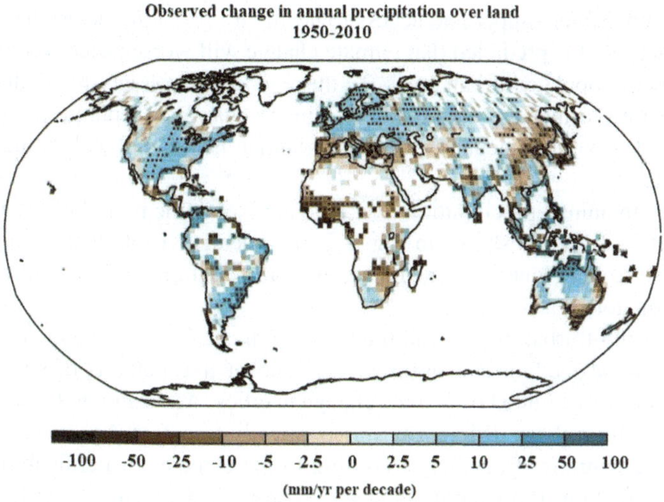

Fig. 4.3 Observed changes in annual precipitation over land area (*Source* IPCC)

Changes in average precipitation in the future (2081–2100) compared to historic values (1986–2005), as reported by IPCC, are shown in Fig. 4.1. It is seen in the above Figures that average precipitation would be substantially modified by the end of the twenty-first century, especially at the poles. There would also be a drastic increase in average precipitation.

Changes in surface temperature in the future (2081–2100) compared to historic values (1986–2005), as reported by IPCC, are depicted in Fig. 4.2. Similar to precipitation, changes in the surface temperature would be substantial, and noticeably in the North Pole.

4.2.1.1 Urban Catchments

Surface runoff management due to climate change is one of the main concerns of managers and urban designers. Population growth, land-use change, and urban sprawl increase impervious surfaces in urban areas, which in turn increases the volume and peak discharge due to climate change (Michelle et al. 2013). Traditional methods of urban runoff management to reduce the risk of floods search for ways to remove runoff from flooded areas, such that, in many cases, runoff enters sanitary sewers, and its capacity to recharge aquifer is lost. In recent years significant changes have been made in urban flood management methods by considering climate change, which emphasizes infiltrating runoff on site instead of discharging it outside city boundaries (Sadeghi et al. 2018).

Moreover, one of the important consequences of climate change is the changes in water quality and access to safe drinking water in urban areas, which is a threat to human health. Water supply and access to health are primarily determined by non-climatic factors. It is predicted that climate change will worsen access conditions in many densely-populated urban areas. To this end, it is necessary to conduct appropriate planning to manage the effects of climate change and public preparedness, in order to deal with its negative impact on human health and well-being in urban communities.

Changes in annual precipitation over land area, over the period of (1950–2010), as reported by IPCC, is shown in Fig. 4.3. It is clearly visible that changes in the region close to the equator are negative, and precipitation increases moving away from the equator.

The control of urban floods and their storage in surface reservoirs and aquifers, through artificial feeding and by gravity, is one of the methods of surface water management in the context of climate change in cities. According to this scheme the output water from these different basins is directly transferred to water collection tanks in a decentralized manner, according to the changes of permeability coefficients, construction, topographic, and the hydrological features of cities. Finally, part of the water that cannot be stored is gradually infiltrated into the subsurface. Stored water is used when needed to irrigate urban green spaces and local parks, for industrial and domestic uses and other non-drinking urban uses, such sports fields, schools, cemeteries, and parks.

Despite the problems and costs of constructing reservoirs governments must consider ways of reducing energy use and the unit cost of water production to minimize the environmental impact of increasing water consumption in cities. Ways of reducing costs of wastewater treatment and containing urban floods are now well established. There is also a need to use systems approach to conceptualize these management issues (Mirchi et al. 2012).

4.2.1.2 River Basin

The amount of water available from a surface or shallow groundwater source depends on seasonal and annual changes in water resources (Chen et al. 2019). More reliable water sources are usually determined by low water demand. Higher temperatures reduce runoff during the summer in basins where it snows frequently, and experience spring peaks (Saifullah et al. 2019). As the earth's climate warms the peak flow change is expected to occur earlier. Winter precipitation is expected to be mainly in the form of rain (Choubin et al. 2019). In many mountainous areas, where glaciers and snow are the main source of runoff, the volume of water stored in glaciers would decrease during the dry season (Pathak et al. 2018). On the other hand, the volume of runoff would increase during the hot and dry season; the volume of this runoff would decrease significantly as the glaciers continue to melt.

A change in sea ice extent during the twentieth century is shown in Fig. 4.4 as reported by IPCC. It has an overall falling trend over the period. It is seen in Fig. 4.4 that, there is a slight increase between 1930 and 1950, but the declining trend has

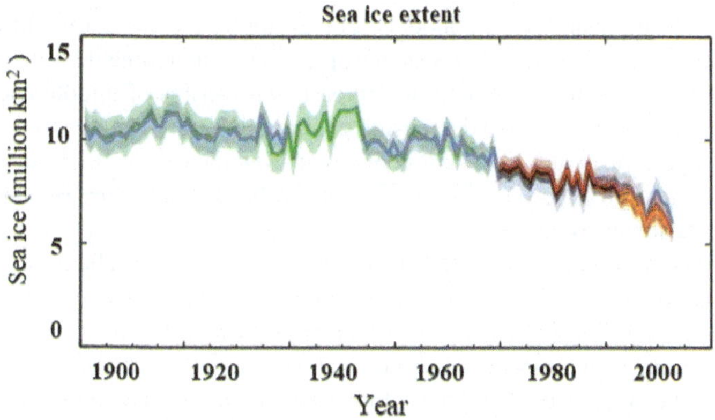

Fig. 4.4 Changes in sea ice extent (*Source* IPCC)

strengthened over the latest 20 years, which has reached about 6 million km^2 of the sea-ice area.

Drought-affected areas are projected to increase (Choubin et al. 2014), with heavy rain likely to increase in both intensity and frequency, which could raise the risk of flooding in river basins (Nyaupane et al. 2018). It is estimated that about 20% of the world's population living near river basins would be affected by flood risk (Mosquera-Machado and Ahmad 2007) by 2080 due to climate change. Freshwater resources, especially in arid and semi-arid regions, are exposed to the consequences of climate change. Many of these areas, including Australia, the western United States, the Mediterranean, and South Africa, are projected to face declining water resources due to climate change.

Once the per capita amount of water available in a basin falls below one thousand cubic meters per year, or the ratio of water withdrawn to the long-term average runoff of that basin is more than 0.4, that basin is under water stress. Humans and natural ecosystems are exposed to reduced rainfall and more precipitation fluctuations in areas affected by water stress caused by climate change. Such basins are located in North Africa, the Mediterranean region, the Middle-East, South Asia, northern China, Australia, Mexico, northeastern Brazil, parts of the United States, and western South America (Bates et al. 2008). The population living in these basins varies between 1.4 and 2.2 billion (Oki et al. 2003; Arnell 2004).

Increases in temperature, and subsequently increases in water temperature, may increase the intensity of rainfall, and raise the concentration of water pollutants in river basins. These changes would affect human health (Impoinvil et al. 2007), the state of natural ecosystems, the reliability of water systems, and their operating costs. In such basins, where runoff is projected to decrease, there are projected increases in seasonal runoff variability, water quality, and flood risks, as the water supply decreases and rainfall changes.

Predictions of future rainfall changes, with soil changes and runoff formation at the basin scale, are not very reliable in many areas because there is no general agreement on the results of the models. For example, increased rainfall and river flow in high latitude basins, as well as in tropical ones (including densely populated areas in Southeast Asia), are likely to increase. The basins of middle and tropical latitudes would have endured a decreasing precipitation trend. Forecasts for changes in rainfall and river water flow and changes in water levels at the basin scale are uncertain. Therefore, the decision to plan at the basin scale must be made in the context of these uncertainties.

Numerous studies have examined the trends of river discharge changes at the basin scale during the twentieth century in the context of climate change. Some of these studies have shown statistically significant relationships between river discharge trends and changes in temperature and rainfall (Pathak et al. 2017). However, many studies have been reported that did not find a specific trend, and could not detect the effects of temperature and precipitation changes on streamflow. Several statistical tests calculate varying intensities of the impacts of climate change. This hinders the certainty of projections about the extent of the impact of climate change in specific basins.

Human intervention in many basins has changed the flow regime in river basins over time (Sagarika et al. 2014). There there is evidence that there is an interconnected pattern of annual runoff changes globally (Thakur et al. 2020). For example, in high latitudes and parts of the United States, there is an increasing trend; and in parts of West Africa, southern Europe, and parts of South America, runoff has (Milly et al. 2005).

Annual changes in streamflow in many basins around the world have been reported under the influence of large-scale climate change patterns such as ENSO,[1] NAO[2] and PNA[3] patterns (Bhandari et al. 2018). The trend of runoff changes in catchments is not always consistent with changes in rainfall because of data limitations, water infrastructure (reservoirs, diversion), or changes in land use. The interaction of rainfall and temperature changes is complex (Bates et al. 2008).

Changes in flood-risk production or generation over the oceans in the future (2051–2060) compared to the historic period (2001–2010), as reported by IPCC, are depicted in Fig. 4.5. According to the figure, there is not any reliable data for the North Pole area, and the highest flood risk is expected in the equatorial seas.

Changes of streamflow in basins under the influence of climate change largely depend on changes in the types of snowfall or rainfall, the times of precipitation, and their topographic features. Projections indicate an increase in runoff at higher latitudes, and a decrease in runoff in areas with medium and lower latitudes in most basin-scale hydrological studies that have simulated runoff under several climate change scenarios. These projections reflect the fact that global warming is affecting the seasonal flow of rivers. Streamflow increases in the spring due to snowmelt, and declines in areas with much more rain (instead of snow) in the winter. This is the case in areas such as the European Alps, Scandinavia, the Himalayas, and the western and

[1] ENSO: Elnino-Southern Oscillation.

[2] NAO: North Atlantic Oscillation.

[3] PNA: Pacific North America.

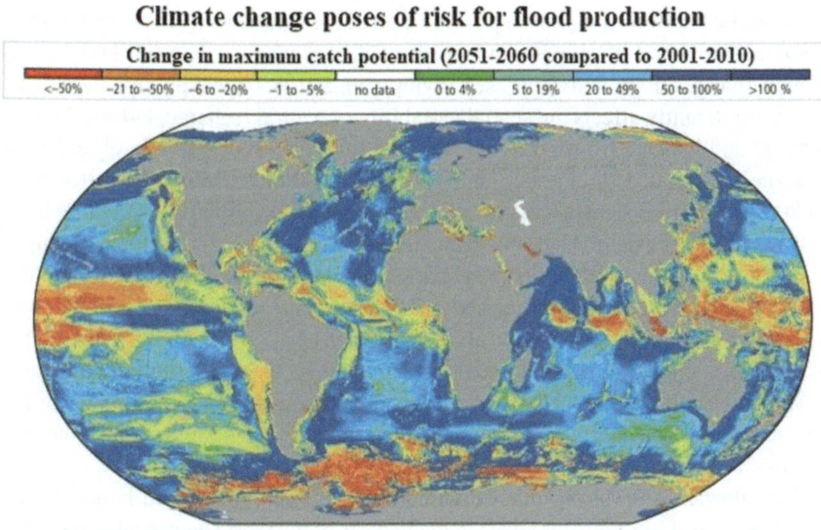

Fig. 4.5 Changes in flood risk production (*Source* IPCC)

central regions of North America. Snowmelt occurs earlier under climate change, and peak river discharges are predicted to occur at least one month earlier by the middle of the twenty-first century. Additionally, in areas where snowfall is very low or it does not snow at all, the changes in river runoff depend on changes in rainfall rather than temperature changes. In glacial regions, such as the highlands of Asia and South America, the melting of glaciers would lead to an increase in water flow in many rivers in the short term (next 10 years), but the major impact is likely to take hold in the coming decades (30 years ahead or beyond).

There is a coherent pattern in the annual change of runoff that has been experienced in many areas. Various studies have been conducted on the effects of rising temperatures and rising runoff in different parts of the world, including (Labat et al. 2004), which predicted a 4% increase in the runoff on a global scale due to regional changes during the twentieth century. However, this claim is undermined by the impact of non-climatic factors on runoff.

4.2.2 *Groundwater*

Groundwater is a source of relatively high quality fresh water supply. The effect of climate change on groundwater resources is of central relevance for humans and the environment (see, e.g., Loáiciga 2003a, b; Rahaman et al. 2019). In many aquatic ecosystems, especially during a meteorological drought, Natural groundwater discharges contribute base flow to surface water resources (Taylor et al. 2013).

Groundwater flow in shallow aquifers is part of the natural water cycle that changes under the influence of climate change through replenishment by infiltration, and through changes in groundwater use (Taylor et al. 2013). Global precipitation distribution significantly affects the spatial variability of natural recharge. Diffuse rain-fed recharge and leakage from surface waters are two main types of natural replenishment that are related to climate, land cover, and geology (Döll and Fiedler 2008; Wada et al. 2010). For instance, groundwater storage diminished considerably in the Murray–Darling basin in Australia from 2000 to 2007 (during multi-annual drought) in response to a fast reduction in recharge (Leblanc et al. 2009). Heavy rainfall in Africa and semiarid regions caused disproportionate and temporal surface water recharge (Small 2005; Favreau et al. 2009; Owor et al. 2009; Taylor et al. 2012). In addition to changing rainfall under climate change, at high latitudes, changes in the snowmelt regime under global warming such as earlier snow melting or rain on snow tend to reduce aquifer recharge (Tague and Grant 2009; Sultana and Coulibaly 2011).

Groundwater is enduring intense withdrawal in many areas, and is undergoing drastic changes. Groundwater levels in many parts of the world have declined sharply in recent decades, but this declining trend is due more to the overdraft of groundwater aquifers (such as in the North China Plain (Chen 2010), northwest India (Rodell et al. 2009) the US high plains (Longuevergne et al. 2010; Scanlon et al. 2012; Loáiciga 2017), Brazil (Foster et al. 2009) and Bangladesh (Shamsudduha et al. 2012) than to the effects of climate change on them. Land use change (LUC) under climate change, significantly influences terrestrial hydrology and the relationship between climate and groundwater. For instance, LUC to cropland caused increasing groundwater storage in the African Sahel (Leblanc et al. 2008).

There are regions where the increase in groundwater withdrawal is not only due to increased demand, but also reduced groundwater recharge, due to the effects of climate change, such as in southwestern Australia. Groundwater systems respond relatively slowly to climate change. This calls for long-term monitoring of groundwater conditions. Interestingly, climate-change effects on groundwater affect surface water resources and may increase the water demand. The reduction of base flow to support streamflow constitutes a direct hydraulic connection through which the effect of climate change on groundwater is reflected on surface water. Additionally, the storage of reservoirs is depleted during drought, and springs dry up or are diminished during drought, which is projected to become more severe in a warmer climate in many regions. As groundwater levels fall, in addition to the disappearance of dependent vegetation, seawater intrusion is exacerbated in coastal aquifer, and excessive irrigation with saline water is likely to contribute to desertification (see Zektser et al. 2005).

4.2.3 *The Oceans*

Climate change has increased the average sea level on a global scale, so that in the second half of the twentieth century the annual sea level rise range between 1.8 and 2.3 mm. Studies have shown that sea-level rise was continuous from the mid-nineteenth century to the end of the twentieth century (Bindoff et al. 2007). It is noteworthy that the roles of several factors driving sea-level change involve uncertainty, but the understanding of knowledge on this issue has increased significantly in recent years. For example, for the decade of 1993–2003, the effect of ocean thermal expansion amounted to an annual sea-level rise of 1.6–1.2 mm, and the melting of glaciers in Greenland's ice sheets contributed an annual sea level between 2.8 and 5 mm, which is estimated to be about 3 mm per year (Bindoff et al. 2007).

Among the primary factors that increase the sea level in the face of climate change are the expansion of ocean water due to rising water temperatures, melting glaciers and shrinking ice masses in Greenland and Antarctica. Other factors that may also contribute to sea level rise are the drainage and discharge of wetlands into the sea, increased deforestation, and land use changes.

Changes in average sea level in the future (2091–2100) compared to the historic period (1986–2005), as reported by IPCC, are shown in Fig. 4.6. It can be concluded that most of the world's seas will face a sea level rise in coming decades. There appears to be broad consensus on the projection of sea level rise base on global-scale satellite measurements, although the actual rise, say, by 2100, remains speculative.

The IPCC projections on the effects of climate change indicate that current models predict sea level rise through the twenty-first century. This increase varies with geographical location. In the United States, for example, sea level along the Atlantic coast rises by about 2–3 mm a year. This increase along the Louisiana coast is between 10 mm per year to a few centimeters per decade; whereas sea level rise in parts of Alaska is affected by subsidence or rising ground (IPCC 2007). The IPCC also predicts that global sea levels will rise by 180–590 mm on a global scale by the end of the twenty-first century compared to the last two decades of the twentieth century (IPCC 2013).

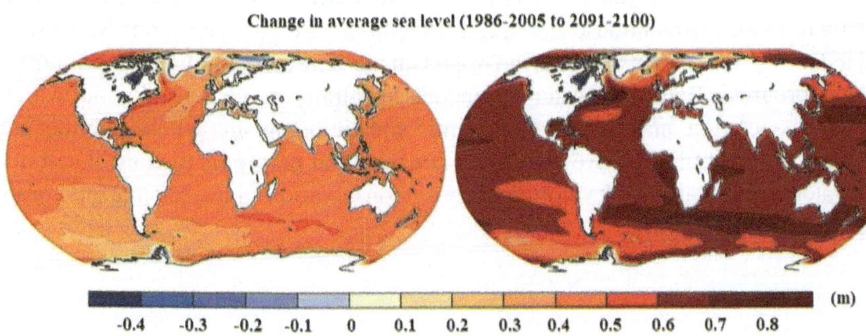

Fig. 4.6 Changes in average sea level (Left 1986–2005, Right 2091–2100, Source IPCC)

These estimates assume that the melting of ice in Greenland and Antarctica will continue at the same rate as previously observed. Under this assumption a high amplitude of the predicted sea-level rise by 2100 ranges between 480 and 790 mm (IPCC 2007).

Rising sea levels increase the salinity of surface and subsurface water resources through the interference of saline and freshwater, which endangers the life of aquatic plants and organisms dwelling in the shores and river deltas. The impact of rising sea levels on water resources includes the gradual flooding of natural systems, flooding of coastal infrastructure, and less efficient draining of coastal areas. The effects in the coastal areas include the following: alteration of coastal wetlands and deltas; more intense erosion of coastal regions; redesign of water quality facilities due to poor drainage to the sea; disruption of wastewater treatment; relocation of infrastructure and populations to higher ground. The impact in coastal areas, such as deltas and coastal cities, and the change of their complex ecosystems, will heighten flood hazards and storm surge damages (see, e.g., Garcia and Loáiciga 2014).

To manage sea-level rise water resources policy makers must take a comprehensive approach to land-use change by protecting coastal areas and offshore facilities, managing wetlands, river deltas, and aquifers, and managing water use. Besides, attention must be paid to the long-term predictions of sea-level rise to prevent seawater intrusion into freshwater aquifers when designing drinking water and sewage systems.

The rising carbon dioxide concentration in the atmosphere by human-induced emissions of greenhouse gases causes more of this gas to dissolve in ocean water, albeit the rising ocean surface temperature (Daryayehsalameh et al. 2021). Increasing the level of carbon dioxide in seawater renders it more acidic (see, e.g., Loáiciga 2006). Atmospheric CO_2 concentration was about 280 ppmv in the mid eighteenth century. It is currently near 417 ppmv (as of April 2021), and continues to rise unabatedly. The impacts of seawater acidification on marine organisms are not well known, but it is believed by many experts that the reproduction of coral reefs and phytoplankton, for instance, would be adversely affected. According to the IPCC rising carbon dioxide in the oceans has reduced the number of marine species by about one-tenth since 1975.

The responses of marine ecosystems to ocean warming and acidification caused by human-induced greenhouse gas emissions remain the topic of research. Moreover, understanding the response of marine organisms to seawater acidification remains a work in progress. Rising seawater has increased salinity in the coastal zone, which damages vegetation, and endangers endemic aquatic organisms whose habitats are altered. This calls for adaptive watershed conservation programs to control salinity and water acidity, and to develop and revise conservation plans for marine corals against ocean pollutants.

4.2.4 Lakes and Wetlands

Increased rainfall intensity over long periods of river flow causes a variety of lake-water pollution, including nutrient accumulation, increased dissolved carbon, pathogens, pesticides, soluble salts, and heat pollution. This increases the density of bacteria in water, which affects river ecosystems and human health, and raises the maintenance costs of water resources systems Increasing surface temperatures may reduce the water quality of lakes (Rusuli et al. 2015), reduce oxygen concentration, and cause the release of phosphorus from sediments. On the other hand, rising temperatures may cause ice to break faster, increase the oxygen level in the water, and reduce winter fish deaths. Increased rainfall due to soil erosion increases the concentration of suspended particles in lakes, causing water pollution, and the transmission of pathogens. Increased erosion (Melesse et al. 2011) releases adsorbed pollutants, such as phosphate and heavy metals, which increases the concentration of pollutants in wetlands.

There is currently no one-size-fits-all trend in global warming due to climate change. For example, the water levels of lakes in China, Australia, Africa, North America, and Europe have decreased due to droughts and global warming, and human activities, but the water levels of some lakes in the mountainous regions of China and northern Europe have increased due to the increased melting of snow and ice (Smith et al. 2005).

Higher water temperatures have been recorded in lakes. Shortening of the glacial period and reduction of the ice thickness of lakes is also a matter of concern. Water temperatures have risen between 0.2 and 2 °C in lakes in Asia and North America since the 1960s. Following the warming of the surface layer of water in lakes their deeper layers have become between 0.2 and 0.7 °C warmer in the twentieth century. Thus, rising temperatures, and longer non-glacial seasons due to climate change have affected the temperature distribution and internal hydrodynamics of lakes and their ecosystems. Water evaporation from lakes has increased in recent years, due to warming of the surface layers of the water, and to summer stratification occurring earlier in the season and thermoclines being shorter. This has changed the temperature layering period to 22 days (starting earlier) in many lakes in North America and Europe, and has extended their thermal stability by up to 3 weeks.

Increasing the stability of lakes reduces the movement of water, which reduces the circulation of oxygen and reduces nutrient availability. This reduces the concentration of nutrients in the surface layer of water and increases it in the deeper layers, which is due to the thermal stability in the lakes. The concentrations of silicates, calcium, and magnesium sulfates, the concentrations of base cations and carbonates in the water, and the alkalinity and electrical conductivity of the water have increased in many lakes. On the other hand, rising temperatures also affect chemical exchange processes in lakes. With the increasing production of phytoplankton there is an associated reduction in dissolved inorganic nitrogen, and an increase in alkalinity and pH of lake water. Decreased solubility due to higher temperatures has significantly reduced the concentration of aluminum by about 13% and increased the concentration of mercury in fishes (Buck et al. 2019).

4.3 Water Use Patterns

Water use management under climate change conditions is key to achieve reliable supply to the municipal and industrial, agricultural, and other sectors. There must be a change in the pattern of water use to reduce the effects of climate change on water resources, and remedies such as overdraft of aquifers and water supply through water transfers may prove short lived. The following is a discussion on water use patterns for the agricultural, domestic, and energy sectors.

4.3.1 Agriculture

The water required for irrigation increases with increasing temperature and decreasing rainfall. There is currently no strong evidence to show a trend in changes of agricultural water consumption in relation to climate change. The reason is that water consumption is always expressed by non-climatic parameters, and there is a lack of sufficient and accurate information.

Irrigation water supply depending on surface water resources and groundwater wells varies seasonally, primarily with the variation of streamflow. In basins where streamflow depends on snowmelt rising temperature reduce river flow and water supply in the summer. Water use has also increased in many countries due to population growth, economic development, and lifestyle changes, and with the development of water supply systems to increase water use for agricultural production (Flörke et al. 2018). About 70% of the world's water resources are used for irrigation, which constitutes more than 90% of the available water (Brocca et al. 2018). In addition, the area of irrigated land has increased from about 140 million hectares in 1960 to 270 million hectares in 2000. The application of nitrogenous fertilizers worldwide will reach 140 million tons by 2050, while their use in 2000 was about 90 million tons. The rising reliance on fertilizer means that more attention must be devoted to protecting water quality with respect to eutrophication (Bates et al. 2008).

The demand for irrigation increases due to rising temperature assuming that rainfall remains constant during the growing season. For example, in China and India, which have the largest irrigation networks in the world, irrigated lands are projected to increase from 2 to 20% in China and from 2 to 7% in India by 2070 (Douville et al. 2002). The IPCC prescribes that a combination of cropping pattern modification with crop rotation would help carbon storage, and improve the functioning of agro-ecosystems (Smith et al. 2007).

Reducing soil operations and minimizing plowing, which keeps plants on the soil's surface, would prevent water evaporation losses, and is being increasingly used worldwide because soil turbidity is exacerbated by increased crop resorption, and erosion releases carbon from the soil to the air. Plowing may also affect nitrogen oxide (N_2O) emissions (Ma et al. 2015). Reducing the energy intensity of agricultural machinery would also be effective in reducing carbon dioxide emissions. It is noteworthy that the conversion of floodplains and wetlands to agricultural land could diminish the ecological function of nature, reduce aquifer recharge, and alter the nutrient cycle.

Because of population growth there is a need to produce more food, thus raising the application o fertilizers and pesticides, which would affect water quality in the future. Moreover, increasing water use by agriculture would raise the concentration of pollutants in water resources, which highlights the need for increasing use of water treatment plants in the future under climate change conditions.

The increasing frequency of droughts and floods would reduce crop yields. These consequences are expected to be seen more often and sooner than expected. An increased frequency of droughts and floods has a greater impact on the yield of local and indigenous crops, especially agricultural products that are produced in lower latitudes. The effect of climate change and irrigation water use must be considered. The use of new reservoirs for irrigation would partially reduce the deficit of water supply for irrigation; yet, the use of reservoirs is not always and everywhere practical or desirable, while these reservoirs themselves will face reduced performance in the long run due to sedimentation, not to mention the loss of inundated land, the displacement of population and animal species, and the submergence valuable habitat and cultural and archeological treasures. Farmers in different parts of the world must adapt to climate change by modifying crop patterns whenever possible, altering the planting time of crops, using improved technology, and redesigning water supply and irrigation systems (Ghumman et al. 2018). It is also suggested that to combat water scarcity in the face of climate change the world must increase the reuse of treated wastewater for irrigation and watering urban resources such as parks, cemeteries, golf courses, and other green areas (Loáiciga 2015). The growing use of water by agriculture and the worsening scarcity due to climate change militate to intensify the competition for water resources. Comprehensive cooperation by international and regional agreements to reach water and food security that meet human needs is necessary to overcome the threats posed by climate change to the agricultural sector.

4.3.2 Domestic Use (Urban)

The urban water system is vulnerable to climate change, and without proper foresight and strategic planning, irreparable consequences may occur (Tamaddun et al. 2018). The harmful effects of climate change usually affect the most vulnerable segments of society. Human settlements must always be prepared to deal with the effects of climate change. Decision makers must use local expertise and knowledge to better manage the climate uncertainty and take effective steps to adapt to its effects. The urban water system alone cannot adapt to climate change; therefore, integrated water management involving the energy, transportation, health, and water-treatment sectors must be implemented (IUWM) (Closas et al. 2012; Furlong et al. 2016; Oral et al. 2020). This means taking into account population growth, land use change, and pollution in conjunction with the measured and projected impacts of climate change.

Urban regions may be exposed to floods and droughts over relatively short periods of time; these local effects of climate change highlight the need for cities to develop accurate water management scenarios.

Urban population growth and suburbanization will accelerate in developing countries, which most likely will result in the exacerbation of misery and poverty, including the rising vulnerability of cities, and intensifying the reliance on already-stressed water resources (Brandt et al. 2016; Babovic et al. 2018; Grafakos et al. 2020). It is predicted that about 68% of the world's population will live in urban areas by 2050, and urban cannot be supported without achieving water security.

The rising demand for domestic and urban water use in the face of climate change is evident, including an increase in the need for electricity to cool buildings and a greater need for water to cool thermal power plants. For example, daily per capita water use at temperatures above 25 °C increases to 11 L per day for New Yorkers (Protopapas et al. 2000; Karim et al. 2021).

Many nations are facing water shortages due to the phenomenon of climate change, and need improved management programs to manage urban water consumption in the short, medium, and long terms, including the following: (1) educational and public awareness programs (Nussbaum et al. 2015) to recognize the threat of climate change and assess the optimal water use; (2) the development of water conservation programs; (3) exaction of fines on those who do not use water efficiently; (4) increase the reuse of municipal and industrial treated wastewater; and (5) retention of urban flood waters in surface and subsurface storage; and (6) increased managed aquifer recharge (MAR).

Many cities have been implemented specific adaptation measures, while most have adopted or at least initiated adaptation strategies (Qaiser et al. 2013; Kandissounon et al. 2018). Different urban areas face a wide range of climate change challenges, and their responses vary according to their circumstances. A number of cities have given priority to one type of adaptation response, such as strengthening governance in Durban, or improving infrastructure in Samarang; other cities have adopted a set of solutions that include the use of economic benefits and natural systems (Loftus et al. 2011; Habibifar et al. 2021b). In general, there is often a strong emphasis on the importance of stakeholder participation in all stages of the process (Vedwan et al. 2008), and in taking into account quality-of-life considerations in IUWM across urban regions in Asia (Reed et al. 2015), Africa (Jørgensen et al. 2014), Europe (Wihlborg et al. 2019), and Australia (Horne 2018).

Flexible options and technologies are integrated with soft solutions such as stakeholder engagement, information, and training, instead of simply focusing on infrastructure projects to cope with climate change in urban areas.

4.3.3 Energy Production and Industries

There is a very close and important connection between water and energy. Water resources management is directly related to energy use for water transfer and purification activities such as pumping, domestic heat generation, and drinking water and wastewater treatment (Shayesteh et al. 2019; Azimian et al. 2021). A source of energy is hydroelectric powerplants, steam turbines, and fossil fuels (Moumouni et al. 2014). The connection between these sources has gradually increased, especially when water resources are rapidly declining and climate change is affecting these conditions. At this time, when many cities are trying to reduce greenhouse gas emissions, high energy use by water resources systems intensifies the effects of climate change on energy production (Wilbanks et al. 2008; Akhmat et al. 2014; Labriet et al. 2015; Van Ruijven et al. 2019).

Various mechanisms for energy production increase greenhouse gas emissions (Shrestha et al. 2011). Several technologies have been proposed to reduce greenhouse gas emissions from fossil fuels in water resources. For example, energy production from biofuels, and carbon capture and storage are options that have implemented, although their application at large scales remains a daunting challenge (Loáiciga 2011, 2013; Salehi et al. 2019; Cao et al. 2020; Sardari et al. 2020). Renewable wind and solar energy is making gains by declining costs and government incentives (Ghasemi et al. 2018). But phasing out of fossil fuels would take some time (Habibifar et al. 2021a, b).

Corn cultivation in North America is partly dedicated to producing ethane; soybean cultivation and extensive operational measures have been taken to produce vegetable diesel oil (Yadav et al. 2020). The sustainable production of biomass must be accompanied by a careful study of its effects on water and soil resources, in the displacement of food production, and as a biodiversity disruptor (Pandey et al. 2019; Elisa et al. 2020). These measures could play effective roles in reducing the effects of climate change by absorbing carbon and storing underground. Other methods of carbon sequestration and storage include the injection of carbon in the deep ocean, or by the biological storage of carbon in forests through the development of afforestation. Stopping reducing deforestation, and sustainable forest management, can significantly reduce carbon emissions, protect resources, prevent floods, reduce runoff peak, control erosion, reduce river sedimentation to protect fish habitat, and enhance biodiversity would reduce energy use and contribute to combating climate change.

4.4 Summary

Global warming, rising temperatures and declining rainfall are among the dangerous climate changes that have affected water resources, human life and natural organisms. There are different types of effects for climate change on water resources which include water resources quality and quantity and water use patterns in this chapter. The former is divided into four main areas of surface water, groundwater, oceans, lakes and wetlands. Also, the latter classified into three main subjects of agricultural, domestic, and industrial (energy production). Increasing temperature and decreasing precipitation have reduced the discharge of surface runoffs in different parts of the world which leads to a decline in the level of groundwater aquifers. In the agriculture section, according to vast climatic and agricultural conditions, solutions have been presented to preserve water resources accordingly. Finally, fossil fuels can be replaced by biofuels to reduce pollution and mitigate climate change conditions in industrial and domestic energy consumption.

References

Ahmad S, Prashar D (2010) Evaluating municipal water conservation policies using a dynamic simulation model. Water Resour Manag 24(13):3371–3395

Ahmad S, Simonovic SP (2000) System dynamics modeling of reservoir operations for flood management. ASCE J Comput Civ Eng 14(3):190–198

Akhmat G, Zaman K, Shukui T, Sajjad F (2014) Does energy consumption contribute to climate change? Evidence from major regions of the world. Renew Sustain Energy Rev 36:123–134

Arnell NW (2004) Climate change and global water resources: SRES emissions and socio-economic scenarios. Glob Environ Chang 14(1):31–52

Azimian M, Amir V, Habibifar R, Golmohamadi H (2021) Probabilistic optimization of networked multi-carrier microgrids to enhance resilience leveraging demand response programs. Sustainability 13(11):5792

Babovic F, Mijic A, Madani K (2018) Decision making under deep uncertainty for adapting urban drainage systems to change. Urban Water J 15(6):552–560

Bates B, Kundzewicz Z, Wu S (2008) Climate change and water. Intergovernmental panel on climate change secretariat

Bhandari S, Kalra A, Tamaddun K, Ahmad S (2018) Relationship between ocean-atmospheric climate variables and regional streamflow of the conterminous United States. Hydrology. https://doi.org/10.3390/hydrology5020030

Bindoff NL, Willebrand J, Artale V, Cazenave A, Gregory JM, Gulev S, Shum CK (2007) Observations: oceanic climate change and sea level

Brandt L, Lewis AD, Fahey R, Scott L, Darling L, Swanston C (2016) A framework for adapting urban forests to climate change. Environ Sci Policy 66:393–402

Brocca L, Tarpanelli A, Filippucci P, Dorigo W, Zaussinger F, Gruber A, Fernández-Prieto D (2018) How much water is used for irrigation? A new approach exploiting coarse resolution satellite soil moisture products. Int J Appl Earth Obs Geoinf 73:752–766

Buck DG, Evers DC, Adams E, DiGangi J, Beeler B, Samánek J, ... , Johnson S (2019) A global-scale assessment of fish mercury concentrations and the identification of biological hotspots. Sci Total Environ 687:956–966

Cao Y, Doustgani A, Salehi A, Nemati M, Ghasemi A, Koohshekan O (2020) The economic evaluation of establishing a plant for producing biodiesel from edible oil wastes in oil-rich countries: case study Iran. Energy 213:118760

Carrier C, Kalra A, Ahmad S (2016) Long-range precipitation forecasts using paleoclimate reconstructions in the Western United States. J Mt Sci 13(4):614–632. https://doi.org/10.1007/s11629-014-3360-2

Carrier C, Kalra A, Ahmad S (2013) Using paleo reconstructions to improve streamflow forecast lead-time in the Western United States. J Am Water Resour Assoc (JAWRA) 49(6):1351–1366. https://doi.org/10.1111/jawra.12088

Chen C, Ahmad S, Kalra A, Xu Z (2017) A dynamic model for exploring water-resource management scenarios in an inland arid area: Shanshan County, Northwestern China. J Mt Sci 14https://doi.org/10.1007/s11629-016-4210-1

Chen C, Kalra A, Ahmad S (2019) Hydrologic responses to climate change using downscaled GCM data on a watershed scale. J Water Clim Change 10(1):63–77

Chen JY (2010) Holistic assessment of groundwater resources and regional environmental problems in the North China Plain. Environ Earth Sci 61:1037–1047

Cheng Y-F, Wang G-X, Xi H-Y, Wang J-D (2007) Variations of land evapotranspiration in the plain of the middle reaches of Heihe River in the recent 35 years. J Glaciol Geocryol 29:406–412

Choubin B, Alamdarloo EH, Mosavi A, Hosseini FS, Ahmad S, Goodarzi M, Shamshirband S (2019) Spatiotemporal dynamics assessment of snow cover to infer snowline elevation mobility in the mountainous regions. Cold Reg Sci Technol. https://doi.org/10.1016/j.coldregions.2019.102870

Choubin B, Khalighi-Sigaroodi S, Malekian A, Ahmad S, Attarod P (2014) Drought forecasting in a semi-arid watershed using climate signals: a neuro-fuzzy modeling approach. J Mt Sci 11(6):1593–1605. https://doi.org/10.1007/s11629-014-3020-6

Closas A, Schuring M, Rodriguez D (2012) Integrated urban water management: lessons and recommendations from regional experiences in Latin America, Central Asia, and Africa (No 75043). The World Bank

Collins M, Knutti R, Arblaster J, Dufresne JL, Fichefet T, Friedlingstein P, Booth BB (2013) Long term climate change: projections, commitments and irreversibility. In: Climate change 2013-the physical science basis: contribution of working group I to the fifth assessment report of the intergovernmental panel on climate change. Cambridge University Press, pp 1029–1136

Daryayehsalameh B, Nabavi M, Vaferi B (2021) Modeling of CO_2 capture ability of [Bmim][BF4] ionic liquid using connectionist smart paradigms. Environ Technol Innov 22:101484

Döll P, Fiedler K (2008) Global-scale modeling of groundwater recharge.

Douville H, Chauvin F, Planton S, Royer JF, Salas-Melia D, Tyteca S (2002) Sensitivity of the hydrological cycle to increasing amounts of greenhouse gases and aerosols. Clim Dyn 20(1):45–68

Dow C, Ahmad S, Stave K, Gerrity D (2019) Evaluating the sustainability of IPR and DPR: a Southern Nevada case study. AWWA Water Sci

Elisa P, Alessandro P, Andrea A, Silvia B, Mathis P, Dominik P, Thomas S (2020) Environmental and climate change impacts of eighteen biomass-based plants in the alpine region: a comparative analysis. J Clean Prod 242:118449

Favreau G, Cappelaere B, Massuel S, Leblanc M, Boucher M, Boulain N, Leduc C (2009) Land clearing, climate variability, and water resources increase in semiarid Southwest Niger: a review. Water Resour Res 45(7)

Flörke M, Schneider C, McDonald RI (2018) Water competition between cities and agriculture driven by climate change and urban growth. Nat Sustain 1(1):51–58

Foster S, Hirata R, Vidal A, Schmidt G, Garduño H (2009) The Guarani Aquifer initiative—towards realistic groundwater management in a transboundary context. The World Bank

Furlong C, De Silva S, Guthrie L (2016) Planning scales and approval processes for IUWM projects; lessons from Melbourne, Australia. Water Policy 18(3):783–802

Garcia ES, Loáiciga HA (2014) Sea-level rise and flooding in coastal riverine floodplains. Hydrol Sci J 59(1–2):204–220

Ghasemi A, Heidarnejad P, Noorpoor A (2018) A novel solar-biomass based multi-generation energy system including water desalination and liquefaction of natural gas system: thermodynamic and thermoeconomic optimization. J Clean Prod 196:424–437

Ghumman AR, Iqbal M, Ahmad S, Hashmi HN (2018) Experimental and numerical investigations for optimal emitter spacing in drip irrigation. Irrig Drain 67:724–737. https://doi.org/10.1002/ird.2284
Global food policy report (IFPRI 2019)
Grafakos S, Viero G, Reckien D, Trigg K, Viguie V, Sudmant A, ..., Dawson R (2020). Integration of mitigation and adaptation in urban climate change action plans in Europe: a systematic assessment. Renew Sustain Energy Rev 121:109623
Greve P, Gudmundsson L, Seneviratne SI (2018) Regional scaling of annual mean precipitation and water availability with global temperature change. Earth Syst Dyn 9(1):227–240
Gu X, Zhang Q, Li J, Singh VP, Liu J, Sun P, Cheng C (2019) Attribution of global soil moisture drying to human activities: a quantitative viewpoint. Geophys Res Lett 46(5):2573–2582
Habibifar R, Khoshjahan M, Saravi VS, Kalantar M (2021a, February) Robust energy management of residential energy hubs integrated with power-to-X technology. In: 2021a IEEE texas power and energy conference (TPEC). IEEE, pp 1–6
Habibifar R, Ranjbar H, Shafie-Khah M, Ehsan M, Catalão JP (2021b) Network-constrained optimal scheduling of multi-carrier residential energy systems: a chance-constrained approach. IEEE Access
Hansen J, Sato M, Ruedy R, Lo K, Lea DW, Medina-Elizade M (2006) Global temperature change. Proc Natl Acad Sci 103(39):14288–14293
Horne J (2018) Resilience in major Australian cities: assessing capacity and preparedness to respond to extreme weather events. Int J Water Resour Dev 34(4):632–651
Impoinvil DE, Ahmad S, Troyo A, Keating J, Githeko AK, Mbogo CM, Kibe L, Githure JI, Gad AM, Hassan AN, Orshan L, Warburg A, Calderón-Arguedas O, Sánchez-Loría VM, Velit-Suarez R, Chadee DD, Novak RJ, Beier JC (2007) Comparison of mosquito control programs in seven urban sites in Africa, the Middle East, and the Americas. Health Policy 83(2–3):196–212
IPCC (2007) Climate change 2007: the physical science basis. Contribution of working group I to the fourth assessment report of the intergovernmental panel on climate change [Solomon S, Qin D, Manning M, Chen Z, Marquis M, Averyt KB, Tignor M, Miller HL (eds)]. Cambridge University Press, Cambridge, United Kingdom and New York, NY, USA
IPCC (2013) Climate change 2013: the physical science basis. Contribution of working group I to the fifth assessment report of the intergovernmental panel on climate change [Stocker TF, Qin D, Plattner G-K, Tignor M, Allen SK, Boschung J, Nauels A, Xia Y, Bex V, Midgley PM (eds)]. Cambridge University Press, Cambridge, United Kingdom and New York, NY, USA, 1535 pp
Janssens C, Havlík P, Krisztin T, Baker J, Frank S, Hasegawa T, Maertens M (2020) Global hunger and climate change adaptation through international trade. Nat Clim Chang 10(9):829–835
Jeuland M, Whittington D (2014) Water resources planning under climate change: assessing the robustness of real options for the Blue Nile. Water Resour Res 50(3):2086–2107
Jørgensen G, Herslund LB, Lund DH, Workneh A, Kombe W, Gueye S (2014) Climate change adaptation in urban planning in African cities: the CLUVA project. In: Resilience and sustainability in relation to natural disasters: a challenge for future cities. Springer, Cham, pp 25–37
Kandissounon GA, Kalra A, Ahmad S (2018) Integrating system dynamics and remote sensing to estimate future water usage and average surface runoff in Lagos, Nigeria. Civ Eng J 4(2):378–393. https://doi.org/10.28991/cej-030998
Karim SHT, Tofiq TA, Shariati M, Rad HN, Ghasemi A (2021) 4E analyses and multi-objective optimization of a solar-based combined cooling, heating, and power system for residential applications. Energy Rep 7:1780–1797
Kundzewicz ZW, Mata LJ, Arnell NW, Döll P, Jimenez B, Miller K, Shiklomanov I (2008) The implications of projected climate change for freshwater resources and their management. Hydrol Sci J 53(1):3–10
Labat D, Goddéris Y, Probst JL, Guyot JL (2004) Evidence for global runoff increase related to climate warming. Adv Water Resour 27(6):631–642

Labriet M, Joshi SR, Vielle M, Holden PB, Edwards NR, Kanudia A, Babonneau F (2015) Worldwide impacts of climate change on energy for heating and cooling. Mitig Adapt Strat Glob Change 20(7):1111–1136

Leblanc MJ, Favreau G, Massuel S, Tweed SO, Loireau M, Cappelaere B (2008) Land clearance and hydrological change in the Sahel: SW Niger. Global Planet Change 61(3–4):135–150

Leblanc MJ, Tregoning P, Ramillien G, Tweed SO, Fakes A (2009) Basin-scale, integrated observations of the early 21st century multiyear drought in Southeast Australia. Water Resour Res 45(4).

Loáiciga HA (2003a) Climate change and Groundwater. Ann Assoc Am Geogr 93(1):33–45

Loáiciga HA (2003b) Modern-age CO_2 and its effect on seawater acidity and salinity. Geophys Res Lett 33:L10605. https://doi.org/10.1029/2006GL026305

Loáiciga HA (2011) Challenges to phasing out fossil fuels as the major source of the world's energy. Energy Environ 22(11):659–679

Loáiciga HA (2015) Managing municipal water supply and use in water-starved regions: looking ahead. J Water Resour Plan Manag 141(1). https://doi.org/10.1061/(ASCE)WR.1943-5452.0000487

Loáiciga HA (2017) The safe yield and climatic variability: implications for groundwater management. Groundw J 55(3):334–345. https://doi.org/10.1111/gwat.12481

Loftus AC, Anton B, Philip R (2011) Adapting urban water systems to climate change: a handbook for decision makers at the local level. Adapting urban water systems to climate change: a handbook for decision makers at the local level

Longuevergne L, Scanlon BR, Wilson CR (2010) GRACE hydrological estimates for small basins: evaluating processing approaches on the High Plains Aquifer, USA. Water Resour Res 46(11).

Ma J, Li LH, Guo LP, Bai L, Zhang JR, Chen ZH, Ahmad S (2015) Variation in soil nutrients in grasslands along the Kunes River in Xinjiang, China. Chem Ecol 31(2):111–122. https://doi.org/10.1080/02757540.2014.917170

Marvel K, Bonfils C (2013) Identifying external influences on global precipitation. Proc Natl Acad Sci 110(48):19301–19306

Marvel K, Cook BI, Bonfils CJ, Durack PJ, Smerdon JE, Williams AP (2019) Twentieth-century hydroclimate changes consistent with human influence. Nature 569(7754):59–65

Melesse AM, Ahmad S, McClain ME, Wang X, Lim YH (2011) Suspended sediment load prediction of river systems: an artificial neural networks approach. Agric Water Manag 98(5):855–866

Michelle TH et al (2013) Global river discharge and water temperature under climate change. Glob Environ Chang 23(2013):450–464

Milly PC, Dunne KA, Vecchia AV (2005) Global pattern of trends in streamflow and water availability in a changing climate. Nature 438(7066):347–350

Mirchi A, Madani K, Watkins D, Ahmad S (2012) Synthesis of system dynamics tools for holistic conceptualization of water resources problems. Water Resour Manage 26(9):2421–2442. https://doi.org/10.1007/s11269-012-0024-2

Mosquera-Machado S, Ahmad S (2007) Flood hazard assessment of Atrato river in Colombia. Water Resour Manage 21(3):591–609

Moumouni Y, Ahmad S, Baker RJ (2014) A system dynamics model for energy planning in Niger. Int J Energy Power Eng 3(6):308–322. https://doi.org/10.11648/j.ijepe.20140306.14

Nazari-Sharabian M, Ahmad S, Karakouzian M (2018) Climate change and eutrophication: a short review. Eng Technol Appl Sci Res 8(6):3668–3672

Nelson GC, Valin H, Sands RD, Havlík P, Ahammad H, Deryng D, Willenbockel D (2014) Climate change effects on agriculture: Economic responses to biophysical shocks. Proc Natl Acad Sci 111(9):3274–3279

Notter B, Hurni H, Wiesmann UM, Abbaspour KC (2012) Modeling water provision as an ecosystem service in a large east African river basin. Hydrol Earth Syst Sci 16(1):69–86

Nussbaum EM, Owens MC, Sinatra GM, Rehmat AP, Cordova JR, Ahmad S, Harris FC, Dascalu SM (2015) Losing the Lake: simulations to promote gains in student knowledge and interest about climate change. Int J Environ Sci Edu 10(6):789–811. https://doi.org/10.12973/ijese.2015.277a

Nyaupane N, Thakur B, Kalra A, Ahmad S (2018) Evaluating future flood scenarios using CMIP5 climate projections. Water 10:1866

Oki T, Agata Y, Kanae S, Saruhashi T, Musiake K (2003) Global water resources assessment under climatic change in 2050 using TRIP. Int Assoc Hydrol Sci Pub 280:124–133

Oral HV, Carvalho P, Gajewska M, Ursino N, Masi F, Hullebusch EDV, ..., Zimmermann M (2020) A review of nature-based solutions for urban water management in European circular cities: a critical assessment based on case studies and literature. Blue-Green Syst 2(1):112–136

Owor M, Taylor RG, Tindimugaya C, Mwesigwa D (2009) Rainfall intensity and groundwater recharge: empirical evidence from the Upper Nile Basin. Environ Res Lett 4(3):035009

Padrón RS, Gudmundsson L, Decharme B, Ducharne A, Lawrence DM, Mao J, Seneviratne SI (2020) Observed changes in dry-season water availability attributed to human-induced climate change. Nat Geosci 13(7):477–481

Pandey A, Larroche C, Gnansounou E, Khanal SK, Dussap CG, Ricke S (eds) (2019) Biomass, biofuels, biochemicals: biofuels: alternative feedstocks and conversion processes for the production of liquid and gaseous biofuels. Academic Press

Pathak P, Kalra A, Ahmad S (2017) Temperature and Precipitation changes in the Midwestern United States: implications for water management. Int J Water Resour Dev 33(6). https://doi.org/10.1080/07900627.2016.1238343

Pathak P, Kalra A, Lamb KW, Miller WP, Ahmad S, Amerineni R, Ponugoti DP (2018) Climatic variability of the Pacific and Atlantic oceans and Western US snowpack. Int J Climatol 38(3):1257–1269. https://doi.org/10.1002/joc.5241

Piao S, Friedlingstein P, Ciais P, de Noblet-Ducoudré N, Labat D, Zaehle S (2007) Changes in climate and land use have a larger direct impact than rising CO_2 on global river runoff trends. Proc Natl Acad Sci 104(39):15242–15247

Protopapas AL, Katchamart S, Platonova A (2000) Weather effects on daily water use in New York City. J Hydrol Eng 5(3):332–338

Puri S, Stephen H, Ahmad S (2011) Relating TRMM precipitation radar land surface backscatter response to soil moisture in the Southern United States. J Hydrol 402(1–2):115–125

Qaiser K, Ahmad S, Johnson W, Batista JR (2013) Evaluating water conservation and reuse policies using a dynamic water balance model. Environ Manag 51(2):449–458

Qiao L, Hong Y, McPherson R, Shafer M, Gade D, Williams D, ... , Lilly D (2014) Climate change and hydrological response in the trans-state Oologah Lake watershed–evaluating dynamically downscaled NARCCAP and statistically downscaled CMIP3 simulations with VIC model. Water Resour Manag 28(10):3291–3305

Rahaman MM, Thakur B, Kalra A, Ahmad S (2019) Modeling of GRACE-derived groundwater information in the Colorado river basin. Hydrology 6(1):19

Ravazzani G, Ghilardi M, Mendlik T, Gobiet A, Corbari C, Mancini M (2014) Investigation of climate change impact on water resources for an alpine basin in northern Italy: implications for evapotranspiration modeling complexity. PloS One 9(10):e109053

Reed SO, Friend R, Jarvie J, Henceroth J, Thinphanga P, Singh D, ..., Sutarto R (2015) Resilience projects as experiments: implementing climate change resilience in Asian cities. Climate Dev 7(5):469–480

Rodell M, Velicogna I, Famiglietti JS (2009) Satellite-based estimates of groundwater depletion in India. Nature 460:999–1002

Rusuli Y, Li L, Ahmad S, Zhao X (2015) Dynamics model to simulate water and salt balance of Bosten Lake in Xinjiang, China. Environ Earth Sci 74(3):2499–2510. https://doi.org/10.1007/s12665-015-4257-2

Sadeghi KM, Loáiciga HA, Kharaghani S (2018) Stormwater control measures for runoff and water quality management in urban landscapes. J Am Water Resour Assoc, 1–10https://doi.org/10.1111/1752-1688.12547

Sardari PT, Babaei-Mahani R, Donald G, Yasseri S, Moghimi MA, Bahai H (2020) Energy recovery from domestic radiators using a compact composite metal Foam/PCM latent heat storage. J Cleaner Prod 257:120504, ISSN 0959-6526. https://doi.org/10.1016/j.jclepro.2020.120504, https://www.sciencedirect.com/science/article/pii/S0959652620305515

Sagarika S, Kalra A, Ahmad S (2014) Evaluating the effect of persistence on long-term trends and analyzing step changes in streamflows of the continental United States. J Hydrol 517:36–53. https://doi.org/10.1016/j.jhydrol.2014.05.002

Saher R, Stephen H, Ahmad S (2021) Urban evapotranspiration of green spaces in arid regions through two established approaches: a review of key drivers, advancements, limitations, and potential opportunities. Urban Water J 18(2):115–127

Saifullah M., Liu S., Tahir AA, Zaman M, Ahmad S, Adnan M, Chen D, Ashraf M, Mehmood A (2019). Development of threshold levels and a climate-sensitivity model of the hydrological regime of the high-altitude catchment of the Western Himalayas, Pakistan. Water 11(7):1454. https://doi.org/10.3390/w11071454

Salehi A, Karbassi A, Ghobadian B, Ghasemi A, Doustgani A (2019) Simulation process of biodiesel production plant. Environ Prog Sustain Energy 38(6):e13264

Sattari MT, Mirabbasi R, Jarhan S, Sureh FS, Ahmad S (2020) Trend and abrupt change analysis in water quality of Urmia Lake in comparison with changes in lake water level. Environ Monit Assess 192:623

Scanlon BR, Faunt CC, Longuevergne L, Reedy RC, Alley WM, McGuire VL, McMahon PB (2012) Groundwater depletion and sustainability of irrigation in the US High Plains and Central Valley. Proc Natl Acad Sci 109(24):9320–9325

Shamsudduha M, Taylor RG, Longuevergne L (2012) Monitoring groundwater storage changes in the highly seasonal humid tropics: validation of GRACE measurements in the Bengal Basin. Water Resour Res 48(2)

Shayesteh AA, Koohshekan O, Ghasemi A, Nemati M, Mokhtari H (2019) Determination of the ORC-RO system optimum parameters based on 4E analysis; Water–energy-environment nexus. Energy Convers Manag 183:772–790

Shrestha E, Ahmad S, Johnson W, Shrestha P, Batista JR (2011) Carbon footprint of water conveyance versus desalination as alternatives to expand water supply. Desalination 280(1–3):33–43

Small, E. E. (2005). Climatic controls on diffuse groundwater recharge in semiarid environments of the Southwestern United States. Water Resour Res 41(4)

Smith LC, Sheng Y, MacDonald GM, Hinzman LD (2005) Disappearing arctic lakes. Science 308(5727):1429–1429

Smith P, Martino D, Cai Z, Gwary D, Janzen H, Kumar P, Towprayoon S (2007) Policy and technological constraints to implementation of greenhouse gas mitigation options in agriculture. Agr Ecosyst Environ 118(1–4):6–28

Sultana Z, Coulibaly P (2011) Distributed modelling of future changes in hydrological processes of Spencer Creek watershed. Hydrol Process 25(8):1254–1270

Tague C, Grant GE (2009) Groundwater dynamics mediate low-flow response to global warming in snow-dominated alpine regions. Water Resour Res 45(7)

Tamaddun K, Kalra A, Ahmad S (2018) Potential of rooftop rainwater harvesting to meet outdoor water demand in arid regions. J Arid Land 10(1):68–83. https://doi.org/10.1007/s40333-017-0110-7

Tamaddun K, Kalra A, Bernardez M, Ahmad S (2017) multi-scale correlation between Western U.S. snow water equivalent and ENSO/PDO using wavelet analyses, water resources management. https://doi.org/10.1007/s11269-017-1659-9

Tamaddun KA, Kalra A, Ahmad S (2019) Spatiotemporal variation in the continental US streamflow in association with large-scale climate signals across multiple spectral bands. Water Resour Manag 33(6):1947–1968

Taylor RG, Martin CT, Kongola L, Maurice L, Nahozya E, Sanga H (2012) Dependence of groundwater resources on intense seasonal rainfall: evidence from East Africa. Nat Clim Change

Taylor RG, Scanlon B, Döll P, Rodell M, Van Beek R, Wada Y, Treidel H (2013) Ground water and climate change. Nat Clim Chang 3(4):322–329

Thakur B, Kalra A, Ahmad S, Lamb KW, Lakshmi V (2020) Bringing statistical learning machines together for hydro-climatological predictions-case study for Sacramento San Joaquin River Basin, California. J Hydrol Reg Stud 27:100651

van Ruijven BJ, De Cian E, Wing IS (2019) Amplification of future energy demand growth due to climate change. Nat Commun 10(1):1–12

Vedwan N, Ahmad S, Miralles-Wilhelm F, Broad K, Letson D, Podesta G (2008) Institutional evolution in Lake Okeechobee management in Florida: characteristics, impacts, and limitations. Water Resour Manag 22(6):699–718

Venkatesan AK, Ahmad S, Johnson W, Batista JR (2011) Salinity reduction and energy conservation in direct and indirect potable water reuse. Desalination 272(1–3):120–127

Wada Y, Van Beek LP, Van Kempen CM, Reckman JW, Vasak S, Bierkens MF (2010) Global depletion of groundwater resources. Geophys Res Lett 37(20)

Wang Z, Ficklin DL, Zhang Y, Zhang M (2012) Impact of climate change on streamflow in the arid Shiyang river basin of Northwest China. Hydrol Process 26(18):2733–2744

Wihlborg M, Sörensen J, Olsson JA (2019) Assessment of barriers and drivers for implementation of blue-green solutions in Swedish municipalities. J Environ Manag 233:706–718

Wilbanks T, Bhatt V, Bilello D, Bull S, Ekmann J, Horak W, Scott MJ (2008) Effects of climate change on energy production and use in the United States. US Department of Energy Publications, 12

WPP, United Nations, Department of Economic and Social Affairs, Population Division (2019) World population prospects. Press Release

Yadav VG, Yadav GD, Patankar SC (2020) The production of fuels and chemicals in the new world: critical analysis of the choice between crude oil and biomass vis-à-vis sustainability and the environment. Clean Technol Environ Policy, 1–18

Zektser IS, Loáiciga HA, Wolf J (2005) Environmental impacts of groundwater overdraft: selected case studies in the Southwestern United States. J Environ Geol 47(3):396–404

Zhang X, Zwiers FW, Hegerl GC, Lambert FH, Gillett NP, Solomon S, Nozawa T (2007) Detection of human influence on twentieth-century precipitation trends. Nature 448(7152):461–465

Zhang Y, Su F, Hao Z, Xu C, Yu Z, Wang L, Tong K (2015) Impact of projected climate change on the hydrology in the headwaters of the Yellow River basin. Hydrol Process 29(20):4379–4397

Zuo D, Xu Z, Peng D, Song J, Cheng L, Wei S, Yang H (2015) Simulating spatiotemporal variability of blue and green water resources availability with uncertainty analysis. Hydrol Process 29(8):1942–1955

Summary

Developing an understanding of climate change is in close connection with the knowledge of the Earth system, particularly the Geosphere including the Lithosphere, Atmosphere, Hydrosphere, Cryosphere. The climate change as acomplex atmospheric-oceanic phenomenon in the Earth system is radically affected by greenhouse gases' increase in the atmosphere, their interaction. The earth experienced major glaciations. Fluctuation in CO_2 concentration-stemmed of volcanic activities, metamorphism emissions, plate limestone formation- plays acrucial role in glaciation, ice-free state. The reason behind the Climate Change studies is its effect on several sectors, such as flora, fauna, water bodies, soil, air, health, agriculture, energy, society.

Climate change caused by various natural factors and anthropogenic ones, in which the natural effects are considerably lower than manmade. To make a clear understanding and comparison between factors, scientists introduce the parameter known as radiative forcing. GHGs with high confidence takes the first place in increasing the Earth received energy. However, some of the GHGs participate in cooling effect through aerosols' creation. As it is said earlier, the CO_2 is the most prominent GHG, intensifying the global warming. Among the natural factors, the solar energy fluctuations and volcanic activities will change the Earth's energy balance by warming and cooling the surface, respectively. The cumulative changes in the radiative forcing and the Earth's energy have grown consistently, but in various intensity since the 1950s.

The critical changes from a water expert viewpoint are temporal and spatial distribution of precipitation and its type (solid or liquid), and evapotranspiration rate. These are followed by changes in the pattern of climate-related natural disasters such as floods and droughts, changes in the characteristics of surface streams, groundwater recharge and water quality. Increasing temperature and decreasing precipitation have reduced the discharge of surface runoffs in different parts of the world which leads to a decline in the level of groundwater aquifers. In the agriculture section, according to vast climatic and agricultural conditions, solutions have been presented to preserve water resources accordingly. Finally, fossil fuels can be replaced by biofuels to reduce pollution and mitigate climate change conditions in industrial and domestic energy consumption.

Part II
Climatic Scenarios and Practical Analysis

Introduction

The Intergovernmental Panel on Climate Change (IPCC) was set up in 1988 under the support of the United Nations Environment Program and the World Meteorological Organization to survey the scientific data pertinent to comprehend the danger of human-actuated environmental change. It does neither do new research nor does it screen environment-related information. It puts together its appraisal with respect to distributed and peer investigated specialized literature. The objective of these evaluations is to advise global arrangement and exchange on environment-related issues. The goal of IPCC is to provide academic and regular reports about all aspects of climate change to governments to ensure that the development of climate change policy is conducted with due regard to rigorous evidence. These assessments reveal the issues that may require further research. Since 1988, five assessment reports have been developed. They are the most extensive academic reports about climate change provided worldwide and provide a range of methodological and specific reports and technical articles in response to the applications of the United Nations Convention on the Climate Change.

In the continuation of this section, climatic models are examined. Climate models are mathematical representations of the Earth's climate system. Besides representing the fundamental controls on Earth's climate in the form of solar radiation, surface and atmospheric albedo, and terrestrial infrared radiation, climate models are able to simulate the interactions and feedback processes of the physical, chemical, and biological features of the climate system at varying levels of detail. Running on powerful computers, they can be employed to predict future climate conditions. The sixth chapter aims to synthesize knowledge of climate models by discussing several categories of models, including Energy Balance Models (EBMs), radiative-convective models, and General Circulation Models (GCMs). As the most comprehensive climate models, GCMs have been widely applied for future climate change projections using different scenarios of population growth, greenhouse gas emissions, land-use changes etc.

Global climatic changes can cause noticeable changes in zonal water accessibility. The significant role of water in society and nature emphasizes the importance of realizing how global climate change can influence tutorial water supplies. Global Climate Models (GCMs) used for climate studies and climate projections are typically run at large spatial resolutions (about 150--200 Km). As they are limited in their ability to resolve crucial subgrid-scale features, GCM-based projections may not be decisive for local impact studies. To conquer this problem, downscaling methods are developed from regional-scale atmospheric variables provided by GCMs.

Chapter 5
Review on IPCC Reports

Anis Hasani, Omid Bozorg-Haddad, Scott Baum, Steven Lucas, and Amin Soltani

The Intergovernmental Panel on Climate Change (IPCC) was set up in 1988 under the support of the United Nations Environment Program and the World Meteorological Organization to survey the scientific data pertinent to comprehend the danger of human-actuated environmental change. It does not do new research, nor does it screen environment-related information. It puts together its appraisal with respect to distributed and peer investigated specialized literature. The objective of these evaluations is to advise global arrangement and exchanges on environment-related issues (Viner and Howarth 2014).

The goal of IPCC is to provide academic and regular reports about all aspects of climate change to governments to ensure that the development of climate change policy is conducted with due regard to rigorous evidence. These assessments reveal

A. Hasani
Water Resources Engineering, Department of Irrigation and Reclamation Engineering, University of Tehran, Karaj, Iran
e-mail: Anis.hasani@ut.ac.ir

O. Bozorg-Haddad (✉)
Faculty of Agricultural Engineering and Technology, Department of Irrigation and Reclamation Engineering, College of Agriculture and Natural Resources, University of Tehran, Karaj, Tehran, Iran
e-mail: OBHaddad@ut.ac.ir

S. Baum
School of Environment and Science at Griffith University, Brisbane, QC, Australia
e-mail: s.baum@griffith.edu.au

S. Lucas
The University of Newcastle, Newcastle, NSW 2308, Australia
e-mail: Steven.Lucas@newcastle.edu.au

A. Soltani
School of Civil, Environmental and Mining Engineering, The University of Adelaide, Adelaide, SA 5005, Australia
e-mail: Amin.Soltani@adelaide.edu.au

O. Bozorg-Haddad (ed.), *Climate Change in Sustainable Water Resources Management*, Springer Water, https://doi.org/10.1007/978-981-19-1898-8_5

the issues that may require further research. IPCC reports are prepared and considered in several stages and are developed with regard to objectivity and clarity. IPCC reports are politically neutral, and are considered necessary in international negotiations related to Climate Change. Since 1988, five assessment reports have been developed. They are the most extensive academic reports about Climate Change provided worldwide and provide a range of methodological and specific reports and technical articles in response to the applications of the United Nations Convention on the Climate Change (IPCC web site, http://www.ipcc.ch/as viewed 29 March 2021).

5.1 Organizational Structure of IPCC

The IPCC is divided into three Working Groups and a Task Force. Working Group 1 (WG1) works on the academic aspects of the climate system as well as assessment of Climate Change, WG2 considers effects of Climate Change, compatibility, vulnerability, and its consequences, and WG3 reflects concepts such as reduction of Climate Change and restriction of Greenhouse gas emissions (Levermore 2008). Each working group is led by at least two chairpersons, typically one from a developing country and one from a developed country and an executive unit located in the foregoing developed country (TSU). The IPCC Bureau comprises the IPCC chairpersons, three vice-chairmen, working groups' chairpersons, and each working groups' vice-chairperson, which members of IPCC assign for a specific period. Working groups observe the process of preparing reports comprehensively. They also lead periodic and expertise meetings as well as required training workshops for providing mentioned reports. Writers and editors of IPCC reports are selected from the most distinguished scientists in the world, and whom the IPCC members to the board of directors are introduced. Agents of the governments who are IPCC members participate at least once a year in the general meeting of IPCC Plenary, which is held in one of the member countries. In addition to the members, some specialist centers and specific governmental and nongovernmental research institutes and some international units participate in these meetings as observers. The selection of the board of directors, chairmen of WG1,2,3 and working groups related to the list of greenhouse gas, planning and blueprint of IPCC, budget allocation, the procedure of spending money, and many other issues are appointed in the general meeting of the IPCC. The approval and oversights of assessment and specialized reports, policy direction, new scenarios, the procedures and functions of workgroups, location and subject of workshops, and required background meetings are all determined in the general meeting. Except for the general meeting, which is held periodically, other meetings such as the board of directors' meetings, report writers' specialized meetings and workshops, scoping meetings are conducted according to this structure. Each of these meetings is held in one of the member countries or Headquarters of the Secretariat of IPCC, Geneva, at specific times for specific goals. Since the foundation of IPCC, every assessment report directly moved into the international policymaking of Climate Change.

IPCC published its first assessment report (FAR) in 1990. This report highlights the importance of Climate Change as a challenge for global consequences and the need for international cooperation. This report has a determinant role in establishing the United Nations Framework Convention on Climate Change (UNFCCC) to reduce global warming and cope with the consequences of Climate Change. The Second Assessment Report (SAR) was published in 1995. This report comprised essential issues for the approval of the Tokyo Protocol in 1997 for governments. The Third Assessment Report (TAR), published in 2001, centered on the effects of Climate Change and adjusting to these changes. The Fourth report (AR4) was published in 2007. This report was a prerequisite for the agreement after the Tokyo Protocol, which focused on reducing global warming to two centigrade. The Fifth Assessment Report (AR5) was published from 2013 to 2014. This report provided the primary information of the Paris Treaty.

In the forty-third session in April 2016, the IPCC decided to prepare three specific reports, one methodological report, and a Sixth Assessment Report. According to the Paris Treaty, the first report from these specific reports referred to as global warming to 1.5°, which was applied by governments worldwide and was finalized in October 2018. The second specific report concerned with Climate Change (SRCCL), which was finalized in August 2019, and the third report has to do with ocean and cryosphere in a changing climate (SROCC), which was finalized in September 2019. The Sixth Assessment Report (AR6) is expected to be prepared by 2022 (IPCC web site, http://www.ipcc.ch/as viewed 29 March 2021).

5.2 FAR[1]

5.2.1 *Climate Change Assessment in the FAR Report*

According to the first report of IPCC, which was published in 1990, the earth's climate is directly dependent on the radiative equilibrium in the atmosphere, with the sun being the driving force. There are several factors, which can alter the balance between the absorbed and emitted energy by the earth; the most notable factor is the change in the energy output of the earth's surface, which is influenced by greenhouse gases. Radiative equilibrium in the atmosphere is also dependent on input radiation and greenhouse gases such as CO_2, CH_4, CFCs, N_2O, and O_3 all of which are increasing due to human activities. The increase in these gases contributes to global warming. It has stated in the FAR that temperature has increased by 3°, compared to 100 years, ago; sea level has also increased by 10–20 cm. According to this report, global warming and increase in sea level were faster than past 100 years due to the extended emission of greenhouse gases. Most ecosystems are influenced by climate

[1] First Assessment Report.

change, which might even lead to the extinction of some ecosystems. Concerning these issues, the IPCC follows these goals in its first report:

- Recognizing and considering factors that might affect the climate change of the next century (especially human activities);
- Atmospheric, ocean, earth, and ice system's response to climate change.
- Considering the abilities of existing models;
- Investigating past climate and change in the present time.

This report maintains that:

- The effect of each of the greenhouse gases on climate change is not the same. For instance, carbon dioxide causes an increase in the concentration of more than 50% of greenhouse gases and is likely to do so in the future.
- The concentration of long-lived greenhouse gases (CO_2, CFCs, N_2O) changes only through reducing their emission. Emission of these gases at the present rate leads to the increase of their concentration in the future.
- In order to stabilize the concentration of greenhouse gases, it is necessary to decrease the emission of these gases, which stem from human activities, by 60%.

In the first report of IPCC, the following predictions were mentioned:

- In the first report, the average amount of global warming was measured as 1.8 °C. This temperature increase was unprecedented in the past 10,000 years. According to the used model in the first report under the title of "scenario A", it was predicted that the average global temperature would increase by 3 °C per decade. Furthermore, according to scenario B, C, and D, the temperature increase would be as follows: 02 °C, a little more than 01 °C, and about 01 °C.
- Earth's surface heats up more quickly than oceans, and high northern latitudes heat up more than the global average in winters.
- The report predicted that the average sea level would increase by about 6 cm per decade (with an uncertainty of 3–10 cm). This increase is due to thermal expansion of oceans and the melting of existing ices in the land.
- Global climate change differs from regional climate change. For instance, the increase in temperature in the south of Europe and the center of North America is predicted to be higher than the global average.

Some uncertainties accompanied the IPCC predictions. The most significant uncertainties highlighted in this report were as follows:

- Oceans, which affect the time and patterns of climate, change.
- Clouds that highly affect the amount of climate change.
- Polar ice sheets affect the prediction of sea-level rise.
- Sources of greenhouse gas emissions that affect future predictions of climate change (FAR 1990).

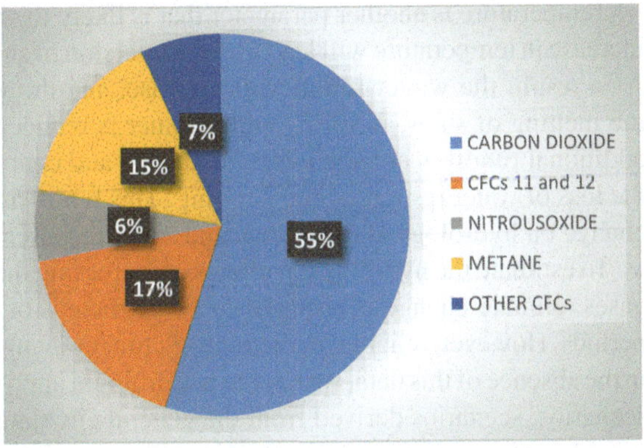

Fig. 5.1 Emission of greenhouse gases due to human activities

5.2.2 *Effective Parameters on Global Climate Change*

The report stated that the concentration of greenhouse gases is increasing. The rise rates of CO_2, CH_4, CFC 11, CFC 12 and N_2O gases in 1990 were expressed by 0.5%, 9%, 4%, 4%, and 0.25% respectively.

Computer models were used to calculate the concentration of greenhouse gases. In order for our climate to stabilize, the models estimated that the emission of these gases must be reduced. In FAR, carbon dioxide gas reduction rate was more than 60%, methane 15 to 20%, nitrogen oxide 70 to 80%, CFC 11 70 to 75%, CFC 12 75 to 85% and HCFC 22 40 to 50%. The amount of greenhouse gases emitted by humans from 1980 to 1990 is illustrated in Fig. 5.1.

5.2.3 *Evaluation of Climate Change Effects on Water Resources in FAR*

Due to the increase in greenhouse gas emissions, climate change has led to a change in temperature, land resources, water resources, and change in time and place of water distribution, hydrological water cycle, water quality, and water supply systems in different regions. Quantitative and qualitative estimation of the hydrological effects of climate change on water resources and investigating and solving its problems is necessary because water resources, directly and indirectly, affect industry, power generation, agriculture, transportation, environmental protection and the removal of current and future water needs. The most critical parameter mentioned in the first report, which is likely to change, is the local precipitations, which, unfortunately, was impossible to predict accurately. Changes in rainfall have important implications for hydrology and water resources.

Temperature is another parameter that is likely to change by climate change. The increase in temperature will lead to the conversion of snow to rain in the snow regions. As a result, the winter runoff will increase, and the water flow, which was due to the melting of snow, in spring and summer is reduced. In addition, in cases where additional runoff is not adequately managed and controlled in winter, it will lead to the loss of water resources and flooding. This report addresses the effects of climate change on hydrological conditions and water resources in different countries.

To estimate the hydrological effects of increasing the concentration of greenhouse gases, climate change is required to be predicted for different regions in different periods. However, reliable predictions of climate change are not available. Therefore, in the absence of this data, we have to use different approaches, including hypothetical scenarios, scenarios derived from the general circulation models of the atmosphere (GCM) and scenarios based on historical reconstructions. Each of these approaches has assumptions. In the first approach, temperature rose from 0.5 to 4 °C and rainfall changes (decrease or increase) are considered to be around 10 to 15%. In the second approach, a double increase for CO_2 was considered. In the third approach, future climate analogues were considered based on the restructuring of the last warm periods, in which the amount of CO_2 was higher than the current value (FAR 1990).

5.2.3.1 Effects of Climate Change in Annual and Seasonal Conditions

Since the late 1970s, changes in annual and seasonal runoff were widely investigated. One of the most important hydrological results of global warming is an increase and decrease in runoff. Its sudden rise will cause flooding, either in management or control, and its decrease will lead to a shortage of water resources, consequently to a drought. The estimation of runoff conditions is fundamental because the hydraulic structures and projects are designed and implemented according to the data observed in the past. If there is a quantitative estimate of meteorological parameters, there can be more reliable results from runoff status in different world regions. Watershed areas in arid and semi-arid areas are susceptible to runoff changes. In areas where snow is considered a water resource, annual runoff and distribution are significant. For instance, it is expected that in the latitudes of the northern hemisphere, by heating 1–2 °C, the winter runoff caused by snow melting will increase, and the runoff will be reduced in the spring.

5.2.3.2 Effects of Climate Change on Water Demand

Considering global climatic conditions, demand for water resources is expected to change in many parts of the world. The actual consumption values depend on climatic factors, weather constraints, and the extent of the development of water consumer sectors. Given the predicted changes in the future climate, water planning is predicted to be complicated and difficult.

Much research has been undertaken on the damages and adaptations of the water system and weather fluctuations (Peterson and Keller 1990; Nemec 1989). Using this information enables us to predict and manage climatic effects on water management and demand and protection against flooding and drought. Peterson and Claire studied the effects of temperature and precipitation changes in the western areas of the United States. They concluded that the increase in temperature at three degrees Celsius and a reduction of 10% of rainfall would reduce crops by 30% (6). the Food and Agriculture Organization (FAO) carried out studies related to the effects of climate change on irrigation use with the British Institute of hydrology in South Africa. Studies revealed that the demand for irrigation water increases by 65% by doubling CO_2 (Nemec 1989).

5.2.3.3 Effects of Climate Change on the Level of Water and Lakes

The hydrological cycle of large lakes affects climate diversity in large areas, including lake basins and surrounding areas. In the future, global warming, due to the increase in gas concentration in the atmosphere, will lead to a change in the lake's water equilibrium (precipitation, evaporation, inlet and outlet).

The hydrological consequences of global warming are not limited to changes in the water runoff and water balance. Other consequences including change in the amount and level of water, erosion in river basins and riverbed, change of water quality, a rise of sea levels and runoff changes can lead to increased flooding of lowland coastal areas, increased erosion in the beach, change in delta processes, change in rivers salinity, and groundwater pollution through saltwater intrusion. The investigation of the studies suggests that the estimation of future water demands must be undertaken by considering the characteristics of each region (FAR 1990).

5.2.4 IPCC Expressed Its Significant Findings in the First Report as Follows

- Climate change is a global issue and requires a global effort to deal with it.
- Industrialized and developing countries share a joint responsibility to deal with the problems caused by climate change.
- The industrialized countries are obligated to take all necessary measures to reduce climate change. They also need to cooperate with developing countries to manage environmental problems due to climate change.
- Greenhouse gases emission is increasing in developing countries. These countries must take the necessary measures to prevent the excessive emission of these gases.
- The economic balance between economic and environmental objectives should be established in all countries.
- Limitation and coping strategies need to supplement each other.

- The decision-making approaches to climate should be determined considering each nation.
- The consequences of climate change are so severe that even in case of uncertainty from possible changes, response strategies to these changes need to be provided.

5.3 SAR[2]

5.3.1 Climate Change Assessment in the SAR Report

After conveying the first report, IPCC proceeded with its second assessment report in 1995. In addition to the reinforcement of issues discussed in the first report, this report has widely considered other social and economic issues. The final goal of UNFCCC: is the stability of concentration of greenhouse gases in the aerospace, which impedes human dangers' interference with the climate system. This level must be done appropriately for the ecosystem to be naturally adapted to climate change. The second report of IPCC was stated for the removal of these challenges.

According to these reports, indications revealed an increase in greenhouse gas concentrations and aerosols before the industry time (since 1750) caused the temperature increase, positive radiative forcing, and general climate change. The atmospheric concentration of greenhouse gases such as CO_2, CH_4, N_2O has risen by 30%, 145%, and 15%, respectively (these numbers belong to 1992) (SAR Synthesis Report 1995). Positive radiative forcing stemming from these, which last long, is estimated at $+2.45\ Wm^{-2}$. Findings also expressed that the concentration of the ozone layer in the northern hemisphere has increased due to human activities compared to the pre-industrial age. The increase in concentration of these gases resulted in positive radiative forcing, which consequently caused global warming. The average global temperature has risen by 0.3–0.6 centigrade since the late nineteenth century. These are gases,which are long-lasting gases and can remain in the atmosphere for 10 to 100 years and affect the radiative energy emitted from the earth. In this report, since 1860, recent years were the warmest years of the past 100 years.

In this report, sea levels have risen 10–25 cm during the past 100 years.

Human aerosols (microscopic airborne particles) originating from burning fossil fuels, biomasses, and other resources result in negative radiative forcing for about $0.5\ Wm^{-2}$, which can be effective in climate change. Despite the greenhouse gases, which are long-lasting, aerosols remain shorter in the atmosphere. Minor changes in the temperature, rainfall and non-linear changes in the evaporation, perspiration, and moisture of soil lead to significant changes in runoffs, specifically in arid and semi-arid regions (SAR 1995).

In this report, there is not sufficient data to judge with certainty whether climate has changed, but on a regional scale, changes have been made in the climate parameters. For instance, ice loss has increased, and rainfall has increased or decreased in some

---

[2] Second Assessment Report.

areas. Climate change is not equal all over the world. For instance, the maximum warming in spring and summer belongs to continents located in the middle latitudes. The maximum rainfall refers to the areas located in the northern hemisphere, specifically in cold seasons. Moreover, the average night temperature has likely risen more than the average daily temperature (SAR 1995).

5.3.2 *Prediction of Climate Change in SAR*

IPCC has produced a wide range of IS92a-f scenarios according to the production of greenhouse gases and aerosols. According to the suppositions regarding population and economic growth, land use, availability of energy in the periods of 1990 to 2100. The predictions of climate, which was provided in the second report of IPCC, were listed below. The models predicted the increase in the temperature and sea level until 2010 compared to 1990. This figure shows that these quantities are for the global average and might be different for various regions. Findings indicate that according to all scenarios, the rate of increase is beyond the past 10,000 years (Table 5.1).

All the simulation of predictive models (whether considering the increase in both greenhouse gases and aerosols or the increase in the greenhouse gases only) included the following results:

- The highest temperature on the earth's surface belongs to the northern latitudes and appears in the winters.
- Earth's surface heats up more than seas in winters.
- Arctic warming has increased in summers.

Table 5.1 Prediction of climate change as reported by the IPCC in the second report

Production scenario	Characteristics of scenario	Temperature change	Change in sea level
IS92a	The average climate sensitivity to the increase in the concentration of aerosols	Temperature increase by about 2 °C	Sea level rise by 50 cm
IS92c	The minimum climate sensitivity to the increase in the concentration of aerosols	Temperature increase by about 1 °C	Sea level rise by 15 cm
IS92e	The maximum climate sensitivity to the increase in the concentration of aerosols	Temperature increase by about 3.5 °C	Sea level rise by 95 cm

- Rainfall and consequently soil moisture have decreased in higher latitudes in the winter.
- The power of thermal circulation has decreased in the North Atlantic.

Large investments are needed in order to respond to the uncertainties regarding predictions, Some of the uncertainties, which exist in the assessment of the effects of climate change on water resources are as follows:

- Uncertainty in General Circulation Model (GCM) and uncertainty in regional characteristics of areas in which consequences of climate change will happen.
- Lack of sufficient knowledge concerning climate change in the future, which is regarded as the main element of water management.
- Uncertainty in estimating changes in watershed water supply due to the changes in the vegetation and situation.
- Uncertainty in future demands in different parts and uncertainty in socio-economic, environmental effects originating from water shortage due to climate change (SAR 1995).

5.3.3 Effects of Climate Change on Water Resources in the Second Report

Water resources (surface water and groundwater) are regarded as an essential and undeniable portion of social, economic, and environmental parts in the world to which all the agriculture, hydroelectric production, environmental, recreational, drinking activities and domestic consumption are dependent. The critical point of the second report of IPCC regarding water resources was that climate change influences global water resources. The goal of water resources management due to climate change is to control and recover the severe effects of climate change and supplying reliable water resources.

The supply of water needs is regarded as one of the most critical challenges of this century. The global water cycle indicates that resources of existing renewable freshwater are about ten times more than the present demands of humans. However, unequal distribution of these spatio-temporal resources causes water challenges (SAR 1995). It seems that developing countries are highly vulnerable to climate change and its effects on water resources since most of these areas are located in arid and semi-arid regions of the world (SAR 1995).

The most important effects of global warming in the context of water supply are as follows:

- Change in the amount of river runoff
- Change in the nourishment of groundwater
- Change in the quality of water
- Increase in sea levels
- Change in the evaporation and perspiration.

Having access to high-quality water specifically for drinking uses is one of the most critical issues of the present world. Climate change influences the quality of water and even water ecosystems. Climate change influences the quality of waters by changing resources, biochemical reactions, and biological effects of pollutants (Xia et al. 2014). In some areas, global warming will lead to severe droughts and in some other areas, it will cause floods. Changes in the amount, intensity, and frequency of rainfalls influence the runoff and intensity of torrents and droughts.

The quality and quantity of water supply are regarded as essential and vital issues of each region. That is why countries that are located in arid regions face severe problems. Climate changes, as occurred in the, past are unexpected, fast, and unpredictable. These changes exclusively originate from the non-linear essence of the climate system and predictions that are more reliable can be achieved by studying and considering non-linear procedures and subsets of climate systems.

5.3.4 *Analysis of Climate Change on Water Resources*

The effects of climate change are related to each region's water resources management conditions and water supply system. In some areas, the effects of climate change on water resources can be minimized by consistent management. In some areas, these changes will impose many social, economic, and cultural costs to the people of that region. In order to prevent climate change from impacting the water supply, consistent and efficient management is required to respond to the demand changes, hydrological information, technologies, economic condition, and social perspectives about the economy and the environment. This management must employ the following approaches:

- New investments for the capacity expansion of existing systems and design and implementation of resources of needed supply
- Optimized use of existing systems
- Maintenance and repair of systems
- Construction and improvement of substructures regarding water resources
- Reinforcement of predictive systems
- Repair and reconstruction of watershed water supply (SAR 1995).

5.4 TAR[3]

5.4.1 *Climate Change Assessment in the TAR Report*

The third assessment report by working group one IPCC was published in 2001 according to the previous assessment reports, their improvement, and new results

[3] Third assessment report.

regarding Climate Change and predictions for the future. Based on this report, the increase in temperature in the late nineteenth century was 0.6 $\mp$ 0.2 °C which was beyond the SAR's prediction. The analysis of the proxy data in the northern hemisphere revealed that the increase in temperature in the twentieth century was probably the maximum rate of increase in temperature in the past 1000 years. Satellite data expressed that snow cover reduced by 10% by the end of 1960. Besides, ground data revealed that Ice Cover of lakes, rivers, and arctic sea reduced in the mid and high latitude of the northern hemisphere (SAR 1995).

Data revealed that sea level increased by 0.1–0.2 in the twentieth century. Meanwhile, the temperature of ocean water has increased since late 1950. Rainfall had probably increased in the high latitude of the northern hemisphere by 0.5–1% in the twentieth century.

Since 1750, the density of CO_2, CH_4, N_2O and O_3 has increased by 31%, 151%, 17%, and 36%, respectively. The increase resulted in positive radiative forcing and consequently global warming.

5.4.2 Climate Change Prediction

Working Group one IPCC provided predictions for the future in the third report. Climate models, which are based on physics, are used for the exact estimate of feedback and characteristics of a region. Although these models cannot simulate all climate aspects yet, and there is a lack of confidence about clouds and their interaction with radiation and Aerosol, confidence in the ability of these models provide helpful predictions regarding future climate. In the twentieth century, improvements regarding the function of meteorology models were made. These included improving our understanding of meteorology models from steam and ocean heat.

Some of the parameters have been predicted in TAR using global simulation and a wide range of scenarios for years between 1990 to 2100. The outcome was for the 35 scenarios and was based on several climate models. Predictions include:

- Increase in temperature globally by 1.4–5.8 centigrade
- Increased sea level by 0.09–0.88 m, which is mainly due to the heat expansion and melting glaciers.
- Increase in rainfall in the higher latitudes in winter and summer, increase in mid-latitude of the northern hemisphere in winter and reduction in the south of Africa, Antarctic, western and southern areas of Asia, Australia, and South America.
- Reduction of snow cover in Northern hemisphere and reduction of sea ice and glaciers (TAR 2001).

5.4.3 *Effects of Climate Change on Water Resources*

The effects of climate change were submitted in the report of working group two of IPCC. Effects of Climate Change on the hydrological parameters are estimated using the definition of scenarios for altering climate entries to the hydrological from producing general hydrological circulation models and water resources (GCMs). After publishing SAR, many studies have been conducted that have altered hydrological characteristics over time, finding relationships between climate models and altering different hydrological behaviors.

Studies have revealed that about 1.7 billion people (One-third of the world's population) live in countries, which that are involved in water stress. Depending on the population growth rate, it has been predicted that about 5 billion people will increase this population by 2025. The predicted climate change can lead to a reduction in water stress in countries. Generally, changes in water management procedures will have a remarkable impact on the effects of climate change. In the following sections, the effects of climate change on some hydrological parameters will be discussed.

Rainfall: Generally, due to climate change, an increase of rainfall in the northern hemisphere and higher geographic latitudes (specifically in fall and winter) and a relative reduction of rainfall in tropical and subtropical zones in both hemispheres has been predicted. Although there were many differences between climate models, most rainfall changes were generally found in higher geographic latitudes, tropical zones, and South East Asia. Even though deducing changes in the frequency of rainfall using global climate models is a demanding task, there were signs (TAR 2001) that demonstrated that intensive rainfall events have increased with global warming.

Ground Water: Groundwater is the primary source of water supply in many parts of the world, specifically in arid and semi-arid regions. Groundwater tables usually fill with effective rainfalls. Changes in rainfall depth effectively cause changes groundwater recharge rates. The groundwater recharge is also affected by the amount of evaporation. Studies suggest that some of the instances of the effect of climate change on the groundwater tables, which are not charged, occurred due to an increase in the amount of evaporation and a remarkable reduction in their level in France, Kenya, Tanzania Island, Texas, New York, and Caribbean (Bouraoui et al. 1999). Meanwhile, climate change affects the sea level; the increase in the sea level causes saline water to absorb groundwater resources, which leads to reduced freshwater.

River Flows: River flows express different incremental and decremental processes in different regions. Thus, it cannot be attributed to the effects of climate change with certainty since there are elements such as physical characteristics of an area that can influence this process.

Glaciers: glaciers and small ice valleys show long-term storage of water. Glaciers supply many river flows in the summer. It has been confirmed that along with the temperature increase, glaciers and ice valleys have melted over time.

Peak streamflow: Peak streamflow has moved from spring to winter due to climate change in several areas in which snowfall a vital component of the water balance. The temperature increase, which is due to climate change caused potential snowfall during the winter months to develop as rain only.

Quality of water: The data collected from the long term research station and ecosystem in North America and the outcomes of the simulations with interpretive models also expressed that climate change (the rate of rainfall and temperature) can have a remarkable influence on the quality of surface waters. Change in water quality, melting snow, and periods of temperature increase or drought can create a situation, which goes beyond the ecosystem tolerance and consequently destroys water quality (Bouraoui et al. 1999). Generally, the quality of water decreases along with the increase in the water temperature. The effect of temperature on the quality of the water improves with the change in the volume flow that is dependent on the change of direction in the flow volume; it might intensify or reduce the temperature effect. An increase in the water temperature modifies the speed of biochemical processes' function. Above all, it reduces the concentration of dissolved oxygen in the water, which influences the ecosystem of that region.

Demand: one of the potential effects of climate, change is the demand for water supply. The change in demand for water is different in different areas and usages. For instance, in agriculture, climate change in the farm will change the amount and time of the watering. Temperature increase leads to the increase in evaporation and consequently soil moisture. The increase in CO_2 also reduces the stomata, and all of these will increase water demand for farming due to climate change.

Water Resources: The effect of climate change on water resources (Arnell 1998) depends on the change in volume, time, recharging current quality, and system's characteristics, change system resilience, the process of completion of system management, and compatibility with the climate change. Non-climate changes might have more influence on water resources than climate changes. Water resource systems must constantly develop in order to respond to the challenges faced with management changes. Consistent management can be regarded as a tool, which is utilized increasingly for agreement and change in usage and different water demands, and it seems that it is more flexible than the management of ordinary water resources. The improved ability, which is used to predict coming weeks and months, remarkably increased water management and the ability, which is used for the prediction of events of future weeks and months, remarkably increased water management and the ability to cope with changes in the hydrological varieties.

5.5 AR4

5.5.1 Climate Change Assessment in the AR4 Report

The Fourth report of IPCC was published in 2007, six years after publishing the third report. In the Fourth report of IPCC, improvements were established regarding understanding the humane and natural stimulating factors which influence climate change. Changes that occurred after the third report (TAR) caused the latter reports on climate change to be more comprehensive and detailed concerning new data, more complex analysis of data, improvement in understanding processes, their simulation in the models, and comprehensive investigation into the range of uncertainty. When the fourth report was published, global warming was confirmed, as the signs such as an increase in the temperature of oceans, melting glaciers, and an increase in the sea levels were observed. Change in the frequency of greenhouse gases and aerosols will change the solar radiation, characteristics of the ground, and energy balance model of climate.

Studies indicated that the density of most of the greenhouse gases has increased. For instance, the density of gases such as CO_2, CH_4, N_2O reached 379 ppm, 1732 ppb, and 319 ppb respectively in 2005, which were beyond the average amount and considered probable that the increase in the emission of these gases is mainly associated with human activities. The contribution of other parts in the CO_2 emission is identified in Fig. 5.2. Radiative forcing stemming from the increase in gases like CO_2, CH_4, N_2O was $+2.3$ W m^{-2}, has been at an unprecedented amount in past 10,000 years.

The data revealed that years 1995 to 2006 were the warmest years since 1850. The linear trend of temperature data revealed that the temperature increase was 0.74 °C in the past 100 years. That is, it was beyond the amount predicted by TAR for these years.

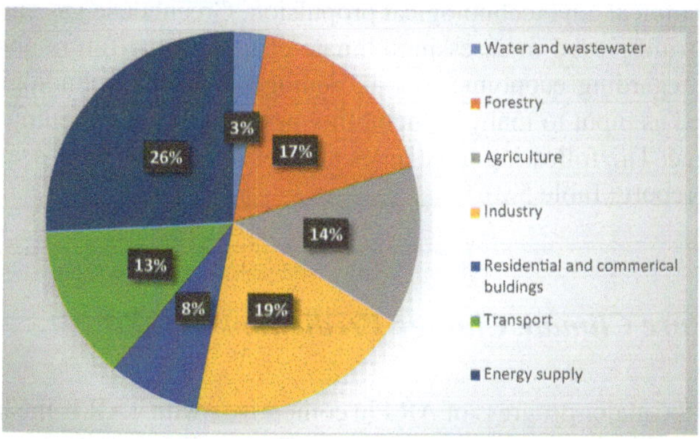

Fig. 5.2 The contribution of different sectors in the publication CO_2

Observations since 1961 showed that the average temperature of the oceans had increased at least to the depth of 3000 (oceans absorb more than 80% of extra heat added to the climate).

Evidence revealed that glaciers and snow cover had reduced generally in both hemispheres. The extensive reduction of glaciers increased sea levels.

The average pace of global sea-level rise is 1.8 mm per year, which is significant compared to 1961–2003. This amount was even faster through the years of 1993–2003. Sea level rise was estimated at 0.17 m in the twentieth century (AR4 2007). Since 1970, more intense and more prolonged droughts were witnessed in broader areas such as tropical and subtropical zones. An increase in the temperature and a reduction in rainfall are important components of these droughts. Change in the temperature of sea level, wind pattern, and reduction of the snow cover, all intensified these droughts. The frequency of heavy rainfalls increased in most areas, which was compatible with warming and increased water vapor in the atmosphere. The data analysis indicated the probability of long term climate change.

However, some parameters remained unchanged, such as reductions in daily temperature ranges (DTR), and the day and night temperature increased at the same rate. For determining existing procedures in Meridional Overturning Circulation (MOC), there were insufficient data on the global oceans or small scale phenomena such as tornados, thunderstorms and dust storms. Therefore, they were not considered in the fourth report.

5.5.2 SRES Scenarios in the Fourth Report

SRES refers to scenarios described in the IPCC special report on the propagation scenarios. SRES scenarios are classified into four scenarios (A1,[4] A2, B1 and B2), which examine alternative development paths and cover a wide range of demographic, economical, and technological propulsion. Greenhouse gas emission anticipations are widely used in assessing climate change in the future and their basic assumptions regarding economic, social, demographic and technological changes are considered as input to many vulnerability assessments and the impact of recent climate change. The following table gives an overview of the main scenarios of the fourth IPCC report (Table 5.2).

5.5.3 Future Climate Change Predictions in AR4

The cause of the major progress of AR4 in comparison with TAR is the large number of simulations with a broader range of models. The predictions done based on a broad range of models in this report were as follows:

[4] Scenario A1 consists of three sub scenarios: A1FI, A1T, and A1B.

Table 5.2 SRES scenarios in the fourth report

SRES scenarios in the fourth report	More emphasis on economic aspects	More emphasis on environmental aspects
The converging world (globalization)	A1	B1
	Rapid economic growth (A1T, A1FI , A1B) temperature increase (1.4–6.4 °C)	Sustainable environmental development temperature increase (1.1–2 °C)
The diverging world (regional)	A2	B2
	Regional economic development temperature increase (2–4.5 °C)	Regional sustainable environmental development temperature increase (4.1–8.3 °C)

Table 5.3 Prediction of average global temperature rise and sea level rise at the end of the twenty-first century

Predicted range for sea level rise	Predicted range for temperature rise	Case
0.18–0.38	1.1–2.9	B1 scenario
0.20–0.45	1.4–6	A1 scenario
0.20–0.43	1.4–3.8	B2 scenario
0.21–0.48	1.7–4.4	A1B scenario
0.23–0.51	2.0–5.4	A2 scenario
0.26–0.59	2.4–6.4	A1FI scenario

- Arctic and Antarctic sea ice are likely to melt. Arctic sea ice will almost disappear in summer, in the late 21st century.
- Future tropical storms are likely to be accompanied by peak speed and higher precipitation.
- The observations showed that the continued emission of greenhouse gases at the same rate, or a higher rate than the current rate, would cause further warming and changes in the global climate system during the 21st century, most likely to be greater than what was observed in the 20th century.
- The amount of rainfall in high latitudes will probably increase, while in most subtropical areas the amount of rainfall was predicted to decrease (AR4 2007).
- Temperature changes and sea levels are listed under 6 SRES scenarios in the following table (Table 5.3).

5.5.4 *The Investigated Effects of Climate Change in AR4*

The evaluation regarding the scientific understanding of the effects of climate change on natural and human systems, adaptability capacity of these systems, and vulnerability were carried out by working group II in the fourth assessment. Since TAR, more than 100 studies have been published regarding the effects of climate change on the flow of rivers. Compared with TAR, there is higher reliability in the predicted patterns of warming and other features on a regional scale, including change in wind patterns, precipitation, and sea change process. Since TAR, studies have made a more orderly sense of time and the amount of effects associated with the different amount and rates of climate change.

However, the focus of studies was mainly on Europe, North America and Australia, and there were few studies carried out in developing countries. Almost all studies had used hydrological model based on the simulation-based scenarios of weather models and some of them used SRES-based scenarios. The reliability of the changes estimated by the models depends on the degree of the flexibility of the weather model for runoff simulation. In the following, the effects of climate change on some water resources parameters have been investigated.

Surface Water: Due to climate change until the mid-twentieth century, average annual rivers runoff and availability of water at high latitudes and tropical humid areas were predicted to increase %10–40 and decrease %10–30 in some arid areas at intermediate latitudes, arid tropical areas, Mediterranean Sea and south of Africa. In some places and especially in particular seasons, the changes were different from these annual figures (Nohara et al. 2006).

Groundwater: Demand for groundwater due to increased global water consumption is likely to rise in the future. Climate change affects the amount of groundwater recharge. However, little information was found about the amount of recharge in developed and developing countries. Studies revealed that changes on the river levels affect groundwater level more than groundwater recharge (Allen et al. 2003).

Some influences of climate change on many of aquifers in the world:

- Groundwater recharge changes from spring to summer (it reduces in summer).
- At high latitudes, melting of glaciers causes a change in level and quality of the groundwater.
- Climate change may lead to changes in vegetation that affect the groundwater recharge.
- With increase in the frequency and magnitude of floods, groundwater recharge may increase, especially in arid and semi-arid regions where severe rainfall and flooding are one of the primary sources of groundwater recharge. Accordingly, the impact of climate change on groundwater recharge should include the effects of rain change.
- For the four scenarios investigated, the groundwater recharge calculated in north-east Brazil, southwest Africa and along the southern coast of the Mediterranean Sea noticeably decreased by more than 70%. It was while in areas like northern China, Siberia and the west of the United States increased by more than 30%.
- In addition, another impact of climate change on water resources is salinization of aquifers. Increase of evaporation and sweating in arid and semi-arid areas and the rise of sea levels may lead to salinization of shallow aquifers.

Quality of Water: Temperature increase, and the changes in the runoff might make undesirable changes in the quality of water, which affects human health and the ecosystem (Hurd et al. 2004). The decreasing water level in rivers and lakes will cause the suspension of sediment, which will negatively affect water resources. Heavy rainfalls will increase the level of suspended solids in lakes and water storage tanks. High temperature on the water surface will give rise to the algae (Kumagai et al. 2003). It has been predicted that waters to be salty in some areas due to the reduction of runoffs. In circumstances in which water quality decreases due to climate change, some strategies for removing wastewater and reuse and recycling of water must be considered (Patrinos et al. 2005).

Flood and Drought: Increase in climate variation increases the danger of flood and drought. A great deal of climate and non-climate motives influence the effects of flood and drought. It is probable that areas that are involved with drought to be expanded. Meanwhile rainfalls, which have been predicted to occur due to climate change, increase the flood danger.

Evidence gathered from all the continents expressed that a large number of natural systems are influenced by regional climate change, specifically temperature increase. There is a probability that other regional climate change effects will increase in the natural and human environments. However, recognizing most of them is demanding due to the compatibility and non-climate motives.

Measures that can be taken for compatibility with climate change in water resources include rainwater investment, water reuse, irrigation efficiency, consistent management of water resources. Many effects stemming from climate change can be reduced or delayed by constant and consistent management. Efforts and investments will have a remarkable impact on achieving sustainable levels within the next few decades (AR4 2007).

5.6 AR5

5.6.1 Climate Change Assessment in the AR5 Report

After the fourth assessment report, in the fifth assessment of Intergovernmental Panel on Climate Change in 2013, the first IPCC working group presented a comprehensive assessment of the basis of the physical sciences of climate change, based on new evidence of climate change, scientific analyses of these observations, and the study of the past climate.

Changes Observed in the Climate System.

The fifth report stated that warming of the climate system had had consequences such as climate and ocean warming, snow and ice reduction, sea-level rise, and rise in the concentration of greenhouse gases. These observed changes were unprecedented over the past millennium.

Observations revealed that the average temperature of the earth's and oceans' surface has increased by 0.85 °C from 1880 to 2012. Also, observations for the years 1901 to 2012, the most extended period in which calculations of regional trends are complete, demonstrated that warming has almost occurred on the whole planet. According to global-scale observations, the warming of the upper parts of the oceans (closer to the surface) is higher. On a global scale, the most warming was observed in the upper 75 m of the ocean, approximately 0.11 °C in each decade from 1971 to 2010. At the time of the publication of AR5, there was not enough information available from the temperature variations below 2000 m. In general, more than %60 of temperature rise in the oceans is stored in the upper ocean (0–700 m), and about 30% is stored in the ocean below 700 m.

Over the last two decades, Greenland and Antarctic ice sheets have lost their ice packs. Almost all glaciers worldwide are reduced, and ices in the Arctic Ocean and snow cover extent in the Northern Hemisphere continue to decline. The extent of sea-level rise from 1901 to 2012 was 0.19 m which were more than the average rise over the last two thousand years. It is most likely that the sea level rise from 1901 to 2010 has been 1.7 mm per year.

In the case of precipitation, the same trend was not observed in all parts of the world. In the latitudes of the Northern Hemisphere, the precipitation has increased since 1901, and in other latitudes, it has probably been reduced.

Evidence indicated that atmospheric concentrations of greenhouse gases have increased. The increase in carbon dioxide, methane and nitrogen oxide is 40%, 150%, and 20%, respectively, concerning the pre-industrial age. In 2011, the concentrations of these gases were ppb 324, ppb 1803, and ppm 391, respectively. It was noted that the average rise rate of greenhouse gases concentration over the last century has been unprecedented in the last 22,000 years.

5.6.2 *Climate Change Assessment in AR5*

In the last decades, climate change has led to change in natural and human systems in all continents and oceans. The effects of climate change, regardless of its cause, represent the sensitivity of natural and human systems to climate change. Understanding the changes in the climate system can be concluded from the combination of observations, the study of feedback processes and simulation of models. Assessment of the ability of climate models in the simulation of recent changes requires considering the status of all components of the model of the climate system model at the beginning of the simulation and the natural and human effects of these models. Compared to AR4, more detailed and longer observations were available to present in the fifth report. They improved climate models enabled the attribution of a human contribution to detected changes in more climate system components. Human influence on the climate system is apparent. This is evident from the increasing greenhouse gas concentrations in the atmosphere, positive radiative forcing, observed warming, and understanding of the climate system. The evidence suggests that human influence on climate change has grown since AR4. Human activities are likely the dominant cause of the observed warming since the mid-twentieth century.

5.6.3 *Future Climate Change Predictions in AR5*

The persistent emissions of greenhouse gases make more warming and changes in all parts of the climate system. Reducing climate change requires a significant and sustainable reduction in greenhouse gases emissions. In the fifth report, the IPCC used the RCP (Representative Concentration Pathways) scenarios, which were constructed in 2010 by a scientific committee under the auspices of the Intergovernmental Panel on Climate Change aimed at providing a set of data that can be used to trace the underlying factors of climate change. Its results can be applied to climate models. These scenarios are more detailed scenarios than the SRES scenarios. One problem with the SRES scenarios was that they do not incorporate carbon emissions controls. These scenarios take into account climate change mitigation policies to limit emissions. In these scenarios, in addition to issues related to the control of carbon dioxide emission, other issues such as social and economic issues, land use, aerosols, and ozone layer are also considered.

In the RCP8.5 scenario, CO_2 concentration was estimated to be 1370 PPM by 2100, and the effect of greenhouse gases on radiative forcing was estimated to be up to 8.5 W/m^2.

In the RCP6 scenario, CO_2 concentration was estimated to be 850 PPM by 2100, and the effect of greenhouse gases on radiative forcing was estimated to be up to 6 W/m^2.

In the RCP4.5 scenario, CO_2 concentration was estimated to be 650 PPM by 2100, and the effect of greenhouse gases on radiative forcing was estimated to be up to 4.5 W/m^2.

In the RCP2.6 scenario, CO_2 concentration was estimated to be 490 PPM by 2100, and the effect of greenhouse gases on radiative forcing was estimated to be up to 2.6 W/m^2.

The predictions presented in the fifth report using these scenarios are as follows:

- The average global temperature for 2016–2035, relative to 1980–2005, is likely to range from 0.3 to 0.7 °C. Relative to 1850–1900, global surface temperature for the end of the twenty-first century is predicted to exceed 1.5 °C for all RCO scenarios, except for RCP2.6. More than about 2 °C change for RCP6.0 and RCP8.5 and less than about 2 °C for RCP4.5 was predicted.
- Changes in the global water cycle in response to the warming over the twenty-first century will not be uniform. Conflict in precipitation between moist and dry regions will increase, but there may be regional exceptions. High latitudes and the Tropical Pacific Ocean in the RCP8.5 scenario are likely to increase the average annual precipitation by the end of the twenty-first century. In many arid and subtropical regions, the average precipitation is more likely to decrease. In many humid regions of mid-latitudes, average precipitation in the PCP8.5 scenario is likely to increase by the end of this century.
- The maximum amount of ocean's temperature rise was predicted for the tropical regions of the Northern Hemisphere by the end of the twenty-first century.

- The Atlantic Meridional Overturning Circulation (AMOC) is likely to become weak over 21st Century.
- It is very likely that the Arctic sea ice cover will continue to shrink and thin. Northern Hemisphere spring snow cover will decrease during the twenty-first century as global mean surface temperature rises. Also, the global volume of glaciers is predicted to decrease.
- It was predicted that the sea level would continue to rise during the twenty-first century. Considering the RCP scenarios, the sea-level rise rate is more likely to be increased due to the oceans' overheating and increase in mass loss from glaciers. The average global sea-level rise for 2100–2081 relative to 1986–2005 is likely to range from 0.26–0.55 m in the RCP2.6 scenario, 0.32–0.63 m in the RCP4.5 scenario, 0.33–0.63 m in the RCP6.0 scenario and 0.45–0.82 m in the RCP8.5 scenario.
- Climate change affects the carbon cycle processes so that CO2 rise was predicted under all four scenarios. The more carbon absorption by the ocean, the more ocean acidification (AR5 2013).

5.6.4 *Investigated Effects of Climate Change in AR5*

Investigating effects, compatibility, and vulnerability against climate change which was surveyed by working group two in the fifth assessment report of IPCC, revealed how danger patterns and potential resources have changed due to climate change since 2007, presented in the fourth assessment report. Dangers regarding climate change and freshwater increase remarkably along with the increases in the concentration of greenhouse gases. Concerning the results of several models, it is predicted that about 7% of the world's population are exposed to the reduction of 20% of renewable water sources for each degree of global warming. Next, the effects of climate change on some of the water resource parameters are considered. Human effects on the climate system are evident, and the expansion of greenhouse gases through human activities reaches the highest recorded levels in history. Recent climate changes have had broad impacts on human and natural systems. Climate change dangers on the freshwater remarkably increase with the increase in the concentration of greenhouse gases. Moreover, the population of people who face water shortage or dangers of massive floods due to the temperature increase rises in the twenty-first century.

Surface water: It is predicted that renewable surface water resources will decrease in many arid subtropical regions. The reduction of water resources intensifies the race for water in agriculture, ecosystems, towns, industry, and energy production. It might remarkably have an impact on the water, energy, and food security of the region.

Groundwater: It is predicted that renewable groundwater resources to be decreased in many arid subtropical regions. Land-use change and groundwater harvesting have made complex observed changes in the groundwater level, storage or discharge due to the climate changes. The observed procedures mainly originate from these effects. The amount of groundwater harvesting influenced by the effects of climate change is not specified now. The observed reduction in the discharge of springs nourished by groundwater in Cashmere (India) was attributed to the observed rainfall reduction since 1980. A groundwater nutrition assessment based on modeling in Spain expressed that due to the reduction of rainfall, refeeding of four aquifers that are exploiting in the region decreased during the 20 century (Arndt et al. 2010).

Flood: There are no comprehensive observations regarding changes in the size and frequency of floods due to climate changes, but according to predictions, the frequency of floods changes. It is predicted that flood risks to be increased in the southern regions, southeast and northeast of Asia. Meanwhile, according to the existing limited evidence, it is probable that global flood risk to be increased due to the condition changes in the future.

Drought: Climate changes under the RCP8.5 scenarios might increase the frequency of meteorological agriculture and hydrological droughts by the end of the twenty-first century. According to the regional observations, most of the droughts and extreme rainfall events from 1990 to 2000 were regarded as the worst droughts and rainfalls since 1950. Most of these changes were concerned with temperature changes (Stahl et al. 2010).

Quality of Water: it is predicted that climate changes reduce water quality. Even with standard water treatment, some dangers threaten drinking water. The maximum amount of reported change refers to the increases of algae in the high temperature regarding lakes and reservoir. The absorption and storage of carbon can reduce the quality of groundwater.

Streamflow: Identified procedures in the streamflow are generally compatible with observed regional changes in the amount of rainfall and temperature since 1950. For instance, in the south and east of Europe, northwest of the Pacific Ocean, and southern regions of the Atlantic Ocean, following the reduction of rainfall, a decrease in the streamflow was observed (Kalra et al. 2008). However, in regions such as North America in the Mississippi, the opposite happened; an increase in the streamflow was observed in the Young river (Forty–sixth session of the IPCC 2017). In regions where snow is regarded as part of their water supply, streamflow of runoffs moves from spring to winter due to the temperature increase and rapid melting of snows.

Glaciers: All the prediction of the twenty-first century lead to the reduction of glaciers. With the reduction of glaciers, runoff peak discharge will occur in the spring. A comparative approach for water management can control the uncertainty originating from climate changes. Comparative techniques include planning, scenario making, and empirical approaches, consisting of learning from experiments and flexibility development which are resistant against uncertainty.

5.7 AR6

At the moment, IPCC is preparing the sixth assessment report (AR6). Since the publication of the fifth report, three specific reports and a methodology report on the expansion of greenhouse gases have presented. Like other reports of IPCC, AR6 consists of three reports gathered through the collaboration of working groups I, II, III and a combined report provided through the combination of presented reports during the cycle.

In its 41st session, held in Nairobi, Kenya, in February of 2015, the IPCC decided to publish its assessment reports each 5–7 years. In its 43rd session in Nairobi, Kenya, which was in April 2016, the IPCC decided on the topics of the specific reports in the assessment cycle. In addition to making decisions for preparing AR6, IPCC accepted the UNFCCC's invitation for preparing a specific report on the effects of global warming of 1.5 °C in its 43rd session. Moreover, it was determined that two specific reports to be prepared, one must be about oceans and cryosphere in a varied climate, and the other must be on the climate changes in the arid areas.

All the IPCC reports are considered in two stages; (1) a first-rate draft which experts study, and (2) a second-rate draft which governments and experts survey.

It has been predicted that the first part of the sixth assessment report, which is related to the basics of climate change physics and will be prepared with the collaboration of working group I, will be finalized in 2021. Issues addressed in the sixth assessment report of IPCC by working group one are similar to other reports of IPCC and include a technical summary and a summary for the use of politicians and headlines, which are mentioned in the following sections:

- Materials and methods and considering the fifth report and its relationship with the sixth report
- Change in the climate system condition
- Effects of human activities on the climate system
- Predictions based on the scenario and short-lived information from future climate
- Surveying global expansion of carbon and other biochemical cycles and their feedback
- Short-lived climate forces
- Earth energy storage, climate feedback and sensitivity to the climate
- Changes in the water cycle
- Changes in the ocean, cryosphere, and sea level
- Effects of global climate change on the regional climate change
- Severe climate conditions (flood, drought) due to the climate change
- Regional climate change data and risk assessment (AR6 2021).

The second part of the sixth report by working group III provides an assessment of compatibility effects and climate change vulnerability. The first-rate draft of this report was examined by working group 3 in march 2020, and it is under second-rate investigation by experts and governments now and will be finalized by 14 March 2021.

Some of the overall parts of this report are as follows:

Introduction and overall construction- the procedure of the expansion of greenhouse gases and effective stimulators on their expansion- methods of compatibility with climate change in the long run- directions of reduction and expansion in the short run to the medium-term- demand, services, and social aspects of climate change-effects of climate change on energy systems, agriculture, forestry and other land uses, urban system and other residences, buildings, transportation, industry, periodic landscapes, politics and national units, international cooperation, innovation, technology development and its transmission, acceleration in transmission in the field of sustainable development (Girod et al. 2009).

The third part of the sixth assessment report, which considers issues of reduction in the effects of climate change, will be presented by working group 2. Experts investigated its first draft from 18 October to 13 December 2019, and its second draft is under investigation by experts and governments. Moreover, the contribution of working group 2 in the sixth assessment report will be completed by 2021. Generally, some of the issues of this report include:

- The importance of climate and regional dangers for the natural and human systems and their interaction with culture, values, morals, identity, and behavior
- Investigation of climate change risks, reduction of uncertainty
- Recognition and investigation of climate effects and methods of compatibility
- Understanding of dynamic climate dangers from reflexive scenarios
- Scientific, technical, and economic aspects- effects of climate changes
- Limitations of compatibility and suitable conditions for practical compatibilities, such as sovereignty, units, economic aspects
- Climate change responses and their interactions with sustainable development
- Investigation of climate change on the continents of Asia, Australia, South and Central America, Europe, North America, Africa, and Isles
- Opportunities for an increase in the directions of sustainable development against climate change (IPCC Special Report on Emissions Scenarios 2009).

A synthetic report (SYR) is a combination of assessment and specific reports which is based on the contents of reports of three working groups: Basics of climate change, effects of compatibility and vulnerability, and the reduction of climate change effects will be presented by working groups 1, 2, and 3, respectively. Working group 3 will also provide three specific reports, which include a global warming of 1.5 °C, climate changes in the earth, and changes in cryosphere due to climate change. According to predictions, the SYR report will be the final product of AR6 and finalized by 2022.

5.8 Scenarios' Evolution from SRES to RCPs

Scenarios are used as the entries for implementing climate models for the assessment of possible effects of climate. Emission scenarios of IPCC are the basis for most of the long-term predictions of climate changes. Many elements in the prediction of global warming in the future must be taken into consideration. For better comparison among various studies and easier connection among models' results, shared scenarios in the scientific community are encouraged. Scenarios are substitutional pictures of how the future emerges and are suitable tools for analyzing how stimulator forces effect on the results expansion of greenhouse gases in the future and are appropriate devices for assessing related uncertainties. They are efficient in analyzing climate changes, such as climate modeling, effect assessment, compatibility, and reduction (Nakicenovic et al. 2000).

IPCC published the first emission scenarios (IS92) in 1992 for use in the global circulation models. These scenarios were the first global scenarios, which provided estimations for the emission of greenhouse gases. These scenarios helped understand the possible emission of greenhouse gases, and how they affect climate, but they failed to consider many issues. The emission of greenhouse gases in the future is the product of complicated dynamic systems, which are determined by stimulator forces such as population increase, economic development, social development, and technical changes. As a result, IS92 scenarios were reassessed in 1995; The assessments were indicators of remarkable increases greenhouse gases, which must be considered in the scenarios. This led to making decisions by the general meeting of IPCC in 1996 to prepare a new complex of scenarios for the third assessment report of IPCC. IPPC published a new complex of scenarios in 2000 for use in the third assessment report of (SRES). SERES scenarios have been created to assess future advances globally, considering a thorough assessment of production and expansion of greenhouse gases. New scenarios have also been presented to assess climate and environmental consequences of the expansion of greenhouse gases and assess reduction and compatibility strategies. They included original expansion lines and last information about economic reconstruction all around the world. They considered differences and technological changes and expanded various economic development directions, such as increasing the income gaps between developed and developing countries (Nakicenovic et al. 2000).

SRES scenarios make use of succession approaches for determining scenarios if a change was made in each previous scenario. These changes were also influential in subsequent scenarios. This caused the continued approach to be time-consuming. Political actions or legislation did not affect these scenarios.

These scenarios were a complex between economic and environmental values and another complex between increasing globalization and regionalism.SRES scenarios are detailed in the following sections.

A1 Scenario: This scenario describes a rapidly developed world along with interactions and strong convergences among regions in which people's annual income is more uniform. The world's population will peak by 2050 and then will decrease, and then more new and efficient technology will be introduced. The remarkable characteristics of the A1 scenario include rapidly economic development, 9 billion population by 2050 and then reduction of this population, the rapid expansion of new and efficient technology and the convergent world (income and way of life among regions will be convergent and development of social and cultural interactions will occur rapidly in the world.). There are three different sub-branches for group A1 according to the used technology in the twenty-first century: emphasis on the intensive use of fossil fuels (A1F1), emphasis on the use of non-fossil fuels (A1T), and moderate use of fossil and non-fossil fuels (A1B).

A2 Scenario: This scenario describes a different world in which countries act independently are relying on themselves. In that population, some regions have reinforced the emphasis on family and familial traditions and, consequently, will have more growth than the A1 scenario (population increases continuously). Per capita economic growth is changeable (regional centred economic growth), and technological development is slow.

B1 Scenario: In this scenario, the world is integrated and environmentally friendly. Population condition in B1 is similar to A1, except that this scenario emphasizes using clean energy, innovative technologies, and reducing pollutants. This scenario emphasizes the global solutions for environmental, economic, and social sustainability and equal human rights among communities. Like the A1 scenario, economic development is fast and based on providing services and information in this scenario.

B2 Scenario: In this scenario, like the A2 scenario, the world is divergent, except that it is considered environment friendly like the B1 scenario. Economic development were moderate and technological changes were fast, but these changes were less than A1 and B1 scenarios. This scenario emphasizes continued population growth, and the rate of population growth increases continuously in this scenario, but the speed of growth is less than the A2 scenario. Moreover, this scenario highlighted regional solutions for economic, social, and ecosystem stability instead of global solutions.

Despite the appropriate function of SRES scenarios, a new complex of scenarios was required for examining climate changes periodically to consider scientific developments in the understanding of climate system and combination of updated data on the expansion of greenhouse gases, reduction of climate changes, effects, compatibility, and vulnerability.

The research community stated that new scenarios are needed to have more detailed information to fulfil climate models based on what provided from the previous complex of scenarios (SRES). Meanwhile, scenarios needed to reflect the effects of various climate policies and the non-climate policy scenarios reflected so far. Such scenarios facilitate the possibility of assessment of climate's long-term goals. The necessity of new scenarios caused IPCC to ask scientific communities to prepare complex scenarios for the ease of climate change assessment in the future. IPCC also decided that such scenarios not be compiled as part of IPCC and left new scenario development to the research community. Eventually, new scenarios were utilized as climate modeling and research for the fifth assessment report (AR5) of IPCC in 2014. RCPs are directions of concentration of greenhouse gases. Four directions were used in the climate modeling and other research for the fifth assessment report. The directions describe different climate futures for the volume of emitted greenhouse gases in the future. One of the positive aspects of RCP scenarios is that each of them describes an emission direction and concentration by 2100.

RCPs were introduced to facilitate interactions between scientific communities working on climate changes, compatibility, and reduction.

Despite SRES, RCPs facilitate the possibility of assessment of expenses and advantages of long-term climate goals. SRESs were not included in any policy for restricting climate changes and thus did not consider reducing climate changes; this problem was solved in RCP scenarios. Despite SRES scenarios, RCP scenarios utilize parallel approaches in the development of scenarios. Parallel approaches allow policy changes to be made. RCP was an indicator of an essential step in the development of new scenarios for climate research. RCPs employs a process in which modeling is less time-consuming, more flexible and cost-effective. Four directions of RCPs scenarios are listed below:

In the RCP8.5 scenario, the amount of concentration of CO_2 is estimated at 1370PPM by 2100, and the greenhouse effect on radiative forcing is estimated at 8.5 W/m^2.

In the RCP6 scenario, the amount concentration of CO_2 has estimated 850 PPM by 2100, and 6 W/m^2 estimate the greenhouse effect on radiative forcing.

In the RCP4.5 scenarios, the amount of concentration of CO_2 has estimated 650 PPM by 2100, and the greenhouse effect on radiative forcing is estimated at 4.5 W/m^2.

In the RCP2.6 scenarios, the amount of concentration of CO_2 has estimated 490 PPM by 2100, and the greenhouse effect on radiative forcing is estimated at 2.6 W/m^2.

Bibliography

Allen DM, Mackie DC, Wei M (2003) Groundwater and climate change: a sensitivity analysis for the Grand Forks aquifer, southern British Columbia. Canada. Hydrogeol Journal. 12:270–290

AR6 Climate Change (2021) Mitigation of climate change—IPCC

Arndt DS, Baringer MO, Johnson MR (2010) State of the climate in 2009. Bull Am Meteor Soc 91(7):S1–S222. https://doi.org/10.1175/BAMS-91-7-StateoftheClimate

Arnell NW (1998) Climate change and water resources in Britain. Springer 39:83–110

Bouraoui F, Vachaud G, Li LZX, Le T, Chen T (1999) Evaluation of the impact of climate changes on water storage and groundwater recharge at the watershed scale. Clim Dyn 15:153–161

Climate Change IPCC (2013) The scientific basis-contribution of working Group I to the fifth assessment report of the intergovernmental panel on climate change: Cambridge University Press. Cambridge University Press, Cambridge, United Kingdom and New York, NY, USA, p 1535

Girod B, Wiek A, Mieg H, Hulme M (2009) The evolution of the IPCC's emissions scenarios. Environ Sci Policy 12(2):103–118

Hurd BH., Callaway M, Smith J, Kirshen P (2004) Climatic change and U.S. water resources: from modeled watershed impacts to national estimates. J Amer Water Resour Assoc 40(1):129–148

IPCC. Climate Change (1995a) Synthesis report Report contribution of working Group I, II and iii to the second assessment report of the intergovernmental panel on climate change .Cambridge United, New York, NY, USA, Oakleigh, melborn. Australia

IPCC. Climate Change (1995b) The science of climate change contribution of working Group I to the second assessment report of the intergovernmental panel on climate change. Cambridge United, New York, NY, USA, Oakleigh, Melborn. Australia

IPCC. Climate Change (1995c) Impacts, adaptations and mitigation of climate change: scientific-technical analyses contribution of working Group II to the second assessment report of the intergovernmental panel on climate change. Cambridge United, New York, NY, USA, Oakleigh, Melborn. Australia
IPCC. Climate Change (1995d) synthesis report contribution of working Group I, II and III to the second assessment report of the intergovernmental panel on climate change. Cambridge United, New York, NY, USA, Oakleigh, Melborn. Australia
IPCC. Climate Change (2001a) The scientific basis-contribution of working Group I to the third assessment report of the intergovernmental panel on climate change: Cambridge University Press
IPCC. Climate Change (2001b) Impacts, adaptation, and Vulenerability. Contribution of working Group II to the third assessment report of the intergovernmental panel on climate change. Geneva, Switzerland
IPCC. Climate Change (2007a) The physical science basis contribution of working Group I to the fourth assessment report of the intergovernmental panel on climate change. Cambridge United Kingdom and New York, NY, USA
IPCC. Climate Change (2007b) Synthesis report working group contributions to the fourth assessment Report of the intergovernmental panel on climate change Valencia, Spain, 12–17 November
IPCC web site, http://www.ipcc.ch/as viewed 29 March 2021
IPCC. Forty—sixth session of the IPCC Montreal. (2017), Canada, 6–10
IPCC. Climate change, (1990) FAR climate change. Australian Government Publishing Service Canberra, The IPCC Impacts Assessment
IPCC. Climate change (1990) FAR climate change, scientific assessment of climate change. Cambridge, new York, port chester, melborn and Sydney. Cambridge University Press
IPCC. Special Report on Emissions Scenarios (2009) A special report of working Group III of tiie intergovernmental panel on climate change. Cambridge University Press
Kalra A, Piechota TC, Davies R, Tootle GA (2008) Changes in U.S. streamflow and western U.S. snowpack. J Hydrol Eng 13(3):156–163
Kumagai M, Ishikawa K, Chunmeng J (2003) Dynamics and biogeochemical significance of the physical environment in Lake Biwa. Wiley Online Library. https://doi.org/10.1046/j.1440-1770.2002.00201.x
Levermore GJ (2008) A review of the IPCC assessment report four, Part 1: the IPCC process and greenhouse gas emission trends from buildings worldwide. Build Serv Eng Res Technol 29(4):349–361
Nakicenovic N, Alcamo J, Davis G (2000) Special report on emissions scenarios: a special report of working Group III of the intergovernmental panel on climate change, Cambridge University Press, Cambridge, U.K., 599 pp. Available online
Nemec J (1989) Impact of climate variability and change on water resources management in agriculture. Conference on climate and water. Helsinki, pp 357–371
Nohara D, Kitoh A, Hosaka M, Oki T (2006) Impact of climate change on river discharge projected by Multimodel ensemble. J Hydrometeorol 7(3):1076–1089
Patrinos A, Bamzai A (2005) Policy needs robust climate science. Nature 438(285)
Peterson DF, Keller A (1990) Effects of climate change on U.S. Irrigation. J Irrig Drainage Eng 116(2). https://doi.org/10.1061/(ASCE)0733-9437(1990)116:2(194)
Stahl K, Hisdal H, Hannaford J, Tallaksen LM, van Lanen HAJ, Sauquet E, Demuth S, Fendekova M, Jodar J (2010) Streamflow trends in Europe: evidence from a dataset of near-natural catchments. Hydrol Earth Syst Sci 14(12):2367–2382. https://doi.org/10.5194/hess-14-2367-2010
Viner D, Howarth C (2014) Practitioners' work and evidence in IPCC reports. Nature 4:848–850
Xia XH, Wu Q, Mou XL, Lai YJ Lai (2014) Potential impacts of climate change on the water quality of different water bodies. J Environ Inform

Chapter 6
Introduction to Key Features of Climate Models

Mahsa Jahandideh Tehrani, Omid Bozorg-Haddad, Santosh Murlidhar Pingale, Mohammed Achite, and Vijay P. Singh

Climate models are mathematical representations of the Earth's climate system. Besides representing the fundamental controls on Earth's climate in the form of solar radiation, surface and atmospheric albedo, and terrestrial infrared radiation, climate models are able to simulate the interactions and feedback processes of the physical, chemical, and biological features of the climate system at varying levels of detail. Running on powerful computers, they can be employed to predict future climate conditions. This chapter aims at discussing several categories of climate

The original version of this chapter was revised: The author name correction have been incorporated throughout the chapter. The correction to this chapter is available at
https://doi.org/10.1007/978-981-19-1898-8_11

M. J. Tehrani
Australian Rivers Institute, Nathan Campus, Griffith University, Brisbane, QLD 4111, Australia
e-mail: m.jahandideh-tehrani@griffith.edu.au

O. Bozorg-Haddad (✉)
Department of Irrigation and Reclamation Engineering, Faculty of Agricultural Engineering and Technology, College of Agriculture and Natural Resources, University of Tehran, Karaj, Tehran, Iran
e-mail: OBHaddad@ut.ac.ir

S. M. Pingale
Hydrological Investigations Division, National Institute of Hydrology, Roorkee, Uttarakhand 247667, India
e-mail: pingalesm.nihr@gov.in

M. Achite
Faculty of Nature and Life Sciences, Laboratory of Water and Environment, University Hassiba Benbouali of Chlef, 02180 Chlef, Algeria
e-mail: m.achite@univ-chlef.dz

V. P. Singh
Department of Biological and Agricultural Engineering and Zachry Department of Civil Engineering, Texas A&M University, College Station, TX 77843-2117, USA
e-mail: vsingh@tamu.edu

O. Bozorg-Haddad (ed.), *Climate Change in Sustainable Water Resources Management*, Springer Water, https://doi.org/10.1007/978-981-19-1898-8_6

models, including Energy Balance Models (EBMs), radiative-convective models, and General Circulation Models (GCMs). Being most comprehensive, GCMs have been widely applied for future climate change projections using different scenarios of population growth, greenhouse gas emissions, and land use changes. Sub-divisions of GCMs, including Atmospheric GCM (AGCM), Ocean GCM (OGCM), and Coupled Atmospheric-Ocean GCM (AOGCM) are also presented. Providing a background of the climate system and climate models, the chapter discusses the features of EBMs, radiative-convective models, and GCMs, and outlines the Coupled Model Inter-comparison Project (CMIP) and the corresponding six phases. Some examples of GCMs taking part in CMIP5 are introduced and their key features are presented. Finally, the key climate projections of the multi-model mean projections of CMIP5 are summarized.

6.1 Climate Models

The Earth's climate is generally controlled by solar radiation, albedo of the Earth–atmosphere system, and terrestrial infrared radiation. The climate system absorbs solar radiation in the atmosphere and at the Earth's surface, and the remainder of solar radiation is reflected back to space, where the amount of this reflection is set by planetary albedo (North and Kim 2017). Figure 6.1 summarizes the solar energy entering the Earth and its conversion into different forms in the Earth-atmosphere system.

Climate models are mathematical representations of the Earth's climate system and are employed to predict future climate changes (IPCC 2007; CSIRO and BOM 2015). These models can be run for different scenarios of Greenhouse Gas (GHG) emissions, land use patterns, and particulate air pollutants (IPCC 2014). GHG emissions are controlled by uncertain factors, such as population growth, lifestyle changes, economic growth, energy and land-use changes, and climate policy (IPCC 2014). Climate models consider the interactions and feedback processes of the climate system's physical, chemical, and biological properties (IPCC 2007). The climate system is represented by models with different levels of complexity. This complexity forms a spectrum or hierarchy of models with different combinations of component(s), number of spatial dimensions, the extent of explicit indications of physical, chemical or biological processes, and the level of empirical parameterization (IPCC 2007).

The climate system consists of five main physical components: atmosphere, oceans, cryosphere, land surface, and biomass. The climate models simulate the interactions of these components at different levels of complexity. The radiation, dynamics, surface processes, chemistry, and resolution in both time and space are the main five items considered when building a climate model (Shine and Henderson-Sellers 1983; McGuffie and Henderson-Sellers 2005). Radiation refers to the process

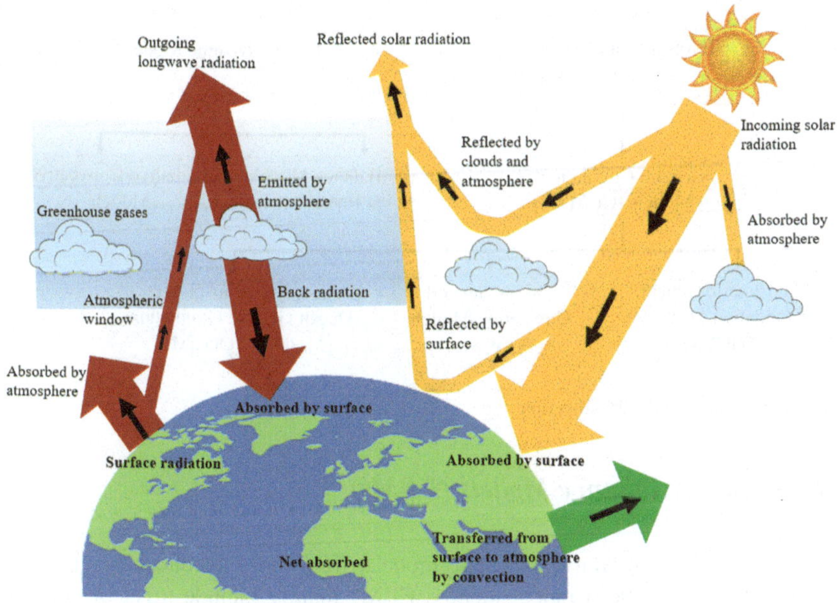

Fig. 6.1 Schematic representation of energy flows in the globally averaged Earth–atmosphere system. Arrows indicate flux densities

of controlling the input and absorption of solar radiation by the surface and atmosphere and the emission of infrared radiation. Dynamics determine energy movements by winds and ocean currents horizontally, and by minor small-scale turbulence and convection vertically. Surface processes refer to patterns and changes in sea and land ice, vegetation, snow, and the resulting impacts on emissivity, albedo, and interchange of energy and moisture between surface and atmosphere. Chemistry describes the atmospheric composition and its interaction with land and the ocean (e.g., exchange of carbon between atmosphere, land, and ocean). Lastly, the resolution is the indication of applied time step and horizontal and vertical scales (temporal and spatial resolution) (McGuffie and Henderson-Sellers 2005).

Depending on the inclusion of conversion between kinetic energy and total potential energy, climate models can be broadly classified into (i) hydrodynamic and (ii) thermodynamic models (Fig. 6.2). Thermodynamic models can be subclassified into (i) Energy Balance Models (EBMs) and (ii) radiative-convective models. Hydrodynamic models are General Circulation Models (GCMs), which can be sub-categorized into (i) Atmospheric General Circulation Models (AGCMs); (ii) Ocean General Circulation Models (OGCMs); and (iii) coupled Atmospheric- Ocean General Circulation Models (AOGCMs) (Schlesinger 1983). Each of these types of climate models, which are shown in Fig. 6.2, is discussed in the subsequent sections.

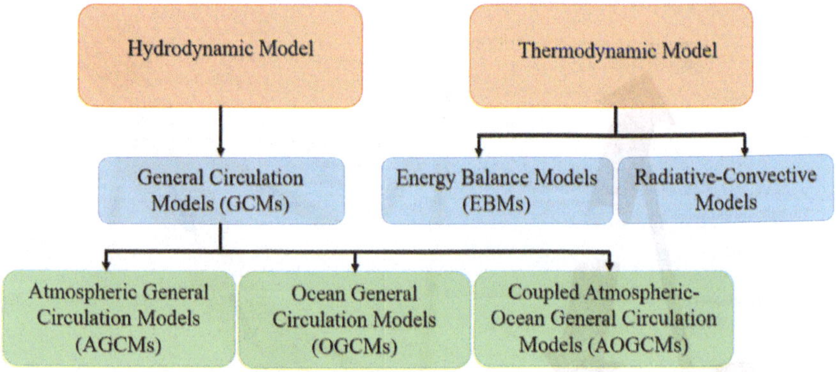

Fig. 6.2 Climate model classification

6.1.1 Energy Balance Models (EBMs)

EBMs, which are the simplest climate models, were first introduced by (Budyko 1969) and Sellers (1969) independently. EBMs mainly include a set of coupled equations to generate a space–time average of the Earth's surface temperature (North and Kim 2017). In a typical one-dimensional EBM model, the average annual sea level temperature in 10° latitude belts is defined as a dependent variable which is a function of solar constant, planetary albedo, turbulence exchange coefficient for the atmosphere and the ocean, and the transparency of atmosphere to the infrared radiation (Budyko 1969). Therefore, the main key feature of EBMs is that these models include only one dependent variable (the Earth's near-surface air temperature), and eliminate the dependency of climate on the full complexity of the wind field (Roques et al. 2014).

In the zero-dimensional EBMs, the most basic version of EBMs, the radiative equilibrium is modelled such that the infrared radiation (outgoing long wave radiation) emitted from the Earth to space is equal to solar radiation (incoming short wave radiation) absorbed by the Earth (Schlesinger 1983; Roques et al. 2014). The formulation of the zero-dimensional EBMs (radiative equilibrium) is as below (Schlesinger 1983):

$$\epsilon_p \sigma T_p^4 4\pi a^2 = S(1-\alpha_p)\pi a^2 \tag{6.1}$$

where ϵ_p= the effective emissivity of the Earth; σ = the Stefan-Boltzmann constant; σT_p^4 = the radiation emitted by the Earth with the effective radiative Earth temperature T_p; $4\pi a^2$ = the Earth's surface area with radius a; S = the solar constant; α_p = the Earth's planetary albedo (reflectivity); and πa^2 = the cross-sectional area of the Earth.

Zero-dimensional EBMs describe only the vertical radiation balance, whereas one-dimensional EBMs also consider the meridional energy transport (Rypdal et al.

2015). Satellite data confirm that much more solar radiation is absorbed in tropical latitudes than cooling through infrared radiation to space, while in the polar regions, the situation is reversed. Therefore, one-dimensional EBMs define the poleward heat transport mechanism in the atmosphere–ocean system (North and Kim 2017). In one-dimensional EBMs, the planetary albedo is defined as a function of polar icecap size by considering the temperature distribution in the north–south (meridional) direction. This spatial temperature distribution is defined as a relationship between the atmosphere–ocean system's meridional (longitudinal) transport of heat (Schlesinger 1983). The two hemisphere's zonally averaged seasonal cycles are different (north hemisphere and south hemisphere consist of approximately 40% and 20% land, respectively), and the two-dimensional EBMs have been developed (North and Kim 2017) which describe zonal transport in addition to vertical radiation balance and meridional energy transport (Schlesinger 1983). These models consider the land and ocean borders to be along meridians with seasonal forcing and idealized geography (North and Kim 2017).

The main limitation of EBMs is that they are unable to describe the temperature field above the boundary layer of the atmosphere (near-surface environment). Additionally, precipitation cannot be modelled well using simple EBMs as precipitation is correlated with the atmospheric and oceanic circulation (North and Kim 2017).

6.1.2 Radiative-Convective Models

The initial studies of modelling heat transfer to the atmosphere by convection were first conducted by Manabe and Wetherald (1967) who first investigated the theory of correlation between surface temperature and radiative forcing. Radiative-convective models compute only the vertical distribution of globally and seasonally averaged surface and atmospheric temperatures, and horizontal dynamics are ignored in these models (Ramanathan and Coakley 1978). These models describe the equilibrium temperature distribution for a column of atmosphere and its underlying surface by considering the solar radiation, atmospheric composition, and surface albedo (Schlesinger 1983). Radiative-convective models determine the mechanism of climatic interaction between atmosphere's surface temperature and H_2O amount (Ramanathan and Coakley 1978). Therefore, these models consist of sub-models to determine the solar transfer and terrestrial radiation, heat exchange between atmosphere and Earth's surface, vertical redistribution of heat in the atmosphere, and the amount of water vapour and clouds in the atmosphere (Schlesinger 1983).

Early studies of radiative-convective models were based on single-column models (e.g., Ramanathan and Coakley 1978). At the early stages of applying radiative-convective models, the main limitation of these models was their inability for determining the feedback mechanism between surface temperature and cloud cover as cloud top temperature, and its altitude was assumed constant (Ramanathan and Coakley 1978; Shine and Henderson-Sellers 1983). However, many limiting assumptions, such as the specification of the clouds and water vapour distributions, have currently been addressed by evolving the early radiative-convective models to cloud-resolving models (Tompkins and Craig 1998; Popke et al. 2013). Such an evolution (applying cloud-resolving models) addresses many issues, including scaling laws for moist convection, aerosol-cloud interactions, and the distribution of convective mass fluxes (Popke et al. 2013; Becker and Wing 2020). Thus, RCMs are still useful for studying in detail local atmospheric dynamics, including cloud processes, and their response to factors such as changing concentrations of GHGs.

6.1.3 General Circulation Models (GCMs)

GCMs have been widely used for different purposes, including simulation of climate system evolution, the study of interactions between components and processes of the climate system, and provision of climate change projections under different scenarios of population growth, GHG emissions etc. (Idso et al. 2013). GCMs apply three-dimensional approximations of the laws of physics, including conservation of mass, energy, and momentum to simulate surface pressure and vertical distributions of water vapour, density, temperature, and velocity as a function of time (Schlesinger 1983; CSIRO and BOM 2015). GCMs represent the atmosphere and ocean as three-dimensional grids to determine the large-scale synoptic features of the atmosphere (e.g., progression of the high and low-pressure systems associated with weather changes). As of the 2000s and the 2010s, GCMs are used to do long global climate simulations which typically have an atmospheric resolution of almost 250–600 km, and 10–20 vertical layers (CSIRO and BOM 2015). As shown in Fig. 6.3, GCMs simulate many aspects of climate, such as atmospheric and oceanic temperatures, winds, precipitation, clouds, and sea-ice extent (IPCC 2014). The movement of materials and energy in each three-dimensional grid cell is calculated using mathematical equations, which are based on the fundamental laws of physics, fluid motion, and chemistry. Additionally, the vertical and horizontal mass and energy transfer between adjacent grid cells is computed (Edwards 2011). Higher resolution (more grid cells) leads to more accurate and detailed simulations, but increases computation time and storage requirements (Edwards 2011).

GCMs are sub-divided into AGCM, OGCM, and AOGCM, and are discussed in the following sub-sections.

6.1.3.1 Atmospheric General Circulation Models (AGCMs)

AGCMs are numerical models that describe the physical and dynamical process (e.g., moist convection, turbulent mixing, and radiation) of the atmosphere, including land–atmosphere coupling but excluding ocean dynamics (Schlesinger 1983). These models simulate the energy transport by the atmospheric flow through fundamental equations (fluid dynamics and thermodynamics equations), which govern the transport of energy, momentum, water mass, and chemical species in the atmosphere

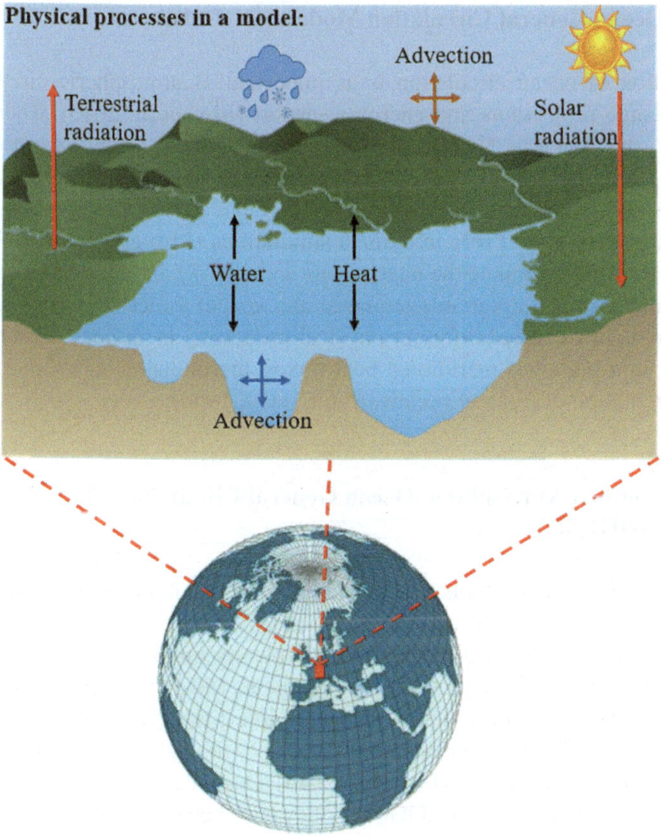

Fig. 6.3 Schematic representation of GCMs

(Idso et al. 2013; Njoku 2014). The flow equations are discretized on a spherical spatial grid with many vertical levels and includes representations of turbulence, cloud formation, dynamic heating (due to the interaction between solar, infrared radiation, atmospheric gases, clouds, and aerosols) (Idso et al. 2013). Within this context, the AGCMs include two main building blocks: (i) model dynamics (determining the large scale motions of the atmosphere) and (ii) model physics (representing a suite of parametrizations of sub-grid-scale physical processes) (Njoku 2014). Large-scale motions are described by solving partial differential equations in the dynamic modules. The cooling, heating, and mixing process are determined as forcing terms in the governing equations in the physics module (Njoku 2014). Both modules (dynamics and physics) include approximations that affect the accuracy of climate predictions (Njoku 2014).

6.1.3.2 Ocean General Circulation Models (OGCMs)

Consideration of ocean circulation is as important as atmospheric circulation in climate change projections and energy transport (Idso et al. 2013). Therefore, in addition to the AGCMs, GCMs include OGCMs that determine ocean's circulation (Idso et al. 2013). Given the dynamic thermal reservoir feature (this feature contributes to the climate system evolution through energy exchange) of oceans, OGCMs also play a vital role in climate simulations (Idso et al. 2013). These are numerical tools that simulate the interactions between oceans and other components of the climate system at variable temporal and spatial scales (Madec et al. 1998). These models investigated different physical processes, including turbulent eddies in oceanic-mid-latitudes, meridional heat fluxes, thermohaline circulation, and the general circulation of tropical oceans (Madec et al. 1998).

6.1.3.3 Coupled Atmospheric-Ocean General Circulation Models (AOGCMs)

AOGCMs are the primary tools that provide a simplified numerical indication of the entire climate system as they evolve at timescales up to several thousand years in a three-dimensional global grid. These models typically simulate complex interactions between the four main climate system components (land, atmosphere, sea ice, and ocean) over time based on four physical principles: (i) conservation of mass (dry mass and water mass), (ii) conservation of energy, (iii) conservation of momentum, and (iv) kinetic theory of gases (Idso et al. 2013). Despite the required vast amount of time and computer memory for AOGCMs, they have been extensively used in all five IPCC reports (Lupo and Kininmonth 2013). For example, as shown in Fig. 6.4, in order to link the interactions between the four main components (land, atmosphere, sea ice, and ocean), a coupler is defined in AOGCMs (Gent 2012).

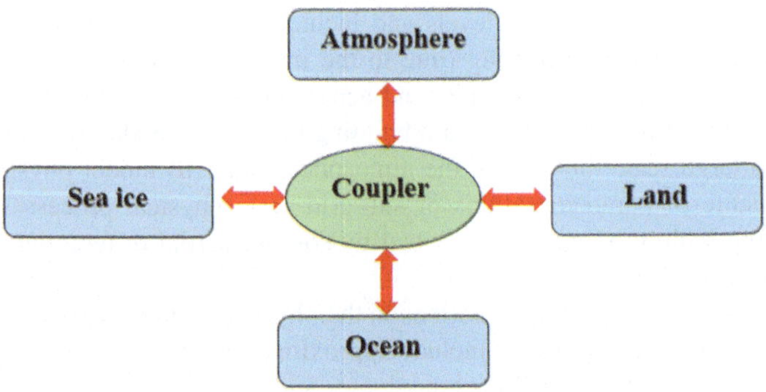

Fig. 6.4 Schematic representation of AOGCMs main components

According to Fig. 6.4, the coupler is a link that transfers information between different components in the fully coupled AOGCMs (Gent 2012). Additionally, in the case of having non-fully coupled AOGCMs, one or more of the components can be replaced by its data equivalent (observed data) rather than performing as a model (Gent 2012). For instance, the numerical ocean and ice components can be replaced by observed time series of surface, ocean, and sea ice temperatures. A coupler can transfer such observations to land and atmosphere models, effectively mimicking an AGCM.

A standardised set of model simulations (experiments) is essential for supporting a systematic evaluation of climate models and their simulated climate change projections. Within this context, over the past two decades, the international Model Intercomparison Projects (MIPs) have provided such an experimental setup, as discussed in the following sections (CSIRO and BOM 2015).

6.2 Coupled Model Inter-Comparison Project (CMIP)

In climate science, inter-comparison projects are based on the idea of running a set of numerical climate models under similar conditions to compare their simulation results (Touzé-Peiffer et al. 2020). Following this, the Coupled Model Intercomparison Project (CMIP) was created in the mid-1990s, and has evolved with different phases. The different phases of CMIP have played an important role in climate research and the synthesis reports of the IPCC, as they provided comprehensive insights into climate models (Touzé-Peiffer et al. 2020). The key advantages of applying standardized experiments are listed below (CSIRO and BOM 2015):

- Comparison of the simulations of the past with observations (assessment of model ability);
- Comparison of projections of different models for different future periods and forcing scenarios (assessment of the range of future climates);
- Identification of strengths and weaknesses of each model; and
- Identification of systematic errors that many or all models have.

In 1990, the Atmospheric Model Inter-comparison Project (AMIP) (the first major experiment) was created to compare the simulation results of AGCMs at seasonal and inter-annual scales (Gates 1992). Model results were analysed and compared under the same boundary conditions (e.g., standardized values of solar constant, CO_2 concentration, observed sea ice distributions, and observed mean sea surface temperature) over the 1979–1988 simulation period. Overall, AMIP aimed to systematically compare and validate the AGCMs (on seasonal and interannual scales) by identifying the differences between different AGCMs and comparing them with observations (Gates 1992).

After the successful validation of the 31 participating AGCMs in the AMIP experiment and recognition of the ability of such experiments in diagnosis, validation, and inter-comparison of global atmospheric models, subsequent inter-comparison projects, specifically CMIPs, have been carried out (Gates et al. 1999; Touzé-Peiffer et al. 2020). In 1996, CMIP1 was initiated to evaluate the AGCMs with the aim of applying the main objectives of AMIP to coupled models (AOGCMs) (Touzé-Peiffer et al. 2020). In 1997, CMIP 2 was created to evaluate the performance of coupled climate models in climate change simulations under a 1% increase of CO_2 annually (Houghton et al. 1992). As the importance of applying adaptation and mitigation strategies became more widely appreciated, CMIP3 was developed, considering future climate change projections under different emission scenarios (Touzé-Peiffer et al. 2020) which included three emission scenarios (B1, A1B, and A2) from the Special Report on Emission Scenarios (SRES). The CMIP3 outcomes were released in the IPCC Fourth Assessment Report (AR4), and also, for the first time among CMIP phases, open access was given to the CMIP3 data, so that many different research groups could analyze them from different perspectives (Meehl et al. 2007). After the success of CMIP3 in climate change projections, CMIP5 was introduced as the most extensive among CMIPs, which enhanced a standard set of model simulations for (i) evaluating the accuracy of climate models in simulating the recent past; (ii) predicting future climate change in the long-term (out to 2100 and beyond) and near-term (out to about 2035) time scales; and (iii) identifying the uncertainty in model projections (Jahandideh-Tehrani et al. 2019). The main difference between CMIP3 and CMIP5 is the inclusion of four new pathways, called Representative Concentration Pathways (RCPs) that replaced the SRES emission scenarios (IPCC 2014; Jahandideh-Tehrani et al. 2019). RCPs were introduced, based on scientific collaboration between climate modellers, emission experts, and ecosystem modellers that considered a low-emission mitigation scenario for the first time (Knutti and Sedláček 2013; IPCC 2014; Jahandideh-Tehrani et al. 2021). The term "representative" represents that RCPs are indicative of a larger set of scenarios in the available scientific literature. The term "concentration pathway" emphasizes that RCPs can be potentially enhanced in terms of fully integrated socio-economic, emission, and climate projections. Radiative forcing levels of + 8.5, + 6, + 4.5, and + 2.6 W/m^2 in the year 2100 compared to pre-industrial values were determined for (1) RCP 8.5, (2) RCP 6, (3) RCP 4.5, and (4) RCP 2.6, respectively (van Vuuren et al. 2011).

CMIP6 is the next generation of climate models that will be introduced in the upcoming 2021 IPCC sixth assessment report (AR6). It is directly based on societal aspects regarding climate change mitigation, adaptation, or impacts. In the upcoming CMIP6, a new set of emissions and land use scenarios will be applied with Integrated Assessment Models (IAMs) that are based on new future pathways of societal development. In CMIP6, the RCPs will be replaced by "Shared Socioeconomic Pathways" (SSPs), SP1-2.6, SSP2-4.5, SSP4-6.0, and SSP5-8.5, where the names still refer to 2100 radiative forcing (O'Neill et al. 2016). Therefore, the main differences between CMIP5 and CMIP6 are that CMIP6 includes updated versions of climate models, SSP-based scenarios with updated versions of IAM, and recent emission trends (Eyring et al. 2016). SSPs 1 and 5 represent relatively optimistic predictions for human development (investment in education and health), rapid economic growth, and well-operated institutions. SSP1 indicates an increasing trend for sustainable practices, while SSP5 refers to a fossil-based economy. SSP2 represents the continuation of historical patterns without significant changes. SSP 3 and 4 indicate more pessimistic developments, with a fast growing population and less human development (less investment in education and health) (O'Neill et al. 2016).

6.3 Examples of CMIP5 Climate Models

This section gives examples of common CMIP5 climate models and their features. Table 6.1 indicates some examples of CMIP5 climate models, including the institutes and countries of origin, and atmospheric and oceanic resolution.

A summary of features of each model will be discussed in the following subsections.

6.3.1 BCC-CSM1.1

The BCC-CSM1.1 stands for the Beijing Climate Center Climate System Model, developed by the Beijing Climate Center (BCC) and China Meteorological Administration (CMA) (Table 6.1). This model is a fully coupled climate-carbon cycle model, which simulates dynamic vegetation and global terrestrial, and oceanic carbon cycles (Wu et al. 2014). BCC-CSM1.1 is supposed to have good performance in simulating the global carbon cycle, as it properly simulates the upper tropospheric jet streams and corresponding transient vortex activity in East Asia (Xiao and Zhang 2012; Wu et al. 2014). According to the decadal prediction experiments performed under CMIP5, this model can accurately forecast global mean surface air temperature at decadal scales. Additionally, this model well-predicted decadal climate in mid and high latitudes (the Indian Ocean in Southern Hemisphere), the tropical Atlantic, and the tropical West Pacific (Gao et al. 2012). Zhao et al. (2013) claimed that BCC-CSM1.1 could potentially capture the basic characteristics of the atmospheric circulation, as the correlation between simulated and observed precipitation is more than 0.8. Wu et al. (2014) demonstrated that BCC-CSM1.1 could also capture trends in global mean temperature as the degree of warming simulated by this model (0.45 °C) is close to the multi-model mean (0.48 °C) over 2000–2005. Overall, this model has the capability to simulate inter-annual changes, long-term trends, and mean climate state under anthropogenic carbon emission scenarios (Wu et al. 2014).

6.3.2 BNU-ESM

The Beijing Normal University—Earth System Model (BNU-ESM) is an atmosphere–land–sea-ice fully coupled model for evaluating the mechanism of ocean–atmosphere interaction, carbon-climate interaction, and natural climate variability at

Table 6.1 Examples of CMIP5 models with the institute and country of origin, and oceanic and atmospheric horizontal resolution

CMIP5 model	Institute and country of origin	Resolution	
		Atmosphere (°LAT × °LON)	Ocean (°LAT × °LON)
BCC-CSM1.1	Beijing Climate Center, China Meteorological Administration (BCC, CMA), China	1.0 × 1.0	2.8 × 2.8
BNU-ESM	Beijing Normal University (BNU), China	0.9 × 1.0	2.8 × 2.8
CCSM4	National Center for Atmospheric Research (NCAR), USA	1.1 × 0.6	1.2 × 0.9
CanESM2	Canadian Centre for Climate Modelling and Analysis (CCCMA), Canada	1.4 × 0.9	2.8 × 2.8
CNRM-CM5	Centre National de Researchers Météorologiques-Centre Européen de Recherche et Formation Avancée en Calcul Scientfique (CNRM-CERFACS), France	1.0 × 0.8	1.4 × 1.4
CSIRO-Mk3.6.0	Commonwealth Scientific and Industrial Research Organisation-Queensland Climate Change Centre of Excellence (CSIRO-QCCCE), Australia	1.9 × 0.9	1.9 × 1.9
FGOALS-g 2	The State Key Laboratory of Numerical Modeling for Atmospheric Sciences and Geophysical Fluid Dynamics/Institute of Atmospheric Physics (LASG/IAP), China	2.8 × 2.8	1.0 × 1.0
FIO-ESM 1.0	First Institute of Oceanography-State Oceanic Administration (FIO—SOA), China	1.1 × 0.6	2.8 × 2.8
GFDL CM3	National Oceanic and Atmospheric Administratio—Geophysical Fluid Dynamics Laboratory (NOAA—GFDL), USA	1.0 × 1.0	2.5 × 2.0

(continued)

Table 6.1 (continued)

CMIP5 model	Institute and country of origin	Resolution	
		Atmosphere (°LAT × °LON)	Ocean (°LAT × °LON)
GISS-E2-H	National Aeronautics and Space Administration/ Goddard Institute for Space Studies (NASA/GISS), NY, USA	2.5 × 2.0	2.5 × 2.0
GISS-E2-H-CC	NASA/GISS, NY, USA	1.0 × 1.0	1.0 × 1.0
GISS-E2-R	NASA/GISS, NY, USA	2.5 × 2.0	2.5 × 2.0
GISS-E2-R-CC	NASA/GISS, NY, USA	1.0 × 1.0	1.0 × 1.0
HadGEM2-AO	National Institute of Meteorological Research—Korea Meteorological Administration (NIMR-KMA), Korea	1.0 × 1.0	1.9 × 1.2
MPI-ESM-LR	Max Planck Institute (MPI), Germany	1.5 × 1.5	1.9 × 1.9
MPI-ESM-MR	MPI, Germany	0.4 × 0.4	1.9 × 1.9
MPI-ESM-P	MPI, Germany	1.5 × 1.5	1.9 × 1.9
NorESM1-M	Norwegian Climate Centre (NCC), Norway	1.1 × 0.6	2.5 × 1.9

both inter-annual to inter-decadal time scales. This model can simulate the annual climatological cycle of precipitation and surface air temperature and the annual cycle of tropical Pacific sea surface temperature (Ji et al. 2014). Overall, this model can simulate annual climatological cycles of surface air temperature and precipitation, while annual temperature and annual precipitation are underestimated and over-estimated, respectively, at a global scale (excluding Antarctica) (Ji et al. 2014). This model can also capture some tropical intra-seasonal oscillation features (e.g., a quadratic relationship between precipitation and zonal wind). However, Lin (2007) explained that BNU-EMS shows precipitation biases over the tropical ocean that require improvement.

6.3.3 CCSM4

The fourth version of the Community Climate System Model (CCSM4.0) was developed by the National Center for Atmospheric Research (NCAR). This model was mainly developed to address the shortcomings of CCSM3 over El Niño–Southern Oscillation (ENSO) periodicity as ENSO was dominated by 2-year variability rather than the 3–7 year period seen in observations (Gent et al. 2011). Therefore, the deep convection parameterization in the atmosphere component was modified to improve the performance of the CCSM4.0 model over the ENSO period (Richter and Rasch 2008; Gent et al. 2011). As a result of such improvement, precipitation extreme event statistics over tropical land became more realistic (Lawrence et al. 2012). Additionally, CCSM4 improved the representation of surface air temperature and extreme events compared to CCSM3.0 (Infanti and Kirtman 2016), while Mei and Wang (2012) claimed that CCSM4.0 had less efficient land–atmosphere coupling than CCSM3.0. This poor performance of CCSM4.0 in land–atmosphere coupling affects the prediction skills of variables, including temperature and precipitation (Infanti and Kirtman 2016).

In another study by Lawrence et al. (2012), it has been demonstrated that both CCSM3.0 and CCSM4.0 show relatively reasonable performance in simulations. They indicated that CCSM4.0 improved simulations of the annual cycle of surface air temperature over Alaska and India, while less accurate simulations were observed over Europe than by the CCSM3.0. In terms of average annual cycles of precipitation, Lawrence et al. (2012) showed that simulation of wet season precipitation and monsoon rainfall had improved in the Amazon and India, respectively. However, significant wet biases have been observed in central and southern Africa and Australia (Lawrence et al. 2012).

6.3.4 CanESM2

The Canadian Earth System Model version 2 (CanESM2) was developed by Canadian Centre for Climate Modelling and Analysis (CCCMA) by combining the Canadian Centre for Climate Modelling and Analysis Version 4 (CanCM4) model and the terrestrial carbon cycle (Chylek et al. 2011). This model can adequately represent ENSO events and represents well sea surface temperature (CSIRO and BOM 2015). However, Sheffield et al. (2013) indicated that this model can properly simulate the observed variability over North America at intra-seasonal and decadal time scales.

6.3.5 CNRM-CM5

The Centre National de Recherches Météorologiques—Climate Model version 5 (CNRM-CM5) has been developed by Centre National de Researchers Météorologiques- Centre Européen de Recherche et Formation Avancée en Calcul Scientfique (CNRM-CERFACS) in France. This model has been developed to address the shortcomings of its previous version (CNRM-CM3) (Voldoire et al. 2013). The main deficiency of the CNRM-CM3 model is its bias in simulating tropical sea surface temperature, and the mean precipitation is significantly overestimated in the Inter-Tropical Convergence Zone (ITCZ) (Waliser et al. 2007). Additionally, ENSO patterns were not realistic in version 4, which have been improved in version 5 (Salas et al. 2011). In the CNRM-CM5 climate model, the horizontal resolution has been increased compared to version 3. Additionally, a new dynamical core, a new radiative scheme, water conservation, and an improved treatment for ozone and aerosols have been added to version 5 of this model. Therefore, a significant reduction of biases in terms of surface mean temperature, sea level pressure was observed. Another improvement of CNRM-CM5 was adding the Caspian Sea (the largest inland water body in the world) to the model (Voldoire et al. 2013). Despite these improvements in version 5, this model still shows precipitation biases in many regions (Salas and Melia et al. 2011).

6.3.6 CSIRO-Mk3.6.0

The Commonwealth Scientific and Industrial Research Organisation (CSIRO) and Queensland Climate Change Centre of Excellence (QCCCE) developed the CSIRO-Mk3.6.0 climate model. This model includes dynamical soil canopy and sea ice components with designated vegetation properties (Gordon et al. 2002). This model is updated from its recent predecessors Mk3.0 and Mk3.5. In Mk3.6, an interactive aerosol scheme has been added to explicitly treat sulfate, sea salt, dust, and carbonaceous aerosol (Syktus et al. 2011). Additionally, the radiation scheme and atmospheric physical components have been updated. That enables this model to investigate the impact of aerosol on climate (Syktus et al. 2011). Within this context, Rotstayn et al. (2011) reported that mineral dust played a key role in simulating rainfall variability that corresponded with ENSO in Australia.

6.3.7 FGOALS-g2

Flexible Global Ocean–Atmosphere-Land System Model: Grid-point Version 2 (FGOALS-g2) was developed by the State Key Laboratory of Numerical Modeling for Atmospheric Sciences and Geophysical Fluid Dynamics/Institute of Atmospheric Physics (LASG/IAP) in China. This model better simulates the precipitation frequency of the tropical land and East Asian Monsoon (Li et al. 2013). Additionally, simulations of the seasonal cycle of East Asian Monsoon and ENSO have improved through the implementation of an updated cumulus parameterization scheme. The associated uncertainties with some key parameters in shallow and deep convection schemes, cloud fraction, cloud macro/microphysical processes and the boundary layer scheme in the atmospheric model have been addressed. Similar to many other CMIP5 models, the simulations of the annual cycle of sea surface temperature and ENSO in the Tropical Pacific have improved (Li et al. 2013).

6.3.8 FIO-ESM 1.0

The First Institute of Oceanography-State Oceanic Earth System Model version 1.0 (FIO-ESM 1.0) was developed by the First Institute of Oceanography-State Oceanic Administration (FIO–SOA) in China. This model, which includes coupled physical climate and carbon cycle sub-models, used the theory of non-breaking surface wave-induced vertical mixing to investigate ocean surface wave impacts (Qiao et al. 2010, 2013; Song et al. 2020). The model is able to predict sea surface temperature and precipitation over the North Pacific with high accuracy, which improved precipitation simulations through the air-sea interaction (Chen et al. 2016). In addition, the model simulated the overall pattern of surface air temperature, precipitation, and ENSO, although it can be improved (Qiao et al. 2013).

6.3.9 GFDL CM3

The Geophysical Fluid Dynamics Laboratory Climate Model version 3 (GFDL CM3) was developed by the National Oceanic and Atmospheric Administration—Geophysical Fluid Dynamics Laboratory (NOAA—GFDL) in the USA. In this model, the aerosol-cloud interactions, chemistry-climate interactions, and links between the troposphere and stratosphere have been improved. The CM3 version indicated more accurate sea surface temperature simulations in high latitudes compared to the previous version (CM2.1) (Griffies et al. 2011). Additionally, implementation of more physically realistic sea ice albedo in CM3 enables the model to improve the Arctic Sea level pressure patterns and shortwave heating increase in the North Pacific (Donner et al. 2011; Griffies et al. 2011). In CM3, the bias in simulating sea surface temperature of South America has decreased due to wind stresses leading to more coastally concentrated upwelling (Griffies et al. 2011).

6.3.10 GISS-E2

The Goddard Institute for Space Studies-E2 (GISS) model was developed by the National Aeronautics and Space Administration/Goddard Institute for Space Studies (NASA/GISS) in the USA. This model consists of three versions for the atmospheric model, a non-interactive version (NINT) and two TCAD versions that include fully interactive aerosols, whole-atmosphere chemistry, and parameterized first indirect aerosol effect on clouds. Each atmospheric model version can be coupled with one of two ocean models, the Russell ocean model and the HYbrid Coordinate Ocean Model (HYCOM) (Nazarenko et al. 2015). Therefore, R and H in GISS-E2-R and GISS-E2-H models denote the Russell ocean model and HYCOM, respectively (Schmidt et al. 2014). Concerning that the GISS-E2-H model overestimates the seasonal cycle of the Arctic sea ice, smaller initial sea ice conditions were observed in the Arctic for all RCP scenarios. Therefore, GISS-E2-H model simulated rapid sea ice melting and an ice-free summer Arctic. In addition, E2-H model simulated warmer surface air temperature resulting in reduced ocean heat uptake compared to E2-R. Overall, GISS models (e.g., GISS-E2-R) considered a too warm condition with a deficiency in Antarctic Sea ice in Southern oceans (Nazarenko et al. 2015).

6.3.11 HadGEM2-AO

The Hadley Centre Global Environmental Model version 2—Atmosphere–Ocean (HadGEM2-AO) was developed by the National Institute of Meteorological Research—Korea Meteorological Administration (NIMR-KMA) in Korea. This model has been developed to address the shortcomings of HadGEM1 model, such as poor ESNO simulation (lack of Southern Oscillation (SO) response), poor tropical variability of sea surface temperature, and land-surface temperature biases in northern continents. In the atmosphere model, to address the stated deficiencies, an adaptive detrainment parameterization was implemented in the convection scheme to improve the tropical convection simulation in the Tropical Pacific. Additionally, in the ocean model, the background ocean diffusivity profile in the thermocline has been changed after identifying a 1-year drift in the global sea surface temperature. As a result of such improvements, the El-Nino variability and Southern Ocean response in the atmosphere become more realistic (Collins et al. 2011).

6.3.12 MPI-ESM (MPI-ESM-LR, MPI-ESM-MR, and MPI-ESM-P)

The Max Planck Institute (MPI), Germany, developed the Max Planck Institute Earth System Model (MPI-ESM) model. This model is the coupled version of the European Centre Hamburg Model (ECHAM6) and Max Planck Institute Ocean Model (MPIOM), and marine biogeochemistry, land, and vegetation models (Giorgetta et al. 2013). In order to improve the MPI-ESM, a carbon cycle has also been added over the model coupling process (Jungclaus et al. 2006). MPT-ESM includes different configurations in terms of resolution, including MPI-ESM-LR (low resolution), MPI-ESM-MR (mixed resolution), and MPI-ESM-P (Paleo). The LR version has been widely used in CMIP5. In contrast, the more computationally intensive MR version has been implemented in CMIP5 with fewer start dates for decadal predictions, and is not used for experiments driven by CO_2 emissions. The P version was implemented for CMIP5 paleo experiments and long-term core experiments (Giorgetta et al. 2013). This model can simulate the tropical variability for the Madden-Julian Oscillation and ENSO patterns with high accuracy.

6.3.13 NorESM1-M

Norwegian Climate Centre (NCC), Norway developed Norwegian Climate Center's Earth System Model version 1 (NorESM1-M) with intermediate resolution. This model was developed, based on the Community Climate System Model version 4 (CCSM4) (Iversen et al. 2013). NorESM1-M underestimates the sensible heat flux in most of the African continent south of the Sahara, in the west coast of India, in Australia, and the western part of the United States. At the same time, it overestimates the sensible heat flux in the extreme eastern part of South America. Like many other climate models, the NorESM1-M reduces precipitation biases in the tropics (along the equator in the Pacific) (Bentsen et al. 2013).

6.4 Climate Projections

According to the fifth IPCC report (2014), the surface temperature is projected to increase over 2081–2100 compared to 1986–2005. However, the rate of future temperature changes significantly depends on the future anthropogenic emissions, and the choice of emission scenarios. The likely range of changes in mean surface temperature and sea-level rise is shown in Tables 6.2 and 6.3, respectively.

As shown in Table 6.2, the global temperature is likely to increase more than 1.5 °C by the end of the twenty-first century under RCP4.5, RCP6.0, and RCP8.5. According to Table 6.3, the sea-level rise is projected to likely exceed 0.30 m over 2081–2100 (by the end of the twenty-first century) under RCP4.5, RCP6.0 and RCP8.5 (IPCC 2014). Moreover, these sea-level rise projections may be underestimates due to the lack of ice sheet dynamics in the CMIP5 models.

Table 6.2 The likely changes of global mean surface temperature (°C) relative to 1986–2005

Scenario period	2046–2065	2081–2100
RCP2.6	0.4–1.6	0.3–1.7
RCP4.5	0.4–1.6	1.1–2.6
RCP6.0	0.4–1.6	1.4–3.1
RCP8.5	0.4–1.6	2.6–4.8

Table 6.3 The likely changes of sea level (m) relative to 1986–2005

Scenario period	2046–2065	2081–2100
RCP2.6	0.17–0.32	0.26–0.55
RCP4.5	0.19–0.33	0.32–0.63
RCP6.0	0.18–0.32	0.33–0.63
RCP8.5	0.22–0.38	0.45–0.82

The multi-model mean projections (CMIP5) of changes in temperature, precipitation, and sea level for the period 2081–2100 under RCP2.6 and RCP8.5 are presented in Fig. 6.5.

As shown in Fig. 6.5, the Arctic region is likely to experience particularly rapid warming (very high confidence). Additionally, land shows higher mean warming than the ocean (IPCC 2014). Figure 6.5 also represents that precipitation changes are

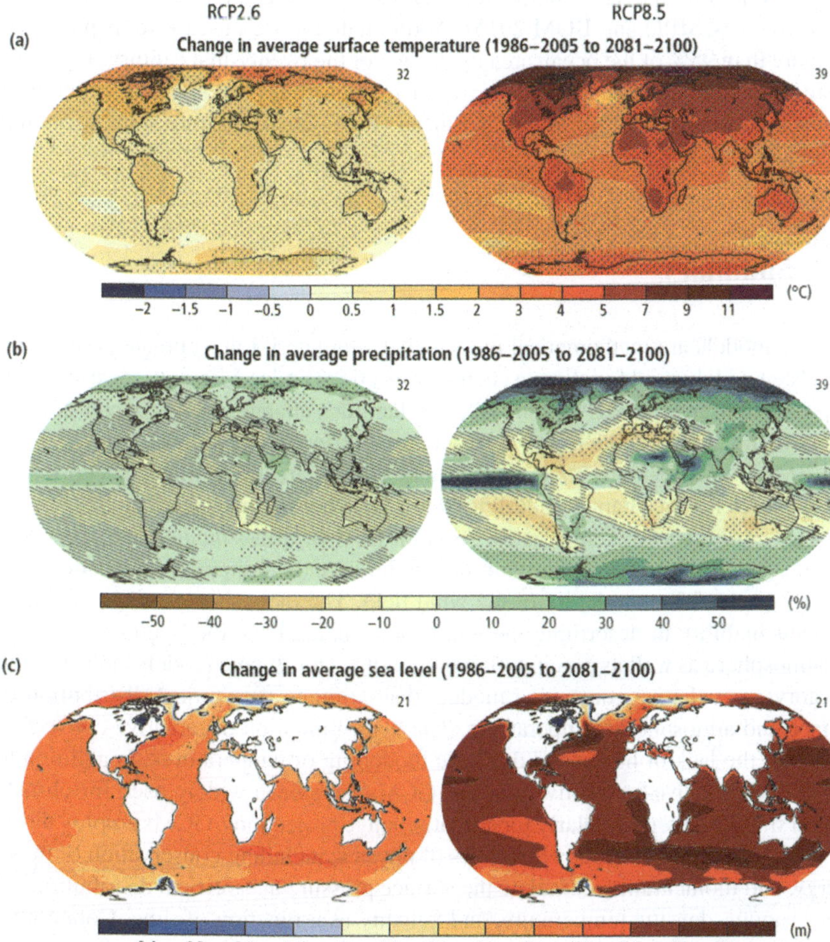

Fig. 6.5 Multi-model mean projections (CMIP5) over 2081–2100 under RCP2.6 (left) and RCP8.8 (right) for **a** annual mean surface temperature changes (°C), **b** annual mean precipitation changes (%), and **c** average sea level (m) over 2081–2100 relative to 1986–2005. On the upper right corner of each row, the number of used CMIP5 is shown. Stippling (dots) denotes regions with large projected changes (greater than two standard deviations of internal variability in 20-year means) and 90% of model changes. Hatching (diagonal lines) denotes regions with low projected changes (less than one standard deviation of natural internal variability in 20-year means) (*Source* IPCC 2014)

not uniform in high latitudes, and the equatorial Pacific is projected to experience increased annual mean precipitation by 2100 under RCP8.5. However, mid-latitude and subtropical regions are likely to experience decreased mean precipitation under RCP8.5. Overall, all RCPs represent that regions with dominated monsoon systems will increase precipitation, and more intensive precipitation is projected due to ENSO patterns (IPCC 2014). This means that the Tropical Pacific will remain dominated by ENSO, and global warming can potentially intensify El Niño-driven drying and El Niño-driven rainfall over western equatorial Pacific, and central and eastern Pacific, respectively (CSIRO and BOM 2015). Additionally, a rise in sea level is projected for more than 95% of the ocean area by the end of the twenty-first century (Fig. 6.5). More specifically, the sea level of 70% of the coastlines are likely to experience changes between ±20% of the global mean (IPCC 2014), despite some local and regional variations.

6.5 Summary

Climate models are mathematical models that simulate Earth's climate system and have been widely used for climate change projections under different scenarios, such as population growth, GHG emission, and land use changes. This chapter aims at synthesizing climate models and corresponding different categories.

EBMs, radiative-convective models and GCMs are three main categories of climate models. The chapter underlines that EBMs consists of a set of coupled equations for Earth's surface temperature generation. Also, EBMs include one dependent variable (the Earth's near-surface air temperature), and eliminate the dependency of climate on the full complexity of the wind field. The main shortcomings of EBMs are their inability in describing the temperature field above the boundary layer of the atmosphere as well as precipitation modelling. The climate models in the second category are radiative-convective models that compute the vertical distribution of surface and atmospheric temperatures. The main limitation of radiative-convective models is the lack of horizontal dynamic modelling of temperatures. In addition to EBMs and radiative-convective models, GCMs have been widely used for climate system simulations, particularly for climate change projections. GCMs employ three-dimensional approximations of the laws of physics, including conservation of mass, energy, and momentum to simulate the surface pressure and vertical distributions of water vapour, density, temperature, and velocity as a function of time. Concerning the wide application of GCMs in predicting future climate change conditions, this chapter also discusses three sub-classifications of GCMs, including AGCM, OGCM, and AOGCM. Within this context, AGCMs, OGCMs, and AOGCMs are numerical models that describe physical and dynamical process of the land–atmosphere, ocean circulation, and four main climate systems (land, atmosphere, sea ice, and ocean), respectively. Additionally, in order to systematically evaluate the climate models and climate change projections, CMIPs have provided experimental setup to run a set of numerical climate models under similar conditions with the aim of comparing model results. Different phases of CMIP (e.g., CMIP3, CMIP5, and CMIP6) played a critical role in climate search and IPCC synthesis reports. The key benefits of CMIP phases are: (i) comparison of historical climate simulations with observations; (ii) comparison of projections of different models; (iii) identification of strengths and weaknesses of each model; and (iv) identification of systematic errors of models. The chapter ends with a description of CMIP5 climate model examples and key overall global climate change projections.

References

Becker T, Wing AA (2020) Understanding the extreme spread in climate sensitivity within the radiative-convective equilibrium model intercomparison project. J Adv Model Earth Syst 12(10). https://doi.org/10.1029/2020MS002165

Bentsen M, Bethke I, Debernard JB, Iversen T, Kirkevåg A, Seland Ø, Drange H, Roelandtm C, Seierstad IA, Hoose C, Kristjánsson JE (2013) The Norwegian Earth System Model, NorESM1-M–Part 1: description and basic evaluation of the physical climate. Geosci Model Dev\ 6(3):687–720. https://doi.org/10.5194/gmd-6-687-2013

Budyko MI (1969) The effect of solar radiation variations on the climate of the Earth. Tellus 21(5):611–619. https://doi.org/10.3402/tellusa.v21i5.10109

CSIRO and BOM (2015) Climate change in Australia information for Australia's natural resource management regions: Technical Report, CSIRO and Bureau of Meteorology, Australia. https://www.climatechangeinaustralia.gov.au/media/ccia/2.1.6/cms_page_media/168/CCIA_2015_NRM_TechnicalReport_WEB.pdf

Chen H, Yin XQ, Bao Y, Qiao FL (2016) Ocean satellite data assimilation experiments in FIO-ESM using ensemble adjustment Kalman filter. Sci China Earth Sci 59(3):484–494. https://doi.org/10.1007/s11430-015-5187-2

Chylek P, Li J, Dubey MK, Wang M, Lesins G (2011) Observed and model simulated 20th century ARCTIC TEMPERATURE VARIABILIty: Canadian earth system model CanESM2. Atmos Chem Phys Discuss 11(8):22893–22907. https://doi.org/10.5194/acpd-11-22893-2011

Collins WJ, Bellouin N, Doutriaux-Boucher M, Gedney N, Halloran P, Hinton T, Hughes J, Jones CD, Joshi M, Liddicoat S, Martin G, O'Connor F, Rae J, Senior C, Sitch S, Totterdell I, Wiltshire A, Woodward S (2011) Development and evaluation of an earth-system model-HadGEM2. Geosci Model Dev 4(4):1051–1075. https://doi.org/10.5194/gmd-4-1051-2011

Donner LJ, Wyman BL, Hemler RS, Horowitz LW, Ming Y, Zhao M, Golaz JC, Ginoux P, Lin SJ, Daniel Schwarzkopf M, Austin J, Alaka G, Cooke WF, Delworth TL, Freidenreich SM, Gordon CT, Griffies SM, Held IM, Hurlin WJ, Klein SA, Knutson TR, Langenhorst AR, Lee HC, Lin Y, Magi BI, Malyshev SL, Milly PCD, Naik V, Nath MJ, Pincus R, Ploshay JJ, Ramaswamy V, Seman CJ, Shevliakova E, Sirutis JJ, Stern WF, Stouffer RJ, Wilson RJ, Winton M, Wittenberg AT, Zeng F (2011) The dynamical core, physical parameterizations, and basic simulation characteristics of the atmospheric component AM3 of the GFDL global coupled model CM3. J Clim 24(13):3484–3519. https://doi.org/10.1175/2011JCLI3955.1

Edwards PN (2011) History of climate modeling. Wiley Interdisciplinary Reviews: Climate Change 2(1):128–139

Eyring V, Bony S, Meehl GA, Senior CA, Stevens B, Stouffer RJ, Taylor KE (2016) Overview of the coupled model intercomparison project phase 6 (CMIP6) experimental design and organization. Geosci Model Dev 9(5):1937–1958. https://doi.org/10.5194/gmd-9-1937-2016

Gates WL (1992) An AMS continuing series: global change–AMIP: the atmospheric model intercomparison project. Bull Am Meteor Soc 73(12):1962–1970. https://doi.org/10.1175/1520 0477(1992)073%3c1962.ATAMIP%3e2.0.CO;2

Gent RP (2012) Coupled climate and earth system models. In Rasch P (ed) Climate change modeling methodology. Springer, New York, NY

Gates WL, Boyle JS, Covey C, Dease CG, Doutriaux CM, Drach RS, Fiorino M, Gleckler PJ, Hnilo JJ, Marlais SM, Phillips TJ, Potter GL, Santer BD, Sperber KR, Taylor KE, Williams DN (1999) An overview of the results of the atmospheric model intercomparison project (AMIP I). Bull Am Meteor Soc 80(1):29–56. https://doi.org/10.1175/1520-0477(1999)080%3c0029:AOOTRO%3e2.0.CO;2

Gent PR, Danabasoglu G, Donner LJ, Holland MM, Hunke EC, Jayne SR, Lawrence DM, Neale RB, Rasch PJ, Vertenstein M, Worley PH, Yang ZL, Zhang M (2011) The community climate system model version 4. J Clim 24(19):4973–4991. https://doi.org/10.1175/2011JCLI4083.1

Giorgetta MA, Jungclaus J, Reick CH, Legutke S, Bader J, Böttinger M, Brovkin V, Crueger T, Esch M, Fieg K, Glushak K, Gayler V, Haak H, Hollweg HD, Ilyina T, Kinne S, Kornblueh L, Matei D, Mauritsen T, Mikolajewicz U, Mueller W, Notz D, Pithan F, Raddatz T, Rast S, Redler R, Roeckner E, Schmidt H, Schnur R, Segschnerider J, Six KD, Stockhause M, Timmreck C, Wegner J, Widmann H, Wieners KH, Claussen M, Maroktzke J, Stevens B (2013) Climate and carbon cycle changes from 1850 to 2100 in MPI-ESM simulations for the coupled model intercomparison project phase 5. J Adv Model Earth Syst 5(3):572–597. https://doi.org/10.1002/jame.20038

Gordon HB, Rotstayn LD, McGregor JL, Dix MR, Kowalczyk EA, O'Farrell SP, Waterman LJ, Hirst AC, Wilson SG, Collier MA, Watterson IG, Elliott TI (2002) The CSIRO Mk3 climate system model. CSIRO Atmospheric Research Technical Paper No. 60

Griffies SM, Winton M, Donner LJ, Horowitz LW, Downes SM, Farneti R, Gnanadesikan A, Hurlin WJ, Lee HC, Liang Z, Palter JB, Samuels BL, Wittenberg AT, Wyman BL, Yin J, Zadeh N (2011) The GFDL CM3 coupled climate model: characteristics of the ocean and sea ice simulations. J Clim 24(13):3520–3544. https://doi.org/10.1175/2011JCLI3964.1

Gao F, Xin X, Wu T (2012) A study of the prediction of regional and global temperature on decadal time scale with BCC−CSM1.1 model. Chin J Atmos Sci 36(6):1165–1179. (in Chinese). https://doi.org/10.3878/j.issn.1006-9895.2012.11243

Houghton J, Callander B, Varney S (eds) (1992) Climate change 1992: the supplementary report to the IPCC scientific assessment. Cambridge University Press, Cambridge

IPCC (2007) Climate Change 2007: Synthesis Report. Contribution of Working Groups I, II and III to the Fourth Assessment Report of the Intergovernmental Panel on Climate Change [Core Writing Team, Pachauri RK, Reisinger A (eds)]. IPCC, Geneva, Switzerland, p 104

IPCC (2014) Contribution of working groups I, II and III to the fifth assessment report of the intergovernmental panel on climate change. Climate change 2014: synthesis report [Core Writing Team Pachauri RK, Meyer LA (eds)]. Cambridge University Press, Cambridge, IPCC, Geneva, Switzerland, p 151

Idso CD, Carter RM, Singer SF (eds) (2013) Climate change reconsidered II: physical science. The Heartland Institute, Chicago, IL

Infanti JM, Kirtman BP (2016) Prediction and predictability of land and atmosphere initialized CCSM4 climate forecasts over North America. J Geophys Res 121(21):12690–12701. https://doi.org/10.1002/2016JD024932

Iversen T, Bentsen M, Bethke I, Debernard JB, Kirkevåg A, Seland Ø, Drange H, Kristjansson JE, Medhaug I, Sand M, Seierstad IA (2013) The Norwegian earth system model, NorESM1-M – Part 2: climate response and scenario projections. Geosci Model Dev 6(2):389–415. https://doi.org/10.5194/gmd-6-389-2013

Jahandideh-Tehrani M, Helfer F, Jenkins G (2021) Impacts of climate change and sea level rise on catchment management: a multi-model ensemble analysis of the Nerang River catchment. Australia Sci Total Environ 777. https://doi.org/10.1016/j.scitotenv.2021.146223

Jahandideh-Tehrani M, Zhang H, Helfer F, Yu Y (2019) Review of climate change impacts on predicted river streamflow in tropical rivers. Environ Monitor Assess 191(12). https://doi.org/10.1007/s10661-019-7841-1

Jungclaus JH, Keenlyside N, Botzet M, Haak H, Luo JJ, Latif M, Marotzke J, Mikolajewicz U, Roeckner E (2006) Ocean circulation and tropical variability in the coupled model ECHAM5/MPI-OM. J Clim 19(16):3952–3972. https://doi.org/10.1175/JCLI3827.1

Ji D, Wang L, Feng J, Wu Q, Cheng H, Zhang Q, Yang J, Dong W, Dai Y, Gong D, Zhang R, Wang X, Liu J, Moore JC, Chen D, Zhou M (2014) Description and basic evaluation of BNU-ESM version 1. Geosci Model Dev Discuss 7(2):1601–1647. https://doi.org/10.5194/gmdd-7-1601-2014

Knutti R, Sedláček J (2013) Robustness and uncertainties in the new CMIP5 climate model projections. Nat Clim Chang 3(4):369–373. https://doi.org/10.1038/nclimate1716

Lupo A, Kininmonth W (2013) Global climate models and their limitations. Climate Change Reconsidered II: Physical Science, pp 7–148

Lawrence DM, Oleson KW, Flanner MG, Fletcher CG, Lawrence PJ, Levis S, Swenson SC, Bonan GB (2012) The CCSM4 land simulation, 1850–2005: assessment of surface climate and new capabilities. J Clim 25(7):2240–2260. https://doi.org/10.1175/JCLI-D-11-00103.1

Li L, Lin P, Yu Y, Wang B, Zhou T, Liu L, Liu J, Bao Q, Xu S, Huang W, Xia K, Pu Y, Dong L, Shen S, Liu Y, Hu N, Liu M, Sun W, Shi X, Zheng W, Wu B, Song M, Liu H, Zhang X, Wu G, Xue W, Huang X, Yang G, Song Z, Qiao F (2013) The flexible global ocean-atmosphere-land system model, Grid-point Version 2: FGOALS-g2. Adv Atmos Sci 30(3):543–560. https://doi.org/10.1007/s00376-012-2140-6

Lin JL (2007) The double-ITCZ problem in IPCC AR4 coupled GCMs: ocean-atmosphere feedback analysis. J Clim 20(18):4497–4525. https://doi.org/10.1175/JCLI4272.1

Meehl GA, Covey C, Delworth T, Latif M, McAvaney B, Mitchell JFB, Stouffer RJ, Taylor KE (2007) The WCRP CMIP3 multimodel dataset: a new era in climate change research. Bull Am Meteor Soc 88(9):1383–1394. https://doi.org/10.1175/BAMS-88-9-1383

Madec G, Delecluse P, Imbard M, Lévy C (1998) OPA 8.1 Ocean general circulation model reference manual. Note du Pôle de modélisation, Institut Pierre-Simon Laplace (IPSL), France, N°11

McGuffie K, Henderson-Sellers A (2005) Energy balance models. A climate modelling primer, vol 1. https://doi.org/10.1002/0470857617.ch3

Manabe S, Wetherald RT (1967) Thermal equilibrium of the atmosphere with a given distribution of relative humidity. J Atmos Sci 24(3):241–259. https://doi.org/10.1175/1520-0469(1967)024<0241:TEOTAW>2.0.CO;2

Mei R, Wang G (2012) Summer land-atmosphere coupling strength in the United States: comparison among observations, reanalysis data, and numerical models. J Hydrometeorol 13(3):1010–1022. https://doi.org/10.1175/JHM-D-11-075.1

Njoku G (2014) Encyclopedia of remote sensing. Springer, New York, NY. https://doi.org/10.1007/978-0-387-36699-9

North GR, Kim K (2017) Energy balance climate models. John Wiley & Sons. https://doi.org/10.1002/9783527698844.ch1

Nazarenko L, Schmidt GA, Miller RL, Tausnev N, Kelley M, Ruedy R, Russell GL, Aleinov I, Bauer M, Bauer S, Bleck R, Canuto V, Cheng Y, Clune TL, Del Genio AD, Faluvegi G, Hansen JE, Healy RJ, Kiang NY, Koch D, Lacis AA, LeGrande AN, Lerner J, Lo KK, Menon S, Oinas V, Perlwitz J, Puma MJ, Rind D, Romanou A, Sato M, Shindell DT, Sun S, Tsigaridis K, Unger N, Voulgarakis A, Yao MS, Zhang J (2015) Future climate change under RCP emission scenarios with GISSModelE2. J Adv Model Earth Syst 7(1):244–267. https://doi.org/10.1002/2014MS000403

O'Neill BC, Tebaldi C, Van Vuuren DP, Eyring V, Friedlingstein P, Hurtt G, Knutti R, Kriegler E, Lamarque JF, Lowe J, Meehl GA, Moss R, Riahi K, Sanderson BM (2016) The Scenario Model Intercomparison Project (ScenarioMIP) for CMIP6. Geosci Model Dev 9(9):3461–3482. https://doi.org/10.5194/gmd-9-3461-2016

Popke D, Stevens B, Voigt A (2013) Climate and climate change in a radiative-convective equilibrium version of ECHAM6. J Adv Model Earth Syst 5(1):1–14. https://doi.org/10.1029/2012MS000191

Qiao F, Song Z, Bao Y, Song Y, Shu Q, Huang C, Zhao W (2013) Development and evaluation of an earth system model with surface gravity waves. J Geophys Res Oceans 118(9):4514–4524. https://doi.org/10.1002/jgrc.20327

Qiao F, Yuan Y, Ezer T, Xia C, Yang Y, Lü X, Song Z (2010) A three-dimensional surface wave-ocean circulation coupled model and its initial testing. Ocean Dyn 60(5):1339–1355. https://doi.org/10.1007/s10236-010-0326-y

Ramanathan V, Coakley JA (1978) Climate modeling through radiative-convective models. Rev Geophys 16(4):465–489. https://doi.org/10.1029/RG016i004p00465

Richter JH, Rasch PJ (2008) Effects of convective momentum transport on the atmospheric circulation in the community atmospheric model, Version 3. J Clim 21(7):1487–1499. https://doi.org/10.1175/2007JCLI1789.1

Roques L, Chekroun MD, Cristofol M, Soubeyrand S, Ghil M (2014) Parameter estimation for energy balance models with memory. Proceed Royal Soc A 470(2169). https://doi.org/10.1098/rspa.2014.0349

Rotstayn LD, Collier MA, Mitchell RM, Qin Y, Campbell SK, Dravitzki SM (2011) Simulated enhancement of ENSO-related rainfall variability due to Australian dust. Atmos Chem Phys 11(13):6575–6592. https://doi.org/10.5194/acp-11-6575-2011

Rypdal K, Rypdal M, Fredriksen HB (2015) Spatiotemporal long-range persistence in earth's temperature field: analysis of stochastic-diffusive energy balance models. J Clim 28(21):8379–8395. https://doi.org/10.1175/JCLI-D-15-0183.1

Seller WD (1969) A global climatic model based on the energy balance of the Earth atmosphere system. J Appl Meteorol Climatol 8(3):392–400. https://doi.org/10.1175/1520-0450(1969)008%3c0392:AGCMBO%3e2.0.CO;2

Schlesinger ME (1983) A review of climate models and their simulation of CO2-induced Warming. Int J Environ Stud 20(2):103–114. https://doi.org/10.1080/00207238308710024

Schmidt GA, Kelley M, Nazarenko L, Ruedy R, Russell GL, Aleinov I, Bauer M, Bauer SE, Bhat MK, Bleck R, Canuto V, Chen YH, Cheng Y, Clune TL, Del Genio A, Fainchtein RD, Faluvegi G, Hansen JE, Healy RJ, Kiang NY, Koch D, Lacis AA, LeGrande AN, Lerner J, Lo KK, Matthews EE, Menon S, Miller RL, Oinas V, Oloso A, Perlwitz JP, Puma MJ, Putman WM, Rind D, Romanou A, Sato M, Shindell DT, Sun S, Syed RA, Tausnev N, Tsigaridis K, Unger N, Voulgarakis A, Yao MS, Zhang J (2014) Journal of advances in modeling earth systems contributions to the CMIP5 archive. J Adv Model Earth Syst 6:141–184. https://doi.org/10.1002/2013MS000265

Sheffield J, Camargo SJ, Fu R, Hu Q, Jiang X, Johnson N, Karnauskas KB, Kim ST, Kinter J, Kumar S, Langenbrunner B, Maloney E, Mariotti A, Meyerson JE, Neelin JD, Nigam S, Pan Z, Ruiz-Barradas A, Seager R, Serra YL, Sun DZ, Wang C, Xie SP, Yu JY, Zhang T, Zhao M (2013) North American climate in CMIP5 experiments. Part II: Evaluation of historical simulations of intraseasonal to decadal variability. J Clim 26(23):9247–9290. https://doi.org/10.1175/JCLI-D-12-00593.1

Shine KP, Henderson-Sellers A (1983) Modelling climate and the nature of climate models: a review. J Climatol 3(1):81–94. https://doi.org/10.1002/joc.3370030107

Salas y Melia, D., Sanchez, E., Decharme, B., Fernandez, E., Cassou, C., Chevallier, M., Geoffroy, O., Senési, S., & Voldoire, A. (2011) Contributing to CMIP5 with CNRM-CM5: model evaluation and simulated climate future climate change. American Geophysical Union, Fall Meeting

Song Y, Zhao Y, Yin X, Bao Y, Qiao F (2020) Evaluation of FIO-ESM v1.0 Seasonal Prediction Skills Over the North Pacific. Front Marine Sci 7(July):1–10. https://doi.org/10.3389/fmars.2020.00504

Syktus J, Jeffrey S, Rotstayn L, Wong K, Toombs N, Dravitzki S, Collier M, Moeseneder C (2011) The CSIRO-QCCCE contribution to CMIP5 using the CSIRO Mk3.6 climate model. MODSIM 2011 - 19th International Congress on Modelling and Simulation-Sustaining Our Future: Understanding and Living with Uncertainty, (December), pp 2782–2788

Touzé-Peiffer L, Barberousse A, Treut HL (2020). The coupled model intercomparison project: history, uses, and structural effects on climate research. Adv Rev 11(4). https://doi.org/10.1002/wcc.648

Tompkins AM, Craig GC (1998) Radiative-convective equilibrium in a three-dimensional cloud-ensemble model. Q J R Meteorol Soc 124(550):2073–2097. https://doi.org/10.1002/qj.49712455013

van Vuuren DP, Edmonds J, Kainuma M, Riahi K, Thomson A, Hibbard K, Hurtt GC, Kram T, Krey V, Lamarque JF, Masui T, Meinshausen M, Nakicenovic N, Smith SJ, Rose SK (2011) The representative concentration pathways: an overview. Clim Change 109(1):5–31. https://doi.org/10.1007/s10584-011-0148-z

Voldoire A, Sanchez-Gomez E, Salas y Mélia D, Decharme B, Cassou C, Sénési S, Valcke S, Beau I, Alias A, Chevallier M, Déuqé M, Deshayes J, Douville H, Fernandez E, Madec G, Maisonnave E, Moine MP, Planton S, Saint-Martin D, Szopa S, Tyteca S, Alkama R, Belamari S, Braun A, Coquart L, Chauvin F (2013) The CNRM-CM5.1 global climate model: Description and basic evaluation. Clim Dyn 40(9–10):2091–2121. https://doi.org/10.1007/s00382-011-1259-y

Waliser D, Seo KW, Schubert S, Njoku E (2007) Global water cycle agreement in the climate models assessed in the IPCC AR4. Geophys Res Lett 34(16):2–7. https://doi.org/10.1029/2007GL030675

Wu T, Song L, Li W, Wang Z, Zhang H, Xin X, Zhang Y, Zhang L, Li J, Wu F, Liu Y, Zhang F, Shi X, Chu M, Zhang J, Fang Y, Wang F, Lu Y, Liu X, Wei M, Liu Q, Zhou W, Dong M, Zhao Q, Ji J, Li L, Zhou M (2014) An overview of BCC climate system model development and application for climate change studies. J Meteorol Res 28(1):34–56. https://doi.org/10.1007/s13351-014-3041-7

Xiao C, Zhang Y (2012) The East Asian upper-tropospheric jet streams and associated transient eddy activities simulated by a climate system model BCC−CSM1.1 Acta Meteorologica Sinica 26(6):700–716. https://doi.org/10.1007/s13351-012-0603-4

Zhao S, Zhi X, Zhang H (2013) A primary assessment of the simulated climatic state by a coupled aerosol-climate model BCC−AGCM2.0.1−CAM. Clim Environ Res 19(3):265–277. (in Chinese). https://doi.org/10.3878/j.issn.1006-9585.2012.12015

Chapter 7
Downscaling Methods

Arash Yoosefdoost, Omid Bozorg-Haddad, Jie Chen, Kwok Wing Chau, and Fahmida Khan

7.1 Introduction

Climatic changes can cause noticeable changes in water accessibility in different zone across the world. The researchers in water emphasize the importance of finding the way that climate change can affect water supplies. In most climate studies and climate projections, Global "Climate Models (GCMs)" are applied (Broderick 2012). These models are typically run at large spatial resolutions (about 150–200 km). These models also are limited in resolving crucial sub grid-scale features, so in local impact studies, GCM based projections may not be crucial. Downscaling methods are developed to solve this problem using regiona hich GCMs provide (Chithra and Thampi 2017).

A. Yoosefdoost
School of Engineering, University of Guelph, Guelph, ON, Canada

O. Bozorg-Haddad (✉)
Faculty of Agricultural Engineering and Technology, Department of Irrigation and Reclamation Engineering, College of Agriculture and Natural Resources, University of Tehran, Karaj, Tehran, Iran
e-mail: Obhaddad@ut.ac.ir

J. Chen
Department of Hydrology and Water Resources, School of Water Resources and Hydropower Engineering, Wuhan University, Wuhan, People's Republic of China
e-mail: jiechen@whu.edu.cn

K. W. Chau
Civil and Hydraulic Engineering, Hong Kong Polytechnic University, Hunghom, China
e-mail: dr.kwok-wing.chau@polyu.edu.hk

F. Khan
Department of Chemistry, National Institute of Technology Raipur, Raipur, India
e-mail: fkhan.chy@nitrr.ac.in

O. Bozorg-Haddad (ed.), *Climate Change in Sustainable Water Resources Management*, Springer Water, https://doi.org/10.1007/978-981-19-1898-8_7

7.2 Downscaling Methods

General Circulation Models (GCM) usually consider as accompanying numerical models. The atmospheric processes in these models are described via mathematical equations. These models were the main tools in most climate change studies in previous years. "various earth systems, such as the atmosphere, oceans, land surface, and sea ice" (Counsell 2018) are represented in GCMS. They also showed significant potential for climate change studies (Liu et al. 2016).

GCMs accepted noticeable change at large scales during the last decades. These models can easily simulate a global climate. These simulations are valuable to most areas (National Research Council (US) Committee on Climate 2001). The same models present weaker results in more minor temporal, local scales. GCM outputs are not reliable at sub-grid box scales and individual grids. Because their spatial resolution grids are too large to sort out many main sub grid-scale processes (Liu et al. 2016; Masson and Knutti 2011), it is essential to modify the bias to produce weather forecasts. The bias correction (BC) approach uses the difference in average and variability among observations and GCMS in the period of study to correct the projected raw daily GCM output (Fig. 7.1; Navarro-Racines and Tarapues 2015).

Raw model output in the future era is applied by BC. It corrects this data by using the differences (Δ) among historical reference data from observations and models. (Navarro-Racines and Tarapues 2015).

In (Fig. 7.1), OREF and T_{REF} are observations and GCM output in the historical reference period from the historical reference period in the historical reference time interval. T_{RAW} and T_{BC} are raw and bias-corrected GCM outputs for the historical or future era.

If we had a chance to equal the variability for GCMs and observations, change in daily date occur with average bias in the reference period (Hawkins et al. 2013; Navarro-Racines and Tarapues 2015) therefore:

$$T_{BC}(t) = T_{RAW}(t) - \overline{T_{REF}}$$

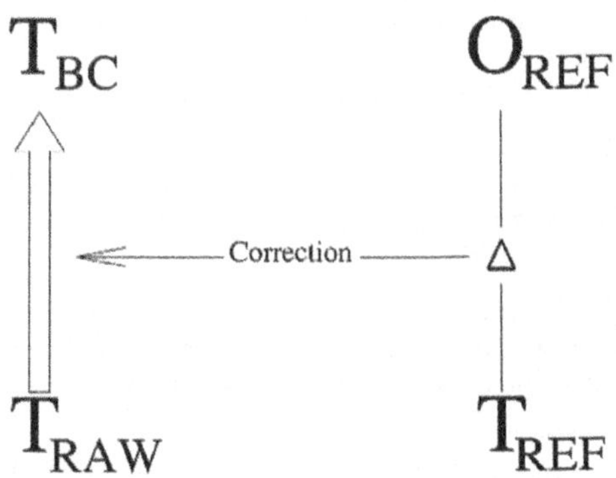

Fig. 7.1 Methodology of the bias correction

To correct the average values and the temporal variability of the output, a more common way of bias correction is possible to use according to (Hawkins et al. 2013; Navarro-Racines and Tarapues 2015).

$$T_{BC}(t) = \overline{O_{REF}} + \frac{\sigma_{0,REF}}{\sigma_{T,REF}}\left(T_{RAW}(t) - \overline{T_{REF}}\right)$$

where $\sigma_{T,REF}$ and $\sigma_{0,REF}$ illustrate standard deviation in the reference and observations period of the daily GCM output. it is important to attend to this for the GCM output; this bias correction method can correct historical and future periods (Liu et al. 2016).

Downscaling the GCM's output can consider as a solution for better space/time accuracy. In the next step, downscaled climate model simulation can manage local area water with a combination of hydrological models (Liu et al. 2016).

Statistical and dynamic downscaling methods are the two main approaches. Dynamic downscaling is about using regional climate models or limited area models (RCMs or LAMS). To produce a higher resolution of climate outputs, it uses lateral boundary conditions and large scale from GCM (Davies 2014). These approaches will be discussed further in the following (Liu et al. 2016). Most downscaling studies to improve the validation of the consequences of the observation use several RCMs to create a multi-model set (Pal et al. 2007a, b). To create high-resolution for climate change scenarios, various climate change assessment projects were held. These projects are large-scale and always are held between several countries and institutes. They are the primary references of regional projections, additional data on RCMs methods and even regional climatic characteristics (Liu et al. 2016). Table 7.1 summarizes these projects (Pal et al. 2007a, b).

7.2.1 Statistical Downscaling Method

The primary source of information in studies on the climate change impact is the climate model projections. However, these models' coarse resolution and bias cannot be used directly in hydrological models (Chen et al. 2018a). Therefore, to address the effects of climate change at the catchment scale, statistical downscaling is considered a necessity. (Chen et al. 2018a; Sunyer et al. 2015a, b).

The statistical downscaling technique tries to links the status of predictors, which are large scale, to the state of predictands, which should be on a smaller scale. Suppose we compare dynamical and statistical downscaling methods. We can note that the statistical approach is cheaper and easier to apply computationally (Chen et al. 2018a). Stochastic weather generator (SWG), model output statistics (MOS) and excellent prognosis (PP) are considered as The most common statistical downscaling approaches (Maraun et al. 2010b), as in Table 7.2 is shown Chen et al. (2018a).

Table 7.1 Information about several regional climate change evaluation projects

Project	Region	Main goal	Technique
PRUDENCE,[a] 2001–2004	Europe	• Prepare high-resolution climate change scenarios for 21st-century in Europe	8 RCMs
ENSEMBLES(ENSEMBLE),[b] 2004–2009	Europe	• To quantitative danger assessment and create a united scenario of climate change in future periods, an ensemble prediction system is developed	RCM ensemble
CLARIS,[c] 2008–present	South America	• Predict climate changes and their socio-economic effect	5 RCMs
NARCCAP,[d] 2006–present	North America	• Prepare climate change scenario data for Canada, Mexico and the United States • To identify the combined and separate uncertainties in zonal climate simulations that were achieved from different GCMs and RCMs	6 RCMs
CORDEX,[e] 2009–present	Africa	• Scientists simplify analysis and "end-user at the local level of climate changes" (Trzaska and Schnarr 2014), studies area • Cultivate international downscaling coordination	10 RCMs

(continued)

Table 7.1 (continued)

Project	Region	Main goal	Technique
AMMA,[f] 2009–present	West Africa	• Improve ability and understanding to predict WAM (Trzaska and Schnarr 2014) • Relate variability of the WAM to several sectors (Trzaska and Schnarr 2014) • Integrate multidisciplinary research with decision-making and prediction	Multi-RCM
STARDEX,[g] 2002–2005	Europe	• Recognize produced future scenarios of extremes by strong downscaling methods for the end of the twenty-first century	Statistical and dynamical

[a]Prediction of Regional scenarios and Uncertainties for Defining European Climate Change Risks and Effects (Bhuvandas et al. 2014)
[b]Based Predictions of Climate Change and their effects
[c]CLimate Change Assessment and Impact Studies
[d]North American Regional Climate Change Assessment Program
[e]Coordinated Regional Climate Downscaling Experiment
[f]African Monsoon Multidisciplinary Analyses
[g]Statistical and Regional Dynamical Downscaling of Extremes for European Regions (Bhuvandas et al. 2014)

Table 7.2 Attributes and constraints of three downscaling approaches (Chen et al. 2018b)

Approach	Characteristics	Limitations
Perfect prognosis (PP)	Estimates a statistical relationship between large-scale predictors and local or site-specific predictands	There is usually no strong relationship between predictors and predictand, especially for precipitation It is incapable of downscaling precipitation occurrence and generating a proper temporal structure of daily precipitation The relationship estimated for the historical period may not hold for the future period The relationship established for reanalysis predictors in the historical period may not hold for GCM data in the future period
Model output statistics (MOS)	Estimates a statistical relationship between a climate model variable and the same variable of observations	They are incapable of correcting the sequence of precipitation occurrence The bias of climate model outputs may not be stationary between historical and future periods Bias correction methods may change the climate change signal provided by climate models The delta change method assumes that precipitation occurrence will not change in the future period The delta change method assumes that the climate change signal at the GCM grid scale is the same as that at the point scale
Weather generator (SWG)	Adjusts stochastic weather generator parameters based on climate change signals projected by climate models.	Most SWG-based methods do not take into account the correction of precipitation occurrence The adjustment of precipitation occurrence is usually based on a delta change method Precipitation extremes are generally not specifically downscaled (Chen et al. 2018b)

The PP method estimates the statistical linear or nonlinear relationship by using observed climate data. This relationship is among site-specific predictands or local and large-scale predictors (Chen et al. 2012, 2018a; Chen and Brissette 2014a; Wilby et al. 2002; Wilby and Wigley 2000). The regression method is a typical type of pp downscaling; it is a Statistical Downscaling Model and is very famous for calcite climate change impact (SDSM, Wilby et al. 2002). These relationships for daily precipitation are usually not strong enough. For the most part, The climate predictors predict daily precipitation variance of something <30% (Chen et al. 2012b; Chen and Brissette 2014a; Wilby et al. 2002). The method, based on regression, is not suitable for crop simulation studies since it is not good at downscaling rain occurrence and generates the suitable time structure of daily rain (Chen et al. 2012b; Mullan et al. 2016a, b; Wilby et al. 2002). The PP creates a connection among predictors and predictions in the previous era, and in continue, this relationship is applied to future periods (Chen et al. 2018a; Wilby et al. 2002). This formula possibly does not stay correct in changing climate for future (Chen et al. 2012a; Chen and Brissette 2014). In general, relations among (predictors, predictands) pair are estimated by applying reanalysis predictors in continue applied to GCM predictors according to the presumption that "reanalysis and GCM predictors are perfectly simulated in grid-scale (Chen et al. 2011; Dibike and Coulibaly 2005; Wilby et al. 2002)" (Chen et al. 2018b). These backings the utilization of huge scope indicators for measurable downscaling, particularly thinking that large-scale flow features include are noticeably precisely reenacted than surface factors. While reanalysis and information from GCM share a few likenesses, these parameters are totally independent (Chen et al. 2018a). Reanalysis targets are showing this actual world, though GCMs work in virtual. Reanalysis information depends "on the assimilation of the actual world, observed information toward a climate model, while GCMs are active in their virtual area in the scale that restricts the similarity among the two universes. (Essou and Sabarly 2016; Essou et al. 2017). Past examinations (Gleckler et al. 2008; Johnson and Sharma 2009, 2012) illustrated that the efficiency of the GCMs in reproducing large-scale variables depends on a model" (Chen et al. 2018b), location and variable (Chen et al. 2018a). All in all, both GCM factors, with reanalysis, are biased; they might be biased in an unexpected way. This clarifies in huge part why GCMs show moderately massive changes in temperature and rain. So, according to these conditions, it is far-fetched that connections are set up for reanalysis indicators hold by GCM information under future and current conditions (Chen et al. 2018a).

The MOS view includes assessing a factual connection "(statistical relationship) between climate model variable and similar variable from observations" data to accurate the outputs of model (Christensen et al. 2008; Mpelasoka et al. 2009; Piani et al. 2010; Themeßl et al. 2012; Johnson and Sharma 2012; Chen et al. 2011, 2013a, b, 2018a, b). The generally utilized change factor techniques (likewise known as the delta change strategy) that use climate change signal assessed applying climate simulation to noticed time series, additionally represented as a MOS strategy in a few investigations (e.g., Chen et al. 2013b). The MOS view has accompanying downsides (Chen et al. 2018a, b; Sangelantoni et al. 2019).

- Although some convoluted bias remedy strategies can address the wet-day recurrence, they are unequipped for amending the "sequence of a precipitation" event, which can impressively influence the "hydrological modelling" and simulation of "plant growth" (Checn et al. 2013a, 2018a, b).
- All bias revision techniques depend on the understanding that "the biases" of outputs of the "climate model are steady" (Chen et al. 2018b) during the long haul. Nonetheless, late investigations have shown that this supposition is not legitimate for specific locales, primarily for precipitation, because of regular "climate fluctuation" (Chen et al. 2015). In this manner, a bias revision strategy that is legitimate for a recorded duration is not possible to be exact for a future interval. A few examinations have illustrated that bias correction strategy can help change the "climate change signal given by models (Hagemann et al. 2011; Maraun 2013; Maurer and Pierce 2014)". For instance, "Maurer and Pierce (2014)" found that alterations of "GCM-extended patterns" by a quantile planning strategy can be just about as broad as the main changes of "GCM-projected" (Chen et al. 2018a).
- The delta change technique accepts that the precipitation event will not change during the future interval (Chen et al. 2011; Teutschbein and Seibert 2012). Be that as it may, this isn't the situation in reality.
- Delta change strategy uses signals of "climate change at the GCM" network "scale to" observation points (Chen et al. 2011, 2018b; Teutschbein and Seibert 2012). Notwithstanding, at a point scale, the climate change signal may not be equivalent to that at a grid-scale, particularly in uneven locales (Chen et al. 2018a).

For some, SWG is utilized as a downscaling device; the climate change effect concentrates on the original article by Wilks (1992). Downscaling utilizing SWG techniques includes changing SWG boundaries dependent "on climate signals" produced "by models (Wilks 1992; Zhang 2005; Qian et al. 2010; Chen et al. 2012)". "Downscaling SWG" boundaries cause it became conceivable to deliver "climate scenario" effect for any acceptable length for considering the effects of the "hazard-based climate" phenomenon (Chen et al. 2018b). Downscaling utilizing SWG strategies includes changing SWG boundaries dependent on climate signals produced by models (Jones and Thornton 1993, 2000). Even though the event change is considered by a couple of researchers, the change is typically found on a "delta change strategy (e.g., Wilks 1992; Chen et al. 2012)"; Chen et al. (2018a) also acquires its downsides. For instance, Wilks (1992) downscaled the precipitation event by changing an unqualified likelihood of precipitation event and a dependency Variable utilizing climate data projected by models. Also, Chen regulates two transitional probabilities of a precipitation event in expansion. Rain exceeding is normally not explicitly downscaled when utilizing techniques based on SWG. All techniques accept that outrageous occasions rain have similar changes as other occasions rain. That means the conceivable tempest increase in future isn't thought of, as revealed in the "Fifth Assessment Report of IPCC (IPCC's Fifth Assessment Report (AR5) 2013)" (Chen et al. 2018b). The parameter connecting to the extreme of the precipitation isn't explicitly adjusted (Chen et al. 2018a).

A factual downscaling technique dependent on a stochastic climate generator (CLINGEN) was created (Zhang 2005). In previous years, this technique has seriously tried utilizing different areas in different climates. In particular, the procedure has been assessed and approved for stationary, non-stationary climate changes and varieties utilizing verifiable climate information in different stations scattered in the United States, the united kingdom, Brazil, Australia and Canada (Zhang et al. 2012, 2013; Mullan et al. 2016a, b). The strategy has effectively applied to GCM projections which are downscaled for sway investigations of climate changes in China, Japan, both the United States and the united kingdom (Zhang and Li 2007). Has shown this refined downscaling way, compared with a traditional downscaling view, simulates occasions excessive daily precipitation agreeable because of its different treatment of temporal, spatial varieties. This technique has been utilized to appraise precipitation erosive alterations under climate change for northeastern China (Zhang 2007); what's more, the southern Appalachian locale of the United States. (Chen et al. 2018a; Hoomehr et al. 2016). Precipitation erosive is for the most part, a function of precipitation severity. The technique has been utilized to simulate site-explicit effects of climate on soil disintegration, surface water, water assets and harvest creation in China (Chen et al. 2018a; Li et al. 2009, 2011). Zhang et al. (2012) demonstrated the climate impacts on overflow, soil disintegration in Arizona, suggestions for moderation with protection works, and the downscaling method. Zhang et al. (2012) examined the impacts of editing and culture frameworks on wheat creation, soil disintegration, and surface spillover under climate change in focal Oklahoma (Chen et al. 2018a; Mullan et al. 2016a, b) examined the impacts of change in climate on the long-term livability of most active substantial take ice street utilizing the downscaling method. (Mullan et al. 2016a, b) likewise displayed the influence of grass strips in overseeing turbid floods under climate change in Belgium. Also, an equivalent method was utilized to study (Gunawardhana and Kazama 2012) the effects of climate and changeability on aquifer and groundwater left and temperatures in one Asian country (Chen et al. 2018a). All the above assessments and applications uncovered the upsides of downscaling technique over others from different viewpoints. Progress and use of the downscaling method in previous years, the time has come to orchestrate and incorporate this load of accomplishments (Chen et al. 2018a).

More significant part of the modern downscaling methods are ordered into three gatherings (Fowler et al. 2007): weather typing schemes, and weather generators (e.g., LARS-WG), regression models (e.g., SVM, ANN). The majority of the new experience compare different statistical downscaling methods. These examinations have represented those different ways have fluctuating efficiency in a specific region, and the unique process has different performance in various investigation regions. Since various methods can catch different parts of the downscaling, consolidating the consequences of the different methods by weighting methods can perform better than a specific method. (Reid 1968; Bates and Granger 1969; Dickinson 1973). The new mixed methods applied instruments such as an artificial linear regression, neural network and simple model average (Shamseldin et al. 1997; Ajami et al. 2005, 2006). These methods include a bunch of deterministic weights to join multiple model outputs. The significance in such a blend can take any optional negative or positive real values that don't have physical descriptions (Ajami et al. 2007; Chen et al. 2018a).

7.2.1.1 Comparison of Statistical and Dynamic Downscaling

For the most part, statistical downscaling strategies are used to examine anticipated hydrological impacts under climate change scenarios contrasted with the dynamic downscaling procedure, which requires limited area models (LAMs) or territorial climate models (RCMs). The overall hypothesis of statistical downscaling includes setting up trial relations in the midst of antiquated large-scale atmospheric and local climate highlights (Trzaska and Schnarr 2014).

Statistical downscaling contains an inharmonious group of various pertinence and perfections; Statistical downscaling strategies are computationally and economically reasonable than "RCMs that need complex displaying of actual physical processes" (Trzaska and Schnarr 2014). Hence, they are a feasible now and then a supportive option for an organization. Computation capacity and specialized experience are fundamental to dynamical downscaling. In contrast to "RCMs, which produce down-scaled projections at a 20–50 km spatial scale" (Chen et al. 2018b), statistical strategies can plan station-scale climate data. However, statistical downscaling is viable, computationally sensible, and contains different ways, but it holds the following inherent assumption (Chen et al. 2018b):

(1) The statistical connection among the predictand indicator doesn't change during the time. (2) "The predictor bears climate signal. (3) There is a sharp connection among the (predictor, predictand) pairs. (4) GCMs exactly simulate the predictor" (Chen et al. 2018b).

The statistical downscaling procedure has a few inconveniences, like the unsure uncertain hypothesis in a future climate. It is computationally affordable and statistically is consider to be a great companion to dynamical downscaling (Chen et al. 2018b). It is imperative that predictor and predictand can be equivalent parameters on various scales. In any case, statistical techniques can select any factor as a predictor without much of a struggle. Other accessible factual downscaling methods have been utilized in various climate regions (Trzaska and Schnarr 2014; Wilby and Wigley 1997; Xu 1999).

7.2.1.2 Synoptic Weather Typing

The automated synoptic weather typing (or air mass typing) is notable for assessing the effect of climate change on different environmental issues explanation is its capacity to sort a complex arrangement of meteorological factors as a coherent index. This can make easy the investigation of local climate change results from weather typing approach has been effectively examined in the Canadian climate for a few study tasks to evaluate the effect of climate change on a few severe climate occasions like freezing precipitation, heavy precipitation, high-/low-stream occasions, air pollution, and human wellbeing. Likewise, over the previous decade, the automated synoptic typing-view way to deal with surveying the effect of climate change on the environment has become well known only on the grounds that these methods depict the similitudes of dynamic meteorological components in a complete system. Ecological issues that are addressed to utilize these methods contain air quality (Kalkstein and Corrigan 1986; Eder et al. 1994; McGregor and Bamzelis 1995; Lam and Cheng 1998; Cheng et al. 2008), human health (Cheng et al. 2008; Cheng 1991; Cheng and Kalkstein 1997; McGregor et al. 1999), and climate research (Kalkstein et al. 1990; Cheng and Kalkstein 1993, 1997). Furthermore, this method is utilized to assess climate impacts on the different environmental problems (Bear and Cheng 2010). The synoptic weather typing can distinguish a complex set of meteorological factors as a coherent indicator (Perry 1983), Utilizing hourly sea level air pressure and hourly surface and upper-air temperature observations, dew point, dew point, u wind, y wind, and complete overcast cover. As indicated by air mass differentiation, an automated synoptic typing method was applied to allocate every day of the developmental dataset to a particular weather type. Subsequently, the surface climate information at 24 different times and six distinct climatic components makes 144 surface climate factors. The upper-air reanalysis data at 6-h intervals and six atmospheric levels make 96 upper-air weather factors. A complete set of 240 upper-air weather variables and surfaces were used in the synoptic typing method. It generates a temporal synoptic index using principal component components analysis (PCA) and a hierarchical agglomerative clustering method. (Kalkstein and Corrigan 1986; Cheng and Kalkstein 1997; Cheng and Lam 2000).

- **Principal Component Analysis**

"Principal Component Analysis," also known as "PCA, is a dimensionality-reduction method" that is always applied to decline large data sets' "dimensionality," turning a large set of variables into smaller ones, but with a lot of group information (Jaadi 2021). Dealing with the group of variables of an information set generally comes at the cost of correctness. However, the way is to trade a small accuracy for ease. Smaller datasets make information visualization and analysis more straightforward and convenient for machine learning algorithms without processing additional parameters (Jaadi 2021). As a result, the PCA approach declines the number of data set variables while retaining information as much as possible. In the following, the PCA method will be explained step by step.

Step 1: Standardization

The primary purpose of "this step is to standardize the range of the continuous primary variables so that each of them contributes equally to the analysis" (Jaadi 2021). In particular, the main reason for standardization before PCA is essential, meaning that the latter is susceptible to the variance of the main variables (Jaadi 2021). Thus, if there are significant variations among the ranges of primary variables, those with a broader range will dominate those over a smaller range, leading to biased outcomes

(Jaadi 2021). In terms of Mathematic, these are able to do by reducing the average and dividing with standard deviation for each variable:

$$z = \frac{value - Mean}{Standard\ deviation}$$

Step 2: Covariance Matrix Computation

This phase aims to realize the way that input data set variables to vary from average by attention to others or to found a connection among them. Since in some situations, parameters are associate in such a way that they involve redundant data. Therefore, to recognize the relationship, a covariance matrix should be calculated.

"A P × P symmetric matrix in that P is the number of dimensions" (Jaadi 2021). The covariance "associated with all possible pairs of the primary variables" is known as the covariance matrix (Jaadi 2021). For instance, for a three-dimensional data series with three parameters, "x, y and z," the below "3 × 3 matrix" is the covariance matrix (Jaadi 2021):

$$\begin{bmatrix} cov(x, x) & cov(x, y) & cov(x, z) \\ cov(y, x) & cov(y, y) & cov(y, x) \\ cov(z, x) & cov(z, y) & cov(z, z) \end{bmatrix}$$

Step 3: Identify the Principal Components by Computing the Eigenvectors and Eigenvalues of the Covariance Matrix

The concepts of linear algebra from the covariance matrix should be calculated to define the data's main parts, eigenvalues and eigenvectors. Before starting the definition of these concepts, it is necessary to understand the main parts. The main parts are novel parameters that are built as linear mixtures or combinations of the primary parameters. These mixtures are done using the new variables uncorrelated. "Most of the information within the primary parameters is compressed or squeezed toward the initial components" (Jaadi 2021). Therefore, the idea is ten-dimensional information gives you ten main parts. However, "PCA tries to put maximum possible" data in the initial part after the maximum remaining data in the second (Jaadi 2021).

Step 4: Feature Vector

Calculating the eigenvectors, arranging them "by their eigenvalues" in declining order would find the main part of importance (Jaadi 2021). "In this part, whether to keep all these p or discard those" of a less extent "(low eigenvalues)" is determined and formed with the left ones; a "matrix of vectors" is called. Thus, the "feature vector" is easily a "matrix that columns the eigenvectors of the components we elect to keep. This makes it the initial step towards dimensionality decrease. If we prefer to stay only p eigenvectors (components) out of n, the final data set will have only p dimensions" (Jaadi 2021).

Step 5: Recast the Data along with the Axes of the Principal Component

In the former step, without considering standardization, you do not exchange any data with the information. You only chose the main parts to form the feature vector, but input sets usually remain in main axes (i.e., the initial variables). At this point, it is the final step. The main purpose is to reorient information "from the original axes to ones" described by the main ingredients using duplicating the "transpose of the original data set" by transposing the feature vector (Jaadi 2021).

$$Final\ DataSet = Feature\ Vector^T * StandadrizeOriginal\ Data\ Set^T$$

- **Hierarchical Agglomerative Clustering Procedure**

It is worth mentioning that cluster analysis is one of the tools to achieve classification. Most authors have tried to define cluster analysis. Although the approaches and the terminology differ considerably, most authors agree with the initial differentiation into two groups of cluster analysis methods, the hierarchical and nonhierarchical, which both of them can continue in cumulative, divisive, or in another way. The agglomerative methods' clusters are produced by successive clustered sets, when "the divisive" ways continue in steps to "part the given set of objects. Most of the published hierarchical clustering methods are of the agglomerative type" (Lukasová 1979). In this part, a more general approach to the "hierarchical agglomerative clustering procedure" will be introduced (Lukasová 1979).

Dissimilarity Coefficient

Most published hierarchical clustering procedures are similar in their successive merging of created clusters. Let us design such an agglomerative approach. However, the main problem in all clustering procedures is the quantitative evaluation of the between cluster relations, enabling us to decide at which stage of the process to merge the clusters. These relations are usually designated as similarity, dissimilarity, association, distance, etc. For us, the most suitable designation seems to be dissimilarity. To define the designation of the hierarchical agglomerative clustering procedure precisely, the designation of the cluster dissimilarity of the object set's partition must first be explained (Lukasová 1979).

Let $\Omega = \{\mathcal{A}_1, \mathcal{A}_2, \ldots, \mathcal{A}_m\}$ be the partition of the object set "ϑ into m clusters $\mathcal{A}_1, \mathcal{A}_2, \ldots, \mathcal{A}_m$ Let D be the function that applies for every pair $(\mathcal{A}_i, \mathcal{A}_j)$ (for $i, j =$ 1, 2,…, m) of partition clusters a real number, $D(\mathcal{A}_i, \mathcal{A}_j)$ satisfying the following requirements" (Lukasová 1979).

(D1)	$D(\mathcal{A}_i, \mathcal{A}_j) = 0$	for $i = 1,2,\ldots,m$
(D2)	$D(\mathcal{A}_i, \mathcal{A}_j) \geq 0$	for $i = 1,2,\ldots,m; i \neq j$
(D3)	$D(\mathcal{A}_i, \mathcal{A}_j) = D(\mathcal{A}_i, \mathcal{A}_j)$	for $i, j = 1,2,\ldots,m$

D is the cluster dissimilarity coefficient; as long as there is no danger of misunderstanding, we shall designate D shortly as the dissimilarity coefficient or dissimilarity measure (Lukasová 1979).

Types of Dissimilarity Coefficients

"The dissimilarity coefficient of two clusters" is most frequently expressed with the "dissimilarity coefficient of two objects. We sometimes stipulated that the latter coefficient was a metric". Some methods utilizing the coefficient of cluster dissimilarity derived from "object dissimilarity can be found in the present literature. We might mention", for example, nearest neighbor, furthest neighbor, and "centroid methods, among others." Such dissimilarity similarity "coefficients will be designated D/d, where d represents the object dissimilarity coefficient. Another means by which we may introduce the cluster dissimilarity coefficient directly and not based on the dissimilarity coefficient of objects forming a cluster. Hall's inhomogeneity" measure is an example of such a method. The properties of various dissimilarity "coefficients will be investigated in the following sections. We shall introduce the terms uni-measure, measure, and tri-measure for the three dissimilarity measure types termed (i)-measure, (i, j)-measure, and (i, j, k)-measure by Lance and Williams" (Lukasová 1979). Hence:

$$\begin{aligned} &\mathrm{D}(\mathcal{A}_i, \mathcal{A}_j) = F^{\mathrm{I}}(\mathcal{A}_i \cup \mathcal{A}_j) && \text{uni-measure,} \\ &\mathrm{D}(\mathcal{A}_i, \mathcal{A}_j) = F^{\mathrm{II}}(\mathcal{A}_i, \mathcal{A}_j) && \text{bi-measure} \\ &\mathrm{D}(\mathcal{A}_i, \mathcal{A}_j) = F^{\mathrm{III}}(\mathcal{A}_i, \mathcal{A}_j, \mathcal{A}_k = \mathcal{A}_i \cup \mathcal{A}_j) && \text{tri-measure} \end{aligned}$$

where $F^{\mathrm{I}}, F^{\mathrm{II}}, F^{\mathrm{III}}$ For clusters, the suitable functionals determine an actual "number corresponding to 1, 2, and 3 sets" (Lukasová 1979). Hall's inhomogeneity measure represents the uni-measure, while the nearest neighbor, furthest neighbor, and centroid methods proceeding, for example, "from the Euclidean metric stands for the measures; Ward-Wishart's method observing" the information "loss accompanying clustering is an example of the tri-measure" (Lukasová 1979).

Clustering Procedure [*HACP*, *D*]

Let "$\vartheta = \{O_1, O_2, \ldots, O_n)$ $n > 1$, be the set of objects, let D be the dissimilarity coefficient" (Lukasová 1979). Let the sequence of the partition $\{\Omega_0, \Omega_1, \ldots, \Omega_{n-1}\}$ Of the set, ϑ correspond to this set. Let a real nonnegative value h($\mathcal{A}$) correspond to each cluster $\mathcal{A}$ of each partition Ω, and the following holds (Lukasová 1979):

1. $\Omega_0 = \{\mathcal{A}_0, \mathcal{A}_{02}, \ldots, \mathcal{A}_{0n}\}$ where $\mathcal{A}_{0i} = 0_i$, $h(\mathcal{A}_{0i}) = 0$ for $i = 1, 2, \ldots, n$.
2. let $\Omega_1 = \{\mathcal{A}_{i1}, \mathcal{A}_{i2}, \ldots, \mathcal{A}_{i,n-i}\}, 0 \le i \le n-2$ be the *ith* partition of the set ϑ in which a real non-negative number h(.$\mathcal{A}_{ij}$), $j = 1, 2, \ldots, n-i$ corresponds to the cluster $\mathcal{A}_{ij}$ and let

$$\mu_i = \min_{x,y=1,2,\ldots\ldots,n-i,x\neq y} \mathrm{D}(\mathcal{A}_{ix}, \mathcal{A}_{iy}).$$

Choose one pair $(\mathcal{A}_{ix}, \mathcal{A}_{iy})$, for which $\mathrm{D}(\mathcal{A}_{ix}, \mathcal{A}_{iy}) = \mu_i$. Then the partition $\Omega_{i+1} = \{\mathcal{A}_{i+1}, \mathcal{A}_{i+2}, \ldots, \mathcal{A}_{i+1,n-i-1}\}$ shows the following properties:

(a) $\exists l(\mathcal{A}_{iu} \cup \mathcal{A}_{ju} = \mathcal{A}_{i+1,l})$

$l\varepsilon\{1, 2, \ldots, n-i-1\}$
$h(\mathcal{A}_{i+1}, l) = \mu_i$

(b) "There is a bijection among the set of subscripts, $j = 1, 2, \ldots, n-i; j \neq \vartheta, u$; of the clusters of the ith partition and the set of subscripts $k = 1, 2, \ldots, n-i-1$; $k \neq l$; of the clusters of the $(n+1)$ the partition such that" (Lukasová 1979).

$$\mathcal{A}_{i+1,k} = \mathcal{A}_{i,j}, h(\mathcal{A}_{i+1,k}) = h(\mathcal{A}_{i,j})$$

3. $\Omega_{n-1} = \{\mathcal{A}_{n-1,1}\}; h(\mathcal{A}_{in-1,1,}) = \mu_{n-2}$

Using steps 1–3, "the hierarchical agglomerative clustering procedure with the coefficient D" (Lukasová 1979), designated as [HACP, D], was established recursively (Lukasová 1979). The numbers $\mu_0, \mu_1, \ldots, \mu_{n-2}$ are the clustering levels corresponding to the partitions $\Omega_0, \Omega_{1,}, \ldots, \Omega_{n-2}$.

The "[HACP, D] corresponds to a finite set of stratified hierarchical clustering to the set ϑ. The number of such clusterings depends on the number of mutually different pairs of clusters in the individual partitions, all of which yield dissimilarity coefficients that take the minimum values within the framework of the partition. If requesting the procedure [HACP, D] to create always the similarity tree, another requirement must be laid on the coefficient D. (D4) Let Ω be the member of the partition sequence of a set created by the clustering procedure [HACP, D] and let" (Lukasová 1979).

$$\mu_i = \min_{x.y=1.2,\ldots,n-i,x\neq y} \mathrm{D}(\mathcal{A}_{ix}, \mathcal{A}_{iy}).$$

"m is the number of clusters of the partition be the minimum value of the dissimilarity coefficient of the clusters belonging to the partition Ω" (Lukasová 1979).

$$\mathrm{D}(\mathcal{A}_i, \mathcal{A}_j) = \mu holds\ true, then$$

$$\mathrm{D} = (\mathcal{A}_i \cup \mathcal{A}_j, \mathcal{A}_k) \geq \mu forallk, k \neq i, k \neq j$$

We illustrate that the D4 condition is essential and enough if the method [HACP, D] corresponds a "finite set ϑ of the similarity trees to each group of the things" (Lukasová 1979). As is evident from it define, a finite group of stratified hierarchical clustering is achieved. When "applying the procedure [HACP, D]" (Lukasová 1979). It remains necessary to show that for all pairs of clusters $\mathcal{A} \subset \mathcal{B}$ "belonging to the stratified hierarchical clustering created by the procedure [HACP, D] with the coefficients D satisfies (D4), the following holds true" (Lukasová 1979):

$\mathrm{h}(\mathcal{A}) \leq h(\mathcal{B})$. Let us designate as in the definition of the procedure [HACP, D]

$$\mu_i = \min_{x.y=1.2......n-i,x\neq y} D(\mathcal{A}_{ix}, \mathcal{A}_{iy}).$$

If D4 holds and the ith step joins two clusters whose dissimilarity coefficient equals μ_i, one of the minimum values from the cluster's dissimilarity coefficients set corresponding to the ith partition is at the same time eliminated (Chen et al. 2018a). Within the *i*th partition, the dissimilarity coefficient of the new cluster formed by the given joining with another arbitrary cluster takes a value higher or equal to μ_i (Lukasová 1979).

7.2.1.3 Transfer Function Approaches

The statistical downscaling approach assumes that large-scale atmospheric variables strongly affect the local climate. Thus, miniaturization models have been formed concerning local climatic parameters and large-scale atmospheric parameters simulated by GCMs. According to the regression equation, transmission performance models are related to forecasting and forecasting and are widely used for statistical climate downscaling. They are regression-based minimization techniques that are simple and require very little calculation compared to other methods. Regression-based methods are limited to places where the prediction-prediction relationship is estimated correctly. Transfer function techniques can depend on predictor variables, statistical fitting methods, or applied mathematical transfer performance. Methods can be nonlinear and "linear regression, artificial neural networks, canonical correlation, principal component analysis, or independent component analysis and (Jeong et al. 2012)" (Jeong et al. 2012).

Transmission performance models generally are according to the linear regression familiar as a linear model (LM). The relations among (prediction, prediction) pairs are often very complicated, and linear regression often fails to work appropriately (Goyal and Ojha 2012; Ahmadi et al. 2014). Several nonlinear and non-parametric regression-based minimization models have been introduced "(Haylock et al. 2006; Hashmi et al. 2011)". Two "linear regression models" (Jeong et al. 2012), Generalized linear methods (GLM) (McCullagh and Nelder 1989) and generalized additive models (GAM) (Hastie and Tibshirani 1990), are effective in deriving the relationship between abnormal response and predictor parameters. Recently, GLM, GAM has been widely applied to reduce the scale of the climate (McCullagh and Nelder 1989; Hastie and Tibshirani 1990; Salameh et al. 2009; Tisseuil et al. 2010; Hu et al. 2013a, b; Liu and Fan 2013; Kigobe et al. 2014; Farajzadeh et al. 2015; Hertig et al. 2014; Lu and Qin 2014). Salameh et al. (2009) introduced GAM to reduce the scale of climate and reported that GAM is more capable in smaller climates than the linear model. (Tisseuil et al. 2010) used GAM and GLM to minimize large-scale simulation with GCM to streamline future river-scale changes in southwestern France. They mentioned the effectiveness of GLM and GAM in downscaling.

Hu et al. (2013a, b) applied GAM to shrink near-surface wind areas in the northeastern Chinese Qinghai-Tibetan Plateau. The GAM-based statistical downscaling approach simulates accurate, fast, and relatively transparent local scale near the field. Inflatable surface offers. Cases of GLM were reported in some research in previous years. (Liu and Fan 2013) applied GLM to reduce the scale of the daily climate in China. (Kigobe et al. 2014) in the Nile. Both reported the significant effect of GLMs in declining climate scale. In addition, (Farajzadeh et al. 2015) applied GLM and several parametric and non-parametric ways to reduce the temperature in the high hill area of Midwestern Iran. Lu and Qin (2014) applied a single site GLM to reduce daily precipitation in Singapore. (Beecham et al. 2014) calculated the suitability of GLM for multi-site daily precipitation modelling in Australia. (Rashid et al. 2016) applied GLM to minimize multi-site daily precipitation forecasting and CMIP5 GCMs in Australia. Qian et al. (2015) applied GLM to minimize severe temperature-related indicators in China. They all reported the significant influence of GLM in reducing monthly and daily precipitation, extreme temperatures and weather. The s above study clearly shows GAM, GLM and LM models can effectively reduce rainfall. However, the performance of linear and nonlinear transmission performance minimization models depends on the data distribution. In general, LM performance is best during the time that the information distribution is normal or close to be normal. GLM and GAM, support nonlinear connections among response and prediction parameters and increase accuracy during the time that data distribution deviates too much normal. This confirms the requirement to evaluate the performance of different statistical downscaling methods to recognize the most appropriate metod for reducing climate change levels in a particular area.

Linear transmission functions are relatively simpler and more accessible than most nonlinear transmission functions. For this reason, if the small functions of the two models do not differ significantly, a linear transmission "function may be preferred to a complex" nonlinear transmission function (Jeong et al. 2012). In several past downscaling types of research, the relation among AOGCM predictors and zonal predictions with linear transmission functions have been modelled "(e.g., Enke and Spekat 1997; Wilby et al. 1998a, b; Huth et al. 2001; Palutikof et al. 2002; Wilby et al. 2002; Guo et al. 2012; Huang et al. 2011)" (Jeong et al. 2012).

On the other hand, a lot of studies such as "(e.g., Schoof and Pryor 2001; Tripathi et al. 2006; Tolika et al. 2007)" Jeong et al. (2012), have illustrated that linear transmission functions are not successful in establishing the relationship among predictions and predictions in SD issues (Jeong et al. 2012). In such cases, nonlinear transmission functions may have the advantage of being limitless to a linear relation among predictions and predictions (Jeong et al. 2012). Neural networks are often used as non-parametric transmission functions to minimize because they can establish nonlinear solid connections among predictions and predictions "(Hewitson and Crane 1996; Wilby et al. 1998a, b; Trigo and Palutikof 1999; Mpelasoka et al. 2001; Cavazos and Hewitson 2005; Miksovsky and Raidl 2005; Tolika et al. 2007, Huth et al. 2008)" (Jeong et al. 2012). Despite that, ANN has several negative effects, such as the possibility of being stuck in the mindset and at least locally in choosing the way "architecture or even training algorithm. This model always needs a large sample size with great quality data to prevent over-adaptation and accurate reflection of the dynamics of target parameters" (Jeong et al. 2012).

In addition, ANN models are more costly and "difficult to interpret than linear regression transfer functions" (Jeong et al. 2012). Different researches examined and compared different linear or nonlinear views to recognize the most suitable transfer functions. "(Weichert and Bürger 1998)" used a linear regression approach according to CCA and a nonlinear ANN to scale the non-seasonal daily parameters of mean temperature, rain and steam pressure in Germany. Wilby et al. (1998b) compared two ANN transmission performance approaches with two air-generating applying vortex flow information as predictions in "six areas across America. Trigo and Palutikof (1999) compared the two approaches for reducing "the maximum and minimum" temperature scale in Portugal" (Jeong et al. 2012). The researchers used a "linear ANN model" rather than a nonlinear "ANN model". Pryor and Schoof (2020) "compared multiple linear regression methods and ANN as transfer functions in SD models for maximum and minimum daily surface temperature, daily and monthly rain in the United States" (Jeong et al. 2012). Mpelasoka et al. (2001) compared ANN with multivariate statistical approaches (PCA, CCA, and bloated regression analysis) to reduce New Zealand's "monthly temperature," rain and showed that the "multivariate statistical" approach performed well (Jeong et al. 2012). Observe average production, variance, and monthly temperature distribution and precipitation (Jeong et al. 2012). Miksovsky and Raidl (2005) using multilayer ANN perceptron, radial base function, "local linear and MLR" to reduce the daily temperature of 25 stations (Jeong et al. 2012). MLR, ANN, and classification approaches are used to minimize daily temperature. The researchers report that MLR based on a step-by-step screening view performed best (Jeong et al. 2012).

Jeong et al. (2012) contrasted three linear models "multiple linear regression (MLR) with ordinary least squares (OLS) estimates, robust regression, and ridge regression" (Jeong et al. 2012) and one non-linear model (artificial neural networks (ANNs)) to determine the best-suited "transfer function in statistical downscaling (SD) models" (Jeong et al. 2012), for the rain (occurrence and amount of rainfall).and temperature (daily minimum and maximum); (Jeong et al. 2012).

Regression equations are applied to determine predictor-predictand relationships in transfer functions, which used linear connection and non-linear relationship, classified as Non-linear transfer functions and linear transfer functions. (Trigo and Palutikof 1999) expanded transfer functions to predict temperature (maximum and minimum) at local scales from the GCM outputs applying Non-linear ANN models and linear models (Jeong et al. 2012).

The performance of non-linear ANN models was calculated to find out their privileges over non-linear models (Jeong et al. 2012). Determining the appropriate transfer function for a specific downscaling issue is the most significant step in Statistical Downscaling applying transfer functions. Linear transfer functions are wholly simpler than complex non-linear functions (Jeong et al. 2012). For this cause, most statistical downscaling studies use linear transfer functions to model the relationship among AOGCM predictors and local predictands. Moreover, linear transfer functions are calibrated separately for various seasons or months have shown better results than annually calibrated models (Jeong et al. 2012). Contrary to this, several other studies show that linear transfer functions fail to appropriately capture the predictor-predictand in statistical downscaling relationships and suggest using complex non-linear transfer functions. The advantage of the nonlinear transfer function over linear transfer functions is the linearity of the predictor-predictand relation does not limit non-linear transfer functions (Jeong et al. 2012).

The Artificial Neural Networks are non-parametric transfer functions with the significant benefit of determining a strong predictor-predictand relation in downscaling by applying statistical methods (Jeong et al. 2012). At the same time, ANN has disadvantages like getting trapped in the selecting model and local minima architecture/training algorithm. There are chances of overfitting, which can be avoided if sample sizes are large enough. ANNs, however, need high operational costs and are complex (Jeong et al. 2012).

Ghosh and Mujumdar (2008) state downscaling using transfer functions is the most popular method of downscaling. Several methods are used, including linear and nonlinear regressions, artificial neural network, fuzzy rule-based system, support vector machine, analog methods. The methods vary "according to the mathematical transfer function, predictor variable, and statistical fitting procedure" (Jeong et al. 2012).

Chen and Brissctte (2014) used the quantile mapping way (transfer function method) to downscale monthly precipitation in two steps. Initially, the transfer function was derived by fitting the first and third-order polynomials, the precipitation simulated, and the observed station by the climate model for the reference period. Furthermore, the transfer functions were used to downscale the "climate model simulated monthly precipitation for the future period" (Chen et al. 2018b). It is observed that the transfer function's ability to reproduce the probability distribution of the monthly precipitation and correct the bias of the grids and match them (Jeong et al. 2012).

- **Transfer Functions**

In Continue, "three linear models and one non-linear model, specifically multiple linear regression (MLR) with ordinary least squares (OLS) evaluated, robust regression, ridge regression, and artificial neural networks (ANNs)," which generally applied to recognize an "appropriate transfer function in statistical downscaling (SD). In regression-based SD models, the relationship between a regional scale predictand and large-scale predictors can be expressed by" (Jeong et al. 2012).

$$y = F(x) + e,$$

"where $y[n \times 1]$ is an observed predictand vector with n record length, $X[n \times q]$ is a carrier matrix that includes the q AOGCM predictors as independent variables, and e $[n \times 1]$ is a random vector with record length n" (Jeong et al. 2012). The "transfer function, $F(x)$," can be a definite or probable "transfer function conditioned by X" predictions. "It can be modelled from the observed or modelled AOGCM" dataset with a "linear or nonlinear mathematical" view (Jeong et al. 2012). In the "SD model," a stochastic method for "the residual vector, e," is always needed to generate the "natural variability" of a regional prediction because "transmission functions" with "global AOGCM predictors" can only represent part of the observed prediction variability (Jeong et al. 2012). Explain. However, this study focuses only on "deterministic transfer functions" to compare the different linear or nonlinear "transfer functions in SD models" (Jeong et al. 2012).

- **Multiple Linear Regression with OLS Estimate**

The usual and essential performance of "linear transmission" is "MLR and OLS" estimation (Chen et al. 2018b). The predictions and y of a site observed in a regional area can be reduced applying the below MLR formulas (Hessami et al. 2008; Jeong et al. 2012; Zorita and Von Storch 1999).

$$y = x\beta + \varepsilon$$

"Where $\varepsilon[n \times 1]$ is a residual vector of MLR, and the parameter vector $\beta[q \times 1]$ can be estimated by the OLS estimation method below" (Chatterjee and Hadi 2015; Jeong et al. 2012).

$$\widehat{\beta} = \left(X^T X\right)^{-1} X^T Y$$

"The variance–covariance matrix for the vector of coefficients of the MLR with OLS estimate, $\widehat{\beta}$; is given by the following" (Jeong et al. 2012):

$$Var\left(\widehat{\beta}\right) = \sigma^2 \left(X^T X\right)^{-1}$$

where σ^2 is a "variance of the error term of the MLR model. From Eq. previous, the standard error of an estimated parameter, $\widehat{B_m}(m = 1,2,\ldots,q)$"; is given by Chatterjee and Hadi (2015), Jeong et al. (2012)" (Jeong et al. 2012).

$$se\left(\widehat{\beta_m}\right) = \frac{\sigma}{\sqrt{\sum_{i=1}^{n} (X_{mi} - \overline{X}_m)(1 - C_m^2)}}$$

"Where σ is the standard error of the predictand y from the MLR model, and C_m^2 is the R^2 of the regression when X_m is a dependent variable and the other $X_j s (j \neq m)$ are independent variables. If X_m is correlated to the other independent variables, C_m^2 becomes more prominent", and that increases the standard error of $\widehat{B_m}$ (Chen et al. 2012, 2018b).

However, several factors influence $se\left(\widehat{\beta_m}\right)$ other than C_m^2. For instance, "increased standard error (σ) of the MLR model because of poor accuracy also increases the standard error of $\left(\widehat{\beta_m}\right)$. Furthermore, the variability of independent variable Xm, as given by $\sum_{i=1}^{n} \left(X_{mi} - \overline{X}_m\right)$; is inversely related to $se\left(\widehat{\beta_m}\right)$; therefore, if the new

X_{mi} is not equal to $\overline{X}_m$; including new records of X_{mi} by increasing the sample size increases the variability and decreases $se\left(\widehat{\beta_m}\right)$" (Jeong et al. 2012).

- **Robust Regression**

One of the essential Disadvantages of "MLR with OLS" estimation is that thc "csti-mated values" are susceptible to precipices (Jeong et al. 2012). During the time that the "distribution of regression errors" is really "heavy or prone" to bounce, robust "regression" views can give a reliable option for estimating more substantial "regres-sion data than the MLR ways" (Chen et al. 2018b; Jeong et al. 2012). The most famous way between robust regressions is the M estimation given by Hewitson and Crane (1996). The present study used the weighted least squares methods with the proposed square weighting scheme (Beaton and Tukey 1974; Jeong et al. 2012). In that method, regression parameters could repeatedly estimate applying IRLS processes (Jeong et al. 2012). In the first iteration, equal weight is assigned to each point, and the approach coefficients are calculated applying OLS. "At each iteration, t, residuals of the previous (t – 1) iteration, I, independent variables", e_i^{t-1} and associated weights, W_i^{t-1} are "recalculated; points farther from model predictions in the previous iter-ation are given a lower weight" (Jeong et al. 2012). "Model coefficients are then updated using weighted least squares" (Jeong et al. 2012). The new weights, $\beta_{o,}^{t}$, of "robust regression at iteration, t, are calculated using the following equation" (Jeong et al. 2012):

$$\beta_0^t = \left(X^T W^{t-1} X\right)^{-1} X^T W_y^{t-1}$$

where W^{t-1} =diag (W_i^{t-1}). "The iteration continues until the values of the coefficient estimates converge. The square (or weight) weight function proposed by Beaton and Tukey (1974) is as follows" (Jeong et al. 2012):

$$w(r) = r\left(1 - r^2\right)^2 \; for \; |r| \le 1 \; and$$
$$w(r) = 0 \; for \; |r| > 1.$$

The weight function is the value of $e(d.c)$, where e is the vector of the residuals of the previous iteration and (d = 1.4862 MAD) is necessary for the standard deviation of the error term. The MAD represents the absolute deviation of the residual mean from their mean. The constant, 1.4862, neutralizes the estimate for the natural distribution. Parameter c is the setting constant (Chen et al. 2018b) that is selected to adjust process efficiency "for data from the normal distribution and is usually set to 4.685" in the square design (Jeong et al. 2012). For significant residues (for example, when $|r| > 1$. in the above equation) sets the weight to zero to eliminate their effects (Jeong et al. 2012).

- **Ridge Regression**

Another major disadvantage of the "MLR" approaches is that the misalignment (or alignment) of the "predictor parameter can make the OLS estimate of the regression coefficients unstable" (Chen et al. 2018b) (for this selection and its use concerning OLS, please see (Hessami et al. 2008). Thus provide a slight bias for more stable estimators (Hoerl and Kennard 1970; Jeong et al. 2012). When there is consistency, the variables estimated by "ridge regression are more potent than the OLS estimates" (Hoerl and Kennard 1970; Jeong et al. 2012).

"Principal component regression (PCR) is an alternative to correlation in predictor variables" (Chatterjee and Hadi 2015; Jeong et al. 2012). PCR usually applied only a few vital "components with highly explained variance as independent variables" (Chen et al. 2018b). It is often said that key components with low variance are possible to be necessary (Jeong et al. 2012). PCR has equal MLR results during the time that it applied all "significant components as predictors." This study used "ridge regression" to compare it with other "transfer functions" applying similar predictions, while "PCR was beyond the scope of this study" (Chen et al. 2018b; Jeong et al. 2012).

- **Artificial Neural Networks**

An ANN model uses three layers: an input layer, a hidden layer and an output layer. The "three-layer ANN equation is given by the following" (Jeong et al. 2012):

$$Y = g_2\left(\left(\sum_j W_j.g_1\left(\sum_i w_{ji}X_i + W_{j0}\right) + W_0\right)\right)$$

"i and j" show the the "input and hidden nodes" (Jeong et al. 2012), respectively, "X_i $[n \times 1]$" (Jeong et al. 2012), is an"input layer predictor vector," and "Y $[n \times 1]$ is a local predictor vector, and output layer Is. Input layer biases (w_0) and output layer biases (w_{j0})" (Jeong et al. 2012), including allowing intermediate level adjustment at each step (Jeong et al. 2012). In this equation, we have two groups of adjustable weights: "(w_{ji}) controls the connection power between input node i and hidden node j, and w_j controls the connection between hidden node j and output node. g_1 and g_2 are the activation (or transfer) functions for the hidden layer and the output layer" (Jeong et al. 2012), respectively (Jeong et al. 2012). The "activation function" (Jeong et al. 2012), selected as a "continuous and finite linear or nonlinear transmission function" (Jeong et al. 2012). This study used a "hyperbolic tangential sigmoid function" (Jeong et al. 2012), for the "latent layer and a linear function for the output layer" (Jeong et al. 2012). A detailed description of the various "activation functions is provided in Haykin (1994)" (Jeong et al. 2012).

ANN was tested under the "error reprogramming algorithm" (Haykin 1994; Jeong et al. 2012). Jeong et al. (2012), Moré (1978) used the "Levenberg–Marquardt cross-diffusion (LMBP)" method to teach the ANN network. "LMBP," a "second-order nonlinear optimization technique, is faster and more dependable than other rear diffusion techniques" (Jeong et al. 2012).

Another critical problem in "ANN" is determining the number of "hidden nodes" (Jeong et al. 2012). There is not still a complete theory to solve this problem, and it depends more on the situation. Fletcher and (Fletcher and Goss 1993) showed that the "optimal number of hidden nodes" is usually in $(2p^{05} + o) \sim (2p + 1)$, where p and o show the number of predictions and variables (Jeong et al. 2012), respectively, moreover this study choose the "optimal number of hidden nodes using trial and error" (Jeong et al. 2012).

7.2.1.4 Stochastic Weather Generator

The stochastic downscaling view usually involves correcting the variables of common meteorological generators like "CLIGEN, WGEN, LARS-WG or EARWIG." Climate change scenarios are generated randomly using a set of modified parameters appropriate to the host GCM outputs. The main benefit is, it can produce several observed "climatic statistics" and is She is broadly used initially to assess the effects of agriculture. In addition, stochastic "generators enable the efficient production of large groups for risk analysis" (Wilby et al. 2002). The main disadvantages are the low reproductive skills of inter-annual to multi-year climatic variability and rainfall's unforeseen effects on secondary variables such as temperature.

- **CLIGEN**

"CLIGEN was initially developed for WEPP (Water Erosion Prediction Project)" (Chen et al. 2008) to predict inflow, sediment and soil erosion at the hill and watershed scales (Chen et al. 2008; Flanagan et al., n.d.; Laflen et al. 1997). It is applied to generate artificial daily weather information which is statistically similar to the recent climate or by spatially interpolating model parameters from adjacent measurement sites to create daily weather for illegal areas Slowly (Baffaut et al. 1996; Chen et al. 2008). Most crucially, "CLIGEN" is suitable for producing daily meteorological series from the monthly output of the "General Circulation Model (GCM) to evaluate soil erosion and crop yield (Chen et al. 2008; Zhang 2003; Zhang and Liu 2005)" (Chen et al. 2008). Its variables can be easily manipulated to simulate desired variance values and change for "sensitivity analysis" (Chen et al. 2008), or intentionally changed to mimic mean and variance alters as predicted by GCM to evaluate the expected effect (Chen et al. 2008; Zhang 2005).

Numerous studies have been performed to calculate the ability of "CLIGEN" to generate "non-sedimentary" variables like "solar radiation (SR)" (Chen et al. 2008), wind speed (TU), dew point temperature (T_{dp}), and maximum (T_{max}) and minimum (T_{min}) air temperature (Chen et al. 2008). Johnson and Rasolofosaon (1996) evaluated them in full at six climate-scattered places in the United States, and (Headrick and Wilson 1997) carried similar research in five locations (Chen et al. 2008). Both researchers used previous versions of "CLIGEN," which used a "conditional probability" (Chen et al. 2008), weight coefficient to adjust the deviation of the "T_{max}, T_{min}, and SR" (Chen et al. 2008), criteria according to current and past day rain to establish an appropriate correlation among "precipitation, temperature, SR and wind" (Chen et al. 2008). This adjustment declined the temperature variance, and "SR" was after that removed from the model (Zhang et al. 2004). Compute the ability of "CLIGEN (v.5.107) to generate non-rainfall" (Chen et al. 2008), parameters at four Oklahoma stations and evaluated the potential effects of reproduced variables on winter wheat simulation efficiency applying the "WEPP" model (Chen et al. 2008). The results showed that the yield of simulated wheat was sensitive to production temperature but not to SR (Chen et al. 2008). "CLIGEN" (Chen et al. 2008) did not maintain proper serial and day-to-day interactions for and between daily temperature and "SR" according to the independence assumptions (Chen et al. 2008).

The CLIGEN model produces "T_{max}, T_{min}, T_{dp}, SR, u and wind direction daily" (Chen et al. 2008). Also, the "rainfall, duration, peak intensity" (Chen et al. 2008) of the storm, "time to peak intensity," and occurrence of "daily precipitation" (Chen et al. 2008) are based on long-term statistical parameters of the meteorological station. "SR is generated daily, T_{max}, T_{min}, T_{dp}" Chen et al. (2008) are reproduced using natural distributions, and "u" are "generated daily using normalized (skewed)" (Chen et al. 2008), distributed distributions. The monthly average and the deviation of the daily T_{max}, T_{min}, criteria are obtained directly from the station records (Chen et al. 2008).

The average monthly T_{dp} is obtained daily from the site records, but the standard deviation is achieved from the T_{min}. The meteorological station "long-term statistical parameter" file (.par) is used by "CLIGEN" (Chen et al. 2008) to generate daily meteorological collections. Daily "T_{max}, T_{min}" are produced independently in versions prior to v5.2253 (Chen et al. 2008; Flanagan et al., n.d.):

$$T_{max} = \mu_{max} + \sigma_{max} \times X$$

$$T_{min} = \mu_{min} + \sigma_{min} \times X$$

$$T_{dp} = \mu_{dp} + \sigma_{min} \times X$$

In v5.22564, T_{max}, T_{min}, and T_{dp} are generated for each day to communicate with each other (Chen et al. 2008). The minor standard deviation T_{max}, T_{min}, is used as the base, and the selected parameter conditions the other two parameters (Chen et al. 2008). Thus, while the standard deviation of "T_{max} is larger than or same to the standard deviation of T_{min}" (Chen et al. 2018b), the daily temperature can be obtained as follows:

$$T_{min} = \mu_{min} + \sigma_{min} \times X$$

$$T_{max} = T_{min} + (\mu_{max} - \mu_{min}) + \sqrt{\sigma_{max}{}^2 - \sigma_{min}{}^2} \times X$$

$$T_{dp} = T_{min} + (\mu_{dp} - \mu_{min}) + \sqrt{\left|\left(\frac{\sigma_{max} + \sigma_{min}}{2}\right)^2 - \sigma_{min}{}^2\right|} \times X$$

While the "standard deviation of T_{max}" is less than those of "T_{min}, daily temperatures" (Chen et al. 2018b), are obtained as below (Chen et al. 2008):

$$T_{max} = \mu_{max} + \sigma_{max} \times X$$

$$T_{min} = T_{max} + (\mu_{max} - \mu_{min}) + \sqrt{\sigma_{min}{}^2 - \sigma_{max}{}^2} \times X^T$$

$$T_{dp} = T_{max} + (\mu_{max} - \mu_{dp}) + \sqrt{\left|\left(\frac{\sigma_{max} + \sigma_{min}}{2}\right)^2 - \sigma_{min}{}^2\right|} \times X$$

where "μ is the monthly mean of daily temperatures, μ is the standard deviation of daily temperatures, and X is a generated standard normal deviation" (Chen et al. 2008). The "standard deviation" (Chen et al. 2008) for each day is obtained by two random numbers. The second issue today will be decline as the first issue tomorrow (Chen et al. 2008). A domain check is applied in "v5.111 to force the daily T_{min} to be less than the T_{max}" (Chen et al. 2008) because they are produced independently. In "v5.22564" (Chen et al. 2008), this domain check is not necessary because the new conditional scheme ensures that "T_{max} is always higher than T_{min} SR" Chen et al. (2008) is produced daily as follows Chen et al. (2008):

$$SR = \mu_{sr} + \sigma_{sr} \times X$$

where "μ_{sr} and σ_{sr} Are monthly mean and standard deviation of daily SR. A maximum SR (SRmax)", Computed from the site width and day of the year, the following formula is applied to calculate the "standard deviation of the daily SR in v5.22564" and earlier versions (Chen et al. 2008):

$$\sigma_{sr} = \frac{SR_{max} - \mu_{sr}}{4}$$

The "wind direction" is separated into 16 main directions (Chen et al. 2008). It is created by sampling the "cumulative distribution" (Chen et al. 2008) of the percentage that the "wind blows" from each of these ways with a stochastic number in (0–1) (Flanagan et al. 1995):

$$u = \mu_u + \frac{2\sigma_u}{\xi}\left\{\left[\frac{\xi}{6}\left(x - \frac{\xi}{6}\right) + 1\right]^3 - 1\right\}$$

Where "μ_u, σ_u And ξ" are the monthly mean, standard deviation, and skewness coefficient of daily u, respectively, for the generated cardinal direction and month (Chen et al. 2008). The mentioned presentation comes from the CLIGEN common mode, in daily meteorological collections are reproduced utilizing monthly statistics free-from interpolation among months (Chen et al. 2008). To produce a continuous and continuous daily air, several interpolation methods (Chen et al. 2008), such as February, are applied in this method to reduce the monthly "statistics" (Chen et al. 2008) to the daily equivalent. Then the daily low volume parameters in the equation are used to reproduce daily meteorological data (Chen et al. 2008).

- **CLIGEN Storm Generation**

The CLIGEN model is based on long-term statistical parameters of monthly, daily rainfall, "depth, D, IP, peak time and maximum daily values, minimum air temperature, dew point temperature, sunlight, wind speed, and wind direction." Precipitation variables are independent of other air variables. "Precipitation and storm patterns" are shown here only (Chen et al. 2009).

"Mean standard deviation and coefficient of deviation of daily rainfall depth, probability of rainfall (wet day followed by a wet day and wet day after dry day), average 0.5 h maximum rainfall intensity per month and time to peak parameters directly from the record Daily station rainfall is extracted" (Chen et al. 2009). A first-order two-state Markov chain generates rain for a day depending on whether it was wet or dry the day before. While a "random number is taken from the uniform distribution for each day less than the probability of precipitation is given for the situation of the previous day, a precipitation event is predicted. For an expected rainy day, the normal deformed (skewed) distribution is used to generate the daily rainfall depth for each month (Lane 1989)" (Chen et al. 2009):

$$x = \frac{6}{g}\left\{\left[\frac{g}{2}\left(\frac{R-\mu}{s}\right)+1\right]^{1/3}-1\right\}+\frac{g}{6}$$

Where "x is the standard normal deviate; R is the daily precipitation depth (mm); μ, s and g are the mean (mm), SD (mm), and skew coefficient of daily depths for the month," respectively (Chen et al. 2009). Two random numbers are used to generate x (the second number for today is reused as the first number for tomorrow), which is then used in the above equation to compute R (Chen et al. 2009). Assuming rainfall rates during a storm decrease exponentially from the maximum rate, the peak intensity (rp) (mm/h) (Chen et al. 2009) "was generated as Arnold and Williams (1989)" (Chen et al. 2009):

$$rp = -2Rln(1-\alpha_{0.5})$$

where $\alpha_{0.5}$ is the ratio of the maximum 0.5-h rainfall depth to the total depth and is drawn from a gamma distribution (Chen et al. 2009). The gamma distribution shape parameter is set to "6.2832 in CLIGEN (v5.22564)", "mean $\alpha_{0.5}$ for the month ($\alpha_{0.5}mean$)" (Chen et al. 2009) is calcute in the model by:

$$\alpha_{0.5mean} = R_{0.5mean}/R_{mean}$$

where "$R_{0.5\mathrm{mean}}$ and R_{mean} are the means of R0.5 (maximum 0.5-h depth) and R for the month, respectively, the $R_{0.5\mathrm{mean}}$ is related to the mean of annual maxima for each month ($R_{0.5\mathrm{max}}$)" (Chen et al. 2009) by:

$$R_{0.5mean} = -R_{0.5max}/lnF$$

where "$R_{0.5\mathrm{max}}$ is an input parameter, and F is the exceedance probability for $R_{0.5\mathrm{max}}$ and is estimated (Yu 2000; Chen et al. 2009)" (Chen et al. 2009) by:

$$F = 2/(2n + 1)$$

where "n is the average number of monthly rain days, the D is (h)" (Chen et al. 2009) can be obtained as:

$$D = -0.5\Delta/ln(1 - \alpha_{0.5})$$

where Δ takes a value of 3.99 in v5.22564 (Chen et al. 2009). Baffaut et al. (1996) pointed out that the mentioned formulas are not final and subject to change as more historical rain information is analyzed. The "ip is calculated as" (Chen et al. 2009):

$$ip = rpD/R$$

By combining the above formulas, ip can be obtained as below:

$$ip = -2Dln(1 - \alpha_{0.5})$$

Since "there are different storm forms in the environment, even in the equal season in the same region" (Chen et al. 2009), it is challenging to produce domestic storm designs. Three hypotheses were performed to ease the simulation of the storm pattern in CLIGEN. First, there is only one storm occurrence on a wet daytime, which indicates that calculating D requires combining several parts of rainfall in one day, and the maximum D is minor than 24 h. Second, each storm trend has only one peak. Third, all storm forms can be described with a two-dimensional function. Arnold et al. (1990) showed that D is "distributed exponentially" (Chen et al. 2009). So, to test this initial observation, D was also generated utilizing the standard exponential distribution (Chen et al. 2009):

$$D = -\ln(1 - X)D_m$$

Where "X is a uniform random number (0 < X < 1) and Dm (h) is the overall mean D for a station average in all months" (Chen et al. 2009).

- **WGEN**

WEGEN (Weather Generator) is a computer simulation model. This model is designed to produce "daily costs for rain," solar radiation, "max and min temperature." This model follows the method defined by Richardson (1982). Different hypotheses have been made that ease the use of the model.

WGEN produced a daily amount of "precipitation (p), maximum temperature (T_{max}), minimum temperature (T_{min}), and solar radiation (r)" (Richardson and Wright 1984) for one year at a specific Place. Rain on a particular day has a significant effect on the "temperature and sunlight." The approximation applied is the production of rainfall for a specific day depending on other parameters. Maximum and minimum temperatures and sunlight are produced depending on whether it is already a dry day or a wet day. This model is defined to maintain the connection in time, the correlation among parameters and seasonal features for the region (Richardson and Wright 1984; Skiles and Richardson 1998).

Precipitation

The WGEN precipitation section is a gamma-ray Markov model. To create dry or wet days, a first-class Markov chain is applied. During the wet day occurs, a "two-parameter gamma distribution" is utilized to generate "rain" (Richardson and Wright 1984). With the "first-class Markov chain model, the probability of rain" on a given day depends on the dry or wet situation of the past day. Wet days are explained as a day with 0.01 inches or more of rainfall (Richardson and Wright 1984). Let "P_i (W/W) be a wet day in the day, give me a wet day in day $i-1$, and let P_i (W/D) be a wet day in my day, a dry day in the day $i-1$ is given" (Richardson and Wright 1984; Skiles and Richardson 1998). Then,

$$P_i(D/W) = 1 - P_i W/W$$
$$p_i(D/D) = 1 = P_i(W/D)$$

where P_i (D/D) and P_i (W/D) are the probabilities of a dry day on day i−1 on a wet day and the probability of a dry day on day i−1 on a dry day, so, the transfer probabilities concerning P_i (W/D) and P_i (W/W) are fully described (Richardson and Wright 1984). Many probability density functions have been used to explain the amount of rainfall distribution "(Skiles and Richardson 1998; Smith and Schreiber 1974; Woolhiser and Roldan 1982)" (Richardson and Wright 1984).

For this program, a "distribution with minimum parameters" was required to define the parameters for many locations (Richardson and Wright 1984). The results of studies (Richardson's 1982) have shown that the gamma distribution of two parameters is noticeably better for explaining daily rain values than the "exponential distribution" of a simple value (Richardson and Wright 1984). The below equation defines the density function of the gamma distribution of two parameters (Skiles and Richardson 1998).

$$f(p) = \frac{p^{\alpha-1}e^{-p/\beta}}{\beta^{\alpha}\Gamma\alpha}, p, \alpha, \beta > 0$$

In Where e is the basis of natural logarithms, "$\Gamma(\alpha)$ is the gamma function α, β is the distribution parameter, and $f(p)$ is the density function p, α. α and β are shape and scale parameters, respectively. For $0 < \alpha < 1, f(p)$ decreases with increasing p (Skiles and Richardson 1998)" (Richardson and Wright 1984).

The pattern is suitable for position, where e is the basis of natural logarithms. "α and β are distribution parameters, (p) is a density function of p, $\Gamma(\alpha)$ is a gamma function α. α and β are form and scale parameters, respectively" (Richardson and Wright 1984). For "$0 < \alpha < 1$, f (p) decreases with increasing p" (Richardson and Wright 1984). Shape/shape is suitable for precipitation values because low values occur more than large values (Richardson and Wright 1984). "P (W/W), P (W/D), α and β vary continuously via the year for several regions (Richardson and Wright 1984). In the "WGEN" model, all four rain values are fixed for a given month but chain from month to month (Skiles and Richardson 1998; Richardson and Wright 1984).

The parameters are applied to produce daily precipitation values using the "Markov chain generation method and the gamma generation method described by Haan (1977)" (Richardson and Wright 1984).

Solar Radiation and Temperature

The method applied in "WGEN for procreating daily parameters of T_{min}, T_{max}, and r are defined by Richardson (1981). The method is according to the weakly static generating process given by Matalas (1967)" *(Richardson and Wright 1984). The formula is:

$$x_i(j) = A_{X_{i-1}}(j) + B\varepsilon_i(j)$$

where "$x_i(j)$ is a 3 X_i matrix for a day i whose elements are residuals of T_{max} $(j = 1)$, T_{min} $(j = 2)$, and r $(j = 3)$, ε_i is a 3 X_i Matrix of independent components and A and B are 3 × 3 matrices that elements are described such that the new sequences have the desired serial correlation and cross-correlation coefficients (Skiles and Richardson 1998)" (Richardson and Wright 1984). The A & B matrices are given by:

$$A = M_1M_0^{-1}$$

$$BB^T = M_0 - M_1M_0^{-1}M_1^T$$

where the "superscripts −1 and T noted the inverse and transpose of the matrix, M_0 and M_1" (Skiles and Richardson 1998) are given as:

$$M_0 = \begin{bmatrix} 1 & \rho_{0(1,2)} & \rho_{0(1,3)} \\ \rho_{0(1,2)} & 1 & \rho_{0(2,3)} \\ \rho_{0(1,3)} & \rho_{0(2,3)} & 1 \end{bmatrix}$$

$$M_1 = \begin{bmatrix} \rho_1(1) & \rho_{1(1,2)} & \rho_{1(1,3)} \\ \rho_{0(1,2)} & \rho_1(2) & \rho_{1(2,3)} \\ \rho_{0(3,1)} & \rho_{1(3,2)} & \rho_1(3) \end{bmatrix}$$

where "$\rho_0(j, k)$ is the correlation coefficient among variables j and k with variable k lagged one day concerning variable j, ρ_1 (j, k) is the correlation coefficient among variables j and k on an equal day, and ρ_1 (j) is the lag one serial-correlation coefficient for variable j" (Richardson 1982). The spatial and seasonal variation in the correlation coefficients were approximately small. If the slight variations are negative, and the mean values of the "correlation coefficients" given by Skiles and Richardson (1998) are used, the "M_0 and M_1" matrices become (Richardson and Wright 1984):

$$M_0 = \begin{bmatrix} 1.000 & 0.633 & 0.186 \\ 0.633 & 1.000 & -0.193 \\ -0.186 & -0.193 & 1.000 \end{bmatrix}$$

$$M_1 = \begin{bmatrix} 0.621 & 0.445 & 0.087 \\ 0.563 & 0.674 & -0.100 \\ 0.015 & -0.091 & 0.251 \end{bmatrix}$$

By using the above equations, the A and B matrices become Skiles and Richardson (1998):

$$A = \begin{bmatrix} 0.567 & 0.086 & -0.002 \\ 0.253 & 0.504 & -0.050 \\ -0.006 & -0.039 & 0.244 \end{bmatrix}$$

$$B = \begin{bmatrix} 0.781 & 0 & 0 \\ 0.328 & 0.638 & 0 \\ 0.238 & -0.341 & 0.873 \end{bmatrix}$$

The A and B matrices are applied with the above equation in "WGEN" to produce new sequences of the "residuals of T_{max}, time, and r that are sequentially correlated and cross-correlated with the correlations being steady at all locations (Skiles and Richardson 1998)", (Richardson and Wright, 1984).

The final daily reproduced parameters of "T_{max}, time, and r are determined by multiplying the residual elements" caused by a "seasonal standard deviation and adding seasonal mean" (Skiles and Richardson 1998) using the equation (Richardson and Wright 1984):

$$t_i(j) = x_i(j).s_i(j) + m_i(i)$$

where "t_i (j) is the daily value of $t_{max}(j = 1)$, $t_{min}(j = 2)$ and r (j = 3), s_i (j) is the standard deviation and m_i (j) is the mean for a day i. The values of $m_i(j)$ and $s_i(j)$ are conditioned on the dry or wet status as specific from the precipitation part of the model" (Richardson and Wright 1984). By representation "terms of the coefficient of variation (c = s/m) rather than the standard deviation" (Richardson and Wright 1984; Skiles and Richardson 1998), the equation becomes:

$$t_i(j) = m_i(j)\left[x_i(j).c_i(j) + 1\right]$$

The "seasonal change in the average and coefficients of variation may be described by Skiles and Richardson (1998)" (Richardson and Wright 1984):

$$U_i = \overline{u} + C\cos(0.712(i - T)), i = 1, \ldots, 365$$

where "U_i is the value of the $m_i(j)$ or c_i(j) on the day i, u is the average of u_i, C is the harmonic amplitude, and T is the harmonic position in days" (Richardson and Wright 1984). Values of "U, C and T" should be determined for each weather variable's mean and coefficient "(T_{max}, time, and r) and the wet or dry condition" (Richardson and Wright 1984). For more information, see Richardson and Wright (1984).

WGEN Rainfall Parameter Assessment in a Dry Area

In WGEN, it is impossible to compute air production variables with less than two moist days in a month longer than the record duration since at least two information points are needed to calculate the two gamma distribution variables. Therefore, whenever it is wet for less than two days in a month, "P (W/W) and P (W/D)" are certainly a value of zero in "WGEN." Under these conditions, the scale parameters of Figure "(β and α)" take a "nominal value of zero" to indicate that they are not defined, and "WGEN" never rains this month when computing artificial air.

For a "10-year" record, with "10%" of wet days, about "30 wet days" are expected, well above the two required to generate "WGEN" variables.

In arid places, a ten-year history contains a wet day in a particular month is expected only once in ten years, just one wet day. A data point leads to zero WGEN parameters, meaning that even with a 100-year simulation, it is predicted that there will be no rainfall over a month for one decade. This insufficiency of the "WGEN system" is exacerbated by reducing the number of data years available to estimate the parameter or a decrease in precipitation frequency.

During the time that there are more than two humid days in a monthly period over the information record length, there can be problems in calculating agreeable

"WGEN" variables. The shape parameters for a given month "j (α_j)" is computed in the "WGEN system" utilizing the "maximum likelihood estimator ($\widehat{a_j}$)" showed by Greenwood and Durand:

$$\widehat{a_j} = \frac{8.898919 + 9.05995y + 0.9775373y^2}{y(17.79728 + 11.968477y + y^2)} \quad 0.5772 < y \le 17.0$$

$$\widehat{a_j} = \frac{0.5000876 + 0.1648852y - 0.0544274y^2}{y} \quad 0 \le y \le 0.5772$$

$$if \widehat{a_j} \ge 1 \rightarrow \widehat{a_j} = 0.998$$

where y is a "function of the precipitation amount in a month," as shown in the next equation. In this group of formulas, "r_{ji} is the precipitation amount for a wet day i in month j, and n is the number of rainy days in month j".

$$\overline{r_j} = \sum_{i=1}^{n} \frac{r_{ji}}{n}$$

$$\overline{lnr_j} = \frac{\sum_{i=1}^{n} lnr_{ji}}{n}$$

$$y = ln\overline{r_j} - \overline{lnr_j}$$

The "moment estimator for the scale parameter, $\widehat{\beta}$", is computed in the "WGEN system" using below equation:

$$\widehat{\beta_j} = \frac{\overline{r_j}}{\widehat{a_j}}$$

Once these parameters have been computed, it is calculating the scale and shape variables for each month by turning "$\alpha_j \approx \widehat{a_j}$ and $\beta_j \approx \widehat{\beta_j}$".

Correction of Precipitation and Temperature

For several areas, the information produced by these methods has a mean monthly temperature and rain near to the average achieve from the actual information. In some situations, spatial smoothing occurs differently when "present in the pattern or topographic features of the site or other things. Procedures have expanded that supply for correcting these discrepancies if the real average monthly amount is assessable and the user chooses to make these corrections". The "use of correction options provides daily generated" parameters close to the monthly average from actual observations (Richardson and Wright 1984). Applying the correction method needs that the real monthly average of the parameters is entered into the production program for correction. According to the report (Groisman and Legates 1994), the average monthly rainfall and/or temperature is available for those places from several sources, like the US Climate Atlas.

The precipitation correction factor is computed as the average monthly "rainfall" from real information, separated from the average monthly rainfall and theoretically produced by the Markov chain gamma model. The precipitation correction factor doubles the daily precipitation produced for the suitable month to achieve the modified rainfall.

Temperature correction is possible to be according to the real average "monthly temperature" or the average "maximum and minimum temperature," depending on the type of information is accessible for an area (Richardson and Wright 1984). For the average "monthly temperature, the temperature correction factor" is computed as the difference between the actual average "monthly temperature" for the area and the average "monthly temperature" (Richardson and Wright 1984), which is apparently reproduced using site variables. The maximum and minimum daily generated temperature are modified by adding a "correction factor" (Richardson and Wright 1984), to the generated temperature. During the time that "average monthly maximum and minimum temperatures" (Richardson and Wright 1984), are accessible, the "maximum temperature and minimum temperature correction" factors are calculated separately.

WGEAN Program

This program is applied to produce daily amounts of sunlight, precipitation, maximum and minimum temperature. This application includes two essential options. When option one is selected, the program generates daily values of "p (inches), maximum (°F), time (°F) and r (ly)" for the number of years selected by the user. If option two is selected, the program reads the real user-generated rainfall and generates the corresponding values of T_{max}, time and r. Option 2 (Richardson and Wright 1984) is provided since; Frequently, a client will possess a long history of real rainfall information with data on "sunlight or daily temperature" (Richardson and Wright 1984). Some options allow the applicant to modify the rainfall and temperature generated according to real data (Richardson and Wright 1984). The user is possible to choose

(1) Do not make any corrections
(2) Correct temperature and rainfall
(3) Only correct rainfall
(4) Just correct the temperature.

WGEN print the daily mount of the four parameters. "A summary of the monthly and annual amounts is published at the final of the year. At the final days of year 10 in study period" (Richardson and Wright 1984), the average and average monthly values are published at that time.

To reproduce weather information for a location outside of 48 adjacent locations or develop production variables for real data from a special site, the "WGEN PAR program reads the daily parameters of p, max, time, and r" (Richardson and Wright 1984). It sets the production parameters needed by "WGEN" (Richardson and Wright

1984). Author. The number of years of weather information requires to expand variables representing a particular place varies with the weather. Generally, at least two decades of rainfall and one decade of "temperature and radiation" (Richardson and Wright 1984) are needed. Long-term recording of rainfall may be necessary for dry places (Richardson and Wright 1984).

- **LARS-WG**

"LARS-WG is a stochastic weather generator" (Semenov et al. 2002). This mode is used to simulate weather information at a single location under future and recent conditions (Racsko et al. 1991; Semenov 2007; Semenov et al. 1998; Semenov and Brooks 1999). The time-series information that is used in this model is in the shape of daily, for a sequence of climate parameters, namely, "min and max temperature (°C), precipitation (mm), and solar radiation (JMM-2 day-1)" (Semenov et al. 2002).

"Stochastic weather generators" (Semenov et al. 2002), originally are expanded for below main goals:

1. Providing statistical characteristics corresponding to simulating synthetic weather time-series to the observed statistics at one location, this data must be long enough to evaluate risk in agricultural or hydrological applicants (Semenov et al. 2002).
2. Providing an average of "extending the simulation of weather time-series to unobserved locations by using the interpolation of the weather generator parameters to run the models at neighboring of the main sites" (Semenov et al. 2002). It is essential to mention that a "stochastic weather generator" (Semenov et al. 2002). It is a simple means to generate "time-series of synthetic weather statistically similar to the observations" (Semenov et al. 2002). It is not a predictive way that is applied in weather forecasting. One result of climate change studies is the new interest in local stochastic weather simulation (Semenov et al. 2002).

Nowadays, global climate models (GCMs) output from temporal resolution and insufficient spatial and reliability to be applied directly in the effect of models (Semenov et al. 2002). However, a "stochastic weather generator" (Semenov et al. 2002) can serve as a "computationally inexpensive tool to produce multiple-year climate change scenarios at a daily time scale (Semenov and Barrow 1997)". The primary version of the "LARS-WG weather generator was expanded in Budapest in 1990 as a section of Assessment of Agricultural Risk in Hungary, a project funded by the Hungarian Academy of Sciences (Racsko et al. 1991)" (Semenov et al. 2002). This work's focus was to "overcome the limitations of the Markov chain model of precipitation occurrence (Bailey 1990)" (Semenov et al. 2002). This expanded used wat of modelling rainfall occurrence is not almost capable to correctly simulate the max dry spell duration; it is essential for a realistic evaluation of agricultural production in several areas of Hungary in the world (Semenov et al. 2002).

The result is the new series' view because the wet and dry and spell duration simulation is the initial step in the "weather generation process" (Semenov et al. 2002).

"A correct version of this weather generator, now called LARS-WG" (Long Ashton Research Station Weather Generator-the location, it was developed in its current form), was applied in the basis of the climate change scenarios and applied in two European Union-funded study the effects of climate change on agricultural potential in Europe, for example CLAIRE (Harrison et al. 1995) and CLIVARA (Downing et al. 2000). More details of these high-resolution climate change scenarios is possible to found (Semenov and Barrow 1997). The crucial version of LARS-WG (version 3.0 for Windows 9x/NT/2000/XP) has undergone a complete redevelopment to produce a robust metod to producing synthetic weather information for a vast range of climates (Semenov et al. 2002). LARS-WG has compared with other applied stochastic weather generators which used the Markov chain approach (WGEN; (Richardson 1981; Richardson and Wright 1984) at many places representing different climates and shown to do at least as well as, if not better than, WHEN at each of these places (Semenov et al. 1998, 2002).

Model Description

The "LARS-WG" (Semenov et al. 2002) is in line with the series meteorologist described in Racsko et al. (1991) uses quasi-experimental distributions for "dry and wet day series" (Semenov et al. 2002), lengths, daily sunlight, and daily rainfall. Histogram, quasi-experimental "distribution Emp = {a_0, a_i; hi, i = 1,….,10} is with t hi show the number of events from the observed data in the ith interval and en intervals, [a_{i-1}, ai], where $a_{i-1} < a_i$" (Semenov et al. 2002).

Random values are selected from quasi-experimental distributions by the initial selection, period selection (applying the ratio of events in a time "interval as the probability of selection"), and selection of the value at that distance from the equal value from the initial values (Semenov et al. 2002). Such a distribution is flexible and can take approximately different shapes by adding distances "[a_{i-1}, a_i]" (Semenov et al. 2002). The value of this "flexibility is that the distribution requires 21 parameters" (Semenov et al. 2002), compared to, for an instant, the three variables for the exponential mix of distributions applied in the latest version (Semenov et al. 2002). A model for determining the series of "dry and wet days" (Racsko et al. 1991; Semenov et al. 2002).

Distances "[a_{i-1}, a_i]" (Semenov et al. 2002) are chosen according to the expected properties of the air parameters. The distance "[a_{i-1}, a_i]" is evenly spaced among the "minimum and maximum information observed for a solar month" (Semenov et al. 2002). In contrast, for the duration of dry and wet series and rainfall, the size of the distance gradually Improves with increasing I. In the latter two cases, there are usually minimal values and enormous values, and this choice of distance structure means that huge resolution is not used for small values (Semenov et al. 2002). Precipitation simulation simulates modelling as a dry and wet series, where a wet day represents a day with a "participation > 0.0 mm" (Semenov et al. 2002). The duration of each series is randomly selected from the semi-experimental "wet or dry distribution" for the month of the series start. In the definition of distributions, the observed series is also assigned to the fish they are formed (Semenov et al. 2002). On a wet day, the amount of precipitation is generated from the semi-experimental distribution of precipitation for a particular month independent of the length of the wet series or the amount of precipitation in the previous days (Semenov et al. 2002). The minimum and maximum daily temperatures are random processes with daily mean and daily standard deviation that depend on the wet or dry state of the day (Semenov et al. 2002). The method applied in the process simulation is the same as the method used in the study (Racsko et al. 1991).

The seasonal periods of the medium and standard deviations are designed with the Fourier finite series, approximating the normal distribution of the residuals (Semenov et al. 2002). The February series for the norm is proportional to the average values observed for any month. Before working with the Fourier series, the "standard deviation, the experimental standard deviation" (Semenov et al. 2002), is determined every month. Removing the calculated influence of the average changes in a month period estimates the average daily standard deviation (Semenov et al. 2002). This setting is computed applying the "Fourier series" (Semenov et al. 2002), previously calculated for average obtained (Semenov et al. 2002). The observed remainder analyzes the temporal correlation correlations for minimum and maximum temperatures by removing appropriate average values from the experimental information (Semenov et al. 2002). For simplicity, it is assumed that both are constant throughout the year for wet and dry days, and the average amount of data observed applies (Semenov et al. 2002). The minimum and maximum temperature residues have a predetermined correlation of 0.6. Sometimes, the simulated minimum temperature is higher than the simulated maximum temperature, in which case the program changes the small temperature by a maximum of 0.1. Analysis of daily solar radiation in many places showed that the natural distribution of daily solar radiation, generally in other air generators, is unsuitable for specific areas (Chia and Hutchinson 1991).

The distribution of sunlight also varies, especially on dry and wet days (Semenov et al. 2002). Therefore, semi-semi-different irony distributions were used to describe the sun's rays on "wet and dry days." An auto-correlation coefficient was computed for "solar radiation" and was "constant" all over the year. Sunlight is simulated separately from temperature (Semenov et al. 2002). "LARS-WG" adopts sundials as an alternative to "solar radiation" information (Semenov et al. 2002). If sunlight information is not available, then sunshine may be applied. These are manually transformed to solar radiation utilizing the described method (Rietveld 1978).

Scheme of the Stochastic Weather Generating Procedure

Generation synthetic weather information can be separated into three recognizable stages:

1. Site Analysis "Calibration of the observed SATE model analysis to determine their statistical characteristics. This information is stored in dual parameter files" (Semenov et al. 2002).
2. "Validation of the QTEST model The statistical aspects of the observed and artificial data are analyzed to define any statistically noticeable differences" (Semenov et al. 2002).

3. Production of artificial climate information “GENERATOR” variable files derived from observed climate information in the calibration of the model, they produce artificial “data with the same statistical characteristics” (Richardson and Wright 1984), as the originally observed information. Artificial data are correlated with a specific “climate change scenario.” They can be created by creating changes derived from the global climate mode in “solar radiation to LARS-WG” variable files, temperature and precipitation. The “LARS-WG” function is explained in Fig. 7.2.

- **SDSM**

“SDSM was first introduced by Wilby and Dawson (2007)” (Wilby et al. 2002) and is expanding applied in hydrological applications in a variety of climatic scenarios. This model uses multiple regression techniques to provide “station-scale” weather data from the “GCM-scale” (Wilby et al. 2002), output with network resolution. This relationship is created between the variable GCMs (so-called predictor) and the application of the regional scale variable (predictor). SDSM was classified as a “hybrid model” (Wilby et al. 2002). He used a “linear regression method” (Wilby et al. 2002), a random meteorological producer. A period of 30 years as a standard reference is suggested for climate change and the study of climate variation in climatology (Mason et al. 2015).

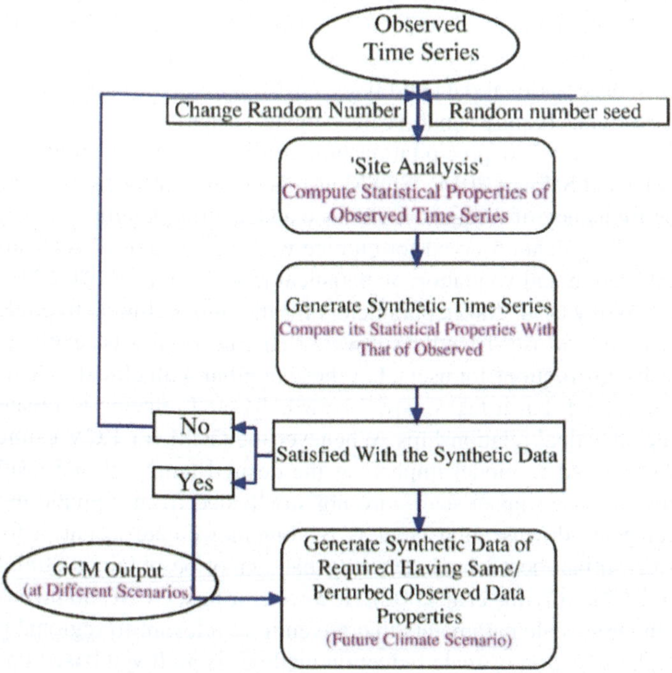

Fig. 7.2 Schematic diagram of LARS WG analysis

This contains two phases: (1) computing whether rainfall happens every day or (2) defining the estimated precipitation per rainy day by considering "GHGs." Consequently, prediction-prediction formulas were developed applying a multilinear regression view for the long-term procuring climate in the region. As a result, the amount of precipitation (y) on day t can be defined by:

$$y_t = F^{-1}[\varphi Z_t]$$

$$Z_t = \beta_0 + \sum\nolimits_{j-1}^{n} \beta_j \widehat{u_t} + \beta_{t-1} + \varepsilon$$

where "F is the experimental function y_t, φ is the natural cumulative distribution function. Z_t is the score z on day t, β is the regression parameter, $\widehat{u_t}$ Is a normal prediction and ε is a variable parameter" (Wilby et al. 2002).

However, "statistical downscaling" has many restrictions (Wangsoh et al. 2017); although the "SDSM model" (Wilby et al. 2002) does not need much computational request to observe the simulation outcomes but can generate "high-quality" (Wilby et al. 2002), prediction outcomes. These benefits made the SDSM a dependable tool for climate mitigation (Samadi et al. 2013; Tukimat and Harun 2015) and were chosen to reduce future climate trends.

SDSM is a user-friendly software package designed to perform statistical minimization methods to develop high-resolution monthly weather data from a simulated large-resolution climate model (GCM). This software applied an air generating approached to generate several realizations (groups) of artificial daily sequences. Using quality observational data and daily GCM outputs for large-scale variables, impact assessments require small-scale climate scenarios. Key Output is "site-specific daily scenarios" for precipitation, humidity, maximum and minimum temperature (Trzaska and Schnarr 2014). SDSM also generates various statistical parameters such as frequency of frequency, variance and spelling length. The key input is the quality of daily data observed for climate variables on a local scale and a large scale for calibration and validation of statistical model (s). Daily "GCM output for large-scale" (Wilby et al. 2002), parameters for the future climate to guide models. This application is a user-friendly software that can usually be explained. It has comprehensive instructions for users. For those familiar with climate science, a little education is needed, but it takes expert knowledge and retrying to create realistic and accurate statistical relationships. Whenever the GCM (or RCM) simulation of the variable (s) used to model impacts at the desired temporal and spatial scales is unrealistic because impact scales are not subdivided from climatic models into model deficiencies, downscaling There is. Scaling may create scenarios for external variables (e.g., urban heat island severity) that cannot be obtained straightly from "GCMs and RCMs" (Wilby et al. 2002). However, the host GCM should have acceptable skills in large-scale parameters that are entirely relevant to regional processes. In practice, the selection of scale reduction methods is archived based on the capability of observational data and GCM since they are needed to create future climate scenarios. "SDSM reduced the task of statistically" (Wilby et al. 2002), minimizing daily meteorological collections to seven separate stages (Wilby et al. 2002):

(1) Check the Quality and transformation of data
(2) Screening of predictor parameters
(3) Calibration the model
(4) Air production
(5) Statistical analysis
(6) Draw the diagram of the result
(7) Scenario generation

SDSM is a described combination of air-generating and random-generating methods in the classification of downscaling techniques. This is according to "large-scale circulation patterns and atmospheric humidity" parameters are applied to situation regional-scale climate-generating variables (e.g., occurrence and precipitation intensity). In addition, random methods are used to artificially inflate the daily slight time-series variance to be more consistent with the observations. The SDSM scale reduction algorithm has been used in a host of meteorological, hydrological and environmental evaluations and a wide variety of geographical areas, contain "Asia, Europe, North America and Africa."

Data Transformation and Quality Control

Few meteorological stations have one hundred percent complete and correct information about an area. The use of lost and incomplete information is almost essential for practical conditions. Routine quantity control checks in "SDSM" allow us to identify underlying errors, prior definitions and missing data codes for calibration. In most cases, it may be appropriate to make predictions or previous predictions and calibrations into a model.

Downscaling Predictor Variables Screening

Understanding the empirical connections among lattice predictions (like an average of sea level pressure) and site predictions (like station rainfall) is vital in all statistical minimization approaches. The primary goal of the operation of display parameters is to help the user select the appropriate predictor of downscaling approaches. It considers one of the most problematic steps in developing any "statistical downscaling" (Wilby et al. 2002) model because the selection of predictors hugely determines the character of the small-scale scenarios. The "decision-making process is complex because the explanatory power of individual predictor variables varies spatially and temporally" (Wilby et al. 2002). Screen variables test seasonal variation in prediction skills.

Calibration the Model

The performance of the calibration model takes a user-defined prediction with a group of prediction parameters. It calculates the variables of multiple regression formula via an optimization algorithm (or two simple ordinary squares).

Specified Model Structure by Users

Seasonal, annual or whether monthly "sub-models are needed (Wilby et al. 2002)". The process is conditional or unconditional. In conditional methods, there is a moderate process among the local climate (e.g., local rainfall blogs about the occurrence of wet days, which belong to local scale predictors like humidity, atmospheric pressure). In "unconditional models" (Wilby et al. 2002), a direct relationship is supposed between forecasters and forecasters (for example, regional wind speed is possible to be a function of local airflow indices).

Weather Generator

The air generating operation creates sets of daily artificial weather sets based on observations of atmospheric predictor variables. This method allows the validation of "calibrated models (using independent information) and the synthesis of artificial time series for existing climatic conditions. Users can easily select a calibrated model, and SDSM" (Wilby et al. 2002) joins all essential predictions to the model's weight. The user should define the recording time and the number of group members. Artificial time series are set to special output files for statistical analysis, charting, and impact modelling.

SDSM includes interrogation of small scenarios and meteorological information viewed with a statistic display and frequency analysis. The user should determine the sub-period, the name of the output file and the selected statistics in both cases. For model output, a group member or average should be defined. Instead, SDSM represents a set of diagnoses, including monthly, seasonal or annual tools, dispersion measures, serial correlation and extremes.

Scenario Generation

Finally, the performance of the generator scenario is a set of daily artificial meteorological sets according to the atmospheric forecasting variables provided with a climate model (for both now and future weather) instead of the observed production forecasts. This function is the same as the climate generator function in all aspects. It is important to determine a "different contract for the model data and the source list for the predictor parameters." Input files for both the "Weather Generator and Scenario Generator options do not require the same length as the files used to achieve the model weight at the calibration steps" (Wilby et al. 2002). Figure 7.3 shows the production scenario of the weather scenario.

- **MAGICC/SCENGEN**

MAGICC/SCENGEN (MAGICC; a model for assessing climate change-related greenhouse gases) is a gas/climate cycle model that drives a space climate scenario (SCENGEN) generator. MAGICC is one of the first models used by the IPCC since 1990 to generate forecasts of average sea levels and future temperature increases (Wigley 2008a,). The climate model at MAGICC is an energy balance, inertia, and diffusion model that produces hemispheric temperatures and the global average r containing the results of ocean thermal expansion (Wigley 2008a). Version 5.3 of

Fig. 7.3 SDSM climate scenario generation process

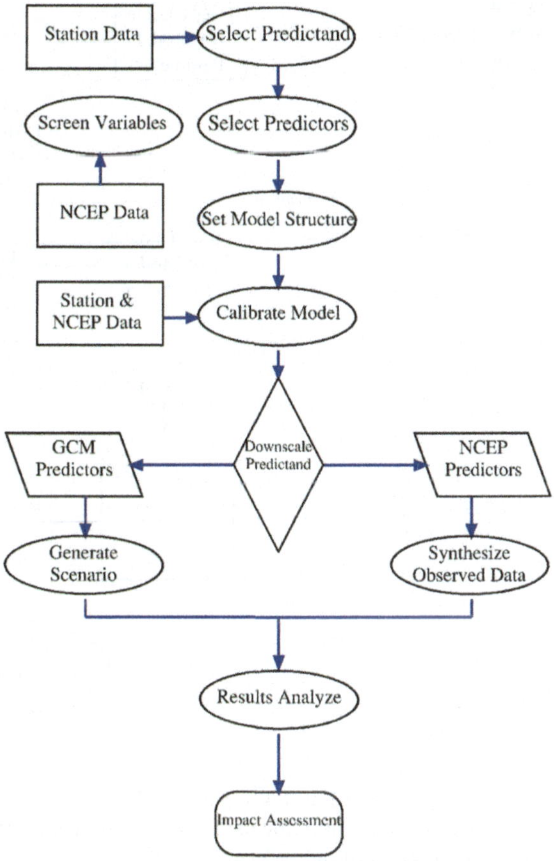

this model is consistent with the IPCC Fourth Assessment Report, Working Group 1 (AR4). Version 4.1 of this software uses the MACICC report of the third IPCC evaluation report, Working Group 1 (TAR) (Wigley 2008a). The MAGICC climate model interacts with a group of gas cycle models that predict significant emissions for greenhouse gas concentrations (Wigley 2008a). Carbon Cycle Climate Feedback Therefore MAGICC is used for SCENGEN driving for global average temperatures (Wigley 2008a). SCENGEN used a copy of the pattern scaling method described in Santer and Wigley (1990) to generate displacement patterns from the Ocean GCM (AOGCM) database of the CMIP3/AR4 archive (Wigley 2008a). The scalability method of the model is based on the separation of global materials and the spatial pattern of other differentials of the second case into greenhouse gas and aerosol components and future climate change (Wigley 2008b).

Spatial patterns are normalized in the database and expressed as changes every one °C at global average temperatures (Wigley 2008a). These normalized aerosol and greenhouse gas components are correctly weighed, augmented, and measured "up to the global average temperature described by MAGICC for a given year, a

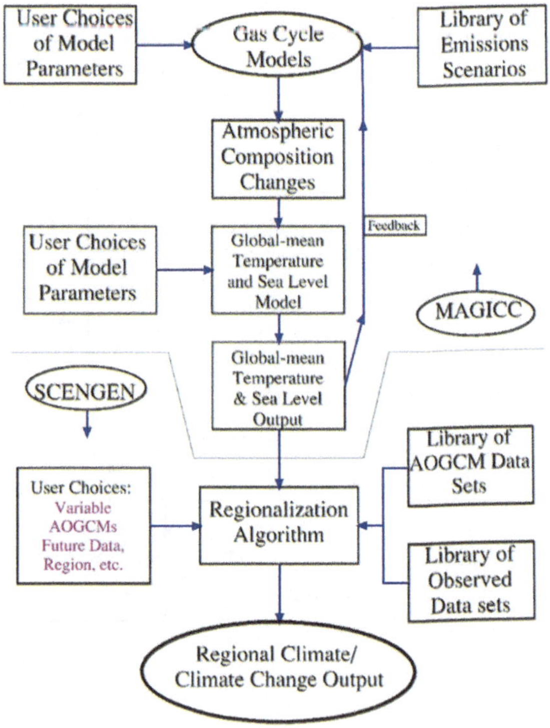

Fig. 7.4 "MAGICC/SCENGEN" diagram

set of climate model parameters and emission scenarios" (Wigley 2008b). For the SCENGEN scale section, the user can choose from different types of AOGCM emissions. The method of using MAGICC/SCENGEN is unchanged from 2000, version 2.4 provided by Giorgi et al. (2001). The part that has changed is the MAGICC 2.4 code (which uses IPCC SAR, Second Evaluation Report, MAGICC version), the AOGCM database used for a more significant number of SCENGEN output options for the user and pattern scaling (Wigley 2008a).

As in the previous stage, the initial step is the implementation of MAGICC (Wigley 2008a). The user starts by selecting a pair of pollution scenarios known as the policy scenario and the reference scenario (Wigley 2008a). The publishing library from which these options were created now contains the latest versions of WRE CO2 stabilization scenarios (Wigley et al. 1996) and complies with SRES scenarios without climate policy (Wigley 2008a). SRES scenarios have a more comprehensive range of emissions than SAR scenarios (Wigley 2008). Therefore, release modes can only be added or edited offline using any editing software the user selects (Wigley 2008a). Reference tags and policies are optional, and the user can easily compare both publishing scenarios with others in the library—the user after selecting a group of parameters of the climate model and the gas cycle. The best set of default estimates may be chosen, or a user set may be prescribed. Both default results and users are transferred to SCENGEN (Wigley 2008b). Figure 7.4 shows the structure of the MAGICC/SCENGEN software (Wigley 2008a).

7.2.2 *Dynamic Downscaling*

Dynamic downscaling notes on the use of GCM-driven RCM to simulate a regional climate (Trzaska and Schnarr 2014). RCM is the same as GCM but possesses "higher resolution" and data outside the area that reflects the regional landscape and possibly regional atmospheric processes (Trzaska and Schnarr 2014). The global method simulates the answer of "global circulation to changes" in the atmosphere's composition via several processes (Trzaska and Schnarr 2014). Some of them are still needed for approximation according to the large resolution of the methods. Also, with a "resolution of 25–50 km" (Trzaska and Schnarr 2014), for parts of the globe, RCM can record several processes on a smaller scale realistically. Atmospheric fields "(e.g. wind, surface pressure, humidity, temperature)" (Trzaska and Schnarr 2014), simulated by the GCM fed at the horizontal and RCM's vertical boundaries. Local equations based on physics and specific data are applied to "process this data and achieve regional climate outputs". The main benefit of "RCMs" is the ability to "explicitly model atmospheric processes and land cover changes" (Trzaska and Schnarr 2014; Methods and Projections 2014).

7.2.2.1 Assumptions and Caveats

Although noticeable progress has been made in the RCMs' technical capability to simulate the region's climate over the past decade, important concerns and challenges remain (Trzaska and Schnarr 2014). Smaller network cells mean more surface data, and usually, more processes participate in RCM. Also, the number of calculations may be higher than GCM, which covers the whole globe. Therefore, RCMs are mathematically demanding, and it is possible to need more processing time than GCM to calculate predictions (Wilby et al. 2009). In addition, they need a significant amount of input, for example, "surface properties and high-frequency GCM" data. Furthermore, realistic simulations often require complex calibration methods (Trzaska and Schnarr 2014). Such a (GCM, RCM)s possess problem in simulating convective precipitation, which is a significant worried about the tropics (Trzaska and Schnarr 2014). Also, most RCMs do not accurately simulate heavy rainfall—with increasing resolution, systematic bias can worsen (Trzaska and Schnarr 2014). To create a better match of model output data with observations data, statistical bias corrections need to be made further (Brown et al. 2008). In some cases, a minor modification to convection designs can enhance the simulated rainfall realism (Trzaska and Schnarr 2014). Still, these settings require considerable expertise and reduce geolocation—that is, they built a version of the model that fits a particular area, but it is possible to perform weakly elsewhere (Trzaska and Schnarr 2014).

In addition, the quality of RCM output is related to GCM driving data, as GCM boundary conditions drive it (Trzaska and Schnarr 2014). For example, if GCM does not change storm routes, errors will occur in the RCM rainfall climatology (Wilby et al. 2009). In addition, various RCMs possess specific dynamic designs and physical variables, meaning that the same GCM RCMs can produce different results (Trzaska and Schnarr 2014).

Finally, the RCM network box size is usually more than ten kilometers, which is still very uneven for agricultural and hydrological effect studies that need more regional or station climatic data (Benestad 2010).To achieve a higher level of resolution outputs, statistical approaches are applied to RCMs, or through statistical methods, RCM output is reduced (Trzaska and Schnarr 2014).

In either case, validation of the model performance across a historical period relative to the simulated parameters of interest (e.g., rainfall, temperature) must be performed (Trzaska and Schnarr 2014). RCM results must not be taken at "face value" (Trzaska and Schnarr 2014).

7.2.2.2 Regional Climate Models

RCM is created by research institutes with sufficient technical expertise and computing capacity (Trzaska and Schnarr 2014). RCMs are usually diverse in physical, numerical and technical issues (Trzaska and Schnarr 2014). The most common RCMs in climate mitigation researches contain the following:

> Canadian Regional Climate Model (CRCM), Hadley Center Regional Climate Model (HadRM) and version 3 (HadRM3), German Regional Climate Model (REMO), Dutch Regional Climate Model (RACMO), Regional Climate Model version 3 (RegCM3), High-Resolution Limited Area Model (HIRLAM), Central Europe-Hamburg (ECHAM).

- **Canadian Regional Climate Model (CRCM)**

CRCM is obtained by pairing the complete set of scale parameters of the "second generation Canadian GCM" network (GCMII, McFarlane et al. 1992) with a highly effective dynamic model (Pal et al. 2007a, b). This model is according to "a semi-Lagrange and semi-implicit integration scheme," the Center for Research Collaboration in Scalability Compression Scale (CCRM-MC2) (Bergeron and Robert 1994; Tanguay et al. 1990)). One-way nesting method is applied to guide CRCM with GCMII information; for example, CRCM receives data from GCMII via its lateral border. Appling "GCMII as a driving model for CRCM" allows both driving and driving models to share a "subnet scale parameter package, which facilitates data transfer between the two models". Another noticeable benefic of applying GCMII subnet parameters is that all geophysical information files "(soil type, vegetation, bare soil albedo, etc.)" required for physics parameterization are being configured globally at the specific level of the sphere model. Lands are thus easier (Pal et al. 2007a, b).

The current configuration of the model includes a 120-by-120-point network that covers the area (5400 km^2) of the Vermont center. Different fields in a "polar-stereographic" network with a resolution of 45–60 km vertical, ten lands with unequal distances with Gal-Chen coordinates (Gal-Chen and Somerville 1975) from the ground near to 30 km are applied. Simulation with version 30 level is underway. GCMII controls the following boundaries: "wind, temperature, pressure and water vapor" components. They are interpolated from GCMII archived data on CRCM lateral boundaries. Preliminary conditions are also taken from GCMII archived information. A spongy area of 10 network points at domain boundary is applied to combine CRCM fields with data received from "GCMII." The nesting approach used by Davies (1976) is inspired by and modified by Yakimiw and Robert (1990). At an interval of ten network points from the "lateral boundaries, "the parameters of the local model are slowly and gently mixed via the (global) driving model." By achieving the "CRCM lateral boundaries," the parameters of the driving model variables are applied. Bypassing through the remains of the network, the parameters of the local model are not influenced by the global model. The dynamics of CRCM are according to Euler's complete non-hydrostatic formulas. So, the dynamic formula does not limit the spatial scales on which the model be performed. The model is also used in millimeter scales to investigate ultrasonic currents, in hundreds of meters for wet convection issues, and synoptic scales for air forecasting. It is cost-effective to integrate the complete Euler equations with the ability of the three-dimensional semi-Lagrange semi-implicit integration scheme. The efficiency of the integration plan also allows a 20-min step with a resolution of 45 km, which is a long step for a "non-hydrostatic numerical" model at this resolution. To compare, the GCMII applies the equal time interval with a Gaussian conversion network with a resolution of approximately 450 kine, and an explicit 45 kine exponential model typically runs at a time of 90s.

Application of CRCM

The dynamics of "CRCM" are according to "Euler's complete non-hydrostatic" formulas. So, the dynamic formula does not limit the spatial scales on which the approach can be executed (Pal et al. 2007a, b). The model is also used in millimeters scales for ultrasonic flow research, several hundred meters for wet convection issues, and synoptic scales for air forecasting (Pal et al. 2007a, b). It is cost-effective to integrate the complete Euler equations with the efficiency of the three-dimensional "semi-Lagrange semi-implicit integration scheme" (Pal et al. 2007a, b). The efficiency of the integration scheme also allows a 20-min step with a resolution of 45 km; it is a long step for a non-hydrostatic numerical approach at this resolution (Pal et al. 2007a, b). To evaluate, the GCMII applies the equal time interval as the Gaussian conversion grid with a resolution of approximately 450 km, and an explicit 45 km resolution O'Leary model typically runs over 90s (Pal et al. 2007a, b).

CRCM has been run over different regions and areas at a wide range of grid spacings between 10 and 100 km. The period of simulation can start "from days until decades to study different scientific" issues (Pal et al. 2007a, b). Several of these applications contain future climate change (Houle et al. 2020; Xie et al. 2017;

Jin and Erhardt 2020; Cook et al. 2020), water resources (Xie et al. 2017; Marcotte et al. 2020; Abbasnezhadi et al. 2020), extreme events (Imberger et al. 2020, Wootten et al. 2017) agriculture (Dayyani et al. 2012; Wagena et al. 2018; Modala et al. 2017; Mehdi et al. 2015); land-cover change (Lotsch et al. 2003), and "biosphere–atmosphere interactions (Sasaki et al. 2006)" (Pal et al. 2007a, b). Also, some initial work has been done by using CRCM for seasonal prediction (Alexandru et al. 2007; Music and Caya 2007). This different "range of applications and locations illustrates the versatility and portability of the RegCM system" (Pal et al. 2007a, b).

- **Hadley Centre's Regional Climate Model (HadRM) and Version 3 (HadRM3)**

The "HadRM3 regional climate model" is based on the atmospheric component, HadAM3 (Pope et al. 2000), from the global companion model HadCM3 (Buonomo et al. 2007; Gordon et al. 2000). HadCM3 changed from HadCM2 in several ways, such as physics parameterization patterns for the atmosphere and ocean. Also, the horizontal resolution of the ocean (1.25°) has increased, while the horizontal resolution of the atmosphere and the vertical surface of "both the atmosphere and the ocean are the same" (Buonomo et al. 2007). A better "description" of the "atmosphere and ocean" eliminates the flux settings "applied at the air-sea interface." To overcome some of the shortcomings in general "atmospheric circulation simulated by HadCM3, a higher-resolution version of its atmospheric component, HadAM3, with improved physics" parameterization, has been built for RCM driving (Buonomo et al. 2007). These changes related to "HadAM3 were the introduction" of a parameter for "critical relative humidity" (Rhcrit) for cloud formation (Cusack et al. 1999), an improved scheme for calculating anvil "radiative properties" (Gregory et al. 1999), a modified section parameterization "The cloud, as a function of specific humidity," involves a radiative connection between vegetation and the ground "surface and frequent" updates of long-wave surface radiative cooling (Buonomo et al. 2007). Moreover, changes were made in "various parameters and thresholds" related to cloud water and ice formation and precipitation. "RCM, HadRM3", was then configured from this "AGCM" (representing HadAM3H) with the same formula except for the implicit horizontal resolution dependence in the Rhcrit scheme and the explicit dependencies entered in the network box assumption of the rainfall assumption fraction (Buonomo et al. 2007).

All of the modifications described earlier, along with the alteration in "atmospheric physics" equations presented in "HadCM3" (the parameterization schemes in Gordon et al. (2000) mentioned for HadCM2, make the diversity among "HadRM3 and HadRM2" However, for convenience, the items that are directly related to precipitation should be listed (Buonomo et al. 2007). Excessive moisture dissipation in these areas can be prevented (Buonomo et al. 2007). "Spurious moisture source," which leads to excessive rainfall over urography (not on wind slopes) (Buonomo et al. 2007). Second, many changes have been made to the extensive rainfall design The water vapor density in "HadRM3" is based on the detected Rhcrit, not vertically dependent surface values, but the horizontal uniform transformation of "cloud water and ice" to precipitation occurs at a lower threshold s in "HadRM3". Rainfall, as well as the rate of freezing rainfall, have increased, the first of which is noticeable progress in realism, and the other is to reduce cloud life/increase rainfall formation act locally (Buonomo et al. 2007). Although significant changes have been made to the convection pattern, these mainly include the onset of convection effects and super convection radiation. There was no change in the formation of convective precipitation (Buonomo et al. 2007).

Application of HadRM3

Several reserves have been done to improve and evaluate the performance or application of HadRM3 in various fields. Among these, the studies are mentioned in Table 7.3.

- **REMO (German Regional Model)**

REMO is a 3D atmosphere model (Pietikäinen et al. 2018). This model was expanded at the Max Planck Meteorological institution in Germany and is currently housed at the "German Climate Service Center (GERICS)" (Pietikäinen et al. 2018). This model is in line with the Europa model, the previous NWP model of the "German Weather Service" (Pietikäinen et al. 2018). Predictive parameters of REMO are surface pressure, horizontal wind ingredients, specific humidity, the temperature of the air, ice and cloud water. Physical packages are derived from the "ECHAM4 global circulation model" (Pietikäinen et al. 2018; Roeckner et al. 1996). In addition, several updates have provided (e.g., Wilhelm et al. 2014a, b; Preuschmann 2012; Pietikäinen et al. 2012 Teichmann 2010; Pfeifer 2006; Semmler et al. 2004; Hagemann 2002).

REMO used a frog jump design with time filtering by Assein (1972) for temporal recognition. Semi-implicit correction is used to allow longer steps. The "vertical atmospheric levels are showed in a hybrid Sigma pressure coordinate system" (Pietikäinen et al. 2018), which is free of the atmospheric model at higher levels and follows the geological level at lower levels. Horizontally, this model has an Arakawa C spherical network (Arakawa 1977) in which all prognostic variables are examined in the center of the network box except the winds. Wind constituents are computed at the edges of the grid boxes.

In the lateral border range, REMO used the relaxation plate created by (Davies 1976) in the lateral border of the model domain. In this plane, the REMO prognostic variables are set to large-scale forcing in eight external network boxes. External coercion in this area decreases exponentially towards the internal model domain. Sea level temperature (SST) and surface distribution are imposed by the lateral boundary forced data set. At ground points, a fraction of the ground surface is applied, and details are able to found for a moment in (Kotlarski 2007) and (Rechid et al. 2009). This method applied a tile view with three tiles: water, earth and ice (Smaller et al. 2004). The bio geophysical specific of the ground tile is explained by surface variables. Leaf area index, fractional green vegetation and lower albedo area Monthly values are prescribed for seasonal vegetation characteristics (Hagemann 2002; Wilhelm et al. 2014a, b). Albedo has a background in other advances in pure vegetation and soil albedo (Rechid 2009; Wilhelm et al. 2014a, b).

Soil temperature profile applying five layers of soil with increased thickness and reaches a depth of ten meters. Soil hydrology is consistent with a simple bucket design

Table 7.3 Application of HadRM3 in different fields

Study filed	Studies	Results
Precipitation	Kundzewicz et al. (2006), Beranová et al. (2018), Lafon et al. (2013), Tryhorn and DeGaetano (2011), Buonomo et al. (2007), Burke et al. (2010), Maraun et al. (2010a), Rivington et al. (2008a), Urrutia and Vuille (2009)	In most areas, this model showed an increase in: • Severe drought • Potential for intense precipitation • Flood hazards • Sea level • Projected number of days with intense precipitation • Maximum Daily precipitation • Evaporation • Maximum and minimum temperatures
Temperature	Ramos et al. (2011), McCarthy et al. (2012), Kundzewicz et al. (2006), Prudhomme et al. (2012), Chow and Levermore (2007), Urrutia and Vuille (2009), Marengo et al. (2009), Kershaw et al. (2010), Rivington et al. (2008b)	
Runoff and water budget	Arnell et al. (2003), Teng et al. (2015), McCarthy et al. (2012), Kundzewicz et al. (2006), Prudhomme and Davies (2009a), Capell et al. (2014), Prudhomme and Williamson (2013), Arnell (2004), Futter et al. (2015), Prudhomme and Davies (2009b)	
Evaporation	Kershaw et al. (2010), Voloudakis et al. (2015), Semenov (2007), Kay et al. (2013), Prudhomme and Davies (2009b), Bell et al. (2011), Prudhomme and Davies (2009a), Prudhomme and Williamson (2013), McCarthy et al. (2012), Tuinenburg et al. (2014)	
Sea level	Ridley et al. (2013), McCarthy et al. (2012), Robins et al. (2014), Lucas-Picher et al. (2011), Prudhomme and Davies (2009a)	
Wind speed	McCarthy et al. (2012), Prudhomme and Williamson (2013), Harrison et al. (2008)	

with an improved accounting of a climatic model box network in the heterogeneity scale of sub-network farm capacities (Hagemann 2002). Due to the turbulence at the subscale scale, the vertical diffusion fluxes are computed for the lowest layer (surface layer) (Louis 1979). Temperature calculation is 2 m to record fluxes (Raddatz et al. 2007).

Snow parameterization is in accordance with the ECHAM4 view. It was later increased for better performance in areas with better latitudes (Semmler et al. 2012). The snow layer is separated into two parts for heat conduction calculations. An upper layer (0.1 m thick) is applied to define the surface temperature of the snow bag and acts as an interface for the atmosphere. The rest of the snow bag climate is computed applying the highest snow layer temperature and soil surface layer temperature. If the temperature of these two layers of snow reaches the melting point, another input energy is used to melt the snow. Snow and heat conduction density is parameterized as snow layer temperature functions (Kotlarski 2007; Roeckner et al. 1996).

Lakes are contained in REMO via the default land cover map: Global Landscape Profile Database (Loveland et al. 2000). Also, the standard floor plan does not tile for lakes and does not allow definitive treatment. Instead, the model applied the closest sea point view to all non-seawater sections to define the temperature of water and conditions of ice. As shown, the most comparative sea point view can significantly distort the simulated climate. These data suggest that REMO benefits from a connection to the process lake model.

Application of REMO

Several researchers have performed to increase and evaluate the performance or application of REMO in various fields. Among these, the following studies can be mentioned in Table 7.4.

- **Dutch Regional Atmospheric Climate Model (RACMO)**

KNMI,[1] in collaboration with the "Danish Meteorological Institute," developed the RACMO study model according to the 1990 "High-Resolution Numerical Prediction Meted" (HIRLAM). In 1993, the UU/IMAU began to model the more complex situation at glacier levels. This early version of RACMO is called RACMO1. In this model, the dynamic core of the HIRLAM model is combined with ECHAM4 physics (van Meijgaard et al. 2008).

The revised polar version of RACMO1 was used for Antarctic ice sheets. RACMO2 is the second version of this model. The center combines the physics of the Integrated Forecasting System (ISF) Meteorological Forecasting Center (ECMWF) with a dynamic core of the HIRLAM model. Versions 2.0 and 2.1 continued the RACMO HIRLAM version 5.0.6 and the ISF CY23r4 cycle, but RACMO 2.3 includes the CY33r1 cycle and HIRLAM version 6.3.7. Due to the rapid increase in computer capacity over the years, these versions of RACMO have been applied in Antarctic and Greenland ice regions and higher resolution in smaller areas such as Dronning Maud Land and Patagonia (van Meijgaard et al. 2008). Figure 7.5 shows the RACMO model applied to ice sheets.

In RACMO, the networks are defined on the equator and then rotated to the desired region. Network distance as a degree of failure. As long as the amplitude is small enough, the results are close to equal points on the grid. It should be noted that this piece is not located on the stereographic polar projection plane. This model uses

[1] Royal Netherlands Meteorological Institute.

Table 7.4 Application of REMO in different fields

Study filed	Studies	Results
Precipitation	Mirakbari et al. (2020), Semmler and Jacob (2004), Saeed et al. (2013), Ahrens et al. (1998), Paxian et al. (2015), Paeth et al. (2011), Mirakbari et al. (2020), Ahrens et al. (1998), Jacob et al. (2012), Kotlarski et al. (2010), Karstens et al. (1996)	It has satisfactory performance in simulating extremes
Temperature	Fotso-Nguemo et al. (2017), Mirakbari et al. (2020), Weinberger and Vetter (2012), Alabastri et al. (2013), Smiatek et al. (2009), Jacob et al. (2012), Kebede et al. (2013), Prömmel et al. (2010), Christensen et al. (2008), Emmanuel et al. (2019)	
Runoff and water budget	Singh et al. (2001), Hagemann and Gates (2003), Karambiri et al. (2011), Jacob and Podzun (1997), Vasyl and Kateryna (2018), Vasyl and Kateryna (2018), Dankers and Middelkoop (2008), Kling et al. (2012), Voudouris et al. (2012), Teichmann et al. (2013)	
Evaporation	Jacob and Podzun (1997), Kerch et al. (2008), Omstedt et al. (1997), Mengelkamp et al. (2006), Di Matteo et al. (2011), Karstens et al. (1996)	
Sea level	Gomis et al. (2008), Soares et al. (2002), Sotillo et al. (2005), Ratsimandresy et al. (2008b), Weisse and Plüβ (2006), Ratsimandresy et al. (2008a), Jacob and Podzun (1997), Tsimplis et al. (2005)	
Wind speed	Winterfeldt and Weisse (2009), Winterfeldt et al. (2011), Weisse and Feser (2003), Koch and Feser (2006), Matzarakis and Endler (2010), Horstmann et al. (2002), Larsén et al. (2010)	

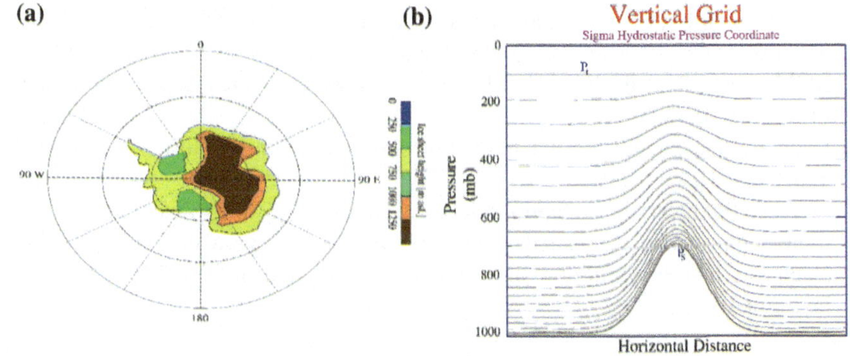

Fig. 7.5 RACMO model. **a** RACMO model applied to ice sheets. **b** Simulation for vertical layers

a hybrid sigma system, which evolves from the near-surface sigma level to the net pressure level at a higher height in the vertical. The real number of horizontal grid points in each method run is different. Research shows that most vertical simulations use 40 vertical layers. RACMO is a local model, so it requires external data at sea level and lateral boundaries. In the lateral boundary area of the model, surface pressure, special humidity, temperature, meridian and regional wind components are calmed every 6 h of the model toward the farms of a global model and sea ice concentration and sea surface temperature. The RMCO interior does not move towards observations and allows it to evolve freely. This model is not necessary on top of the model (van Meijgaard et al. 2008).

Model Alteration

To increase the performance of the methods in glacial regions, different modifications have been made to the model equation. In RACMO1, the parameters of snow albedo, snow properties, initial snow temperature and roughness length were adjusted. Moreover, an additional vertical surface was placed within seven meters of the surface. RACMO2.0 was interactively mixed with four-layer dry snow melted. This model is modified by a multilayer snow model (over 100), which calculates infiltration, melting, re-freezing and meltwater runoff for the Greenland ice sheet.

At the top of the connection to the multilayer snow model, RACMO2.1 also includes an increased surface albedo design according to the lower atmosphere and a snowdrift routine that resembles snowdrift interactions with the surface and a prognosis design for snowflake size and a background-fixed (fixed) ice-albedo from MODIS fields. It should be noted that "RACMO2.2 does not have a separate polar version, and RACMO2.3" applies the same RACMO2.1 snow or ice model.

As a result, the most important new features in setting up RACMO version 2 of the core HIRLAM and ECMWF include the following changes (see Lenderink 2003):

- Introduce a modified relaxation scheme of prognostic parameters in the lateral boundary zone of the method domain

- Reduce small-scale urography variance using a "Raymond" filter to suppress network-scale perturbations by the dynamics solver. In addition, the dissolved RCM urography in the lateral border region is relaxed to match the gross urography of the host model.
- Decrease the specific coefficient of control of horizontal moisture penetration by a factor of 10.
- Increase the layers of the soil model by a factor of 3.5 to allow the model for a more comprehensive seasonal cycle in the values of soil moisture. The associated changes affect the yield of plant water stress and reduce the infiltration rate by a factor of 10.

Application of RACMO

Several studies have been conducted to increase and evaluate the performance or application of RACMO in various fields. Among these, the following studies can be mentioned in Table 7.5.

- **Regional Climate Model Version 3 (RegCM3)**

The development of ICTP[2] RegCM3 involves the efforts of several scientists around the world. EDN scientists have made some contributions because of RegCNET activities like the inclusion of the "Massachusetts Institute of Technology (MIT) convection scheme and a dust module" (Pal et al. 2007a, b). This section provides a "short history of the RegCM system and then describes new increases in RegCM3" (Pal et al. 2007a, b). The current version of RegCM3, also a beta version with test features, is possible to found (Pal et al. 2007a, b).

RegCM History

Table 7.6 lists "the various dynamic and physical packages of consecutive versions of the RegCM system. RegCM3 is a significant integration achieved in version 2.5 (RegCM2.5) as explained in (Giorgi and Gutowski 2015)" (Pal et al. 2007a, b). This progress is in the display of "rainfall physics, surface physics, atmospheric chemistry and airborne particles, model input fields and user interface. Also, the model code has been modified for parallel computing. An essential aspect of RegCM3 is that it is user-friendly and runs on a variety of computer operating systems". To this end, fundamental changes have been made to the post-processing, preprocessing and execution of the model (Pal et al. 2007a, b). In addition, RegCM3 has options for interfaces with different analyzes and "GCM boundary conditions to provide a complete set of simulation options" (Pal et al. 2007a, b).

Dynamical Core

Similar to RegCM2 and 2.5 versions, the RegCM3 dynamic core is "built on the hydrostatic version of the Pennsylvania State University National Center for Space Research (PSU-NCAR) scaling model" (MM5; Grell 1993; Pal et al. 2007a, b). This

[2] International Centre for Theoretical Physics.

Table 7.5 Application of RACMO in different fields

Surface mass balance	Van de Berg et al. (2005), van Lipzig et al. (2002a), Vernon et al. (2013), Van Wessem et al. (2014), Franco et al. (2012), Izeboud et al. (2020), Franco et al. (2012), Van Lipzig et al. (2002), Edwards et al. (2014), Lang et al. (2015), Lenaerts et al. (2018), Fyke et al. (2017), Zou et al. (2020), Lenaerts et al. (2012), van de Berg and Medley (2016), Shean et al. (2017), Wilton et al. (2017), van Lipzig et al. (2002b), Feldmann et al. (2019), Sellevold et al. (2019), Mottram et al. (2017), Sutterley et al. (2013), Krebs-Kanzow et al. (2020), Day et al. (2013), Wang et al. (2015), Sutterley et al. (2016)	High perfornamce in tropical area
Antarctic	Trusel et al. (2013), Van den Broeke and Van Lipzig (2003), Van Lipzig et al. (2004), Frieler et al. (2015), Zwally et al. (2015), Van Den Broeke and Van Lipzig (2004), Eom and Rim (2020), Huss and Farinotti (2014)	
Precipitation	Leander and Buishand (2007), Marshall et al. (2017), Hanel et al. (2009), van Lipzig et al. (2002a), Bisselink and Dolman (2009), Te Linde et al. (2010), Van Meijgaard (1995), Van Den Broeke and Van Lipzig (2004), VanZanten et al. (2011), Klutse et al. (2016)	
Temperature	van Lipzig et al. (2002a), Leander and Buishand (2007), Van Meijgaard et al. (2012), Erik van Meijgaard et al. (2008), Boberg and Christensen (2012), Van Den Broeke and Van Lipzig (2004), Van Lipzig et al. (2002), Christensen et al. (2008), Boers et al. (2019), Manders-Groot et al. (2011)	

(continued)

Table 7.5 (continued)

Runoff and water budget	Karambiri et al. (2011), Xu et al. (2013), Van den Hurk et al. (2002), Van Pelt et al. (2012), Mankoff et al. (2020), Kling et al. (2012), King et al. (2019), Muerth et al. (2013), Mohajerani et al. (2015), Vautard et al. (2019)	
Evaporation	Bisselink and Dolman (2009), Jansen and Teuling (2020), Wipfler (2010), van Meijgaard et al. (2008), Mohamed et al. (2006), Dolman et al. (2001), Mohamed et al. (2005), Dolman et al. (1999), Kotsopoulos et al. (2015)	
Sea level	Quiquet et al. (2013), Spada et al. (2013), Edwards et al. (2014), Graversen et al. (2011), Zammit-Mangion et al. (2015), Meyssignac et al. (2017)	
Wind speed	Van Lipzig et al. (2004), Lenaerts et al. (2012), Lenaerts et al. (2012), Van den Broeke and Van Lipzig (2003, 2004), da Silva Soares (2016), Vautard et al. (2019)	

is a model of elementary, hydrostatic, compressible, sigma-vertical coordinates (Pal et al. 2007a, b). For more details, "we refer the reader to the MM5 documentation (Grell et al. 1994) and the articles describing previous versions of the model" (Giorgi et al. 1993a; Giorgi and Mearns 1999; Pal et al. 2007a, b).

Physics of Model

Several of the "model physcal packages in RegCM3 have been enhanced because of the release of RegCM2.5" (Pal et al. 2007a, b). These contain convective and non-convective sediment designs, atmospheric and aerosol chemistry, surface, ocean flux designs modules (Pal et al. 2007a, b). Although minor adjustments have been made to some of the model parameters, the radiation transmission package and planetary boundary layer design have not yet been substantially modified (Pal et al. 2007a, b).

Precipitation

The precipitation organization in RegCM3 is shown in two ways: resolvable (or large scale) and convective (sub-grid). The resolvable rainfall is usually dependent on large-scale systems that move slightly in the vertical route and are most usual in the winter. Therefore, convective precipitation usually occurs in the summer hemisphere

Table 7.6 Explanation of the advance of the versions of the RegCM system" (Pal et al. 2007a, b)

	RegCM1	RegCM2	RegCM2.5	RegCM3
Primary references	Zeng et al. (1998), Naz et al. (2018)	Giorgi et al. (1993a, b)	Small et al. (1999), Pal et al. (2000)	
Dynamics	MM4 (Anthes et al. 1987)	MM5 (hydrostatic) (Grell 1993)	MM5 (hydrostatic) (Grell et al. 1994)	MM5 (hydrostatic) (Grell et al. 1994)
Radiative transfer	CCM1 (Kiehl et al. 1996)	CCM2 (Briegleb 1992)	CCM3 (Kiehl et al. 2005)	CCM3 (Kiehl et al. 2005)
Boundary layer	Local (Deardorff 1972)	Nonlocal (Holtslag et al. 1990)	Nonlocal (Holtslag et al. 1990)	Nonlocal (Holtslag et al. 1990)
Land surface	BATS 1a (Dickinson et al. 1986)	BATS 1e (Dickinson et al. 1993)	BATS 1e (Dickinson et al. 1993)	SUBBATS (Giorgi et al. 2003)
Convective precipitation	Anthes (1977)	Grell (1993),Anthes (1977)	Zhang and McFarlane (1995), Grell (1993)	MIT (Emanuel 1991; Anthes 1977; Grell 1993)
Resolvable precipitation	Implicit (Dickinson et al. 1989)	Explicit (Hsie et al. 1984)	SIMEX (Qian and Giorgi 1999)	SUBEX Pal et al. (2000)
Aerosols and chemistry	Not available	Not available	Qian and Giorgi (1999) (no dust)	Solmon et al. (2006), Zakey et al. (2006)

and tropics at scales smaller than 1 km. since fine-scale, convective rainfall should be parameterized in almost all climate methods (Pal et al. 2007a, b).

"In RegCM3, the resolvable-scale precipitation is represented using the sub grid explicit moisture (SUBEX) scheme (Pal et al. 2000). SUBEX accounts for the sub grid-scale variability of clouds and includes formulations for the auto conversion of cloud water into rainwater; the cloud droplets accumulate as raindrops fall and raindrops evaporate" (Pal et al. 2007a, b). Cloud fraction coverage is calculated from humidity, clouds during the time that the humidity increased the describe threshold under the cell saturation. Moreover, the threshold of cloudy water for rain data is the amount of cloudy liquid water according to experimental observations in the cloud. The "implementation of this plane has been illustrated to significantly increase the simulation of temperature, rainfall and other cloud-related parameters in the Americas (Pal et al. 2000, 2007a, b)" (Pal et al. 2007a, b).

Convective rainfall parameterization rests one of the major sources of error in climate studies methods. There are three parts in RegCM3 to show aggregation convection:

(1) "Anthes-Kuo repaired scheme" (Anthes 1977; Giorgi 1991; Pal et al. 2007a, b).
(2) The Grell scheme (Grell 1993; Pal et al. 2007a, b).
(3) "The MIT scheme (Emanuel and Živković-Rothman 1999; Emanuel 1991). The designs of Anthes-Kuo and Grell are well documented in articles describing the results of previous versions of the RegCM system" (e.g., Giorgi 1991; Giorgi et al. 1993a; Pal et al. 2007a, b).

MIT is the latest aggregation convection option in RegCM3. For more details, please refer to "Emanuel and Živković-Rothman (1999) and Emanuel (1991)" (Pal et al. 2007a, b). The plane presumes that the combination in the clouds is really heterogeneous and episodic and continues the convection fluxes following an ideal model of subsidence at the scale below the cloud (Pal et al. 2007a, b). Convection is adjusted when the neutral swimming level is more than the base. Among these two surfaces, air rises and the fraction of dense moisture builds until a residual fraction of cloud forms (Pal et al. 2007a, b). It is assumed that the cloud mixes with "air from the environment based on a fixed spectrum of the mixture that ascends or slopes to the corresponding neutral floating surface" (Pal et al. 2007a, b). The upward mass flux of the cloud base is calm relative to the quasi-equilibrium of the sub-cloud layer. Moreover, to achieve a better physical display of convection, the MIT design suggests several different advantages over other RegCM3 convection options (Pal et al. 2007a, b). For example, this includes a formula for automatically converting cloud water to rain within cumulus clouds. Ice processes become temperature-dependent with letting the automatic conversion threshold of water. In addition, "rainfall is added to a low hydrostatic, unsaturated pressure that carries water and heat. Last, the MIT plan noticed the transportation of previous trackers" (Pal et al. 2007a, b).

Biosphere

"Since RegCM2, surface physics calculations have been performed using the Biosphere–Atmosphere Transfer Scheme (BATS) version 1 (Dickinson et al. 1993), which describes the transfer of energy, mass, and motion between the atmosphere and the biosphere (Pal et al. 2007a, b). In RegCM3, additional modifications have been made to BATS to diversify the land cover subnetwork using a mosaic-type approach and topography" (Pal et al. 2007a, b), (called SUBBATS; Giorgi et al. 2003). This repair adapts an orderly microscale surface subnet for every large model grid cell (Pal et al. 2007a, b). Meteorological parameters are separated from the extensive network to fine network due to the topographic difference among the sub-network and large network. BATS computations are performed for every subnet cell separately, and the input surface fluxes are aggregated into the atmospheric model on the large lattice cell (Pal et al. 2007a, b). This parameterization shows a significant improvement in "surface hydrological cycle displacement in mountainous areas" (Giorgi et al. 2003; Pal et al. 2007a, b).

Water Bodies

"In RegCM3, water bodies are able to be classified as open (e.g., oceans) and surrounded (e.g., lakes)". sea energy flux and air from open bodies are calculated from "sea surface temperature" (SST) without two-way interaction (Pal et al. 2007a, b). "That is, the ocean influences the atmosphere, but the atmosphere does not influence the ocean." The energy flux of enclosed water bodies can be calculated applying one of two ways: either the columnar lake meted with two-way interaction (Hostetler and Bartlein 1990) or the SST method prescribed for open bodies (Pal et al. 2007a, b). In RegCM3, we have two parameterization options for calculating fluxes from open water: the BATS formulation and the recently implemented Zeng scheme (Pal et al. 2007a, b; Zeng et al. 1998) (Pal et al. 2007a, b). BATS applied the "Monin–Obukhov standard similarity" (Pal et al. 2007a, b), connections for calcite fluxes without any specific treatment of really stable convective situations. Moreover, the roughness duration is set to a stable value (Pal et al. 2007a, b). "It is not a function of wind and stability" (Pal et al. 2007a, b). For these causes, due to "(Zeng et al. 1998)" (Pal et al. 2007a, b), ocean flux calculations from BATS are interested in overestimating evaporation under both low and high wind situations (Pal et al. 2007a, b). The Zeng plane shows all stability situations and contains a spin speed to calculate the extra flux due to the variation of the boundary layer scale. Experiments illustrate that RegCM3, together with the Zeng plane, further estimates evaporation flows across the South Pacific (Pal et al. 2007a, b).

Chemistry and Atmospheric Aerosols

"Atmospheric aerosols are known to have a significant impact on the climate system, especially at regional scales." An aerosol design for sulphate, organic carbon and carbon black aerosols is included in RegCM3 as described by Fu and Johanson (2005), Solmon et al. (2006), Pal et al. (2007a, b). This scheme briefly describes the trend of atmospheric winds, turbulence propagation, vertical transportation by accumulated deep convection, wet and dry removal processes, and chemical gas and aqueous phase conversion mechanisms (Pal et al. 2007a, b). Wet removal is parameterized by soluble scale precipitation and subgrade "as a function of the conversion rate from cloud to rain" (Pal et al. 2007a, b) and different water embroidery of airborne particles (Pal et al. 2007a, b). "Surface dry deposition" is computed based on the set sediment velocity on the ground and water surface (Pal et al. 2007a, b).

"Direct and indirect aerosol" (Pal et al. 2007a, b), radiation effects are effected in RegCM3 (Giorgi et al. 2003). Direct radiation influences are calculated by determining the optical settings of the aerosol: "extinction coefficient, single scattering albedo and asymmetry" (Pal et al. 2007a, b), variable. Indirect influences are explained assuming the dependence of the cloud droplet radius on the "aerosol mass concentration" (Pal et al. 2007a, b). This model of aerosol-associated chemistry has been applied in the East Asian region (Georgia et al. 2003) and the Europe-Africa region (Pal et al. 2007a, b; Solmon et al. 2006) (Pal et al. 2007a, b).

Moreover, the connection of desert dust parameters to RegCM3 has lately completed applying a design according to primarily on the work of (Laurent et al.

2008) and (Alfaro and Gomes 2001). The "RegCM3-dust" (Pal et al. 2007a, b) mobile model has been examined for dust storm cases and long-term simulations of an "Africa-Europe territory by Pal et al. (2007a, b), Zakey et al. (2006)" (Pal et al. 2007a, b).

Initial and Boundary Conditions

RegCM3 needs time-dependent primarily and time-dependent lateral situations for temperature, wind, water vapor components and surface pressure (Pal et al. 2007a, b). Moreover, SSTs should be identified on the oceans. The interface is designed to quickly transfer different GCM boundary conditions and re-analyze the RegCM3 framework (Pal et al. 2007a, b). until now, different global analytics "products and GCMs have provided boundary conditions for RegCM3" (Pal et al. 2007a, b), such as "the National Environmental" Forecasting Center NCEP-NCAR (NNRP), a 40-year-old European Center for Weather Forecasts for 40-year Medium Range (ECMWF)—"Analysis (ERA-40), NCAR Community Climate Model 3 (CCM3), ECHAM, Hadley Center Climate Model (HadAM3) and fvGCM. In addition, RegCM3 can be incorporated into itself or other RCMs, including providing a regional climate for impact studies (PRECIS; Krichak 2005, Personal Communication) and MM5 (Ashfaq 2005, published manuscript)" (Pal et al. 2007a, b).

Soil moisture usually starts based on vegetation characteristics (e.g., desert = arid) (Dickinson et al. 1989), but if needed, soil moisture is adjusted able based on the Water Prediction Center data set, and air (CPC) started (Giorgi et al. 1999; Fan and Van Den Dool 2004) with the prescribed vertical profile (Pal and Eltahir 2001) or applying soil moisture from the driving model (Pal et al. 2007a, b). Land cover is determined by applying "Global Version 2 Global Landscape Description (GLCC) information provided by the USGS (US Geological Survey)" Land Resource Observation System information Center (Loveland et al. 2000)" (Pal et al. 2007a, b). Like soil texture and moisture, class is usually prescribed base on the characteristics of the vegetation (e.g., desert = sand). Alternatively, there is an interface to the "Food and Agriculture Organization (FAO)" soil texture information (Webb et al. 1993) to the "RegCM3 network" (Pal et al. 2007a, b).

Examples of RegCM Applications

RegCM3 runs several days in different network and simulation courses from a few days to decades to research various scientific issues. Several of these usages contain future climate change (Diffenbaugh et al. 2005; Gao et al. 2006), air quality (Giorgi et al. 2006; Solmon et al. 2006), water resources (Pal and Eltahir 2001), extreme events (Giorgi et al. 2004), agriculture (White et al. 2006), land-cover change (Akinyemi and Abiodun 2019; Gao et al. 2006), and biosphere–atmosphere interactions (Pal and Mather 2003). Also, some initial work has been performed applying RegCM3 for seasonal prediction usages (Pal et al. 2007a, b; Rauscher et al. 2006). This various range of usages and regions demonstrates the performance and portability of the "RegCM system" (Pal et al. 2007a, b).

- **High-Resolution Limited Area Model (HIRLAM)**

The High-Resolution Area Model, known as HIRLAM, is a Numerical Weather Prediction System (NWP) developed by the international HIRLAM program.

In Europe, Operational short-range Limited Area Models (LAM) models are being developed mainly in collaboration with the National Meteorological Service (NMS). HIRLAM (High-Resolution Limited Area Model) was the first limited modelling consortium in Europe, built-in 1985 in Nordic countries and later expanded to other members. HIRLAM collaboration was initially organized into projects, and there were six of them (starting HIRLAM-1 to HIRLAM-6). The members then decided to continue their cooperation in the form of five-year plans, HIRLAM-A (2006–2010), HIRLAM-B (2010–2015) and (currently in operation) HIRLAM-C (2016–2020) (Bengtsson et al. 2017).

HIRLAM's long-term primary goal is to provide its members with advanced practical and concise practical skills and scope of the weather forecasting system. The immediate use of the NWP system is to generate usable weather forecasts for member services, with a particular emphasis on identifying and forecasting severe weather conditions and public safety services (Cats and Wolters 1996; Rontu et al. 2019).

HIRLAM and ALADIN, in 2005, a strategic decision was made for HIRLAM to work closely with the ALADIN consortium. This research collaboration aims to develop and maintain the standard ECMWF/Arpege Integrated Prediction System (IFS) in the mid-scale masterpiece LAM model code for short-range numerical prediction. This standard scale analysis and forecasting system, called HIRLAM HARMONIE, is intended to replace the HIRLAM model in all its applications eventually. Terms of cooperation HIRLAM-ALADIN were approved in the cooperation agreement between ALADIN and HIRLAM consortia in December 2016. The two consortia have agreed to form a single consortium by the end of 2020 at the latest (Batrak et al. 2018).

HIRLAM-ALADIN Cooperstown will continue in the HIRLAM-C program. The main goal of the HIRLAM-C program is to conduct research and development with ALADIN partners to prepare members with a comprehensive NWP system on a large scale as a perfect tool to assist them in beneficial weather forecasting activities. Most of this work is done in close coordination with the ALADDIN program under compatible scientific work programs. More details and details of the program at HIRLAM-C MoU 2016–2020 (Rontu et al. 2019).

Models being developed in HIRLAM-C text:

- HARMONIE-AROME scaling model; in general, run-in local collections with a resolution of 2.5 km in landscape mode and 65 layers in portrait mode.
- HARMONIE-AROME based Mesoscale LAM EPS systems are generally called Harmon EPS. Local HarmonEPS kits for Norwegian, Swedish and Finnish Met Coop operational cooperation (so-called MEPS system), operational for DMI (COMEPS), AEMET (SRPES) and KNMI (KEPS) are operational or under construction.

GLAMEPS Pan-European Hydrostatic Limited Group Prediction System, which is practically implemented in ECMWF
- The HIRLAM hydrostatic model (horizontal resolution 3–15 km) is no longer under active development but is still used operationally in many HIRLAM services.

The HIRLAM program is a collaboration between the following meteorological institutes (Cats and Wolters 1996):

> Danish Meteorological Institute (DMI) (Denmark), Estonian Institute of Meteorology and Hydrology (EMHI) (Estonia), Meteorological Institute of Finland (FMI) (Finland), Meteorological Office of Iceland (IMO) (Iceland), Lithuanian Hydrological and Meteorological Services (LHMS) (Lithuania), Met Éireann (ME) (Ireland), Norwegian Meteorological Institute (MET) (Norway), Royal Netherlands Meteorological Institute (KNMI) (Netherlands), Agencia Estatal de Meteorología (AEMET), formerly INM (Spain), Swedish Institute of Meteorology and Hydrology (SMHI) (Sweden), Also, this organization (Météo-France (France)) is a research partner in the international cooperation of HIRLAM.

Hirlam is known as a "short-range weather forecasting system." Versions 6 and c of the Hirlam system now use the usual weather forecast in most of the participating institutions. Modern weather numerical forecasting systems possess three initial components: an analysis unit, forecasting method and a post-processing processor. It provides an analysis of the basic situations of the model according to recent observations and other sources of data. The prediction method spatially and temporally integrates the classical equations of Newton's second law, mass protection, and thermodynamics. The finite spectral methods are usually applied for spatial discretization and a jump plane for temporal discretization. Temperature, wind and humidity a 3D network is usually used. Surface pressure and temperature are 2D. to forecasts with almost long military service; this model should cover the ground. When the chosen time is two days or less, the calculations are performed in a range of limited horizontal dimensions. The drawback is that a lateral boundary situation should provide. They are always obtained from a previous prediction with a global method. In the post-processing stage, climate-related phenomena (e.g., wind speed at an altitude of 10 m) are computed from the model parameters.

The Hirlam Reference System

The "reference version of Hirlam is maintained at the European Center for Medium-Range Meteorological Forecasts (ECMW) and is regularly updated." All variations in Hirlam are done through the reference system. Hirlam members can download modified versions via their link to ECNMrF or via the system administrator. They can then run the numeric method on regional hardware. Portability is one of the main aspects of reference system design for easy regional installation of recent versions. As with most of these systems, Hiram has three primary components.

Analysis

Analysis of surface pressure, atmospheric temperature, wind and humidity is according to a 3D multivariate optimal interpolation scheme. Sea surface, surface temperature (2 m above sea level), sea ice and snow cover are analyzed with a 2D optimal interpolation scheme. At 01, the model domain is divided into several subdomains called boxes, which depend on the number of observations inside and near the box. For every box, a so-called correlation matrix is formed and inverted. The number of boxes for a "typical Hirlam performance" is 1000, and the mean size of a correlation matrix is "300 × 300". Schecme 1 applied by 13irlam is the same as the method applied in ECMWE. It is adapted for applications in a limited range method, but the system's main features are equal. In this plan, two scans are performed with 01: the initial is to control the quality of the data (identify and reject insufficient data). The second case is for calculating the analyzed variables at network points.

To prepare and exchange observations of the global atmosphere, the world's nations collaborate on the "World Weather Watch program (under the auspices of the World Meteorological Organization)." This global observation network includes surface observations, radio probes, satellites, aircraft reports, and ships and vessels. The Hirlam analysis scheme combines the wind, humidity, pressure, and temperature showed by these observations with the "initial guess," the Hirlam model's six-hour forecast from the previous analysis of six hours. The computer needs of the analysis plan are mediocre: 400 processor seconds on a Cray C90 processor. The code is executed efficiently on parallel memory machines (Cray, Convex)) using the "process generation" strategy: Computations for a box form a separate process. Each similar processor executes the first process in the list of methods to be performed. Since just a few sync points are required (for instance, after data quality check), this parallelization view is practical: the speed on eight processors is more than 7. The analysis has not yet been distributed on the memory device.

Forecast Model

> The forecast model is a hydrostatic grid point model, with a horizontal resolution of 5–5.5 km and a vertical surface of 16–31 km. Vertical diffusion is done through a first-order closure design, dense processes through a Sunquest design, and surface processes through a two-layer design with snow, ice, and soil moisture. For more details, readers are referred elsewhere (Källén 1996).

The prediction method is the most expensive part of the Hirlam system: a 48-h prediction of a C90 processor takes 3600 s. Significant efforts have been made to implement this method in an expanded range of architectures since it can (and will be) applied as a benchmark for the same codes, like the most expensive component predictor model in the system. Hiram is a 48-h forecast on the C90 that requires 3600 s of CPU. Significant efforts have been made to implement this method in an expanded range of architectures since it can (and can be) applied as a benchmark for the same codes, like climate method and regional air pollution dispersion methods on Limited regions.

Post-processing and Other Components

"A post-processing package involves pressure levels and transmits the output known as GRIB files (a standard format for international network data exchange, developed by the WMO)". Hirlam calls the meteorologist at regular intervals to show the predicted products. Further visualization study is ongoing in this region. The reference system includes several other components that do not require a lot of computer resources and are not very important in terms of time but are very valuable: a standard diagnostic package, including a tool for monitoring data quality control procedures. For a package to validate the model against observations; and documents, it contains publications in free literature and articles (see Hirlam's home page at http://www.knmi.nl/hirlam).

Application of HIRLAM

Several types of research have been performed to increase and evaluate the performance or application of Hilram in various fields. These include the following studies in Table 7.7:

- **European Centre-Hamburg "(ECHAM)"**

"ECHAM is a general circulation model (GCM) developed by the Max Planck Institute for Meteorology," a Max Planck Association research organization. It was designed to be used for climate research by correcting global forecasting methods

Table 7.7 Application of HIRLAM in various fields

Application	Example
Air pollutant	Baklanov et al. (2008), Toros et al. (2014), Niska et al. (2005), Rantamäki et al. (2005)
Improve verification and reanalysis of Hirlam	Baklanov et al. (2017), Senkova et al. (2007), Launiainen (2015), Dahlgren et al. (2016), Navascués et al. (2013), De Bruijn et al. (2006)
Direct radiative effect	Toll et al. (2016)
Parametrization of lakes	Rontu et al. (2012), Eerola et al. (2010), Pour et al. (2014)
Cloud studies	de Haan and van der Veen (2014), Stengel et al. (2010)
Analysis of temperature; precipitation; wind	Sekula et al. (2019), Isotta et al. (2014), Lindskog et al. (2004), Landelius et al. (2016), Watson (1994),
Sea breeze thunderstorms	Azorin-Molina et al. (2014)
Digital filter	Lynch and Huang (1992)
Thermodynamic sea ice scheme	Batrak et al. (2018)
SEVIRI infrared radiances	Stengel et al. (2010)
GPS and NWP	The ue of GPS to validate NWP systems: the HIRLAM model
Diabatic digital-filtering	Huang and Lynch (1993)
Urban effects	Mahura et al. (2009), Vu et al. (2013)
Graphics processing unit optimizations	
Chemical weather forecasting	Korsholm et al. (2008)

expanded by the ECMWF. This model is named as a mix of its principle ("EC" stands for "ECMWF") and the location of its parameter package, Hamburg. The model's default form resolves the atmosphere to 10 horsepower but can be configured to 0.01 hPa for use in the lower mesosphere and stratosphere. ECHAM calculates the spread of global weather conditions (wind, cold, temperature, etc.) in time 20–40 min when using a spatial resolution of 300–500 km. Climate-related gases, sea surface temperature, solutions are prescribed (Pietikäinen et al. 2018). This mainly determines the simulated climate. The observed weather conditions are not selected; they are simulated; instead, the characteristic air circulation models such as high-pressure and low-pressure Azores routes present the average locations of Iceland. Over the years, there have been new developments in ECHAM5 in assessing the impact of climate on congestion (prefabricated) routes. A new type of cloud, the Cirrus Density Sequence, is constantly being introduced to the natural clouds in the model. The development of cirrus condensation sequences and their impact on climate has been calculated (Pietikäinen et al. 2018).

ECHAM models are primarily Central European spectral weather forecast models for medium-range weather forecasts (ECMWF, Simmons et al. 1989). The section describing the dynamics of ECHAM is in line with the ECMWF documentation, which has been modified to explain newly implemented features and fundamental changes for climate experiments. The new version of ECHAM is fully portable and runs on all significant high-performance operating systems (Pietikäinen et al. 2018). The restart mechanism runs on top of the netCDF and is therefore independent of the underlying architecture (Pietikäinen et al. 2018). The dynamic part of ECHAM is formulated in spherical harmonics. After comparisons between models by Girard and Jarraud (1982), the shortened expansion in terms of spherical harmonics was used to represent dynamic backgrounds (Pietikäinen et al. 2018). The conversion technique was developed by Eliasen et al. (1970), Orszag (1970). This method is used in which nonlinear terms, including parameters, are computed at a set of almost regularly distributed network points in the Gaussian network. In the vertical position, a flexible coordinate is applied, enabling the model to use the usual ground sigma coordinates (Phillips 1957) or the combined coordinates for which the surface of high-level models is smoothed on sloping terrain and causes rotation of "constant pressure levels in the stratosphere (Simmons and Burridge 1981)" (Pietikäinen et al. 2018). Wet processes in several ways using a mass-saving algorithm to transmit different chemical tracers and water species. Transportation takes place in the Gaussian network (Lin and Rood 1996).

Several versions of ECHAM, primarily the various ECHAM5 settings, are the basis for the multiple releases listed below.

ECHAM2

According to the ECMWF1 numerical weather forecast model, the ECHAM2 spectral atmospheric circulation model was developed "jointly by the Meteorologists Institute der Universität Hamburg and the Max-Planck Institute für Meteorology, Hamburg, Germany." Prognostic variables are rotation, divergence, temperature (logarithm), surface pressure, humidity and cloud water (ice and water phase). This model includes radiation, cloud formation and precipitation, convection and vertical and horizontal diffusion. Ground surface processes are expressed by a five-layer thermal conductivity model and by a hydrological model to determine local evaporation and runoff. Vegetation is prescribed. This model is currently used in two different horizontal resolutions, T21 and T42. This corresponds to a horizontal homogeneous resolution of approximately 8.6 and 4.3 degrees, respectively, for dynamic processes. The corresponding "Gaussian grid for calculating nonlinear and diabetic terms has a resolution of 5.60 and 28°". The method applied 19 vertical layers in a high-altitude hybrid ap "coordinate system" at ten horsepower. Annual and daily solar radiation cycles are available. The annual cycle of sea surface temperature and ice distribution is prescribed as low boundary conditions for offline running. ECHAM2 differs from ECHAMl, used in the simulation of the Hamburg 002 scenario (Cubasch et al. 1992), in orography (ECHAM2 uses a medium urography instead of the envelope) and gravity wave drawing parameters. At T21 resolution, this forced choice ensures air circulation simulation, mostly in the North Atlantic (Sausen et al. 1992). Compared to ECHAM3, the climate simulated with ECHAM2/T21 is a little different, and it will challenge to decide which version simulates the better observation.

ECHAM3

ECHAM3 includes several modifications, mainly in the parameter, to adjust the model for climate simulation. T42 is the reference resolution, but the model is set to use the resolution range T21 to T106. It should be noted that long-term mergers are performed only for T21, T42 and T63.

ECHAM4

Compared to the previous version of ECHAM3, several fundamental changes have been made to the physics and model number.

These contain:

- Semi-Lagrangian transportation plan for cloud water tracking materials, water vapor
- A new formula for vertical diffusion coefficients as turbulent kinetic energy functions
- New ECMWF (Radiation Plan) with changes in water vapor continuity, optical properties of clouds and greenhouse gases
- A new goal for deep convection is based on convection instability instead of moisture convergence.
- A new set of ground-level parameters has been introduced for the new model

It should be noted that minor changes are related to horizontal diffusion parameters, stratigraphic clouds and surface processes. In addition, the model climatology, derived from two extensive AMIP simulations at T42L19 resolution, is compared and documented with ECMWF operational analysis.

Some of the biases listed for the previous version of the model remain the same. For example:

- Regional wind faults are more significant than the 200 hpa level
- The upper polar troposphere and the lower stratosphere are too cold
- Low-frequency variability is still minimal, but errors are reduced by about 50% compared to ECHAM-3.

Zonal wind and tropospheric temperature errors are always smaller than previous models, except for the tropics. The most significant improvement, compared to previous developments, is in the Earth's surface climate. Precipitation and temperature errors are usually more minor than in the past, and the biomass distribution of these parameters is more realistic in ECHAM-4. This improvement can be attributed to the better display of surface radiation currents through the more significant absorption of solar radiation in the atmosphere due to water vapor and clouds (Roeckner et al. 1996).

ECHAM5

Compared to ECHAM4, there have been several fundamental changes in physics and model number. This includes:

- Semi-Lagrangian flux transport scheme for specific positive variables, such as chemical detectors, water components and
- Separate prognostic equations and longwave radiation scheme for cloud ice and liquid cloud water
- Statistical prediction of statistical cloud cover parameters and new microphysical design of cloud and one. The number of spectral distances has been improved in both high wave and short-wave sections of the spectrum
- Changes in the representation of ground surface processes, including an implicit connection between the atmosphere and the surface and the display of urography traction forces.
- A new set of ground-level parameters has been developed for the new model.

It should be noted that horizontal and vertical penetration, summative convection as well as spectral dynamics remain essentially unchanged.

ECHAM6

This version is currently the most advanced version of ECHAM coupling among diabatic processes and large-scale circulations. Eventually, both are driven by radiant coercion. This includes a transport model for scalar quantities dry, spectral-transform dynamical core, other than surface pressure and temperature. Physical parameters are suitable for displaying diabetic activity and boundary data sets for external parameters, such as tabulations of gas absorption optical properties, trace gas and aerosol distributions, temporal variations in spectral solar irradiance, land-surface properties, etc.

The maximum changes relative to ECHAM5 include:

- Show the progress of radiation transmission in the short wave (or solar) part of the spectrum
- A new description of aerosols

– Demonstration of surface albedo progress, including treatment of melting ponds on sea ice, significantly increases the representation of the middle atmosphere as part of the standard model.

It should be noted that the minimum changes in the representation of convection processes have been made by selecting slightly varied vertical discretization in the troposphere and changing the model parameters. (Stevens et al. 2013).

Applications of ECHAM

"ECHAM has run over different grid spacings, regions and simulation periods from days to decades to study various scientific problems" (Pal et al. 2007a, b). Some of these applications include future climate change, air quality water resources, extreme events, agriculture, land-cover change, and biosphere–atmosphere interactions (Pal et al. 2007a, b). Also, some preliminary work has been performed using "RegCM3 for seasonal prediction applications. This diverse range of applications and regions shows the versatility and portability of the ECHAM system" (Pal et al. 2007a, b).

Examples of Applying Different Regional Climate Model

The study (Jacob et al. 2007c), "possible regional climate" change across Europe, was "simulated by ten regional climate models." The primary purpose of this study was to understand how the systematic biases of the original model differ in "different models." Two fundamental aspects of model validation were addressed the ability of simulation: (i) long-term (30–40 years) means climate, (ii) inter-annual diversity. The analysis concentrates on near-surface air temperature and precipitation over land and concentration mainly on winter and summer seasons ("December–January–February (DJF), June–July–August (JJA)"). However, the transition seasons of "March–April–May (MAM) and September–October–November (SON) are also considered" (Jacob et al. 2007b).

The result illustrates that RACMO and Remo have the most predicted compared to other models. Moreover, the result of HIRHAM, REMO, and HadRM is close to the leading data. It is necessary to mention that the output of all modes is less close to the primary data in the second half of the year. In Region 2, the closest result to primary data is related to RACMO; also, HIRAM, REMO, and REGCM have the largest underestimate in the first six months of the year, and the biggest overestimate during the second half of the year.

In Tables 7.8, 7.9, 7.10 and 7.11, different biases related to precipitation estimation, the temperature in summer and winter are examined. As can be seen, the model has the highest bias value among other models in estimating temperature and rainfall parameters in summer. Also, in winter, two-way models have the highest and lowest bias values in assessing temperature and precipitation parameters, respectively.

Table 7.8 Temperature Bias (°C) DJF with connection to the CRU climatology for each of the eight regions. For some sub-regions, fewer models enter the ensemble average because of limited coverage by some models. The last entry provides the corresponding CRU average value. The average bias (left columns) and the inter-annual standard deviation (right columns) are presented for each sub-region. For the ensemble average, the inter-model standard deviation of the 30-year biases is shown instead (Jacob et al. 2007a)

	BI		IP		FR		ME		SC		AL		MD		EA	
	Bias	I.an	Bias	I.an	Bias	I.an	Bias	I.an	Bias	I.an	Bias	I.an	Bias	I.an	Bias	I.an
HIRHAM	0.70	0.97	0.44	0.79	1.32	1.13	1.28	1.69	1.27	1.76	0.29	1.28	−0.27	1.01	1.05	2.09
HIRHAM25	0.91	0.88	0.43	0.74	1.26	1.04	1.35	1.56	1.59	1.69	0.06	1.26	−0.23	0.93	1.27	1.84
HIRHAM12	0.93	0.94	0.41	0.76	1.14	1.20	1.17	1.63	1.56	1.58	-0.34	1.35	−0.32	0.98	1.04	1.81
CHRM	0.23	0.91	−1.19	0.63	0.27	0.94	0.82	1.19	1.53	1.56	0.15	0.90	−1.78	0.84	0.17	1.66
CLM	1.37	0.78	−1.01	0.72	1.19	0.98	1.63	1.24	2.98	1.25	-0.13	1.03	−1.56	0.86	1.15	1.57
HadRM3H	0.98	0.99	0.20	0.80	1.31	1.18	1.57	1.49	1.71	1.66	0.96	1.17	−0.34	0.89	1.15	1.69
RegCM	1.50	0.85	0.07	0.71	1.27	1.01	1.03	1.27	–	–	0.92	1.11	−1.14	0.86	0.36	1.40
RACMO	1.48	0.84	–	0.68	1.25	1.00	1.76	1.31	3.34	1.45	1.47	1.04	−0.37	0.85	1.33	1.51
HIRHAM.no	1.31	0.89	–	–	1.55	1.01	1.63	1.38	2.88	1.75	0.99	1.23	–	–	1.36	1.62
REMO	1.70	0.82	0.57	0.73	1.41	1.11	1.34	1.47	3.46	1.30	1.49	1.10	0.73	0.73	0.97	1.55

Table 7.9 As for Table 7.8 but JJA (Jacob et al. 2007a)

	BI		IP		FR		ME		SC		AL		MD		EA	
	Bias	I.an	Bias	I.an	Bias	I.an	Bias	I.an	Bias	I.an	Bias	I.an	Bias	I.an	Bias	I.an
HIRHAM	0.43	0.72	1.24	0.84	1.09	1.12	1.08	1.06	0.36	0.78	−0.16	1.01	1.86	1.01	2.60	1.19
HIRHAM25	0.32	0.60	0.78	0.81	0.52	0.94	0.74	0.93	0.31	0.68	−0.43	0.88	1.10	0.97	2.58	1.21
HIRHAM12	0.46	0.56	0.74	0.78	0.61	0.89	0.82	0.87	0.32	0.71	−0.36	0.81	0.95	0.92	2.59	1.04
CHRM	−0.43	0.55	−0.58	0.82	0.12	0.98	−0.13	0.73	−0.84	0.69	−0.34	0.91	0.18	0.87	1.10	0.72
CLM	0.26	0.61	−1.28	0.76	−0.07	0.95	−0.06	0.82	−0.52	0.75	−0.80	0.87	−0.33	0.70	1.44	1.02
HadRM3H	0.33	0.70	1.97	0.97	1.53	1.49	1.23	1.38	−0.13	0.76	1.95	1.45	3.12	1.38	3.12	1.40
RegCM	−0.29	0.67	−0.01	0.95	0.25	1.36	0.21	1.08	–	–	0.37	1.27	0.87	1.20	1.97	1.17
RACMO	0.65	0.50	0.39	0.82	0.41	0.76	0.60	0.59	0.93	0.74	0.80	0.74	1.19	0.81	1.79	0.70
HIRHAM.no	0.50	0.77	–	–	0.36	0.98	0.00	1.04	-0.12	1.01	−0.90	0.83	–	–	1.14	0.38
REMO	0.70	0.51	1.30	0.83	0.96	0.87	1.25	0.84	0.53	0.81	1.44	0.91	1.93	0.96	2.50	1.19

Table 7.10 As for Table 7.8 but precipitation (mm/day) (Jacob et al. 2007a)

	BI		IP		FR		ME		SC		AL		MD		EA	
	Bias	I.an	Bias	I.an	Bias	I.an	Bias	I.an	Bias	I.an	Bias	I.an	Bias	I.an	Bias	I.an
HIRHAM	0.06	0.63	−0.32	1.01	0.45	0.74	0.70	0.67	0.17	0.42	−0.25	0.82	−1.37	0.65	0.13	0.26
HIRHAM25	0.11	0.63	−0.22	1.05	0.52	0.72	0.76	0.69	0.30	0.44	−0.01	0.88	−1.17	0.70	0.22	0.30
HIRHAM12	0.16	0.64	−0.10	1.12	0.53	0.74	0.76	0.71	0.30	0.44	0.05	0.97	−1.11	0.73	0.19	0.29
CHRM	−0.51	0.57	−1.05	0.85	0.01	0.68	0.47	0.60	0.33	0.37	−0.62	0.85	−1.69	0.60	0.14	0.25
CLM	1.30	0.92	0.01	1.13	1.36	1.00	1.51	0.94	1.00	0.46	0.38	1.01	−0.71	0.76	0.66	0.34
HadRM3H	0.27	0.73	−0.17	1.02	0.38	0.76	0.79	0.73	0.78	0.44	0.85	1.11	−0.60	0.94	0.37	0.30
RegCM	0.32	0.74	−0.17	1.02	0.70	0.81	0.91	0.71	–	–	0.41	0.90	−0.62	0.87	0.42	0.31
RACMO	0.63	0.73	−0.26	0.99	0.51	0.78	0.73	0.66	0.55	0.42	−0.07	0.98	−1.07	0.77	0.25	0.28
HIRHAM.no	−0.07	0.57	–	–	0.69	0.74	1.16	0.72	0.77	0.43	0.27	0.95	–	–	0.74	0.29
REMO	0.59	0.81	−0.49	0.98	0.29	0.74	0.67	0.69	0.85	0.42	−0.24	0.89	−1.05	0.82	0.25	0.29

Table 7.11 As for Table 7.8 but JJA (Jacob et al. 2007a)

	BI		IP		FR		ME		SC		AL		MD		EA	
	Bias	I.an	Bias	I.an	Bias	I.an	Bias	I.an	Bias	I.an	Bias	I.an	Bias	I.an	Bias	I.an
HIRHAM	0.09	0.51	0.31	0.31	0.17	0.39	0.04	0.46	0.24	0.37	−0.28	0.47	0.14	0.31	−0.44	0.47
HIRHAM25	0.15	0.49	0.41	0.28	0.44	0.36	0.14	0.41	0.33	0.35	−0.07	0.55	0.45	0.41	−0.34	0.52
HIRHAM12	0.14	0.46	0.56	0.32	0.66	0.39	0.40	0.53	0.47	0.27	0.20	0.53	0.61	0.44	−0.13	0.53
CHRM	−0.14	0.57	−0.50	0.27	−0.47	0.40	−0.33	0.48	0.46	0.39	−1.15	0.64	−0.62	0.44	−0.86	0.45
CLM	0.55	0.62	0	0.31	0.47	0.45	0.33	0.52	0.73	0.31	−0.51	0.66	−0.06	0.55	−0.70	0.44
HadRM3H	0.02	0.57	0.01	0.34	−0.02	0.50	0.17	0.65	0.49	0.34	−0.63	0.95	−0.27	0.50	−0.43	0.61
RegCM	0.79	0.67	0.07	0.35	0.51	0.58	0.82	0.64	–	–	0.07	0.72	0.22	0.68	0.16	0.72
RACMO	0.26	0.52	−0.13	0.31	−0.03	0.35	−0.01	0.43	0.49	0.37	−0.28	0.68	−0.44	0.44	−0.56	0.42
HIRHAM.no	−0.32	0.46	–	–	0.01	0.46	−0.08	0.36	0.38	0.44	0.28	0.72	–	–	−0.57	0.40
REMO	0.47	0.62	0.32	0.34	0.50	0.53	0.48	0.55	0.59	0.36	0.53	0.64	0.22	0.58	−0.14	0.64

7.3 Summary

In climate change studies, Downscaling means that turning large-scale data into local-scale forecasters. Each climate model's main goal is to simulate and predict climate processes and climate, respectively. However, by considering the climate change effect, predicting the climate state is almost impossible; an alternative solution is to identify the various possibilities for the climate scenario. Currently, the most reliable tools for generating these scenarios are "General Rotation Models" (GCM) and "Atmospheric-Ocean General Circulation Models" (AOGCM).

Two primary forms of mentioned technique are Statistical and Dynamical Downscaling. In this chapter, each method is described in detail, then the structures and equations are explained and finally, the straightforward application of each model is summarized.

References

Abbasnezhadi K, Rousseau AN, Bohrn S (2020) Mid-21st century anthropogenic changes in extreme precipitation and snowpack projections over Newfoundland. Can Water Resour J/Revue Can. des ressources hydriques 1–21

Abbott MB, Bathurst JC, Cunge JA, O'Connell PE, Rasmussen J (1986) An introduction to the European Hydrological System—Systeme Hydrologique Europeen, "SHE", 1: History and philosophy of a physically-based, distributed modelling system. J Hydrol 87:45–59. https://doi.org/10.1016/0022-1694(86)90114-9

Abrahart Robert J, Anctil F, Coulibaly P, Dawson CW, Mount NJ, See LM, Shamseldin AY, Solomatine DP, Toth E, Wilby RL (2012) Two decades of anarchy? Emerging themes and outstanding challenges for neural network river forecasting. Prog Phys Geogr 36:480–513. https://doi.org/10.1177/0309133312444943

Abrahart Robert J, Mount NJ, Shamseldin AY (2012) Neuroemulation: definition and key benefits for water resources research. Hydrol Sci J-J Des Sci Hydrol 57:407–423. https://doi.org/10.1080/02626667.2012.658401

Ahmad S, Simonovic SP (2005) An artificial neural network model for generating hydrograph from hydro-meteorological parameters. J Hydrol 315:236–251. https://doi.org/10.1016/j.jhydrol.2005.03.032

Ahmadi A, Moridi A, Lafdani EK, Kianpisheh G (2014) Assessment of climate change impacts on rainfall using large scale climate variables and downscaling models—a case study. J Earth Syst Sci 123:1603–1618

Ahrens B, Karstens U, Rockel B, Stuhlmann R (1998) On the validation of the atmospheric model REMO with ISCCP data and precipitation measurements using simple statistics. Meteorol Atmos Phys 68:127–142

Ajami NK, Duan Q, Gao X, Sorooshian S (2006) Multimodel combination techniques for analysis of hydrological simulations: application to distributed model intercomparison project results. J Hydrometeorol 7:755–768. https://doi.org/10.1175/JHM519.1

Ajami NK, Duan Q, Moradkhni H, Sorooshian S (2005) 1.3 Recursive bayesian model combination for streamflow forecasting

Ajami NK, Duan Q, Sorooshian S (2007) An integrated hydrologic Bayesian multimodel combination framework: confronting input, parameter, and model structural uncertainty in hydrologic prediction. Water Resour Res 43. https://doi.org/10.1029/2005WR004745

Akinyemi FO, Abiodun BJ (2019) Potential impacts of global warming levels 1.5 °C and above on climate extremes in Botswana. Clim Change 154:387–400. https://doi.org/10.1007/s10584-019-02446-1

Alabastri A, Tuccio S, Giugni A, Toma A, Liberale C, Das G, De Angelis F, Di Fabrizio E, Zaccaria RP (2013) Molding of plasmonic resonances in metallic nanostructures: dependence of the non-linear electric permittivity on system size and temperature. Materials (Basel) 6:4879–4910

Alexandru A, de Elia R, Laprise R (2007) Internal variability in regional climate downscaling at the seasonal scale. Mon Weather Rev 135:3221–3238

Alfaro SC, Gomes L (2001) Modeling mineral aerosol production by wind erosion: emission intensities and aerosol size distributions in source areas. J Geophys Res Atmos 106:18075–18084

Andrews WH, Riley JP, Masteller MB (1978) Mathematical modeling of a sociological and hydrologic decision system

Anthes RA (1977) A cumulus parameterization scheme utilizing a one-dimensional cloud model. Mon Weather Rev 105:270–286

Anthes RA, Hsie E-Y, Kuo Y-H (1987) Description of the Penn State/NCAR mesoscale model version 4 (MM4). NCAR Boulder, Co

Arakawa A (1977) Computational design of the basic dynamical processes of the UCLA general circulation model. Methods in Computational Physics. Adv Res Appl 17. Gen Circ Model Atmos 337

Arnell NW (2004) Climate-change impacts on river flows in Britain: the UKCIPO2 scenarios. Water Environ J 18:112–117

Arnell NW, Hudson DA, Jones RG (2003) Climate change scenarios from a regional climate model: estimating change in runoff in southern Africa. J Geophys Res Atmos 108

Arnold JG, Srinivasan R, Muttiah RS, Williams JR (1998) Large area hydrologic modeling and assessment. Part I: Model development. J Am Water Resour Assoc 34:73–89. https://doi.org/10.1111/j.1752-1688.1998.tb05961.x

Arnold JG, Williams JR (1989) Stochastic generation of internal storm structure at a point. Trans Am Soc Agric Eng 32:161–167. https://doi.org/10.13031/2013.30976

Arnold JG, Williams JR, Nicks AD, Sammons NB (1990) SWRRB; A basin scale simulation model for soil and water resources management. SWRRB; A basin scale. Simul Model Soil Water Resour Manag, Texas A&M University Press

Assein R (1972) Frequency filter for time integrations. Mon Weather Rev 100:487–490. https://doi.org/10.1175/1520-0493(1972)100%3c0487:ffti%3e2.3.co;2

Authority TV (1972) A continuous daily streamflow model: upper bear creek. Exp Proj Res Pap (8)

Azorin-Molina C, Tijm S, Ebert EE, Vicente-Serrano SM, Estrela MJ (2014) Sea breeze thunderstorms in the eastern Iberian Peninsula. Neighborhood verification of HIRLAM and HARMONIE precipitation forecasts. Atmos Res 139:101–115. https://doi.org/10.1016/j.atmosres.2014.01.010

Baffaut C, Nearing MA, Nicks AD (1996) Impact of cligen parameters on wepp-predicted average annual soil loss. Trans ASAE 39:447–457. https://doi.org/10.13031/2013.27522

Bailey NTJ (1990) The elements of stochastic processes with applications to the natural sciences. Wiley

Baklanov A, Korsholm US, Nuterman R, Mahura A, Nielsen KP, Sass H, Rasmussen A, Zakey A, Kaas E, Kurganskiy A, Sørensen B, González-Aparicio I (2017) Enviro-HIRLAM online integrated meteorology-chemistry modelling system: strategy, methodology, developments and applications (v7.2). Geosci Model Dev 10:2971–2999. https://doi.org/10.5194/gmd-10-2971-2017

Baklanov A, Korsholm U, Mahura A, Petersen C, Gross A (2008) ENVIRO-HIRLAM: on-line coupled modelling of urban meteorology and air pollution

Bates JM, Granger CWJ (1969) The combination of forecasts. J Oper Res Soc 20:451–468. https://doi.org/10.1057/jors.1969.103

Bathurst JC, O'connell P (1992) Future of distributed modelling: the Systeme Hydrologique Europeen [WWW Document]. Hydrol Process. https://scholar.google.com/scholar_lookup?title=FutureofdistributedparametermodelingTheSystemeHydrologiqueEuropeen&journal=HydrologicalProcesses&volume=6&pages=265-277&publication_year=1992&author=Bathurst%2CJ.C.&author=O%27Connell%2CP.E . Accessed 14 May 2021

Batrak Y, Kourzeneva E, Homleid M (2018) Implementation of a simple thermodynamic sea ice scheme, SICE version 1.0-38h1, within the ALADIN–HIRLAM numerical weather prediction system version 38h1. Geosci Model Dev 11:3347–3368

Bear J (2012) Hydraulics of groundwater. Courier Corporation

Bear J, Cheng A (2010) Modeling groundwater flow and contaminant transport, (Vol 23). Springer, Dordrecht. https://doi.org/10.1007/978-1-4020-6682-5

Beasley D (1977) A mathematical model for simulating the effects of land use and management on water quality Purdue University. https://www.proquest.com/docview/302840394

Beaton AE, Tukey JW (1974) The fitting of power series, meaning polynomials, illustrated on band-spectroscopic data. Technometrics 16:147–185

Becker A, Nemec J (1987) Macroscale hydrologic models in support to climate research. Influ Clim Change ClimVar Hydrol Reg Water Resour 431–445

Becker A, Serban P (1990) Hydrological models for water-resources system design and operation. WMO No 740. https://library.wmo.int/doc_num.php?explnum_id=1696

Beecham S, Rashid M, Chowdhury RK (2014) Statistical downscaling of multi-site daily rainfall in a South Australian catchment using a Generalized Linear Model. Int J Climatol 34:3654–3670

Bell VA, Gedney N, Kay AL, Smith RNB, Jones RG, Moore RJ (2011) Estimating potential evaporation from vegetated surfaces for water management impact assessments using climate model output. J Hydrometeorol 12:1127–1136

Benestad RE (2010) Downscaling precipitation extremes. Theor Appl Climatol 100:1–21

Bengtsson L, Andrae U, Aspelien T, Batrak Y, Calvo J, de Rooy W, Gleeson E, Hansen-Sass B, Homleid M, Hortal M (2017) The HARMONIE–AROME model configuration in the ALADIN–HIRLAM NWP system. Mon Weather Rev 145:1919–1935

Beranová R, Kyselý J, Hanel M (2018) Characteristics of sub-daily precipitation extremes in observed data and regional climate model simulations. Theor Appl Climatol 132:515–527

Bergeron G, Robert AJ (1994) Formulation of the mesoscale compressible community (MC2) model. Cooperative Centre for Research in Mesometeorology = Centre coopératif pour ….

Beven KJ (2011) Rainfall-runoff modelling: the primer. Wiley

Beven K (2012a) Rainfall-runoff modelling the primer. 2nd edn

Beven K (2012b) Down to basics: runoff processes and the modelling process. In: Rainfall-runoff modelling. Wiley, Ltd., pp 1–23. https://doi.org/10.1002/9781119951001.ch1

Beven K (2013) So how much of your error is epistemic? Lessons from Japan and Italy. Hydrol Process 27:1677–1680. https://doi.org/10.1002/hyp.9648

Beven KJ, Kirkby MJ (1979) A physically based, variable contributing area model of basin hydrology. Hydrol Sci Bull 24:43–69. https://doi.org/10.1080/02626667909491834

Beven KJ, Kirkby MJ, Kirkby AJ (1979) A physically based, variable contributing area model of basin hydrology/Un modèle à base physique de zone d'appel variable de l'hydrologie du bassin versant) A physically based, variable contributing area model of basin hydrology/Un modèle à base physique de zone d'appel variable de l'hydrologie du bassin versant A physically based, variable contributing area model of basin hydrology. Hydrol Sci J 24:43–69. https://doi.org/10.1080/02626667909491834

Beven K, Westerberg I (2011) On red herrings and real herrings: disinformation and information in hydrological inference. Hydrol Process. https://doi.org/10.1002/hyp.7963

Beven K, Lamb R, Quinn P, Romanowicz R, Freer J (1995) TOPMODEL. Comput Model Watershed Hydrol 627–668

Bhuvandas N, Timbadiya PV, Patel PL, Porey PD (2014) Review of downscaling methods in climate change and their role in hydrological studies. Int J Environ Ecol Geol Mar Eng 8:713–718

Bisselink B, Dolman AJ (2009) Recycling of moisture in Europe: contribution of evaporation to variability in very wet and dry years. Hydrol Earth Syst Sci Discuss 6

Biswas AK (1970) History of hydrology. North-Holland Publishing Company

Boberg F, Christensen JH (2012) Overestimation of Mediterranean summer temperature projections due to model deficiencies. Nat Clim Change 2:433–436

Boers R, Bosveld F, Baltink HK, Knap W, Van Meijgaard E, Wauben W (2019) Observing and modelling the surface radiative budget and cloud radiative forcing at the Cabauw experimental site for atmospheric research (CESAR), The Netherlands, 2009–17. J Clim 32:7209–7225

Bouraoui F, Braud I, Dillaha TA (2002) ANSWERS: a nonpoint source pollution model for water, sediment and nutrient losses. Math Model Small Watershed Hydrol Appl 833–882

Bowden GJ, Maier HR, Dandy GC (2012) Real-time deployment of artificial neural network forecasting models: understanding the range of applicability. Water Resour Res 48:10549. https://doi.org/10.1029/2012WR011984

Bras RL, Rodriguez-Iturbe I (1993) Random functions and hydrology. Courier Corporation

Briegleb BP (1992) Delta-Eddington approximation for solar radiation in the NCAR Community Climate Model. J Geophys Res. Atmos 97:7603–7612

Broderick CJ (2012) Climate Change and Atlantic salmon (Salmo salar): changes in flow and freshwater habitat in the burrishoole catchment national university of Ireland, Maynooth (Ireland). https://mural.maynoothuniversity.ie/3996/1/Ciaran_Broderick_PhD_thesis.pdf

Brown C, Greene AM, Block PJ, Giannini A (2008) Review of downscaling methodologies for Africa climate applications. IRIfCa. Society. https://doi.org/10.7916/D8M04C88

Brunner P, Simmons CT (2012) HydroGeoSphere: a fully integrated, physically based hydrological model. Ground Water https://doi.org/10.1111/j.1745-6584.2011.00882.x

Buonomo E, Jones R, Huntingford C, Hannaford J (2007) On the robustness of changes in extreme precipitation over Europe from two high resolution climate change simulations. Q. J. R. Meteorol. Soc. A. J Atmos Sci Appl Meteorol Phys Oceanogr 133:65–81

Burke EJ, Perry RHJ, Brown SJ (2010) An extreme value analysis of UK drought and projections of change in the future. J Hydrol 388:131–143

Burnash R (1975) The NWS River Forecast System-Catchment Modeling. Singh V (Ed) Computer Models of Watershed Hydrology. In: water resources publication, Colorado

Burnash RJ, Ferral RL, McGuire RA (1973) A generalized streamflow simulation system: Conceptual modelling for digital computers. US Department of Commerce, National Weather Service, and State of California. https://books.google.com.hk/books?id=aQJDAAAAIAAJ%26printsec=frontcover%26hl=zhCN%26source=gbs_ge_summary_r%26cad=0#v=onepage%26q%26f=false

Capell R, Tetzlaff D, Essery R, Soulsby C (2014) Projecting climate change impacts on stream flow regimes with tracer-aided runoff models-preliminary assessment of heterogeneity at the mesoscale. Hydrol Process 28:545–558

Carter T, Parry M, Harasawa H, Nishioka S (1994) IPCC technical guidelines for assessing climate change impacts and adaptations. In: Part of the IPCC special report to the first session of the conference of the parties to the UN framework convention on climate change. Intergovernmental Panel on Climate Change, Department of Geography, University College London, UK and Center for Global

Cats G, Wolters L (1996) The Hirlam project [meteorology]. IEEE Comput. Sci. Eng. 3:4–7

Cavazos T, Hewitson BC (2005) Performance of NCEP-NCAR reanalysis variables in statistical downscaling of daily precipitation. Clim Res 28:95–107. https://doi.org/10.3354/cr028095

Chapuis RP (2012) Predicting the saturated hydraulic conductivity of soils: a review. Bull Eng Geol Environ 71:401–434. https://doi.org/10.1007/s10064-012-0418-7

Chatterjee S, Hadi AS (2015) Regression analysis by example. Wiley

Chen J, Brissette FP (2014) Stochastic generation of daily precipitation amounts: review and evaluation of different models. Clim Res. https://doi.org/10.3354/cr01214

Chen J, Chen H, Guo S (2018a) Multi-site precipitation downscaling using a stochastic weather generator. Clim Dyn 50:1975–1992. https://doi.org/10.1007/s00382-017-3731-9

Chen J, Zhang XJ, Li X (2018b) A weather generator-based statistical downscaling tool for site-specific assessment of climate change impacts. Trans ASABE 61:977–993. https://doi.org/10.13031/TRANS.12601

Chen J, Brissette FP, Leconte R (2011) Uncertainty of downscaling method in quantifying the impact of climate change on hydrology. J Hydrol 401:190–202. https://doi.org/10.1016/j.jhydrol.2011.02.020

Chen J, Brissette FP, Lucas-Picher P (2015) Assessing the limits of bias-correcting climate model outputs for climate change impact studies. Wiley Online Libr 120:1123–1136. https://doi.org/10.1002/2014JD022635

Chen J, Zhang X, Liu W, Li Z (2009) Evaluating and extending CLIGEN precipitation generation for the Loess Plateau of China1. JAWRA J Am Water Resour Assoc 45:378–396. https://doi.org/10.1111/J.1752-1688.2008.00296.X

Chen J, Brissette FP, Leconte R (2012a) Coupling statistical and dynamical methods for spatial downscaling of precipitation. Clim Change 114:509–526. https://doi.org/10.1007/s10584-012-0452-2

Chen J, Brissette FP, Chaumont D, Braun M (2013a) Finding appropriate bias correction methods in downscaling precipitation for hydrologic impact studies over North America. Wiley Online Libr 49:4187–4205. https://doi.org/10.1002/wrcr.20331

Chen J, Brissette FP, Chaumont D, Braun M (2013b) Performance and uncertainty evaluation of empirical downscaling methods in quantifying the climate change impacts on hydrology over two North American river basins. J Hydrol 479:200–214. https://doi.org/10.1016/j.jhydrol.2012.11.062

Chen J, Zhang XC, Liu WZ, Li Z (2008) Assessment and improvement of CLIGEN non-precipitation parameters for the Loess Plateau of China. Trans ASABE 51:901–913. https://doi.org/10.13031/2013.24529

Chen G, Hua W, Fang X, Wang C, Li X (2021) Distributed-framework basin modeling system: II. Hydrologic modeling system. Water 2021 13:744, 13:744. https://doi.org/10.3390/W13050744

Chen J, Brissette F, Research RL-C (2012b) Undefined, n.d. Downscaling of weather generator parameters to quantify hydrological impacts of climate change. https://www.int-res.com

Cheng S (1991) Synoptic climatological categorization and human mortality in Shanghai, China. Proc Middle S Div Assoc Am Geogr 24:5–11

Cheng S, Kalkstein LS (1997) Determination of climatological seasons for the East Coast of the US using an air mass-based classification. Clim Res 8:107–116

Cheng S, Lam K-C (2000) Synoptic typing and its application to the assessment of climatic impact on concentrations of sulfur dioxide and nitrogen oxides in Hong Kong. Atmos Environ 34:585–594

Cheng CT, Wang WC, Xu DM, Chau KW (2008) Optimizing hydropower reservoir operation using hybrid genetic algorithm and chaos. Water Resour Manag 22:895–909. https://doi.org/10.1007/s11269-007-9200-1

Cheng S, Kalkstein LS (1993) An evaluation of climate change in Phoenix using an automatic synoptic climatological approach. World Resour Rev States 5

Chia E, Hutchinson MF (1991) The beta distribution as a probability model for daily cloud duration. Agric For Meteorol 56:195–208

Chiew F, Hydrology TM-J (1994), undefined, n.d. Application of the daily rainfall-runoff model MODHYDROLOG to 28 Australian catchments. Elsevier

Chithra NR, Thampi SG (2017) Downscaling future projections of monthly precipitation in a catchment with varying physiography. ISH J Hydraul Eng 23:144–156

Chow DHC, Levermore GJ (2007) New algorithm for generating hourly temperature values using daily maximum, minimum and average values from climate models. Build Serv Eng Res Technol 28:237–248

Chow VT, Maidment DR, Mays LW (1988) Applied hydrology. MacGraw-Hill. Inc., New York

Christensen JH, Boberg F, Christensen OB, Lucas-Picher P (2008) On the need for bias correction of regional climate change projections of temperature and precipitation. Geophys Res Lett 35

Clark Martyn P, Kavetski D, Fenicia F (2011) Pursuing the method of multiple working hypotheses for hydrological modeling. Water Resour. Res. 47:9301. https://doi.org/10.1029/2010WR009827

Clark MP, Kavetski D, Fenicia F (2011) Pursuing the method of multiple working hypotheses for hydrological modeling. Water Resour Res 47:9301. https://doi.org/10.1029/2010WR009827

Clarke RT (1973) A review of some mathematical models used in hydrology, with observations on their calibration and use. J Hydrol 19:1–20. https://doi.org/10.1016/0022-1694(73)90089-9
Clarke RT (1988) Stochastic processes for water scientists: development and applications. Wiley
Cook LM, McGinnis S, Samaras C (2020) The effect of modeling choices on updating intensity-duration-frequency curves and stormwater infrastructure designs for climate change. Clim Change 159:289–308
Corzo GA, Solomatine DP, Hidayat H, De Wit M, Werner M, Uhlenbrook S, Price RK (2009) Combining semi-distributed process-based and data-driven models in flow simulation: a case study of the Meuse river basin. Hydrol Earth Syst Sci 13:1619–1634. https://doi.org/10.5194/hess-13-1619-2009
Counsell CJA (2018) End-to-end ensemble modelling for water resources planning under uncertainty. Open University (United Kingdom)
Crawford N, Linsley RK (1966) Digital simulation in hydrology' stanford watershed model IV
Cubasch U, Hasselmann K, Höck H, Maier-Reimer E, Mikolajewicz U, Santer BD, Sausen R (1992) Time-dependent greenhouse warming computations with a coupled ocean-atmosphere model. Clim Dyn 8:55–69
Cusack S, Edwards JM, Kershaw R (1999) Estimating the subgrid variance of saturation, and its parametrization for use in a GCM cloud scheme. Q J R Meteorol Soc, 125(560):3057–3076
Dahlgren P, Landelius T, Kållberg P, Gollvik S (2016) A high-resolution regional reanalysis for Europe. Part 1: Three-dimensional reanalysis with the regional High-Resolution Limited-Area Model (HIRLAM). Q J R Meteorol Soc 142:2119–2131
Dankers R, Middelkoop H (2008) River discharge and freshwater runoff to the Barents Sea under present and future climate conditions. Clim Change 87:131–153
da Silva Soares JP (2016) Wind energy utilization in arctic climate–RACMO 2.3 Greenland Climate Runs Project
Darcy H (1856) Les fontaines publiques de la ville de Dijon: exposition et application... Victor Dalmont
Davies H (1976) A lateral boundary formulation for multi-level prediction models. Q J R Meteorol Soc 102:405–418
Davies T (2014) Lateral boundary conditions for limited area models. Q J R Meteorol Soc 140:185–196. https://doi.org/10.1002/qj.2127
Dawdy DR, O'Donnell T (1965) Mathematical models of catchment behavior. J Hydraul Div 91:123–137
Dawdy DR, Lichty RW, Bergmann JM (1972) A rainfall-runoff simulation model for estimation of flood peaks for small drainage basins. US Government Printing Office. https://books.google.com/books?hl=en&lr=lang_en%26id=eSY5vje5k00C%26oi=fnd%26pg=PA10%26ots=DoKYSMFZYO%26sig=bEJHoNpCdqZmgcBMvoTmpZnNZ4#v=onepage%26q%26f=false
Dawson CW, Mount NJ, Abrahart RJ, Louis J (2014) Sensitivity analysis for comparison, validation and physical legitimacy of neural network-based hydrological models. J Hydroinform 16:1–18. https://doi.org/10.2166/hydro.2013.222
Dawson CW, Wilby RL (2001) Hydrological modelling using artificial neural networks. Prog Phys Geogr 25:80–108. https://doi.org/10.1177/030913330102500104
Day JJ, Bamber JL, Valdes PJ (2013) The Greenland Ice Sheet's surface mass balance in a seasonally sea ice-free Arctic. J Geophys Res Earth Surf 118:1533–1544
Dayyani S, Prasher SO, Madani A, Madramootoo CA (2012) Impact of climate change on the hydrology and nitrogen pollution in a tile-drained agricultural watershed in Eastern Canada. Trans ASABE 55:389–401
De Bruijn EIF, De Bruijn C, Van Meijgaard E (2006) Verification of HIRLAM with ECMWF physics compared with HIRLAM reference versions field experiment with a hot-air balloon at Cabauw View project verification of HIRLAM with ECMWF physics compared with HIRLAM reference versions
de Haan S, van der Veen SH (2014) Cloud initialization in the rapid update cycle of HIRLAM. Weather Forecast 29:1120–1133. https://doi.org/10.1175/WAF-D-13-00071.1

Deardorff JW (1972) Parameterization of the planetary boundary layer for use in general circulation models. Mon Weather Rev 100:93–106

Devia GK, Ganasri BP, Dwarakish GS (2015) A review on hydrological models. Aquat Procedia 4:1001–1007. https://doi.org/10.1016/j.aqpro.2015.02.126

Dibike YB, Coulibaly P (2005) Hydrologic impact of climate change in the Saguenay watershed: comparison of downscaling methods and hydrologic models. J Hydrol 307:145–163. https://doi.org/10.1016/j.jhydrol.2004.10.012

Dickinson JP (1973) Some statistical results in the combination of forecasts. J Oper Res Soc 24:253–260. https://doi.org/10.1057/jors.1973.42

Dickinson RE, Errico RM, Giorgi F, Bates GT (1989) A regional climate model for the western United States. Clim Change 15:383–422

Dickinson RE, Henderson-Sellers A, Kennedy PJ, Wilson MF (1986) Biosphereatmosphere transfer scheme (BATS) for NCAR community climate model. NCAR Tech Note Atmos Anal Predict Div Nat Cent Atmos Res, Boulder, CO

Dickinson E, Henderson-Sellers A, Kennedy J (1993) Biosphere-atmosphere transfer scheme (BATS) version 1e as coupled to the NCAR community climate model

Diffenbaugh NS, Pal JS, Trapp RJ, Giorgi F (2005) Fine-scale processes regulate the response of extreme events to global climate change. Proc Natl Acad Sci 102:15774–15778

Di Matteo L, Bigotti F, Ricco R (2011) Compressibility of kaolinitic clay contaminated by ethanol-gasoline blends. J Geotech Geoenvironmental Eng 137:846–849

Dolman AJ, Soet M, Ronda R, Kabat P, de Bruin HAR, Feddes RA, Kok MJT, Verweij W (1999) Representation of the seasonal hydrological cycle in climate and weather prediction models in West Europe. In: Dutch national research programme on global air pollution and climate change: proceedings of the first NRP-II symposium on climate change research, Garderen, the Netherlands, 29–30 October 1998, pp 3–8

Dolman AJ, Soet M, Van den Hurk B, Ijpelaar RJM, Ronda RJ (2001) The representation of the seasonal hydrological cycle in a regional climate model in west Europe. IAHS Publ 11–18

Donigian AS (1977) Agricultural runoff management (ARM) model version II: refinement and testing. Environmental Protection Agency, Office of Research and Development. https://books.google.com/books?hl=en%26lr=lang_en%26id=0e0RzttuK60C%26oi=fnd&pg=PR4%26ots=3JYVrf_St%26sig=EpM6qZ7m7L6B4K6aW1bFxKsbgX8#v=onepage%26q%26f=false

Downing TE, Harrison PA, Butterfield RE, Lonsdale KG (2000) Climate change, climatic variability and agriculture in Europe: an integrated assessment. University of Oxford, Environmental Change Institute

Duan Q, Gupta HV, Sorooshian S, Rousseau AN, Turcotte R (2004) Calibration of watershed models. American Geophysical Union

Eder BK, Davis JM, Bloomfield P (1994) An automated classification scheme designed to better elucidate the dependence of ozone on meteorology. J Appl Meteorol 33:1182–1199

Edwards TL, Fettweis X, Gagliardini O, Gillet-Chaulet F, Goelzer H, Gregory JM, Hoffman M, Huybrechts P, Payne AJ, Perego M (2014) Effect of uncertainty in surface mass balance–elevation feedback on projections of the future sea level contribution of the Greenland ice sheet. Cryosph 8:195–208

Eerola K, Rontu L, Kourzeneva E, Shcherbak E (2010) A study on effects of lake temperature and ice cover in hirlam

Eliasen E, Machenhauer B, Rasmussen E (1970) On a numerical method for integration of the hydrodynamical equations with a spectral representation of the horizontal fields. Kobenhavns Universitet, Institut for Teoretisk Meteorologi

Emanuel KA (1991) A scheme for representing cumulus convection in large-scale models. J Atmos Sci 48:2313–2329

Emanuel KA, Živković-Rothman M (1999) Development and evaluation of a convection scheme for use in climate models. J Atmos Sci 56:1766–1782

Emmanuel LA, Houngué NR, Biaou CA, Badou DF (2019) Statistical analysis of recent and future rainfall and temperature variability in the Mono River Watershed (Benin, Togo). Climate 7:8

Enke W, Spekat A (1997) Downscaling climate model outputs into local and regional weather elements by classification and regression. Clim Res 8:195–207. https://doi.org/10.3354/cr008195
Eom J, Rim H (2020) Recovery of mass changes in antarctic ice-sheet based on the regional climate model. RACMO Econ Environ Geol 53:147–157
Essou GR, Sabarly F, Lucas-Picher P, Brissette F, Poulin A (2016) Can precipitation and temperature from meteorological reanalyses be used for hydrological modeling? J Hydrometeorol, 17(7):1929–1950. https://doi.org/10.1175/JHM-D-15-0138.1
Essou GR, Brissette FP, Lucas-Picher P (2017) The use of reanalyses and gridded observations as weather input data for a hydrological model: comparison of performances of simulated river flows based. J Hydrometeorol
Ewen J, Parkin G, O'Connell PE (2000) SHETRAN: distributed river basin flow and transport modeling system. J Hydrol Eng 5:250–258. https://doi.org/10.1061/(asce)1084-0699(2000)5:3(250)
Fan Y, Van Den Dool H (2004) Climate prediction center global monthly soil moisture data set at 0.5 resolution for 1948 to present. J Geophys Res Atmos 109
Farajzadeh M, Oji R, Cannon AJ, Ghavidel Y, Bavani AM (2015) An evaluation of single-site statistical downscaling techniques in terms of indices of climate extremes for the Midwest of Iran. Theor Appl Climatol 120:377–390
Fares A, El-Kadi AI (2008) Coastal watershed management. WIT Press
Feldman AD (1981) HEC models for water resources system simulation: theory and experience. In: Advances in hydroscience. Elsevier, pp 297–423
Feldmann J, Levermann A, Mengel M (2019) Stabilizing the West Antarctic Ice Sheet by surface mass deposition. Sci Adv 5:aaw4132
Fenicia F, Kavetski D, Savenije HHG (2011) Elements of a flexible approach for conceptual hydrological modeling: 1. Motivation and theoretical development. Water Resour Res 47. https://doi.org/10.1029/2010WR010174
Flanagan D, Rep MN-N (1995) undefined, n.d. USDA-Water Erosion Prediction Project: Hillslope profile and watershed model documentation. https://www.ars.usda.gov
Fletcher D, Goss E (1993) Forecasting with neural networks. An application using bankruptcy data. Inf Manag 24:159–167. https://doi.org/10.1016/0378-7206(93)90064-Z
Flury M, Flühler H, Jury WA, Leuenberger J (1994) Susceptibility of soils to preferential flow of water: a field study. Water Resour Res 30:1945–1954. https://doi.org/10.1029/94WR00871
Fortin J-P, Turcotte R, Massicotte S, Moussa R, Fitzback J, Villeneuve J-P (2001) Distributed watershed model compatible with remote sensing and GIS data. I: description of model. J Hydrol Eng 6:91–99. https://doi.org/10.1061/(asce)1084-0699(2001)6:2(91)
Fotso-Nguemo TC, Vondou DA, Tchawoua C, Haensler A (2017) Assessment of simulated rainfall and temperature from the regional climate model REMO and future changes over Central Africa. Clim Dyn 48:3685–3705
Fowler HJ, Blenkinsop S, Tebaldi C (2007) Linking climate change modelling to impacts studies: recent advances in downscaling techniques for hydrological modelling. Int J Climatol 27:1547–1578. https://doi.org/10.1002/joc.1556
Franco B, Fettweis X, Lang C, Erpicum M (2012) Impact of spatial resolution on the modelling of the Greenland ice sheet surface mass balance between 1990–2010, using the regional climate model MAR. Cryosphere 6:695–711
Frieler K, Clark PU, He F, Buizert C, Reese R, Ligtenberg SRM, Van Den Broeke MR, Winkelmann R, Levermann A (2015) Consistent evidence of increasing Antarctic accumulation with warming. Nat Clim Change 5:348–352
Fu Q, Johanson CM (2005) Satellite-derived vertical dependence of tropical tropospheric temperature trends. Geophys. Res. Lett. 32:1–5. https://doi.org/10.1029/2004GL022266
Futter MN, Whitehead PG, Sarkar S, Rodda H, Crossman J (2015) Rainfall runoff modelling of the Upper Ganga and Brahmaputra basins using PERSiST. Environ Sci Process Imp 17:1070–1081
Fyke J, Lenaerts J, Wang H (2017) Basin-scale heterogeneity in Antarctic precipitation and its impact on surface mass variability. Cryosph 11

Gal-Chen T, Somerville RCJ (1975) On the use of a coordinate transformation for the solution of the Navier-Stokes equations. J Comput Phys 17:209–228

Gao X, Pal JS, Giorgi F (2006) Projected changes in mean and extreme precipitation over the Mediterranean region from a high resolution double nested RCM simulation. Geophys Res Lett 33

Gelhar LW (1986) Stochastic subsurface hydrology from theory to applications. Water Resour Res 22:135S–145S

Ghosh S, Mujumdar PP (2008) Statistical downscaling of GCM simulations to streamflow using relevance vector machine. Adv Water Resour 31:132–146. https://doi.org/10.1016/j.advwatres.2007.07.005

Girard C, Jarraud M (1982) Short and medium range forecast differences between a spectral and grid point model: an extensive quasioperational comparison (Issue 32). European Centre Medium Range Weather Forecasts

Giorgi F (1991) Sensitivity of simulated summertime precipitation over the western United States to different physics parameterizations. Mon Weather Rev 119:2870–2888

Giorgi F, Bi X, Pal J (2004) Mean, interannual variability and trends in a regional climate change experiment over Europe. II: climate change scenarios (2071–2100). Clim Dyn 23:839–858

Giorgi F, Francisco R, Pal J (2003) Effects of a subgrid-scale topography and land use scheme on the simulation of surface climate and hydrology. Part I: Effects of temperature and water vapor disaggregation. J Hydrometeorol 4:317–333

Giorgi F, Gutowski WJ (2015) Regional dynamical downscaling and the CORDEX initiative. Ann Rev Environ Resour 40:467–490. https://doi.org/10.1146/annurev-environ-102014-021217

Giorgi F, Huang Y, Nishizawa K, Fu C (1999) A seasonal cycle simulation over eastern Asia and its sensitivity to radiative transfer and surface processes. J Geophys Res Atmos 104:6403–6423

Giorgi F, Mearns LO (1999) Introduction to special section: regional climate modeling revisited. J Geophys Res Atmos. 104:6335–6352

Giorgi F, Pal JS, Bi X, Sloan L, Elguindi N, Solmon F (2006) Introduction to the TAC special issue: the RegCNET network. Theor Appl Climatol. https://doi.org/10.1007/s00704-005-0199-z

Giorgi F, Marinucci MR, Bates GT (1993a) Development of a second-generation regional climate model (RegCM2). Part I: Boundary-layer and radiative transfer processes. Mon Weather Rev 121:2794–2813

Giorgi F, Marinucci MR, Bates GT, De Canio G (1993b) Development of a second-generation regional climate model (RegCM2). Part II: Convective processes and assimilation of lateral boundary conditions. Mon. Weather Rev 121:2814–2832

Giorgi F, Hewitson B, Christensen J, Hulme M, Storch H, Whetton P, Jones R, Mearns L, Fu C (2001) Regional climate information—evaluation and projections

Gleckler PJ, Taylor KE, Doutriaux C (2008) Performance metrics for climate models. J Geophys Res Atmos 113. https://doi.org/10.1029/2007JD008972

Gleick PH (1986) Methods for evaluating the regional hydrologic impacts of global climatic changes. J Hydrol 88:97–116. https://doi.org/10.1016/0022-1694(86)90199-X

Gomis D, Ruiz S, Sotillo MG, Álvarez-Fanjul E, Terradas J (2008) Low frequency Mediterranean sea level variability: the contribution of atmospheric pressure and wind. Glob Planet Change 63:215–229

Gong W, Gupta HV, Yang D, Sricharan K, Hero AO (2013) Estimating epistemic and aleatory uncertainties during hydrologic modeling: an information theoretic approach. Water Resour Res 49:2253–2273. https://doi.org/10.1002/wrcr.20161

Gordon C, Cooper C, Senior CA, Banks H, Gregory JM, Johns TC, Mitchell JFB, Wood RA (2000) The simulation of SST, sea ice extents and ocean heat transports in a version of the Hadley Centre coupled model without flux adjustments. Clim Dyn 16:147–168. https://doi.org/10.1007/s003820050010

Goyal MK, Ojha CSP (2012) Downscaling of surface temperature for lake catchment in an arid region in India using linear multiple regression and neural networks. Int J Climatol 32:552–566

Granata F, Gargano R, de Marinis G (2016) Support vector regression for rainfall-runoffmodeling in urban drainage: a comparison with the EPA's storm water management model. Water (Switzerland) 8:1–13. https://doi.org/10.3390/w8030069
Graversen RG, Drijfhout S, Hazeleger W, van de Wal R, Bintanja R, Helsen M (2011) Greenland's contribution to global sea-level rise by the end of the 21st century. Clim Dyn 37:1427–1442
Grell GA (1993) Prognostic evaluation of assumptions used by cumulus parameterizations. Mon Weather Rev 121:764–787
Grell GA, Dudhia J, Stauffer DR (1994) A description of the fifth-generation Penn State/NCAR Mesoscale Model (MM5)
Groisman PY, Legates DR (1994) The accuracy of United States precipitation data. Bull Am Meteorol Soc 75:215–228
Gunawardhana LN, Kazama S (2012) Statistical and numerical analyses of the influence of climate variability on aquifer water levels and groundwater temperatures: the impacts of climate change on aquifer thermal regimes. Glob Planet Change 86:66–78
Guo Jing, Chen H, Xu CY, Guo S, Guo Jiali (2012) Prediction of variability of precipitation in the Yangtze River Basin under the climate change conditions based on automated statistical downscaling. Stoch Environ Res Risk Assess 26:157–176. https://doi.org/10.1007/s00477-011-0464-x
Gupta HV, Nearing GS (2014) Debates—the future of hydrological sciences: a (common) path forward? Using models and data to learn: a systems theoretic perspective on the future of hydrological science. Water Resour Res https://doi.org/10.1002/2013WR015096
Gregory ML, Raymond WD, Bell A, Fosler-Lussier E, Jurafsky D (1999) The effects of collocational strength and contextual predictability in lexical production. Chicago Linguistic Society, 35:151–166
Haan CT (1977) Statistical methods in hydrology. The Iowa State University Press
Hagemann S, Chen C, Haerter JO, Heinke J, Gerten D, Piani C (2011) Impact of a statistical bias correction on the projected hydrological changes obtained from three GCMs and two hydrology models. J Hydrometeorol 12:556–578. https://doi.org/10.1175/2011JHM1336.1
Hagemann S, Gates LD (2003) Improving a subgrid runoff parameterization scheme for climate models by the use of high resolution data derived from satellite observations. Clim Dyn 21:349–359
Hagemann S (2002) An improved land surface parameter dataset for global and regional climate models
Hanel M, Buishand TA, Ferro CAT (2009) A nonstationary index flood model for precipitation extremes in transient regional climate model simulations. J Geophys Res Atmos 114
Harrison GP, Cradden LC, Chick JP (2008) Preliminary assessment of climate change impacts on the UK onshore wind energy resource. Energy Sourc Part A 30:1286–1299
Harrison PA, Butterfield RE, Downing TE (1995) Climate change and agriculture in Europe: assessment of impacts and adaptations. Environmental Change Unit, University of Oxford
Hashmi MZ, Shamseldin AY, Melville BW (2011) Statistical downscaling of watershed precipitation using Gene Expression Programming (GEP). Environ Model Softw 26:1639–1646
Hastie TJ, Tibshirani RJ (1990) Generalized additive models. CRC Press
Hawkins E, Osborne TM, Ho CK, Challinor AJ (2013) Calibration and bias correction of climate projections for crop modelling: an idealised case study over Europe. Agric For Meteorol 170:19–31. https://doi.org/10.1016/j.agrformet.2012.04.007
Haykin S (1994) Neural networks: a comprehensive foundation. MacMillan College Publishing Co., New York
Haylock MR, Peterson TC, Alves LM, Ambrizzi T, Anunciação YMT, Baez J, Barros VR, Berlato MA, Bidegain M, Coronel G (2006) Trends in total and extreme South American rainfall in 1960–2000 and links with sea surface temperature. J Clim 19:1490–1512
Headrick MG, Wilson BN (1997) Evaluation of stochastic weather parameters for Minnesota and their impact on WEPP. In: Paper—American Society of Agricultural Engineers

Hertig E, Seubert S, Paxian A, Vogt G, Paeth H, Jacobeit J (2014) Statistical modelling of extreme precipitation indices for the Mediterranean area under future climate change. Int J Climatol 34:1132–1156

Hessami M, Gachon P, Ouarda TBMJ, St-Hilaire A (2008) Automated regression-based statistical downscaling tool. Environ Model Softw 23:813–834

Hewitson BC, Crane RG (1996) Climate downscaling: techniques and application. Clim Res 7:85–95

Hoerl AE, Kennard RW (1970) Ridge regression: biased estimation for nonorthogonal problems. Technometrics 12:55–67

Holton JR, Staley DO (1973) An introduction to dynamic meteorology. Am J Phys 41:752–754. https://doi.org/10.1119/1.1987371

Holtslag AAM, De Bruijn EIF, Pan HL (1990) A high resolution air mass transformation model for short-range weather forecasting. Mon Weather Rev 118:1561–1575

Hoomehr S, Schwartz JS, Yoder DC (2016) Potential changes in rainfall erosivity under GCM climate change scenarios for the southern Appalachian region, USA. Catena 136:141–151. https://doi.org/10.1016/j.catena.2015.01.012

Horstmann J, Koch W, Lehner S (2002) High resolution wind fields retrieved from SAR in comparison to numerical models. In: IEEE international geoscience and remote sensing symposium. IEEE, pp 1877–1879

Horton RE (1933) The rôle of infiltration in the hydrologic cycle. Eos Trans Am Geophys Union 14:446–460. https://doi.org/10.1029/TR014i001p00446

Horton RE (1945) Erosional development of streams and their drainage basins; hydrophysical approach to quantitative morphology. Geol Soc Am Bull, 56(3): 275–370. https://doi.org/10.1130/0016-7606(1945)56[275:EDOSAT]2.0.CO;2

Hostetler SW, Bartlein PJ (1990) Simulation of lake evaporation with application to modeling lake level variations of Harney-Malheur Lake Oregon. Water Resour Res. 26:2603–2612

Houle D, Marty C, Augustin F, Dermont G, Gagnon C, Kaiser K (2020) Impact of climate change on soil hydro-climatic conditions and base cations weathering rates in forested watersheds in Eastern Canada. Front For Glob Change 3:1–12

Hsie E-Y, Anthes RA, Keyser D (1984) Numerical simulation of frontogenesis in a moist atmosphere. J Atmos Sci 41:2581–2594

Hsu P-C, Nguyen C (1995) Theoretical investigation of a class of new planar transmission lines from microwave and millimeter-wave integrated circuits. In: Millimeter and Submillimeter Waves II. SPIE, pp 159–161. https://doi.org/10.1117/12.224222

Hu F, Wang L, Quan B, Xu X, Li Z, Wu Z, Pan X (2013a) Design of a polarization insensitive multiband terahertz metamaterial absorber. J Phys D Appl Phys 46:195103

Hu Y, Maskey S, Uhlenbrook S (2013b) Downscaling daily precipitation over the Yellow River source region in China: a comparison of three statistical downscaling methods. Theor Appl Climatol 112:447–460

Huang X-Y, Lynch P (1993) Diabatic digital-filtering initialization: application to the HIRLAM model. Mon Weather Rev 121:589–603

Huang J, Zhang J, Zhang Z, Xu CY, Wang B, Yao J (2011) Estimation of future precipitation change in the Yangtze River basin by using statistical downscaling method. Stoch Environ Res Risk Assess 25:781–792. https://doi.org/10.1007/s00477-010-0441-9

Huber WC, Dickinson RE (1988) Storm water management model user's manual, version 4. Rep. No. EPA/600/3–88/001a, US Environmental Protection Agency, Athens, GA

Huber WC (1995) Chapter 22: EPA storm water management model SWMM, Computer models of watershed hydrology, Singh, VP ed..

Huggins LF, Monke EJ (1970) Mathematical simulation of hydrologic events of ungaged watersheds. https://docs.lib.purdue.edu/cgi/viewcontent.cgi?article=1014%26context=watertech

Huss M, Farinotti D (2014) A high-resolution bedrock map for the Antarctic Peninsula. Cryosph 8:1261–1273

Huth R, Kliegrová S, Metelka L (2008) Non-linearity in statistical downscaling: Does it bring an improvement for daily temperature in Europe? Int J Climatol 28:465–477. https://doi.org/10.1002/joc.1545

Huth R, Kyselý J, Climate MD-J (2001) undefined, n.d. Time structure of observed, GCM-simulated, downscaled, and stochastically generated daily temperature series. https://journals.ametsoc.org

Imberger M, Larsén XG, Davis N, Du J (2020) Approaches toward improving the modelling of mid-latitude cyclones entering at the lateral boundary corner in the limited area model WRF. Q J R Meteorol Soc. 146(732):3225–3244. https://doi.org/10.1002/qj.3843

IPCC's Fifth Assessment Report (AR5) (2013) Fifth Assessment Report—Climate Change 2013, IPCC

Isotta FA, Vogel R, Frei C (2014) Evaluation of European regional reanalyses and downscalings for precipitation in the Alpine region. Meteorol. Zeitschrift 24:15–37. https://doi.org/10.1127/metz/2014/0584

Izeboud M, Lhermitte S, Van Tricht K, Lenaerts JTM, Van Lipzig NPM, Wever N (2020) The spatiotemporal variability of cloud radiative effects on the Greenland Ice Sheet surface mass balance. Geophys Res Lett 47:e2020GL087315

Jaadi ZJ (2021) A step-by-step explanation of principal component analysis (PCA)|Built In [WWW Document]. https://builtin.com/data-science/step-step-explanation-principal-component-analysis. Accessed 8 Sept 2021

Jackson B, Mcintyre N, Pechlivanidis IG, Jackson BM, Mcintyre NR, Wheater HS (2011) Catchment scale hydrological modelling: a review of model types, calibration approaches and uncertainty analysis methods in the context of recent developments in technology and app Catchment Scale Hydrological Modelling: A Review Of Model Types, Calibration Approaches And Uncertainty Analysis Methods In The Context Of Recent Developments In Technology And Applications, Article in GlobalNEST International Journal

Jacob D, Elizalde A, Haensler A, Hagemann S, Kumar P, Podzun R, Rechid D, Remedio AR, Saeed F, Sieck K (2012) Assessing the transferability of the regional climate model REMO to different coordinated regional climate downscaling experiment (CORDEX) regions. Atmosphere (Basel). 3:181–199

Jacob D, Podzun R (1997) Sensitivity studies with the regional climate model REMO. Meteorol Atmos Phys 63:119–129. https://doi.org/10.1007/BF01025368

Jacob D, Bärring L, Christensen OB, Christensen JH, De Castro M, Deque M, Giorgi F, Hagemann S, Hirschi M, Jones R (2007a) An inter-comparison of regional climate models for Europe: model performance in present-day climate. Clim. Change 81:31–52

Jacob D, Bärring L, Christensen OB, Christensen JH, de Castro M, Déqué M, Giorgi F, Hagemann S, Hirschi M, Jones R, Kjellström E, Lenderink G, Rockel B, Sánchez E, Schär C, Seneviratne SI, Somot S, van Ulden A, van den Hurk B (2007b) An inter-comparison of regional climate models for Europe: model performance in present-day climate. Clim Change 811:81, 31–52. https://doi.org/10.1007/S10584-006-9213-4

Jacob D, Bärring L, Christensen OB, Christensen JH, De Castro M, Déqué M, Giorgi F, Hagemann S, Hirschi M, Jones R, Kjellström E, Lenderink G, Rockel B, Sánchez E, Schär C, Seneviratne SI, Somot S, Van Ulden A, Van Den Hurk B (2007c) An inter-comparison of regional climate models for Europe: model performance in present-day climate. Clim. Change 81:31–52. https://doi.org/10.1007/s10584-006-9213-4

Jansen FA, Teuling AJ (2020) Evaporation from a large lowland reservoir-(dis) agreement between evaporation models from hourly to decadal timescales. Hydrol Earth Syst Sci 24:1055–1072

Jeong DI, St-Hilaire A, Ouarda TBMJ, Gachon P (2012) Comparison of transfer functions in statistical downscaling models for daily temperature and precipitation over Canada. Stoch Environ Res Risk Assess 26:633–653. https://doi.org/10.1007/s00477-011-0523-3

Jin Z, Erhardt RJ (2020) Incorporating climate change projections into risk measures of index-based insurance. North Am Actuar J 1–15

Johnson PA, Rasolofosaon PNJ (1996) Manifestation of nonlinear elasticity in rock: convincing evidence over large frequency and strain intervals from laboratory studies. Nonlinear Process Geophys 3:77–88

Johnson F, Sharma A (2009) Measurement of GCM skill in predicting variables relevant for hydroclimatological assessments. J Clim 22:4373–4382. https://doi.org/10.1175/2009JCLI2681.1

Johnson F, Sharma A, Singh V (2017) Handbook of applied hydrology

Johnson F, Sharma A (2012) A nesting model for bias correction of variability at multiple time scales in general circulation model precipitation simulations. Water Resour Res 48. https://doi.org/10.1029/2011WR010464

Jones PG, Thornton PK (1993) A rainfall generator for agricultural applications in the tropics. Agric For Meteorol 63:1–19. https://doi.org/10.1016/0168-1923(93)90019-E

Jones PG, Thornton PK (2000) MarkSim: software to generate daily weather data for Latin America and Africa. Agron J 92:445–453. https://doi.org/10.2134/agronj2000.923445x

Kalkstein LS, Corrigan P (1986) A synoptic climatological approach for geographical analysis: assessment of sulfur dioxide concentrations. Ann Assoc Am Geogr 76:381–395

Kalkstein LS, Dunne PC, Vose RS (1990) Detection of climatic change in the western North American Arctic using a synoptic climatological approach. J Clim 3:1153–1167

Kampf SK, Burges SJ (2007) A framework for classifying and comparing distributed hillslope and catchment hydrologic models. Wiley Online Libr 43:5423. https://doi.org/10.1029/2006WR005370

Karambiri H, García Galiano SG, Giraldo JD, Yacouba H, Ibrahim B, Barbier B, Polcher J (2011) Assessing the impact of climate variability and climate change on runoff in West Africa: the case of Senegal and Nakambe River basins. Atmos Sci Lett 12:109–115

Karstens U, NOLTE-HOLUBE, R., Rockel, B., (1996) Calculation of the water budget over the Baltic Sea catchment area using the regional forecast model REMO for June 1993. Tellus A 48:684–692

Kavetski D, Clark MP (2010) Ancient numerical daemons of conceptual hydrological modeling: 2. Impact of time stepping schemes on model analysis and prediction. Water Resour Res 46. https://doi.org/10.1029/2009WR008896

Kavetski D, Fenicia F (2011) Elements of a flexible approach for conceptual hydrological modeling: 2. Application and experimental insights of a flexible approach for conceptual hydrological modeling: 2. Application and experimental insights. Elem Water Resour Res 47:11511. https://doi.org/10.1029/2011WR010748

Kay AL, Bell VA, Blyth EM, Crooks SM, Davies HN, Reynard NS (2013) A hydrological perspective on evaporation: historical trends and future projections in Britain. J Water Clim Change 4:193–208

Kebede A, Diekkrüger B, Moges SA (2013) An assessment of temperature and precipitation change projections using a regional and a global climate model for the Baro-Akobo Basin, Nile Basin Ethiopia. J Earth Sci Clim Change 4:133

Kerch G, Rustichelli F, Ausili P, Zicans J, Meri RM, Glonin A (2008) Effect of chitosan on physical and chemical processes during bread baking and staling. Eur Food Res Technol 226:1459–1464

Kershaw T, Sanderson M, Coley D, Eames M (2010) Estimation of the urban heat island for UK climate change projections. Build Serv Eng Res Technol 31:251–263

Kiehl JT, Caron JM, Hack JJ (2005) On using global climate model simulations to assess the accuracy of MSU retrieval methods for tropospheric warming trends. J Clim 18:2533–2539. https://doi.org/10.1175/JCLI3492.1

Kiehl T, Hack J, Bonan B, Boville A, Briegleb P, Williamson L, Rasch J (1996) Description of the NCAR community climate model (CCM3)

Kigobe M, Wheater H, McIntyre N (2014) Statistical downscaling of precipitation in the upper nile: use of Generalized Linear Models (GLMs) for the Kyoga basin. In: Nile River Basin. Springer, pp 421–449

King MD, Howat I, Vijay S (2019) Large scale impacts of surface melt on ice sheet dynamics: empirical evidence of outlet glacier sensitivity to late-season melt events in Greenland. AGUFM 2019:C12A-02

Kite GW (1995) The SLURP model. Comput Model Watershed Hydrol 521–562

Kling H, Fuchs M, Paulin M (2012) Runoff conditions in the upper Danube basin under an ensemble of climate change scenarios. J Hydrol 424:264–277

Klutse NAB, Sylla MB, Diallo I, Sarr A, Dosio A, Diedhiou A, Kamga A, Lamptey B, Ali A, Gbobaniyi EO (2016) Daily characteristics of West African summer monsoon precipitation in CORDEX simulations. Theor Appl Climatol. 123:369–386

Knudsen J, Thomsen A, Refsgaard JC (1986) WATBAL a semi-distributed, physically based hydrological modelling system. Nord Hydrol 17:347–362. https://doi.org/10.2166/nh.1986.0026

Koch W, Feser F (2006) Relationship between SAR-derived wind vectors and wind at 10-m height represented by a mesoscale model. Mo. Weather Rev 134:1505–1517

Kokkonen T, Koivusalo H, Karvonen T (2001) A semi-distributed approach to rainfall-runoff modelling—a case study in a snow affected catchment. Environ Model Softw 16:481–493. https://doi.org/10.1016/S1364-8152(01)00028-7

Korsholm US, Baklanov A, Gross A, Mahura A, Sass BH, Kaas E (2008) Online coupled chemical weather forecasting based on HIRLAM–overview and prospective of Enviro-HIRLAM. HIRLAM Newsl. 54:151–168

Kotlarski S, Paul F, Jacob D (2010) Forcing a distributed glacier mass balance model with the regional climate model REMO. Part I: Climate model evaluation. J Clim 23:1589–1606

Kotlarski S (2007) A subgrid glacier parameterisation for use in regional climate modelling. University of Hamburg. https://pure.mpg.de/rest/items/item_994357/component/file_994356/content

Kotsopoulos S, Nastos P, Lazogiannis K, Poulos S, Ghionis G, Alexiou I, Panagopoulos A, Farsirotou E, Alamanis N (2015) Evaporation, evapotranspiration and crop water requirements under present and future climate conditions at Pinios delta plain. In: 14th international conference on environmental science and technology, Rhodes, Greece. pp 3–5

Kouwen N, Fathi-Moghadam M (2000) Friction factors for coniferous trees along rivers. J Hydraul Eng 126:732–740. https://doi.org/10.1061/(asce)0733-9429(2000)126:10(732)

Kouwen N, Soulis ED, Pietroniro A, Donald J, Harrington RA (1993) Grouped response units for distributed hydrologic modeling. J Water Resour Plan Manag 119:289–305. https://doi.org/10.1061/(asce)0733-9496(1993)119:3(289)

Krebs-Kanzow U, Gierz P Rodehacke CB, Xu S, Yang H, Lohmann G (2020) The diurnal Energy Balance Model (dEBM): a convenient surface mass balance solution for ice sheets in Earth System modeling. Cryosph Discuss 1–36

Kumar P, Folk M, Markus M, Alameda J (2005) Hydroinformatics: data integrative approaches in computation, analysis, and modeling

Kundzewicz ZW, Radziejewski M, Pinskwar I (2006) Precipitation extremes in the changing climate of Europe. Clim Res 31:51–58

Källén E (1996) HIRLAM documentation manual. System 2:122–147

Laflen JM, Elliot WJ, Flanagan DC, Meyer CR, Nearing MA (1997) WEPP-predicting water erosion using a process-based model. J Soil Water Conserv 52

Lafon T, Dadson S, Buys G, Prudhomme C (2013) Bias correction of daily precipitation simulated by a regional climate model: a comparison of methods. Int J Climatol 33:1367–1381

Lall U (2014) Debates-The future of hydrological sciences: a (common) path forward? One water. One world. Many climes Many Souls. Wiley Online Libr 50:5335–5341. https://doi.org/10.1002/2014WR015402

Lam K-C, Cheng S (1998) A synoptic climatological approach to forecast concentrations of sulfur dioxide and nitrogen oxides in Hong Kong. Environ Pollut 101:183–191

Landelius T, Dahlgren P, Gollvik S, Jansson A, Olsson E (2016) A high-resolution regional reanalysis for Europe. Part 2: 2D analysis of surface temperature, precipitation and wind. Q J R Meteorol Soc 142:2132–2142. https://doi.org/10.1002/qj.2813

Lane L (1989) USDA-Water Erosion Prediction Project: hillslope profile model documentation. USDA-ARS, National Soil Erosion Research Laboratory. https://books.google.com/books?hl=fa%26lr=lang_en%26id=UPhXfy6I5g4C%26oi=fnd%26pg=PA1%26dq=Water+Erosion+Prediction+Project:+hillslope+profile+model+documentation.+USDAARS,+National+Soil+Erosion+&ots=ct7Va5PEXk%26sig=diWWYXZetToF2Kudi2KUdHPMxLU#v=onepage%26q=Water%20Erosion%20Prediction%20Project%3A%20hillslope%20profile%20model%20documentation.%2C%20National%20Soil%20Erosion%26f=false

Lang C, Fettweis X, Erpicum M (2015) Stable climate and surface mass balance in Svalbard over 1979–2013 despite the Arctic warming. Cryosphere 9:83–101

Larsén XG, Mann J, Berg J, Göttel H, Jacob D (2010) Wind climate from the regional climate model REMO. Wind Energy Int J Prog Appl Wind Power Convers Technol 13:279–296

Launiainen J (2015) Validation of HIRLAM boundary-layer structures over the Baltic Sea

Laurenson E (1962) Hydrograph synthesis by runoff routing. University of New South Wales. Water Res Lab, Manly Vale, NSW. https://doi.org/10.4225/53/57959065caad3

Laurenson EM (1964) A catchment storage model for runoff routing. J Hydrol 2:141–163

Laurenson EM, Mein RG (1990) RORB-Version 4, Runoff Routing Program: User Manual. Monash University Department of Civil Engineering

Laurenson E, Watershed RM-C models of (1995) undefined, n.d. RORB: Hydrograph synthesis by runoff routing. cabdirect.org

Laurent B, Marticorena B, Bergametti G, Léon JF, Mahowald NM (2008) Modeling mineral dust emissions from the Sahara desert using new surface properties and soil database. J Geophys Res Atmos 113

Leander R, Buishand TA (2007) Resampling of regional climate model output for the simulation of extreme river flows. J Hydrol 332:487–496

Lenaerts JTM, Van Den Broeke MR, Scarchilli C, Agosta C (2012) Impact of model resolution on simulated wind, drifting snow and surface mass balance in Terre Adélie, East Antarctica. J Glaciol 58:821–829

Lenaerts JTM, Ligtenberg SRM, Medley B, Van de Berg WJ, Konrad H, Nicolas JP, Van Wessem JM, Trusel LD, Mulvaney R, Tuckwell RJ (2018) Climate and surface mass balance of coastal West Antarctica resolved by regional climate modelling. Ann Glaciol 59:29–41

Lenderink G, Lenderink G (2003) Simulation of present-day climate in RACMO2: first results and model developments

Li Z, Liu WZ, Zhang XC, Zheng FL (2011) Assessing the site-specific impacts of climate change on hydrology, soil erosion and crop yields in the Loess Plateau of China. Clim Change 105:223–242. https://doi.org/10.1007/s10584-010-9875-9

Li Z, Liu W-Z, Zhang X-C, Zheng, F-L (2009) Impacts of land use change and climate variability on hydrology in an agricultural catchment on the Loess Plateau of China. J Hydrol 377:35–42. https://doi.org/10.1016/j.jhydrol.2009.08.007

Liang S, Li X, Xie X (2013) Land surface observation, modeling and data assimilation. World Scientific. https://books.google.com/books?hl=fa%26lr=lang_en%26id=eyu7CgAAQBAJ%26oi=fnd%26pg=PR5%26dq=Land+surface+observation+modeling+and+data+assimilation+World+Scientific%26ots=dNbJaWgcQ1%26sig=5nxZuwmHh8L1yraV1Z4mQqVIkk8

Lin S-J, Rood RB (1996) Multidimensional flux-form semi-Lagrangian transport schemes. Mon Weather Rev 124:2046–2070

Lindskog M, Salonen K, Järvinen H, Michelson DB (2004) Doppler radar wind data assimilation with HIRLAM 3DVAR. Mon Weather Rev 132:1081–1092

Liu Y, Fan K (2013) A new statistical downscaling model for autumn precipitation in China. Int J Climatol 33:1321–1336

Liu J, Yuan D, Zhang L, Zou X, Song X (2016) Comparison of three statistical downscaling methods and ensemble downscaling method based on Bayesian model averaging in upper Hanjiang River Basin, China. Adv Meteorol 2016. https://doi.org/10.1155/2016/7463963

Lotsch A, Tian Y, Friedl MA, Myneni RB (2003) Land cover mapping in support of LAI and FPAR retrievals from EOS-MODIS and MISR: classification methods and sensitivities to errors. Int J Remote Sens 24:1997–2016

Louis J-F (1979) A parametric model of vertical eddy fluxes in the atmosphere. Boundary-Layer Meteorol 17:187–202

Loveland TR, Reed BC, Brown JF, Ohlen DO, Zhu Z, Yang L, Merchant JW (2000) Development of a global land cover characteristics database and IGBP DISCover from 1 km AVHRR data. Int J. Remote Sens 21:1303–1330

Lu Y, Qin XS (2014) Multisite rainfall downscaling and disaggregation in a tropical urban area. J Hydrol 509:55–65

Lucas-Picher P, Christensen JH, Saeed F, Kumar P, Asharaf S, Ahrens B, Wiltshire AJ, Jacob D, Hagemann S (2011) Can regional climate models represent the Indian monsoon? J Hydrometeorol 12:849–868

Lukasová A (1979) Hierarchical agglomerative clustering procedure. Pattern Recogn 11:365–381. https://doi.org/10.1016/0031-3203(79)90049-9

Lynch P, Huang X-Y (1992) Initialization of the HIRLAM model using a digital filter. Mon Weather Rev 120:1019–1034

Ma X, Cheng W (1996) A modeling of hydrological processes in a large low plain area including lakes and ponds. J Japan Soc Hydrol Water Resour 9:320–329

Ma X, Fukushima Y, Hashimoto E, Hiyama E (1999) Application of a simple SVAT model in a mountain catchment under temperate humid climate. J Japan Soc Hydrol Water Resour

Mahura A, Baklanov A, Petersen C, Nielsen NW, Amstrup B (2009) Verification and case studies for urban effects in HIRLAM numerical weather forecasting. In: Meteorological and air quality models for urban areas. Springer Berlin Heidelberg, pp 143–150. https://doi.org/10.1007/978-3-642-00298-4_14

Manders-Groot AMM, Schaap M, van Ulft B, van Meijgaard E (2011) Coupling of the air quality model Lotus-Euros to the climate model Racmo

Mankoff KD, Noël B, Fettweis X, Ahlstrøm AP, Colgan W, Kondo K, Langley K, Sugiyama S, van As D, Fausto RS (2020) Greenland liquid water discharge from 1958 through 2019. Earth Syst Sci Data 12:2811–2841

Maraun D (2013) Bias correction, quantile mapping, and downscaling: revisiting the inflation issue. J Clim 26:2137–2143. https://doi.org/10.1175/JCLI-D-12-00821.1

Maraun D, Wetterhall F, Ireson AM, Chandler RE, Kendon EJ, Widmann M, Brienen S, Rust HW, Sauter T Themel M, Venema VKC, Chun KP, Goodess CM, Jones RG, Onof C, Vrac M, Thiele-Eich I (2010a) Precipitation downscaling under climate change: recent developments to bridge the gap between dynamical models and the end user. Rev Geophys 48. https://doi.org/10.1029/2009RG000314

Maraun D, Wetterhall F, Ireson AM, Chandler RE, Kendon EJ, Widmann M, Brienen S, Rust HW, Sauter T, Themeßl M, Venema VKC, Chun KP, Goodess CM, Jones RG, Onof C, Vrac M, Thiele-Eich I (2010b) Precipitation downscaling under climate change: Recent developments to bridge the gap between dynamical models and the end user. Rev Geophys 48:RG3003. https://doi.org/10.1029/2009RG000314

Marcotte D, MacDonald RJ, Nemeth MW (2020) Participatory water management modelling in the Athabasca River Basin. Can. Water Resour J/revue Can. Des Ressources Hydriques 45:109–124

Marengo JA, Jones R, Alves LM, Valverde MC (2009) Future change of temperature and precipitation extremes in South America as derived from the PRECIS regional climate modeling system. Int J Climatol A J R Meteorol Soc 29:2241–2255

Marshall GJ, Thompson DWJ, van den Broeke MR (2017) The signature of Southern Hemisphere atmospheric circulation patterns in Antarctic precipitation. Geophys Res Lett 44:11–580

Mason S, Kruczkiewicz A, Ceccato P, Crawford A (2015) Accessing and using climate data and information in fragile, data-poor states

Masson D, Knutti R (2011) Spatial-scale dependence of climate model performance in the CMIP3 ensemble. J Clim, 24(11):2680–2692

Matalas NC (1967) Mathematical assessment of synthetic hydrology. Water Resour Res 3:937–945. https://doi.org/10.1029/WR003i004p00937
Matzarakis A, Endler C (2010) Climate change and thermal bioclimate in cities: impacts and options for adaptation in Freiburg, Germany. Int J Biometeorol 54:479–483
Maurer EP, Pierce DW (2014) Bias correction can modify climate model simulated precipitation changes without adverse effect on the ensemble mean. Hydrol Earth Syst Sci 18:915–925. https://doi.org/10.5194/hess-18-915-2014
May RJ, Dandy GC, Maier HR, Nixon JB (n.d.) Application of partial mutual information variable selection to ANN forecasting of water quality in water distribution systems. Elsevier. https://doi.org/10.1016/j.envsoft.2008.03.008
Mays LW, Tung Y-K (2002) Hydrosystems engineering and management. Water Resources Publication
McCarthy MP, Harpham C, Goodess CM, Jones PD (2012) Simulating climate change in UK cities using a regional climate model, HadRM3. Int J Climatol 32:1875–1888
McCullagh P, Nelder JA (1989) Generalized linear models, vol 37. CRC Press
McFarlane NA, Boer GJ, Blanchet JP, Lazare M (1992) The Canadian Climate Centre second-generation general circulation model and its equilibrium climate. J Clim 5:1013–1044
McGregor GR, Bamzelis D (1995) Synoptic typing and its application to the investigation of weather air pollution relationships, Birmingham, United Kingdom. Theor Appl Climatol 51:223–236
McGregor GR, Walters S, Wordley J (1999) Daily hospital respiratory admissions and winter air mass types, Birmingham UK. Int J Biometeorol 43:21–30
Mehdi B, Lehner B, Gombault C, Michaud A, Beaudin I, Sottile M-F, Blondlot A (2015) Simulated impacts of climate change and agricultural land use change on surface water quality with and without adaptation management strategies. Agric Ecosyst Environ 213:47–60
Mengelkamp H-T, Beyrich F, Heinemann G, Ament F, Bange J, Berger F, Bösenberg J, Foken T, Hennemuth B, Heret C (2006) Evaporation over a heterogeneous land surface. Bull Am Meteorol Soc 87:775–786
Metcalf E (1971) University of Florida and Water Resources Engineers, Inc., Storm water management model, vol I, Final Report. EPA Report 11024 DOC 07/71 (NTIS PB-203289). Environ. Prot. Agency Washington, DC, USA 352
Methods D, Projections CC (2014) A review of downscaling methods for climate change projections
Meyssignac B, Slangen ABA, Melet A, Church JA, Fettweis X, Marzeion B, Agosta C, Ligtenberg SRM, Spada G, Richter K (2017) Evaluating model simulations of twentieth-century sea-level rise. Part II: Regional sea-level changes. J Clim 30:8565–8593
Miksovsky J, Raidl A (2005) Testing the performance of three nonlinear methods of time series analysis for prediction and downscaling of European daily temperatures. Nonlinear Process Geophys 12:979–991. https://doi.org/10.5194/npg-12-979-2005
Mirakbari M, Mesbahzadeh T, Soleimani Sardoo F, Miglietta MM, Krakauer NY, Alipour N (2020) Observed and projected trends of extreme precipitation and maximum temperature during 1992–2100 in Isfahan province, Iran using REMO model and copula theory. Nat Resour Model 33:e12254
Mitchell TM (1997) Does machine learning really work? AI magazine, 18(3):11–11. https://doi.org/10.1609/aimag.v18i3.1303
Modala NR, Ale S, Goldberg DW, Olivares M, Munster CL, Rajan N, Feagin RA (2017) Climate change projections for the Texas high plains and rolling plains. Theor Appl Climatol 129:263–280
Mohajerani Y, Sutterley TC, Velicogna I, van den Broeke MR, Fettweis X (2015) Reducing uncertainties in Greenland surface mass balance using IceBridge and ICESat Altimetry, GRACE data and regional atmospheric climate model outputs. AGUFM 2015:C23C-0803
Mohamed YA, Van den Hurk B, Savenije HHG, Bastiaanssen WGM (2005) Hydroclimatology of the Nile: results from a regional climate model
Mohamed YA, Savenije HHG, Bastiaanssen WGM, Van den Hurk BJJM (2006) New lessons on the Sudd hydrology learned from remote sensing and climate modeling

Moré JJ (1978) The Levenberg-Marquardt algorithm: implementation and theory. In: Numerical analysis. Springer, pp 105–116

Mottram R, Boberg F, Langen P, Yang S, Rodehacke C, Christensen JH, Madsen MS (2017) Surface mass balance of the Greenland ice sheet in the regional climate model HIRHAM5: Present state and future prospects. 低温科学 75:105–115

Mount NJ, Maier HR, Toth E, Elshorbagy A, Solomatine D, Chang FJ, Abrahart RJ (2016) Data-driven modelling approaches for socio-hydrology: opportunities and challenges within the Panta Rhei Science Plan. Hydrol Sci J 61:1192–1208. https://doi.org/10.1080/02626667.2016.1159683

Mpelasoka FS, Mullan AB, Heerdegen RG (2001) New Zealand climate change information derived by multivariate statistical and artificial neural networks approaches. Int J Climatol 21:1415–1433. https://doi.org/10.1002/joc.617

Mpelasoka F, Hydrometeorology FC-J (2009) undefined, n.d. Influence of rainfall scenario construction methods on runoff projections. https://journals.ametsoc.org

Muerth MJ, St-Denis BG, Ricard S, Velázquez JA, Schmid J, Minville M, Caya D, Chaumont D, Ludwig R, Turcotte R (2013) On the need for bias correction in regional climate scenarios to assess climate change impacts on river runoff. Hydrol Earth Syst Sci 17

Mullan D, Chen J, Dynamics XZ-C (2016a) undefined, n.d. Validation of non-stationary precipitation series for site-specific impact assessment: comparison of two statistical downscaling techniques. Springer

Mullan D, Chen J, Zhang XJ (2016b) Validation of non-stationary precipitation series for site-specific impact assessment: comparison of two statistical downscaling techniques. Clim Dyn 46:967–986. https://doi.org/10.1007/s00382-015-2626-x

Mulvany TJ (1850) On the use of self-registering rain and flood gauges in making observations of the rainfall and flood discharges in a given catchment. Trans Min Proc Inst Civ Eng Ireland, Sess 1

Music B, Caya D (2007) Evaluation of the hydrological cycle over the Mississippi River basin as simulated by the Canadian Regional Climate Model (CRCM). J Hydrometeorol 8:969–988

National Research Council (US) Committee on Climate EID and HH (2001) Temporal and spatial scaling: an ecological perspective

Navarro-Racines CE, Tarapues JE (2015) Bias-correction in the CCAFS-climate portal: a description of methodologies

Navascués B, Calvo J, Morales G, Santos C, Callado A, Cansado A, Cuxart J, Díez M, del Río P, Escribà P, García-Colombo O, García-Moya JA, Geijo C, Gutiérrez E, Hortal M, Martínez I, Orfila B, Parodi JA, Rodríguez E, Sánchez-Arriola J, Santos-Atienza I, Simarro J (2013) Long-term verification of HIRLAM and ECMWF forecasts over Southern Europe. History and perspectives of Numerical Weather Prediction at AEMET. Atmos Res. https://doi.org/10.1016/j.atmosres.2013.01.010

Naz BS, Kao S-C, Ashfaq M, Gao H, Rastogi D, Gangrade S (2018) Effects of climate change on streamflow extremes and implications for reservoir inflow in the United States. J Hydrol 556:359–370. https://doi.org/10.1016/J.JHYDROL.2017.11.027

Ng HYF, Marsalek J (1992) Sensitivity of streamflow simulation to changes in climatic inputs. Nord Hydrol 23:257–272. https://doi.org/10.2166/nh.1992.0018

Niska H, Rantamäki M, Hiltunen T, Karppinen A, Kukkonen J, Ruuskanen J, Kolehmainen M (2005) Evaluation of an integrated modelling system containing a multi-layer perceptron model and the numerical weather prediction model HIRLAM for the forecasting of urban airborne pollutant concentrations. Atmos Environ 39:6524–6536. https://doi.org/10.1016/j.atmosenv.2005.07.035

Omstedt A, Meuller L, Nyberg L (1997) Interannual, seasonal and regional variations of precipitation and evaporation over the Baltic Sea. Ambio 484–492

Orszag SA (1970) Transform method for the calculation of vector-coupled sums: application to the spectral form of the vorticity equation. J Atmos Sci 27:890–895

Paeth H, Hall NMJ, Gaertner MA, Alonso MD, Moumouni S, Polcher J, Ruti PM, Fink AH, Gosset M, Lebel T (2011) Progress in regional downscaling of West African precipitation. Atmos Sci Lett 12:75–82

Pal JS, Eltahir EAB (2001) Pathways relating soil moisture conditions to future summer rainfall within a model of the land–atmosphere system. J Clim 14:1227–1242

Pal M, Mather PM (2003) An assessment of the effectiveness of decision tree methods for land cover classification. Remote Sens Environ 86:554–565. https://doi.org/10.1016/S0034-4257(03)00132-9

Pal JS, Small EE, Eltahir EAB (2000) Simulation of regional-scale water and energy budgets: representation of subgrid cloud and precipitation processes within RegCM. J Geophys Res Atmos 105:29579–29594

Pal JS, Giorgi F, Bi X, Elguindi N, Solmon F, Gao X, Rauscher SA, Francisco R, Zakey A, Winter J (2007a) Regional climate modeling for the developing world: the ICTP RegCM3 and RegCNET. Bull Am Meteorol Soc 88:1395–1410

Pal JS, Giorgi F, Bi X, Elguindi N, Solmon F, Gao X, Rauscher SA, Francisco R, Zakey A, Winter J, Ashfaq M, Syed FS, Bell JL, Differbaugh NS, Karmacharya J, Konari A, Martinez D, Da Rocha RP, Sloan LC, Steiner AL (2007b) Regional climate modeling for the developing world: the ICTP RegCM3 and RegCNET. Bull Am Meteorol Soc 88:1395–1409. https://doi.org/10.1175/BAMS-88-9-1395

Palutikof J, Goodess C, ... SW-J (2002) undefined, n.d. Generating rainfall and temperature scenarios at multiple sites: examples from the Mediterranean. https://journals.ametsoc.org

Paxian A, Hertig E, Seubert S, Vogt G, Jacobeit J, Paeth H (2015) Present-day and future Mediterranean precipitation extremes assessed by different statistical approaches. Clim Dyn 44:845–860

Perrin C, Michel C, Andréassian V (2003) Improvement of a parsimonious model for streamflow simulation. J Hydrol 279:275–289. https://doi.org/10.1016/S0022-1694(03)00225-7

Perry A (1983) Growth points in synoptic climatology. Prog Phys Geogr 7:90–96. https://doi.org/10.1177/030913338300700104

Pfeifer S (2006) Modeling cold cloud processes with the regional climate model REMO. University of Hamburg. https://pure.mpg.de/rest/items/item_994658/component/file_994657/content

Phillips NA (1956) The general circulation of the atmosphere: a numerical experiment. Q J R Meteorol Soc 82:123–164. https://doi.org/10.1002/qj.49708235202

Phillips NA (1957) A coordinate system having some special advantages for numerical forecasting. J Meteorol 14:184–185. https://doi.org/10.1175/1520-0469(1957)014%3C0184:ACSHSS%3E2.0.CO;2

Piani C, Haerter JO, Coppola E (2010) Statistical bias correction for daily precipitation in regional climate models over Europe. Theor Appl Climatol 99:187–192. https://doi.org/10.1007/s00704-009-0134-9

Pietikäinen J-P, Markkanen T, Sieck K, Jacob D, Korhonen J, Räisänen P, Gao Y, Ahola J, Korhonen H, Laaksonen A (2018) The regional climate model REMO (v2015) coupled with the 1-D freshwater lake model FLake (v1): Fenno-Scandinavian climate and lakes. Geosci Model Dev 11:1321–1342

Pietikäinen J-P, O'Donnell D, Teichmann C, Karstens U, Pfeifer S, Kazil J, Podzun R, Fiedler S, Kokkola H, Birmili W (2012) The regional aerosol-climate model REMO-HAM. Geosci Model Dev 5:1323–1339

Pinder GF, Gray WG (2013) Finite element simulation in surface and subsurface hydrology. Elsevier

Pirazzini R, Vihma T, Launiainen J, Tisler P (2002) Validation of HIRLAM boundary-layer structures over the Baltic Sea. BOREAL Enviro Res 7(3):211–218. http://www.borenv.net/BER/archive/pdfs/ber7/ber7-211.pdf

Pope VD, Gallani ML, Rowntree PR, Stratton RA (2000) The impact of new physical parametrizations in the Hadley Centre climate model: HadAM3. Clim Dyn, 16(2):123–146

Poston T, Stewart I (2014) Catastrophe theory and its applications. Courier Corporation. In: dover publications, Inc, Mineola, New York.

Pour HK, Rontu L, Duguay C, Eerola K, Kourzeneva E (2014) Impact of satellite-based lake surface observations on the initial state of HIRLAM. Part II: Analysis of lake surface temperature and ice cover, Tellus A Dyn Meteorol Oceanogr 66:21395. https://doi.org/10.3402/tellusa.v66.21395

Preuschmann S (2012) Regional surface albedo characteristics-analysis of albedo data and application to land-cover changes for a regional climate model. Hamburg University. https://pure.mpg.de/rest/items/item_1539450/component/file_1539448/content
Prudhomme C, Dadson S, Morris D, Williamson J, Goodsell G, Crooks S, Boelee L, Davies H, Buys G, Lafon T (2012) Future flows climate: an ensemble of 1-km climate change projections for hydrological application in Great Britain. Earth Syst Sci Data 4:143–148
Prudhomme C, Davies H (2009a) Assessing uncertainties in climate change impact analyses on the river flow regimes in the UK. Part 1: Baseline climate. Clim Change 93:177–195
Prudhomme C, Davies H (2009b) Assessing uncertainties in climate change impact analyses on the river flow regimes in the UK. Part 2: Future climate. Clim Change 93:197–222
Prudhomme C, Williamson J (2013) Derivation of RCM-driven potential evapotranspiration for hydrological climate change impact analysis in Great Britain: a comparison of methods and associated uncertainty in future projections. Hydrol Earth Syst Sci 17:1365–1377
Pryor SC, Schoof JT (2020) Differential credibility assessment for statistical downscaling. J Appl Meteorol Climatol 59:1333–1349. https://doi.org/10.1175/JAMC-D-19-0296.1
Prömmel K, Geyer B, Jones JM, Widmann M (2010) Evaluation of the skill and added value of a reanalysis-driven regional simulation for Alpine temperature. Int J Climatol A J R Meteorol Soc 30:760–773
Pyle D, Cerra DD, Kaufmann M (1999) Data Preparation for data mining, Morgan kaufmann. https://books.google.com/books?hl=fa%26lr=lang_en%26id=hhdVr9FJfAC%26oi=fnd%26pg=PR17%26dq=Data+preparation+for+data+mining.+morgan+kaufmann%26ots=6hdW7QKwbt%26sig=VGuwUGKXqdEWqEKLMgCWJvHBiM#v=onepage%26q=Data%20preparation%20for%20data%20mining.%20morgan%20kaufmann%26f=false
Qian B, Gameda S, De Jong R, Falloon P, Gornall J (2010) Comparing scenarios of Canadian daily climate extremes derived using a weather generator. Clim Res 41:131–149. https://doi.org/10.3354/cr00845
Qian Y, Giorgi F (1999) Interactive coupling of regional climate and sulfate aerosol models over eastern Asia. J Geophys Res Atmos 104:6477–6499
Qian C, Zhou W, Fong SK, Leong KC (2015) Two approaches for statistical prediction of non-Gaussian climate extremes: a case study of Macao hot extremes during 1912–2012. J Clim 28:623–636
Quick MC, Pipes A (1977) UBC watershed model/Le modèle du bassin versant UCB. Hydrol Sci J 22:153–161
Quick MC (1995) The UBC watershed model. Comput Model Watershed Hydrol 233–280
Quiquet A, Ritz C, Punge HJ, Salas y Mélia D (2013) Greenland ice sheet contribution to sea level rise during the last interglacial period: a modelling study driven and constrained by ice core data. Clim Past 9:353–366
Racsko P, Szeidl L, Semenov M (1991) A serial approach to local stochastic weather models. Ecol Modell 57:27–41
Raddatz TJ, Reick CH, Knorr W, Kattge J, Roeckner E, Schnur R, Schnitzler K-G, Wetzel P, Jungclaus J (2007) Will the tropical land biosphere dominate the climate–carbon cycle feedback during the twenty-first century? Clim Dyn 29:565–574
Ramos AM, Trigo RM, Santo FE (2011) Evolution of extreme temperatures over Portugal: recent changes and future scenarios. Clim Res 48:177–192
Rantamäki M, Pohjola MA, Tisler P, Bremer P, Kukkonen J, Karppinen A (2005) Evaluation of two versions of the HIRLAM numerical weather prediction model during an air pollution episode in southern Finland. In: Atmospheric environment. Elsevier Ltd, pp 2775–2786. https://doi.org/10.1016/j.atmosenv.2004.12.050
Rashid MM, Beecham S, Chowdhury RK (2016) Statistical downscaling of rainfall: a non-stationary and multi-resolution approach. Theor. Appl. Climatol. 124:919–933
Ratsimandresy AW, Sotillo MG, Albiach JCC, Fanjul EA, Hajji H (2008a) A 44-year high-resolution ocean and atmospheric hindcast for the Mediterranean Basin developed within the HIPOCAS Project. Coast Eng 55:827–842

Ratsimandresy AW, Sotillo MG, Fanjul EÁ, Albiach JCC, Gómez BP, Hajji H (2008b) A 44-year (1958–2001) sea level residual hindcast over the Mediterranean Basin. Phys Chem Earth Parts A/B/C 33:250–259

Rauscher SA, Seth A, Qian J-H, Camargo SJ (2006) Regional climate model domain choice in the tropics based on process considerations. Theor Appl Clim 86:229–246

Rechid D, Raddatz TJ, Jacob D (2009) Parameterization of snow-free land surface albedo as a function of vegetation phenology based on MODIS data and applied in climate modelling. Theor Appl Climatol 95(1):197–221. https://doi.org/10.1007/s00704-007-0371-8

Rechid D (2009) Land surface scheme of REMO. Internal report, Max Planck Institute for Meteorology, Hamburg. Available at ….

Rechid D, Hagemann S, Jacob D (2009) Sensitivity of climate models to seasonal variability of snow-free land surface albedo. Theoretical and applied climatology, 95(1):97–221. https://doi.org/10.1007/s00704-007-0371-8

Refsgaard JC, Knudsen J (1996) Operational validation and intercomparison of different types of hydrological models. Water Resour Res 32:2189–2202. https://doi.org/10.1029/96WR00896

Refsgaard JC, Storm B, Mike SHE (1995) Computer models of watershed hydrology. Water Resour Publ 809–846

Reid DJ (1968) Combining Three Estimates of Gross Domestic Product. Economica 35:431. https://doi.org/10.2307/2552350

Remson I, Hornberger GM, Molz FJ (1971) Numerical methods in subsurface hydrology. https://books.google.com/books/about/Numerical_Methods_in_Subsurface_Hydrolog.html?id=0PBOAAAAMAAJ

Richardson CW (1981) Stochastic simulation of daily precipitation, temperature, and solar radiation. Water Resour Res 17:182–190

Richardson CW (1982) Dependence structure of daily temperature and solar radiation. Trans ASAE 25:735–739

Richardson CW, Wright DA (1984) WGEN: a model for generating daily weather variables. ARS

Ridley J, Wiltshire A, Mathison C (2013) More frequent occurrence of westerly disturbances in Karakoram up to 2100. Sci Total Environ 468:S31–S35

Rietveld MR (1978) A new method for estimating the regression coefficients in the formula relating solar radiation to sunshine. Agric Meteorol 19:243–252. https://doi.org/10.1016/0002-1571(78)90014-6

Rivington M, Miller D, Matthews KB, Russell G, Bellocchi G, Buchan K (2008a) Downscaling regional climate model estimates of daily precipitation, temperature and solar radiation data. Clim Res 35:181–202

Rivington M, Miller D, Matthews KB, Russell G, Bellocchi G, Buchan K (2008b) Evaluating regional climate model estimates against site-specific observed data in the UK. Clim Change 88:157–185

Robins PE, Lewis MJ, Simpson JH, Howlett ER, Malham SK (2014) Future variability of solute transport in a macrotidal estuary. Estuar Coast Shelf Sci 151:88–99

Rockwood D (1982) Theory and practice of the SSARR model as related to analyzing and forecasting the response of hydrologic systems. Appl Model Catchment Hydrol

Roeckner E, Arpe K, Bengtsson L, Christoph M, Claussen M, Dümenil L, Esch M, Giorgetta MA, Schlese U, Schulzweida U (1996) The atmospheric general circulation model ECHAM-4: Model description and simulation of present-day climate

Rontu L, Eerola K, Kourzeneva E, Vehviläinen B (2012) Data assimilation and parametrisation of lakes in HIRLAM. Tellus A Dyn. Meteorol Oceanogr 64:17611. https://doi.org/10.3402/tellusa.v64i0.17611

Rontu L, Pietikäinen J-P, Martin Perez D (2019)Renewal of aerosol data for ALADIN-HIRLAM radiation parametrizations. Advances in Science and Research, 16:129–136. https://asr.copernicus.org/articles/16/129/2019/

Ross TJ (2010) Fuzzy logic with engineering applications, 3rd edn. Wiley. https://doi.org/10.1002/9781119994374

Saeed F, Haensler A, Weber T, Hagemann S, Jacob D (2013) Representation of extreme precipitation events leading to opposite climate change signals over the Congo basin. Atmosphere (Basel). 4:254–271

Salameh T, Drobinski P, Vrac M, Naveau P (2009) Statistical downscaling of near-surface wind over complex terrain in southern France. Meteorol Atmos Phys 103:253–265

Samadi S, Wilson CAME, Moradkhani H (2013) Uncertainty analysis of statistical downscaling models using Hadley Centre Coupled Model. Theor Applied Climatol, 114(3):673–690

Sangelantoni L, Russo A, Gennaretti F (2019) Impact of bias correction and downscaling through quantile mapping on simulated climate change signal: a case study over Central Italy. Theor Appl Climatol 135:725–740. https://doi.org/10.1007/s00704-018-2406-8

Santer BD, Wigley TML (1990) Regional validation of means, variances, and spatial patterns in general circulation model control runs. J Geophys Res 95:829–850. https://doi.org/10.1029/JD095iD01p00829

Sasaki H, Kurihara K, Takayabu I, Murazaki K, Sato Y, Tsujino H (2006) Preliminary results from the coupled atmosphere-ocean regional climate model at the Meteorological Research Institute. J Meteorol Soc Japan Ser II 84:389–403

Sausen R, Voss R, Ponater M (1992) Orographic forcing in ECHAM. Meteorologisches Institut der Universität Hamburg

Schoof JT, Pryor SC (2001) Downscaling temperature and precipitation: a comparison of regression-based methods and artificial neural networks. Int J Climatol 21:773–790. https://doi.org/10.1002/joc.655

Schulze RE (1997) Impacts of global climate change in a hydrologically vulnerable region: challenges to South African hydrologists. Prog Phys Geogr 21:113–136

Sekula P, Bokwa A, Bochenek B, Zimnoch M (2019) Prediction of air temperature in the Polish Western Carpathian Mountains with the ALADIN-HIRLAM numerical weather prediction system. Atmosphere (Basel). 10:186. https://doi.org/10.3390/atmos10040186

Sellevold R, Van Kampenhout L, Lenaerts J, Noël B, Lipscomb WH, Vizcaino M (2019) Surface mass balance downscaling through elevation classes in an earth system model: application to the Greenland ice sheet. Cryosphere 13:3193–3208

Semenov MA (2007) Development of high-resolution UKCIP02-based climate change scenarios in the UK. Agric For Meteorol 144:127–138. https://doi.org/10.1016/j.agrformet.2007.02.003

Semenov MA, Barrow EM (1997) Use of a stochastic weather generator in the development of climate change scenarios. Clim Change 35:397–414. https://doi.org/10.1023/A:1005342632279

Semenov MA, Brooks RJ (1999) Spatial interpolation of the LARS-WG stochastic weather generator in Great Britain. Clim Res 11:137–148

Semenov M, Brooks R, Barrow E, Richardson C (1998) Comparison of the WGEN and LARS-WG stochastic weather generators for diverse climates. Clim Res 10:95–107. https://doi.org/10.3354/cr010095

Semenov MA, Barrow EM, Lars-Wg A (2002) A stochastic weather generator for use in climate impact studies. User Man Herts UK

Semmler T, Cheng B, Yang Y, Rontu L (2012) Snow and ice on Bear Lake (Alaska)–sensitivity experiments with two lake ice models. Tellus A Dyn Meteorol Oceanogr 64:17339

Semmler T, Jacob D (2004) Modeling extreme precipitation events—a climate change simulation for Europe. Glob Planet Change 44:119–127

Semmler T, Jacob D, Schlünzen KH, Podzun R (2004) Influence of sea ice treatment in a regional climate model on boundary layer values in the Fram Strait region. Mon Weather Rev 132:985–999

Sen Z (2009) Fuzzy logic and hydrological modeling. CRC Press. https://doi.org/10.1201/9781439809402

Senkova AV, Rontu L, Savijärvi H (2007) Parametrization of orographic effects on surface radiation in HIRLAM. Tellus. Ser A Dyn Meteorol Oceanogr 59:279–291. https://doi.org/10.1111/j.1600-0870.2007.00235.x

Shamseldin AY, O'Connor KM, Liang GC (1997) Methods for combining the outputs of different rainfall–runoff models. J Hydrol 197:203–229. https://doi.org/10.1016/S0022-1694(96)03259-3

Shean DE, Christianson K, Larson KM, Ligtenberg SRM, Joughin IR, Smith BE, Stevens CM, Bushuk M, Holland DM (2017) GPS-derived estimates of surface mass balance and ocean-induced basal melt for Pine Island Glacier ice shelf, Antarctica. Cryosph. 11:2655–2674

Simmons AJ, Burridge DM (1981) An energy and angular-momentum conserving vertical finite-difference scheme and hybrid vertical coordinates. Mon Weather Rev 109:758–766

Simmons AJ, Burridge DM, Jarraud M, Girard C, Wergen W (1989) The ECMWF medium-range prediction models development of the numerical formulations and the impact of increased resolution. Meteorol Atmos Phys 40:28–60. https://doi.org/10.1007/BF01027467

Singh VP (2015) Entropy theory in hydraulic engineering: an introduction entropy theory in hydraulic engineering: an introduction. Am Soc Civil Eng (ASCE). https://doi.org/10.1061/978 0784412725

Singh VP (2016) Handbook of Applied Hydrology, Second Edition. McGraw-Hill Education. https://books.google.com/books?id=2kUBswEACAAJ

Singh VP (2018) Hydrologic modeling: progress and future directions. Geosci Lett 5:15. https://doi.org/10.1186/s40562-018-0113-z

Singh VP, Rontu L, Pietikäinen JP, Martin Perez D (Eds) (2019). Renewal of aerosol data for ALADIN-HIRLAM radiation parametrizations. Advances in Science and Research, 16:129–136. https://asr.copernicus.org/articles/16/129/2019/

Singh VP, Woolhiser DA (2002) Mathematical modeling of watershed hydrology. J Hydrol Eng 7:270–292. https://doi.org/10.1061/(ASCE)1084-0699(2002)7:4(270)

Singh V, Frevert D (2002a) Mathematical models of small watershed hydrology and applications

Singh V, Frevert D (2002b) Mathematical models of large watershed hydrology

Singh VP, Frevert DK (2002) Mathematical models of large watershed hydrology. Water Resources Publications. https://books.google.nl/books?hl=fa%26lr=lang_en%26id=abRyo4OLjgYC%26oi=fnd%26pg=PR11%26dq=)+Mathematical+models+of+large+watershed+hydrology%26ots=R_uDQXmVY4%26sig=nITrhsRON2_j0PYWetcvhW25pk%26redir_esc=y#v=onepage%26q=)%20Mathematical%20models%20of%20large%20watershed%20hydrology%26f=false

Singh VP, Zhang L (2018) Copula–entropy theory for multivariate stochastic modeling in water engineering. Geosci Lett. https://geoscienceletters.springeropen.com/articles/10.1186/s40562-018-0105-z

Singh A, Herrmann A, Kaushal MP (2001) AU-REMO: a two dimensional finite element runoff and soil erosion model for agricultural lands. In: Soil erosion. American Society of Agricultural and Biological Engineers, p 334

Singh VP, Jain SK, Tyagi A (2007) Risk and reliability analysis: a handbook for civil and environmental engineers. Am Soc Civil Eng. https://ascelibrary.org/doi/abs/10.1061/978078440 8919

Singh V (1995a) Computer models of watershed hydrology. Water Resources Publications. https://www.wrpllc.com/books/cmwhn.html

Singh VP (1995b) Computer models of watershed hydrology. Water Resources Publications

Singh V (2013) Entropy theory and its application in environmental and water engineering. John Wiley & Sons. https://books.google.nl/books?hl=fa%26lr=lang_en%26id=A8_-5-VWJiAC%26oi=fnd%26pg=PT9%26dq=Entropy+theory+and+its+application+in+environmental+and+water+engineering%26ots=6ZzSqqSKFF%26sig=wIj7pv4d4YWuVG2CuwsHZJDb67w%26redir_esc=y#v=onepage%26q=Entropy%

Singh V (2014) Entropy theory in hydraulic engineering: an introduction. https://doi.org/10.1061/9780784412725

Singh VP (2017) Handbook of applied hydrology, 2nd edn. McGraw-Hill Education, New York

Sitterson J, Knightes C, Parmar R, Wolfe K, Avant B, Muche M (2018) An overview of rainfall-runoff model types. https://scholarsarchive.byu.edu/iemssconference/2018/Stream-C/41/

Sittner WT, Schauss CE, Monro JC (1969) Continuous hydrograph synthesis with an API-type hydrologic model. Water Resour Res 5:1007–1022

Sivakumar B, Berndtsson R (2010) Nonlinear dynamics and chaos in hydrology. Advances in data-based approaches for hydrologic modeling and forecasting, 411–461. https://doi.org/10.1142/9789814307987_0009
Skiles JW, Richardson CW (1998) A stochastic weather generation model for Alaska. Ecol Modell 110:211–232. https://doi.org/10.1016/S0304-3800(98)00061-1
Small EE, Sloan LC, Hostetler S, Giorgi F (1999) Simulating the water balance of the Aral Sea with a coupled regional climate-lake model. J Geophys Res Atmos 104:6583–6602
Smiatek G, Kunstmann H, Knoche R, Marx A (2009) Precipitation and temperature statistics in high-resolution regional climate models: evaluation for the European Alps. J Geophys Res Atmos 114
Smith RE, Schreiber HA (1974) Point processes of seasonal thunderstorm rainfall: 2 Rainfall Depth Probabilities. Water Resour Res 10:418–423
Soares CG, Weisse R, Carretero JC, Alvarez E (2002) A 40 year hindcast of wind, sea level and waves in European waters. In: International conference on offshore mechanics and arctic engineering, pp 669–675
Soil Conservation Service (scs),Computer model for project formulation hydrology, Tech (1965) USDA, Washington
Solmon F, Giorgi F, Liousse C (2006) Aerosol modelling for regional climate studies: application to anthropogenic particles and evaluation over a European/African domain. Tellus B Chem Phys Meteorol 58:51–72
Solomatine DP, Wagener T (2011) Hydrological modeling. In: Treatise on water science. Elsevier, pp 435–457. https://doi.org/10.1016/B978-0-444-53199-5.00044-0
Solomatine D, See LM, Abrahart RJ (2009) Data-driven modelling: concepts, approaches and experiences. Pract Hydroinform 17–30
Solomatine DP (2005) Data-driven modeling and computational intelligence methods in hydrology. In: Encyclopedia of hydrological sciences. Wiley. https://doi.org/10.1002/0470848944.hsa021
Sotillo MG, Ratsimandresy AW, Carretero JC, Bentamy A, Valero F, González-Rouco F (2005) A high-resolution 44-year atmospheric hindcast for the Mediterranean Basin: contribution to the regional improvement of global reanalysis. Clim Dyn 25:219–236
Spada G, Bamber JL, Hurkmans R (2013) The gravitationally consistent sea-level fingerprint of future terrestrial ice loss. Geophys Res Lett 40:482–486
Srinivasulu S, Jain A (2008) Rainfall-runoff modelling: integrating available data and modern techniques. In: Practical hydroinformatics. Springer Berlin, Heidelberg, pp 59–70. https://link.springer.com/chapter/10.1007/978-3-540-79881-1_5
Stengel M, Lindskog M, Undén P, Gustafsson N, Bennartz R (2010) An extended observation operator in HIRLAM 4D-VAR for the assimilation of cloud-affected satellite radiances. Q J R Meteorol Soc 136:1064–1074. https://doi.org/10.1002/qj.621
Stevens B, Giorgetta M, Esch M, Mauritsen T, Crueger T, Rast S, Salzmann M, Schmidt H, Bader J, Block K, Brokopf R, Fast I, Kinne S, Kornblueh L, Lohmann U, Pincus R, Reichler T, Roeckner E (2013) Atmospheric component of the MPI-M earth system model: ECHAM6. J Adv Model Earth Syst 5:146–172. https://doi.org/10.1002/jame.20015
Sugawara M (1974) Tank model and its application to Bird Creek, Wollombi Brook, Bikin River, Kitsu River, Sanaga River and Nam Mune. Res Notes Natl Res Cent Disas Prev 11:1–64
Sugawara M (1995) Tank model. Comput Model Watershed Hydrol. Singh VP (Ed) Colorado: Water Resources Publications, Highlands Ranch, 1130. https://www.wrpllc.com/books/cmwhn.html
Sunyer MA, Hundecha Y, Lawrence D, Madsen H, Willems P, Martinkova M, Vormoor K, Bürger G, Hanel M, Kriaučiūnienė J, Loukas A, Osuch M, Yücel I (2015a) Inter-comparison of statistical downscaling methods for projection of extreme precipitation in Europe. Hydrol Earth Syst Sci 19:1827–1847. https://doi.org/10.5194/hess-19-1827-2015
Sunyer MA, Hundecha Y, Lawrence D, Madsen H, Willems P, Martinkova M, Vormoor K, Bürger G, Hanel M, Kriaučiuniene J, Loukas A, Osuch M, Yücel I (2015b) Inter-comparison of statistical

downscaling methods for projection of extreme precipitation in Europe. Hydrol Earth Syst Sci 19:1827–1847. https://doi.org/10.5194/hess-19-1827-2015a

Suryavanshi S, Pandey A, Chaube UC (2017) Hydrological simulation of the Betwa River basin (India) using the SWAT model. Hydrol Sci J 62:960–978. https://doi.org/10.1080/02626667.2016.1271420

Sutterley TC, Velicogna I, Csatho BM, van den Broeke MR, Wahr JM, Flament T, Rezvan-Behbahani S, Babonis GS (2013) Using GRACE measurements of time variable gravity, elevation changes from ICESat, OIB and ENVISAT and surface mass balance outputs from RACMO to improve ice mass balance estimates. AGUFM 2013:C51A-0515

Sutterley TC, Velicogna I, Fettweis X, van den Broeke MR (2016) Surface mass balance model evaluation from satellite and airborne lidar mapping. AGUFM 2016:C12B-05

Tanguay M, Robert A, Laprise R (1990) A semi-implicit send-lagrangian fully compressible regional forecast model. Mon Weather Rev 118:1970–1980

Tayfur G (2014) Soft computing in water resources engineering: artificial neural networks, fuzzy logic and genetic algorithms, WIT Press. https://books.google.nl/books?hl=fa%26lr=lang_en%26id=OzYrBQAAQBAJ%26oi=fnd&pg=PP1%26dq=Soft+computing+in+water+resources+engineering:+Artificial+neural+networks,+fuzzy+logic+and+genetic+algorithm%26ots=LWQFTbG18F%26sig=

Te Linde AH, Aerts J, Bakker AMR, Kwadijk JCJ (2010) Simulating low-probability peak discharges for the Rhine basin using resampled climate modeling data. Water Resour Res 46

Teichmann C, Eggert B, Elizalde A, Haensler A, Jacob D, Kumar P, Moseley C, Pfeifer S, Rechid D, Remedio AR (2013) How does a regional climate model modify the projected climate change signal of the driving GCM: a study over different CORDEX regions using REMO. Atmosphere (Basel). 4:214–236

Teichmann C (2010) Climate and air pollution modelling in South America with focus on megacities, Hamburg University. https://pure.mpg.de/rest/items/item_993870/component/file_993869/content

Teng J, Potter NJ, Chiew FHS, Zhang L, Wang B, Vaze J, Evans JP (2015) How does bias correction of regional climate model precipitation affect modelled runoff? Hydrol Earth Syst Sci 19

Teutschbein C, Seibert J (2012) Bias correction of regional climate model simulations for hydrological climate-change impact studies: review and evaluation of different methods. J Hydrol 456–457:12–29. https://doi.org/10.1016/j.jhydrol.2012.05.052

The Use of GPS to Validate NWP Systems: The HIRLAM Model. J Atmos Ocean Technol 17(6) (2000) [WWW Document], n.d. https://journals.ametsoc.org/view/journals/atot/17/6/1520-0426_2000_017_0773_tuogtv_2_0_co_2.xml?tab_body=fulltext-display. Accessed 12 December 2020

Themeßl MJ, Gobiet A, Heinrich G (2012) Empirical-statistical downscaling and error correction of regional climate models and its impact on the climate change signal. Clim Change 112:449–468. https://doi.org/10.1007/s10584-011-0224-4

Tisseuil C, Vrac M, Lek S, Wade AJ (2010) Statistical downscaling of river flows. J Hydrol. 385:279–291. https://doi.org/10.1016/j.jhydrol.2010.02.030

Todini E (1988) Rainfall-runoff modeling—past, present and future. J Hydrol 100:341–352. https://doi.org/10.1016/0022-1694(88)90191-6

Todini E (1996) The ARNO rainfall-runoff model. J Hydrol 175:339–382. https://doi.org/10.1016/S0022-1694(96)80016-3

Todini E (1995) New trends in modelling soil processes from hillslope to GCM scales. In: The role of water and the hydrological cycle in global change. Springer Berlin, Heidelberg, pp 317–347. https://doi.org/10.1007/978-3-642-79830-6_11

Todini E (2007) Hydrological catchment modelling: past, present and future. Hydrol Earth Syst Sci. 11(1):468–482. https://doi.org/10.5194/hess-11-468-2007

Tolika K, Maheras P, Vafiadis M, Flocas HA, Arseni-Papadimitriou A (2007) Simulation of seasonal precipitation and raindays over Greece: a statistical downscaling technique based on artificial neural networks (ANNs). Int J Climatol 27:861–881. https://doi.org/10.1002/joc.1442

Toll V, Gleeson E, Nielsen KP, Männik A, Mašek J, Rontu L, Post P (2016) Impacts of the direct radiative effect of aerosols in numerical weather prediction over Europe using the ALADIN-HIRLAM NWP system. Atmos Res 172–173:163–173. https://doi.org/10.1016/j.atmosres.2016.01.003

Toros H, Geertsema G, Cats G (2014) Evaluation of the HIRLAM and HARMONIE numerical weather prediction models during an air pollution episode over greater İstanbul area. CLEAN—Soil Air Water 42:863–870. https://doi.org/10.1002/clen.201200306

Trigo RM, Palutikof JP (1999) Simulation of daily temperatures for climate change scenarios over Portugal: a neural network model approach. Clim Res 13:45–59. https://doi.org/10.3354/cr013045

Tripathi S, Srinivas VV, Nanjundiah RS (2006) Downscaling of precipitation for climate change scenarios: a support vector machine approach. J Hydrol 330:621–640. https://doi.org/10.1016/j.jhydrol.2006.04.030

Trusel LD, Frey KE, Das SB, Munneke PK, Van Den Broeke MR (2013) Satellite-based estimates of Antarctic surface meltwater fluxes. Geophys Res Lett 40:6148–6153

Tryhorn L, DeGaetano A (2011) A comparison of techniques for downscaling extreme precipitation over the Northeastern United States. Int J Climatol 31:1975–1989

Trzaska S, Schnarr E (2014) A review of downscaling methods for climate change projections. Vermont, US: Tetra Tech ARD & Center for International Earth Science Information Network. https://www.ciesin.org/documents/Downscaling_CLEARED_000.pdf

Tsimplis MN, Álvarez-Fanjul E, Gomis D, Fenoglio-Marc L, Pérez B (2005) Mediterranean sea level trends: atmospheric pressure and wind contribution. Geophys Res Lett 32

Tuinenburg OA, Hutjes RWA, Stacke T, Wiltshire A, Lucas-Picher P (2014) Effects of irrigation in India on the atmospheric water budget. J. Hydrometeorol 15:1028–1050

Tukimat NNA, Harun S (2015) Climate change impact on rainfall and temperature in Muda irrigation area using multicorrelation matrix and downscaling method. J Water Clim Change, 6(3):647–660

Tung Y-K, Yen B-C (2005) Hydrosystems engineering uncertainty analysis. Asce

US HEC, US WRSC (1981) HEC-1 flood hydrograph package: users manual. US Army Corps of Engineers, Water Resources Support Center, Hydrologic ….

Uhlenbrook S, Roser S, Tilch N (2004) Hydrological process representation at the meso-scale: the potential of a distributed, conceptual catchment model. J Hydrol 278–296. Elsevier. https://doi.org/10.1016/j.jhydrol.2003.12.038

Urrutia R, Vuille M (2009) Climate change projections for the tropical Andes using a regional climate model: Temperature and precipitation simulations for the end of the 21st century. J Geophys Res Atmos 114

Van den Broeke MR, Van Lipzig NPM (2003) Factors controlling the near-surface wind field in Antarctica. Mon Weather Rev 131:733–743

Van Den Broeke MR, Van Lipzig NPM (2004) Changes in Antarctic temperature, wind and precipitation in response to the Antarctic Oscillation. Ann Glaciol 39:119–126

van de Berg WJ, Medley B (2016) Brief Communication: Upper-air relaxation in RACMO2 significantly improves modelled interannual surface mass balance variability in Antarctica. Cryosph 10:459–463

Van de Berg WJ, Van den Broeke MR, Reijmer CH, Van Meijgaard E (2005) Characteristics of the Antarctic surface mass balance, 1958–2002, using a regional atmospheric climate model. In: Annals of glaciology. Cambridge University Press, pp 97–104. https://doi.org/10.3189/172756405781813302

Van den Hurk BJJM, Graham LP, Viterbo P (2002) Comparison of land surface hydrology in regional climate simulations of the Baltic Sea catchment. J Hydrol 255:169–193

Van Lipzig NPM, Van Meijgaard E, Oerlemans J (2002) The effect of temporal variations in the surface mass balance and temperature-inversion strength on the interpretation of ice-core signals. J Glaciol 48:611–621

Van Lipzig NPM, Turner J, Colwell SR, van Den Broeke MR (2004) The near-surface wind field over the Antarctic continent. Int J Climatol A J R Meteorol Soc 24:1973–1982 van Lipzig NPM, van Meijgaard E, Oerlemans J (2002a) The spatial and temporal variability of the surface mass balance in Antarctica: results from a regional atmospheric climate model. Int J Climatol A J R Meteorol Soc 22:1197–1217

van Lipzig NPM, van Meijgaard E, Oerlemans J (2002b) Temperature sensitivity of the Antarctic surface mass balance in a regional atmospheric climate model. J Clim 15:2758–2774

Van Meijgaard E (1995) Precipitation forecasts from atmospheric models during recent flooding events of the Meuse. Phys Chem Earth 20:497–502

van Meijgaard E, van Ulft LH, Van de Berg WJ, Bosveld, FC, Lenderink G, Siebesma AP (2008a) The KNMI regional atmospheric climate model RACMO version 2.1. Technical report; TR-302 43

van Meijgaard E, Van Ulft LH, Van de Berg WJ, Bosveld FC, Van den Hurk B, Lenderink G, Siebesma AP (2008b) The KNMI regional atmospheric climate model RACMO, version 2.1. KNMI De Bilt, Netherlands

Van Meijgaard E, Van Ulft LH, Lenderink G, De Roode SR, Wipfler EL, Boers R, van Timmermans RMA (2012) Refinement and application of a regional atmospheric model for climate scenario calculations of Western Europe. KVR

Van Pelt WJJ, Oerlemans J, Reijmer CH, Pohjola VA, Pettersson R, Van Angelen JH (2012) Simulating melt, runoff and refreezing on Nordenskiöldbreen, Svalbard, using a coupled snow and energy balance model. Cryosph 6:641–659

Van Wessem JM, Reijmer CH, Morlighem M, Mouginot J, Rignot E, Medley B, Joughin I, Wouters B, Depoorter MA, Bamber JL (2014) Improved representation of East Antarctic surface mass balance in a regional atmospheric climate model. J Glaciol 60:761–770

VanZanten MC, Stevens B, Nuijens L, Siebesma AP, Ackerman AS, Burnet F, Cheng A, Couvreux F, Jiang H, Khairoutdinov M (2011) Controls on precipitation and cloudiness in simulations of trade-wind cumulus as observed during RICO. J Adv Model Earth Syst 3

Vandenberg A (1989) A physical model of vertical integration, drain discharge, and surface runoff for layered soils. National Hydrology Research Institute

Vasyl G, Kateryna M (2018) Forecasting the runoff on rivers of the dnister river basin according to the remo numeric climatic model. World Sci 1

Vautard R, Van Oldenborgh GJ, Otto F, Yiou P, De Vries H, Van Mijgaard E, Stepek A, Soubeyroux J-M, Philip S, Kew S (2019) Human influence on European winter wind storms such as those of January 2018. Earth Syst Dyn 10

Vaze J, Jordan P, Beecham R, Frost A, Summerell G (2011) Guidelines for rainfall-runoff modelling: towards best practice model application

Vernon CL, Bamber JL, Box JE, Van den Broeke MR, Fettweis X, Hanna E, Huybrechts P (2013) Surface mass balance model intercomparison for the Greenland ice sheet. Cryosph 7:599–614

Voloudakis D, Karamanos A, Economou G, Kalivas D, Vahamidis P, Kotoulas V, Kapsomenakis J, Zerefos C (2015) Prediction of climate change impacts on cotton yields in Greece under eight climatic models using the AquaCrop crop simulation model and discriminant function analysis. Agric Water Manag 147:116–128

Voudouris K, Mavromatis T, Krinis P (2012) Assessing runoff in future climate conditions in Messara valley in Crete with a rainfall-runoff model. Meteorol Appl 19:473–483

Vu VT, Cats G, Wolters L (2013) Graphics processing unit optimizations for the dynamics of the HIRLAM weather forecast model. Concurr Comput Pract Exp 25:1376–1393

Wagena MB, Collick AS, Ross AC, Najjar RG, Rau B, Sommerlot AR, Fuka DR, Kleinman PJA, Easton ZM (2018) Impact of climate change and climate anomalies on hydrologic and biogeochemical processes in an agricultural catchment of the Chesapeake Bay watershed, USA. Sci Total Environ 637:1443–1454

Wang Y, Hou S, Sun W, Lenaerts JTM, van den Broeke MR, van Wessem JM (2015) Recent surface mass balance from Syowa Station to Dome F, East Antarctica: comparison of field observations, atmospheric reanalyses, and a regional atmospheric climate model. Clim Dyn 45:2885–2899

Wangsoh N, Wathayu W, Sukawat D (2017) A hybrid climate model for rainfall forecasting based on combination of self-organizing map and analog method. Sains Malays https://doi.org/10.17576/jsm-2017-4612-32

Watson SJ (1994) Application of wind speed forecasting to the integration of wind energy into a large scale power system. IEE Proc Gener Transm Distrib 141:357. https://doi.org/10.1049/ip-gtd:19941215

Webb RS, Rosenzweig CE, Levine ER (1993) Specifying land surface characteristics in general circulation models: soil profile data set and derived water-holding capacities. Glob Biogeochem Cycles 7:97–108

Weichert A, Bürger G (1998) Linear versus nonlinear techniques in downscaling. Clim Res 10:83–93. https://doi.org/10.3354/cr010083

Weinberger S, Vetter M (2012) Using the hydrodynamic model DYRESM based on results of a regional climate model to estimate water temperature changes at Lake Ammersee. Ecol Modell 244:38–48

Weisse R, Feser F (2003) Evaluation of a method to reduce uncertainty in wind hindcasts performed with regional atmosphere models. Coast Eng 48:211–225

Weisse R, Plüß A (2006) Storm-related sea level variations along the North Sea coast as simulated by a high-resolution model 1958–2002. Ocean Dyn 56:16–25

Wheater HS, Jakeman AJ, Beven KJ (1993) Progress and directions in rainfall-runoff modelling.

White M, Diffenbaugh N, Jones G, Pal J, Giorgi F (2006) Increased heat stress in the 21st century reduces and shifts premium wine production in the United States. Proc Natl Acad Sci 103

Wigley TML, Jones PD, Briffa KR, Smith G (1990) Obtaining sub-grid-scale information from coarse-resolution general circulation model output. J Geophys Res 95:1943–1953. https://doi.org/10.1029/JD095iD02p01943

Wigley TML, Richels R, Edmonds JA (1996) Economic and environmental choices in the stabilization of atmospheric CO_2 concentrations. Nature 379:240–243. https://doi.org/10.1038/379240a0

Wigley T (2008a) MAGICC/SCENGEN 5.3: User manual (version 2). Ncar 1–81

Wigley TML (2008b) MAGICC/SCENGEN 5.3: User manual (version 2). NCAR, Boulder, CO 80

Wigmosta MS, Vail LW, Lettenmaier DP (1994) A distributed hydrology-vegetation model for complex terrain. Water Resour Res 30:1665–1679

Wilby RL, Dawson CW, Barrow EM (2002) SDSM—a decision support tool for the assessment of regional climate change impacts. Environ Model Softw 17:145–157. https://doi.org/10.1016/s1364-8152(01)00060-3

Wilby Robert L, Hassan H, Hanaki K (1998a) Statistical downscaling of hydrometeorological variables using general circulation model output. J Hydrol 205:1–19. https://doi.org/10.1016/S0022-1694(97)00130-3

Wilby RL, Wigley TML, Conway D, Jones PD, Hewitson BC, Main J, Wilks DS (1998b) Statistical downscaling of general circulation model output: a comparison of methods. Water Resour Res 34:2995–3008. https://doi.org/10.1029/98WR02577

Wilby RL, Troni J, Biot Y, Tedd L, Hewitson BC, Smith DM, Sutton RT (2009) A review of climate risk information for adaptation and development planning. Int J Climatol A J R Meteorol Soc 29:1193–1215

Wilby RL, Wigley TML (1997) Downscaling general circulation model output: a review of methods and limitations. Prog Phys Geogr. https://doi.org/10.1177/030913339702100403

Wilby RL, Wigley TML (2000) Precipitation predictors for downscaling: observed and general circulation model relationships. Int J Climatol 20:641–661. https://doi.org/10.1002/(SICI)1097-0088(200005)20:6%3c641::AID-JOC501%3e3.0.CO;2-1

Wilby RL, Dawson CW (2007) SDSM 4.2-A decision support tool for the assessment of regional climate change impacts User Manual, Elsevier

Wilderer PA (2010) Treatise on water science. Newnes

Wilhelm C, Rechid D, Jacob D (2014a) Interactive coupling of regional atmosphere with biosphere in the new generation regional climate system model REMO-iMOVE 1093–1114. https://doi.org/10.5194/gmd-7-1093-2014a

Wilhelm C, Rechid D, Jacob D (2014b) Interactive coupling of regional atmosphere with biosphere in the new generation regional climate system model REMO-iMOVE. Geosci Model Dev 7:1093–1114

Wilks DS (1992) Adapting stochastic weather generation algorithms for climate change studies. Clim Change 22:67–84. https://doi.org/10.1007/BF00143344

Wilton DJ, Jowett A, Hanna E, Bigg GR, Van Den Broeke MR, Fettweis X, Huybrechts P (2017) High resolution (1 km) positive degree-day modelling of Greenland ice sheet surface mass balance, 1870–2012 using reanalysis data. J Glaciol 63:176–193

Winterfeldt J, Geyer B, Weisse R (2011) Using QuikSCAT in the added value assessment of dynamically downscaled wind speed. Int J Climatol 31:1028–1039

Winterfeldt J, Weisse R (2009) Assessment of value added for surface marine wind speed obtained from two regional climate models. Mon Weather Rev 137:2955–2965

Wipfler L (2010) Modification and testing of a Land Surface Scheme as adopted by the RCM RACMO. https://library.wur.nl/WebQuery/wurpubs/fulltext/310765

Woolhiser DA, Roldan J (1982) Stochastic daily precipitation models: 2. A comparison of distributions of amounts. Water Resour Res 18:1461–1468

Wootten A, Terando A, Reich BJ, Boyles RP, Semazzi F (2017) Characterizing sources of uncertainty from global climate models and downscaling techniques. J Appl Meteorol Climatol 56:3245–3262. https://doi.org/10.1175/JAMC-D-17-0087.1

Xie ZH, Zeng YJ, Xia J, Qin PH, Jia BH, Zou J, Liu S (2017) Coupled modeling of land hydrology–regional climate including human carbon emission and water exploitation. Adv Clim Change Res 8:68–79. https://doi.org/10.1016/j.accre.2017.05.001

Xu C (1999) Climate change and hydrologic models: a review of existing gaps and recent research developments. Water Resour Manag 13:369–382. https://doi.org/10.1023/A:1008190900459

Xu J, Lv C, Zhang M, Yao L, Zeng Z (2015) Equilibrium strategy-based optimization method for the coal-water conflict: a perspective from China. J Environ Manag 160:312–323. https://doi.org/10.1016/j.jenvman.2015.06.036

Xu Y, Rignot E, Fenty I, Menemenlis D, Flexas MM (2013) Subaqueous melting of Store Glacier, west Greenland from three-dimensional, high-resolution numerical modeling and ocean observations. Geophys Res Lett 40:4648–4653

Yakimiw E, Robert A (1990) Validation experiments for a nested grid-point regional forecast model: research note. Atmosphere-Ocean 28:466–472

Yang B, Li M-H (2011) Assessing planning approaches by watershed streamflow modeling: Case study of The Woodlands; Texas. Landsc Urban Plan 99:9–22

YoosefDoost A, Asghari H, Abunuri R, Sadegh Sadeghian M (2018a) Comparison of CGCM3, CSIRO MK3 and HADCM3 models in estimating the effects of climate change on temperature and precipitation in Taleghan Basin. Am J Environ Prot 6:28–34. https://doi.org/10.12691/env-6-1-5

YoosefDoost A, YoosefDoost I, Asghari H, Sadegh Sadeghian M (2018b) Comparison of HadCM3, CSIRO Mk3 and GFDL CM2.1 in prediction the climate change in Taleghan river basin. Am J Civ Eng Archit 6:93–100. https://doi.org/10.12691/ajcea-6-3-1

Yu B (2000) Improvement and evaluation of cligen for storm generation. Trans ASAE 43:301–307. https://doi.org/10.13031/2013.2705

Zakey AS, Solmon F, Giorgi F (2006) Development and testing of a desert dust module in a regional climate model. Atmospheric Chemistry and Physics Discussions, 6(2):1749–1792

Zammit-Mangion A, Rougier J, Schön N, Lindgren F, Bamber J (2015) Multivariate spatio-temporal modelling for assessing Antarctica's present-day contribution to sea-level rise. Environmetrics 26:159–177

Zeeman EC (1976) Catastrophe theory. JSTOR

Zeng X, Zhao M, Dickinson RE (1998) Intercomparison of bulk aerodynamic algorithms for the computation of sea surface fluxes using TOGA COARE and TAO data. J Clim 11:2628–2644
Zhang XC (2003) Assessing seasonal climatic impact on water resources and crop production using cligen and WEPP models. Trans ASAE 46:685–693. https://doi.org/10.13031/2013.13603
Zhang XC (2005) Spatial downscaling of global climate model output for site-specific assessment of crop production and soil erosion. Agric For Meteorol 135:215–229. https://doi.org/10.1016/j.agrformet.2005.11.016
Zhang XC (2007) A comparison of explicit and implicit spatial downscaling of GCM output for soil erosion and crop production assessments. Clim Change 84:337–363. https://doi.org/10.1007/s10584-007-9256-1
Zhang X-CJ (2012) Cropping and tillage systems effects on soil erosion under climate change in Oklahoma. Soil Sci Soc Am J 76:1789–1797. https://doi.org/10.2136/sssaj2012.0085
Zhang XC (2013) Verifying a temporal disaggregation method for generating daily precipitation of potentially non-stationary climate change for site-specific impact assessment. Int J Climatol 33:326–342. https://doi.org/10.1002/joc.3425
Zhang Q, Li H (2007) MOEA/D: a Multiobjective evolutionary algorithm based on decomposition. IEEE Trans Evol Comput 11:712–731. https://doi.org/10.1109/TEVC.2007.892759
Zhang XC, Liu WZ (2005) Simulating potential response of hydrology, soil erosion, and crop productivity to climate change in Changwu tableland region on the Loess Plateau of China. Agric For Meteorol 131:127–142. https://doi.org/10.1016/j.agrformet.2005.05.005
Zhang GJ, McFarlane NA (1995) Sensitivity of climate simulations to the parameterization of cumulus convection in the Canadian Climate Centre general circulation model. Atmosphere-Ocean 33:407–446
Zhang Q, Sun P, Singh VP, Chen X (2012) Spatial-temporal precipitation changes (1956–2000) and their implications for agriculture in China. Glob Planet Change 82–83:86–95. https://doi.org/10.1016/j.gloplacha.2011.12.001
Zhang G, Xiang X, Tang H (2011) Time series prediction of chimney foundation settlement by neural networks. Int J Geomech 11:154–158. https://doi.org/10.1061/(ASCE)GM.1943-5622.0000029
Zhang C, Zhang C, Member A, Hydrologist R (2004) CLIGEN non−precipitation parameters and their impact on WEPP crop simulation
Zorita E, Von Storch H (1999) The analog method as a simple statistical downscaling technique: comparison with more complicated methods. J Clim 12:2474–2489. https://doi.org/10.1175/1520-0442(1999)012%3c2474:tamaas%3e2.0.co;2
Zou F, Tenzer R, Fok HS, Nichol JE (2020) Mass balance of the Greenland ice sheet from GRACE and surface mass balance modelling. Water 12:1847
Zwally HJ, Li J, Robbins JW, Saba JL, Yi D, Brenner AC (2015) Mass gains of the Antarctic ice sheet exceed losses. J Glaciol 61:1019–1036

Summary

At the beginning of this section, IPCC reports on the issues of water resources are considered. The procedure of global warming is confirmed in all IPCC reports. Each IPCC report has initially considered previous reports and has set new goals in each report for itself. In each report, predictions have been reported regarding future climate changes. Generally, every set of reports consist of 4 subparts: the first subpart generally deals with a comprehensive evaluation of the basics of the physical science of climate changes. Evaluation of effects of climate change on water resources, evaluation of temperature variation, changes in the expansion of greenhouse gases, snow cover, changes in the sea level, changes in the rainfall, are all presented in the evaluative reports. In the second subpart, effects, compatibility, and vulnerabilities influenced by climate changes are taken into regard. The third subpart considers the procedures of reduction of climate changes. IPCC also presents a combined report which is composed of the mentioned three subparts. First report (FAR) to sixth (AR6) report pf IPCC have been discussed in the fifth chapter and the trends of improvement in these reports have surveyed.

Climate models are mathematical models that simulate Earth's climate system. Climate models have been widely used for climate change projections under different scenarios. In the sixth chapter aims to synthesize knowledge of climate models and corresponding different categories. EBMs, radiative-convective models and GCMs are three main categories of climate models. The chapter underlines that EBMs consists of a set of coupled equations for Earth's surface temperature generation. Also, EBMs include one dependent variable (the Earth's near-surface air temperature), and eliminate the dependency of the climate on the full complexity of the wind field. The main shortcomings of EBMs underline their inability in describing temperature field above the boundary layer of the atmosphere as well as precipitation modelling. The second category of climate models are radiative-convective models that compute the vertical distribution of surface and atmospheric temperatures. Main limitation of radiative-convective models is lack of horizontal dynamic modelling of temperatures. In addition to EBMs and radiative-convective models, GCMs have been widely used for climate system simulations, and particularly climate change projections. GCMs employ three-dimensional approximations of the laws of physics, including conservation of mass, energy, and momentum to simulate the surface pressure and vertical distributions of water vapour, density, temperature, and velocity as a function of time. Concerning the wide application of GCMs in predicting future climate change conditions, this chapter also discusses three sub-classifications of GCMs including AGCM, OGCM, and AOGCM.

In climate change studies, Downscaling refers to the process of moving from large-scale to local-scale forecasters. Each climate model tries to simulate the climate processes and predict the climate for the future years. However, as it is impossible to predict the future state of the climate under the climate change influence, an alternative solution is to identify the various possibilities for the climate scenario. Currently, the most reliable tools for generating these scenarios are "General Rotation Models" (GCM) and "Atmospheric-Ocean General Circulation Models" (AOGCM). Two primary forms of mentioned technique are Statistical and Dynamical Downscaling. In the seventh chapter, each method is described in detail, then the structures and equations are explained and finally, the straightforward application of each model is summarized.

Part III
Modeling to Plan Mitigation and Adaptation Measures

Introduction

Climate change is a global challenge that affects humankind in a complex manner across political, economic, social, and environmental dimensions. Based on the available evidence and studies, freshwater resources can be intensely affected by climate change, with extensive consequences for human communities and ecosystems. Climate change has immediate and long-term effects on water resources such as floods, droughts, rising sea levels in estuaries, drying up of rivers, poor water quality in surface and groundwater systems, distortions of water vapor and precipitation patterns, improper distribution Ice is snow and earth and the amount of access and demand for water resources. Vulnerability due to climate change varies according to different countries, geographical location, and capacity to reduce or adapt to change. Due to the heterogeneity in watersheds and the nonlinearity of hydrological and erosion behaviors, it is complex and difficult to grasp the connections between them. The impact of measurement should be considered when assessing and examining these systems; nevertheless, even if it is a valid criterion, it is insufficient to rely on exclusively; as a result, hydrologic models are used. These models are simplified representations of the actual hydrological system that aid in the study of how the basin responds to various inputs and aid in a better understanding of hydrological processes. Mitigation and adaptation as a tool share the ultimate purpose of reducing climate change impacts. In order to adapt to climate change and mitigate its damage to society, extensive research is required to uncover monthly and annual several climatic fluctuations of different regions (with different climatic characteristics), notably in terms of trends in extreme events. Mitigating the effect of the consequences of climate change without a scientific perspective or using applied research results will not be fruitful. In this section, after reviewing the history of the hydrological model and its importance, as well as a brief look at the influence of climate change, the classification methods are reviewed. A description of the structure of some known hydrological models is also provided. In the following section, the risks of climate change, the vulnerability of water systems to it, the concepts of mitigation and adaptation, and their role in sustainable water resource management have all been examined from various perspectives. Finally, the last chapter looks at climate change on six continents: Africa, Asia, Europe, North America, South America, and Antarctica, its effects on water resources, and the policies and strategies that have been implemented.

Chapter 8
Hydrological Models

Icen Yoosefdoost, Omid Bozorg-Haddad, Vijay P. Singh, and Kwok Wing Chau

8.1 Summary

Hydrological modelling can be characterized as the process of abstracting real hydrological features through small-scale physical models, mathematical analogs, or computer simulations (Chen et al. 2021). Hydrologic models can be separated into several classes according to model structures and spatial processes. In this chapter, after a discussion about the history of these models, a concise glance at the impacts of clime change is examined. Finally, in the last part, the structure of some recognized hydrological models is explained.

I. Yoosefdoost
Department of Sciences and Water Engineering, Faculty of Agriculture, University of Birjand, Birjand, Iran
e-mail: icenyd@birjand.ac.ir

O. Bozorg-Haddad (✉)
Department of Irrigation and Reclamation Engineering, Faculty of Agricultural Engineering and Technology, College of Agriculture and Natural Resources, University of Tehran, Karaj, Tehran, Iran
e-mail: OBHaddad@ut.ac.ir

V. P. Singh
Department of Biological and Agricultural Engineering, Zachry Department of Civil and Environmental Engineering, Texas A&M University, College Station, Texas 77843-2117, USA
e-mail: vsingh@tamu.edu

K. W. Chau
Civil and Hydraulic Engineering, Hong Kong Polytechnic University, Hung Hom, Hong Kong
e-mail: dr.kwok-wing.chau@polyu.edu.hk

O. Bozorg-Haddad (ed.), *Climate Change in Sustainable Water Resources Management*, Springer Water, https://doi.org/10.1007/978-981-19-1898-8_8

8.2 Introduction

Due to the heterogeneity in watersheds and the nonlinearity of hydrological and erosion behaviours, it is complex and difficult to grasp the connections between them. Consequently, in assessing and evaluating hydrological systems, we should consider the effect of measurements; however, it is not enough to rely solely upon them, even though they are a reliable criterion. As a result, hydrologic models are used. These models are helpful to simulate the process of rain-runoff and hydraulic models to analyze the runoff flow in the river and study how it spreads. Since they are simplified representations of the actual hydrological system, they help study the function of the basin in response to various inputs and understand hydrological processes more efficiently.

8.3 Hydrological Models

8.3.1 History of Hydrologic and Modelling

Hydrology has a long history (Biswas 1970). Hydrological modelling began in the 1850s with the study by (Mulvany 1850) and continued with (Darcy 1856); nowadays, it is known as Darcy's law. In the 1960s, computer revolution took place, made great strides in hydrological modelling, and caused the birth of a new approach in hydrology, which is called numerical or digital hydrology. It also created another branch rooted in random or statistical hydrology. This branch requires the analysis of large volumes of data. After that, several improvements have been made. For Instance, a dream comes true! Simulation of the whole hydrological cycle. This happened with the development of the Stanford Watershed Management Model (Crawford and Linsley 1966a, b; Mount et al. 2016). This work follows a myriad of basin models developed simultaneously around the world (Singh 1995a, b; Singh and Frevert 2002a, b, 2006), as well as the development of optimization technologies; it is used as a basis for parameter estimation, reservoir management, and river basin simulation and exploitation. Numerous methods have been developed and used to calibrate hydrological models (Beven and Westerberg 2011; Duan et al. 2004). Then, the next advancement due to progress in numerical mathematics happened; it produced two- and three-dimensional models. Using this new achievement, development in two- and three-dimensional groundwater models occurred (Bear 2012; Pinder and Gray 2013; Remson et al. 1971). After that time, the computational power proved to be exponential, and hydrology started to develop and grow in width (horizontally and vertically). Mathematics, statistics, and information theory tools are integral components of hydrology (Bras and Rodriguez-Iturbe 1993; Clarke 1988; Gelhar 1986; Mays and Tung 2002; Singh et al. 2007; Tung and Yen 2005). In recent decades, the expansion of ANNs,[1] FL,[2] GP,[3] and WM[4] has been achieved (Kumar et al. 2005; Ross 2010; Sen 2009; Singh 2018; Tayfur 2014). Recent theories such as entropy theory (Johnson et al. 2017; Singh 2014, 2013, 2015; Singh and Zhang 2018), chaos theory (Singh 2017), network theory (Singh 2017) and Disaster theory (Poston and Stewart 2014; Zeeman 1976) were used in hydrology studies (Singh 2018). New accurate and advanced instruments have been developed to measure various hydrological parameters, like soil moisture, water & air quality variables, and energy flux. In addition, data transfer tools in extracting information from a measurement location, processing, retrieving, and disseminating became powerful (Liang et al. 2013; Singh 2018; Sivakumar and Berndtsson 2010).

8.3.2 Importance of Hydrological Modelling

Hydrological models are commonly used to generate flow data to:

- Fill in the blanks of the available data.
- Estimate current and/or natural flow and efficiency.
- Create long-term records to establish performance reliability estimates and other statistics.
- Provide input data for another research, for example, environmental flow
- Prepare scenario for forecasting.

In most developing countries, we have only one or two long-term flow measurement sites in a catchment area. Many stations may be measured sites, but the data is usually only stored for a few years, so it is obviously impossible to access long-term recorded data to generate long-term returns or trends. To solve this problem, hydrological modelling can be used to estimate the performance of catchments. A hydrological model can produce: estimation of flow and trends over long periods, historical flow records or natural flow. In addition, it can contribute to future returns or scenario modelling.

Predicting future seasonal flow data is required for a wide range of water managers and users, such as urban managers, environmental managers, rural water suppliers, irrigators, and hydropower producers. This forecast can help plan, manage decisions about available water resources, and ensure adequate government access to water security.

[1] Artificial Neural Networks.

[2] Fuzzy Logic.

[3] Genetic Programming.

[4] Wavelet Models.

8.4 Role of Climate Change in Hydrological Analysis

Climate change is recognized as a complex and long-term global atmosphere and ocean phenomenon (Yoosefdoost et al. 2018b). Global warming has caused the average global temperature and sea levels to rise. Since minor changes in hydrological variables can significantly impact water resources, climate change has an undeniable effect on precipitation, evaporation, and surface runoff at the regional and local scales (Yoosefdoost et al. 2018a). Natural processes are quite sensitive to temperature changes, since global temperatures have risen steadily in recent decades. They directly affect water vapour concentrations, clouds, precipitation patterns, and flow patterns, all of which are related to hydrology. Figure 8.1 shows what happens in the water cycle during climate change.

The hydrological change affects all aspects of human well-being. For example, in an area where runoff is expected to grow, larger drainage canals and more reservoir overflows require funding. Global atmospheric general circulation models (GCM) simulate the current climate and predict future climate change (Xu 1999). GCMs are highly efficient on continental and hemispherical spatial scales, but they are

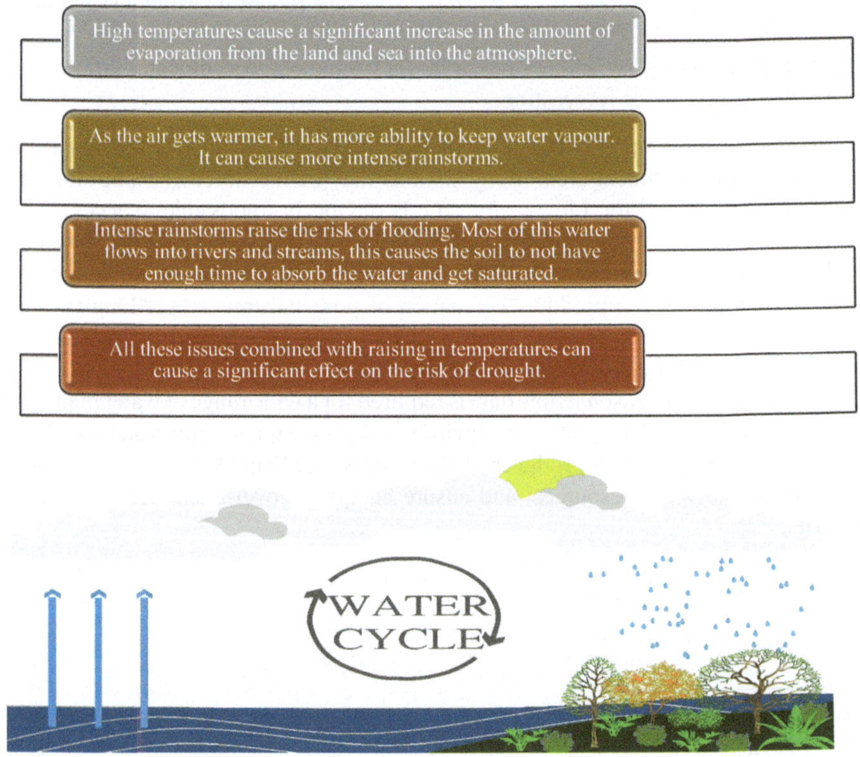

Fig. 8.1 Water cycle in climate change condition

inadequate to constitute local aspects and dynamics at the subgrid-scale (Carter et al. 1994; Wigley et al. 1990; Xu 1999). While observing "global climate change" impacts, the emphasis is fundamentally on societal reactions to local and regional outcomes of large-scale changes (Xu 1999). The contention between GCM execution at regional spatial scales and the needs of regional-scale impact is, for the most part, reliant upon a few components (Xu 1999). Although hydrologists might consider various topics, the following essential problems are considered as below:

- GCMs precision reduces at finer spatial scales; however, their impacts improve with higher resolution in contrast.
- GCM accuracy is reduced from free tropospheric variables to surface variables, while ground surface parameters are used directly in water balance calculations (Xu 1999).
- GCM accuracy decreases from climate-related variables, including wind, temperature, air pressure, humidity to evapotranspiration, precipitation, and soil moisture, while later variables in hydrological regimes are important terms of requirements (Xu 1999). The solutions to the above gaps are.
- Using the dynamic reduction method to generate high-resolution meteorological data, which is required for hydrological models.
- Using statistical scale reduction method to simulate local scale surface parameters from free tropospheric parameters.
- Using hydrological large-scale modelling procedures to simulate the flow of aquifers in large river basins to mitigate the weakness in displaying hydrological processes in GCMs.

8.4.1 Roles of GCMs in Climate Change Study

The GCM was first introduced in (Phillips 1956)'s works; this was originally produced for model average, synoptic-scale (i.e., almost 105 km^2) atmospheric circulation patterns for particular external forcing states. Afterward, a wide range of atmospheric GCMs have been planned to simulate mean large-scale atmospheric circulation (e.g., Holton and Staley 1973; Xu 1999). Over the years, GCMs have been utilized to model climate sensitivity to increasing carbon dioxide concentrations and other necessary variables, as well as to predict future climate change. The most usual use of GCMs are as below.

Models include those from the Canadian Climate Center (CCC), the Dynamic Geophysical Fluid Laboratory (GFDL), the Goddard Institute for Space Studies (GISS), the National Center for Atmospheric Research (NCAR), Oregon State University (OSU) and the U.K. Meteorological Office (UKMO), (Xu 1999). Currently, the only available tools for accurate modelling of future climate evolution are general circulation models. These models are not suitable for responding to hydrological diversity on a regional scale. Since GCMs run on a large spatial scale and the best-modelled GCM time resolution is monthly, the benefits of direct use of

GCM output in practical hydrological applications are limited. Today, the free troposphere is well modelled by coarse GCMs. In contrast, the regional or local features of near-surface or surface climate variables, their changeability, and the likelihood of severe events are not directly accessible to GCM (Xu 1999).

8.4.2 Role of Hydrological Models in Climate Change Studies

The use of regional hydrological simulations to study the climate change effects has several notable features (Gleick 1986; Schulze 1997). Initially, models that have been tested for several weather/physiographic situations are easily accessible and built for use at different spatial scales and displaying dominant processes. This gives us enough flexibility in choosing the best method to measure each specific region. The next is the ability of hydrological models to fit existing data specifications. GCM-derived climatic perturbations could be applied as input data to the model. Thus, different responses to climate change scenarios can be simulated. It is worth mentioning that it is easier to manipulate hydrological models on a regional scale than general circulation simulations.

Another issue could be that the regional model can calculate the sensitivity of a specific domain to both large-scale changes predicted by GCM and hypothetical climate changes. Finally, methods capable of combining both large-scale accurate GCM outputs and regional hydrological features can provide a good place to take advantage of continuous advances in hydrology, regional geography, and segregation of climate models (Xu 1999).

The model selection depends on several factors when the main aim of the study is data and model acceptability (Xu 1999; Gleick 1986; Ng and Marsalek 1992). Complex models are expected to provide sufficient results for different applications in terms of input requirements and structure. When the objective function is appropriate, simple models with a smaller range of applications can result in significant cost reductions. Distinguishing between physics-based and simple parametric models is one of high or low complexity, and is closely related to the main goals of the models that must be applied. Therefore, choosing a suitable model means identifying the situation in which simple models can be used.

8.4.3 Gap Between Hydrological Modelling and Climate Modelling

The atmospheric components of GCMs are complex. (Kite 1995) showed that the land-phase parameters in existing GCMs do not concur in predicting different hydrological variables, even when the atmospheric force is the same (Xu 1999). There are gaps between climate and hydrological modelling given below:

Gap1. Spatial and temporal scale mismatch between hydrological needs and GCM capability:

The GCMS forecasting capability shows a significant reduction from the global catchment scale to regional and local scale and from annual to monthly and daily scale. It is also noteworthy that the validity of hydrological weather forecasts increases from annual to daily and global to local scale.

Gap 2. The vertical level mismatch between hydrological requirements and GCM capability:

GCMs are highly skilled in modelling the free troposphere climate compared to surface climates, although hydrological models use surface variables. Results of several studies showed that GCMS indicated better predictions at higher altitudes but had less correlation with ground surface variables.

Gap 3. The mismatch between the hydrological significance of GCM accuracy and the variables:

GCMs are able to make accurate predictions of several variables, such as temperature, air pressure, and wind field. Rainfall and cloud cover are less forecasted by GCM. The fact is that the variables of other hydrological regimes, such as soil moisture, evapotranspiration, and runoff, cannot be well represented by GCM. According to recent studies, GCM simulation skills are reduced from climatic parameters to hydrological variables, but the hydrological significance expands in the same direction. Another noteworthy point about the GCM is that the simulation of precipitation data is lower than that of temperature. This is because these types of frequent happenings, such as hurricanes, take place at a finer spatial scale than the GCM network. In addition, evapotranspiration is also not appropriately constituted by GCMs, because it happens at boundaries. Also, the runoff estimation of GCM output is less accurate. The main problem is that most GCMs do not have lateral water transfer in the land phase. This type of model carries out vertical water distribution at each time period (at each grid point) utilizing evaporation, transpiration, precipitation and groundwater reservoirs, although any kind of overflow or excess water is easily excluded and removed from the model calculations. If GCMs could model excess water accurately, they would be working with a deficient hydrological cycle.

8.5 Model Classification

8.5.1 Principle of Classification

Classification of hydrological models is useful and even necessary for a wide range of reasons. This helps in some cases, such as choosing the right model for a particular application and evaluating the availability of valuable models. From the users' point of view, the following classification is the most appropriate according to the criteria related to the problem and user-centered (Becker and Serban 1990):

(a) The main aim of simulation, (b) system type, (c) "Hydrological process" or relevant criterion (variable, element) to be examined, (d) fegree of process causality, and (e) discretization of the required time and place.

From this classification, three criteria need to be examined more closely and generally accepted, namely (a), (d) and (e). Therefore, the situation is recognized to understand the ongoing trends and demands for model application (Becker and Serban 1990). It is dominating to portray the contrast between a model and a method (process, procedure): When a model delineates a certain procedure, a methodology is recognized to be an amalgamation of "sub-procedures" containing as sub-routines:

- Processes that control input datasets and convert them to the required form of the model, which is called the sub-routines.
- Computational methods in accordance with the hydrological model or a combination of procedures. A prime example is an updated method for "real-time forecasting."
- Techniques for introducing output in a "user-oriented" appearance (Becker and Serban 1990).

On several occasions, models constitute the required core or at least one essential element of a process. For example, procedures are needed as a "framework" for executing simulations for a particular exercise, and they frequently specify the "availability" or "unavailability" of the model (Becker and Serban 1990). Accordingly, the models are the essential subject of this report, and processes additionally must be thought about. Classification can be done in terms of the type system and hydrological process or variable.

The classification does not pose a fundamental problem in terms of these two criteria. In the case of "coupled systems," it is worth mentioning that under CS[5] water systems, just the CS solely, including lakes, rivers, canals, reservoirs, etc., when "feeding" the land surface components of the river basin or aquifer are not explicitly reviewed. Their output is considered as BCs[6] in the CS. However, they form the core part of the river basin model for mentioned regions. In general, they are defined in many areas such as water quality, moisture, and soil evapotranspiration. The subdivision of this group (water level and discharge) into two subcategories (i. time steps over a day or ii. less than or equal to a day) seems suitable. This currently shows that within the general frameworks determined by the principal types of "hydrological systems" and procedures, any needed alternative or sub-classification could be presented (Becker and Serban 1990).

---

[5] Complex surface.

[6] Boundary conditions.

8.5.2 Causality Degree

"Causality" is demonstrated as cause and effect relationships. They are clearly considered deterministic models, which convert the dependent variables (effects, outputs), y, into a set of independent variables, x, which are: inputs or other variables, including BCs (Becker and Serban 1990).

$$y = f(x, a)$$

In the above equation, coefficients or parameters describing system behaviour are presented in "a," there are several types of deterministic models. Depending on modelling purpose and the modelling procedure, they differ in terms of basic structure, physical "soundness," and dimensionality. However, here are three main categories of definitive models:

(DL) Physics-based models, especially based on hydro-thermodynamics, biology, chemistry, etc., are known as "white-box" models.
(DC) The conceptual simulation reflects relatively simple rules and generally involves a certain degree of empiricism known as gray box models (Becker and Serban 1990).
(DB) These models do not explicitly consider the governing rules and include cause-and-effect relationship of inputs in a completely empirical way. These models are known among users as the black-box model (Becker and Serban 1990).

Stochastic models are other basic hydrological models that do not consider the principle of causality in principle. Probabilistic (SP) models are one of the subcategories of these models, generally characterized by probabilistic distribution functions of hydrological variables, like maximum and minimum discharges, peak and low flow currents, and water levels or storage volumes (Becker and Serban 1990). They are regularly depicted in terms of parameters such as mean, standard deviation, and anomaly coefficients. The main assumption of model formation is that there is no causal relationship between different process elements (variables).

The second subset of the SMs[7] is time series (ST) generation models that retrieve the sequence of variables or recorded happenings by maintaining their statistical variables. The famous ARIMA is categorized in this group (Becker and Serban 1990).

It is obvious that the SMs are generally associated with a specific hydrological procedure observed at the station and only in an "integrated" manner. Therefore, the system type criterion cannot regularly be used in conjunction with stochastic models unless there is a specific relationship with a particular system. Conversely, DMs[8] are definitely associated with specific systems, so they relate to inputs, fixed conditions, and outputs (Becker and Serban 1990).

---

[7] Stochastic Model.

[8] Deterministic Model.

It is worth bearing in mind that hydrological processes, in any case, have random and decisive components. It is valid not only for forecasting, planning, and designing simulations where at least the "forecasting section" frequently requires a random model element as a deterministic component but also for real-time prediction, where the method updates and all manufactured methods predict input and correct. Data, such as precipitation, snowfall, etc., indicate the need for a random model component (Becker and Serban 1990).

Another verifiable truth is that in a definite model of a complicated hydrological system, we are regularly incapable of paying attention to all the parameters that affect the output. Thus, a specific error "EY (N)" occurs, which reduces the number of independent parameters (n) and increases the variables to a certain number. It could be explained by a stochastic model, which embraces two types of error: the "measurement error" and the "model error" (Becker and Serban 1990).

8.5.3 Time and Space Discretization

Selection of a suitable time step is often made, according to the aim of the model program. For erosion, flood and water quality studies, one-day phases are typically needed, while for other motivations, longer periods (up to one month) may be acceptable. Some of the lately evolved models are planned so that distinctive time steps can be applied. It is important to pay attention to this issue in choosing the right model for a particular application (Becker and Serban 1990).

It tends to be expressed that stochastic models are required in a significant part of any study, forecasting, planning, and design. Their role expands significantly in real-time applications with increasing predictive length. In the short-term forecast, they address a component of secondary importance that is embraced in the update method; however, in the medium-term and long-term forecast, they quickly arrive at a similar significance as in managing and planning studies (Becker and Serban 1990). Distribution models in their original format take into consideration the spatial variability of model parameters, inputs, outputs, and state variables, for example, by sub-dividing a land surface region into elementary surface units determined by a regular grid. In micro-scale considerations, grid zones are selected small enough to guarantee the physical soundness and other constitutions.

For the following reasons, the use of distribution models is needed:

- Non-uniformity in the characteristics of the basin space (topography, vegetation, soil, etc.).
- Spatial differences and nonlinearity of "mass and energy transfer" procedures that occur in a catchment.
- The effect of non-uniform distribution of human exercises on the parts of the hydrological cycle.

Lump models, as another utmost case, do not consider the spatial distribution of the specified features, so the models are much simpler and easier to manage. In addition to simplicity, or even in view of effortlessness, these models are generally used for a variety of objectives, such as straightforward plan studies and real-time discharge estimation (Becker and Serban 1990).

Between the two limits, there are compromise solutions that, on the one hand, overcome the restraints of Lump models (L) and, on the other hand, require large amounts of input data and parameter-based demand to prevent computational overhead grid-based distributed models (IG). An old method was to introduce "statistical distribution functions" for the main parameters. For example, a linear model of soil capacity for evapotranspiration and infiltration in the river basin can be used, a method proposed by (Crawford and Linsley 1966a, b) and known as the source of the area or linear model of soil storage capacity (for capillary water). The soil root layer was presented by Becker in 1975 (Becker and Nemec 1987).

The following stage may involve splitting the land surface or river basin into larger subsets (regions) with approximately equal inputs and similar hydrological behaviour. This subgroup (IS) of distributed models is known as semi-distributed models (Knudsen et al. 1986). Currently, these models are increasingly accepted and used. Since they prevent a number of problems related to the application of large-scale distributed network models; these problems have been reported in many studies, for example, related to the proper assessment of feedback, regional diversity, and spatial integration in larger areas (Becker and Nemec 1987). This type can work better overall, given the availability of the data input model. The following criteria should be considered for dividing a land region or river basin into "sub-areas" (IS group):

(I) Local distribution of the most dominant meteorological inputs in specific precipitation and possible evapotranspiration, taking into consideration the existing measurement system and observation information.

(II) The regional distribution of land features that remarkably determine general hydrological regimes, such as "topography," "land use," and "hydrogeology" (Becker and Serban 1990).

In this regard, it should be noted that the division of a river basin into sub-basins should not be interpreted as a semi-distributed model. This is a combination of lump models that are compatible with the respective sub-basins. In surface water systems, the discretization method has been a little adjusted and is basically identified with the model dimension, which can be one, two or three-dimensional (Becker and Serban 1990).

One-dimensional descriptions are generally sufficient for simple river flow routing without considering large floodplains, reservoirs, lakes, and so on. In these models, the river system is split according to the existing measuring stations as well as the confluence of branches with the principal river, while with a concept model or black box, these values for each element in a single model, the use of a distributed hydraulic model makes the river split into shorter "calculations."

Wherever there are wide floodplains, it is better to use a 2D or "semi-distributed" model. Although a good point network is just required for particular research. In many other cases, a semi-distributed conceptual model derived from a "quasi-two-dimensional description" can be used as a very simple but often equally efficient solution. The river channel is solely introduced as a separate primary conceptual model and arches as a sub-secondary conceptual model. They are routing models with different variables for channels and storage components. (Becker and Serban 1990). These sub-models are activated when the specific threshold discharge exceeds the dewatering discharge. The special shapes of these models are called multilayer models. Some of the effective aspects in discretizing river basin space and surface water systems are:

(I) The motivation behind the model exercise and the necessary precision of outputs
(II) The quantity and quality of input information and available information.
(III) Special needs and restrictions that must be observed to use a selected mathematical model.
(IV) The type and impact of control of existing or planned hydraulic structures (Becker and Serban 1990).

8.5.4 Main Types of Hydrological Models

Over the past decades, various structures of hydrological modelling with computer code have been developed, implemented and used in many studies (e.g., see Todini 1988) for a historical review of rainfall-runoff modelling. Classification of these model structures can help better understand the concept of hydrological models. Several classification methods are defined for hydrological models. Such methods as presented in the studies (Chow et al. 1988; Clarke 1973; Refsgaard and Knudsen 1996; Singh 1995a, b; Todini 1988; Wheater et al. 1993). Hydrological models can be classified based on the representation of physical processes, space, time or randomness (Bathurst and O'connell 1992). The classification of hydrological models focuses on the biological process based on the input and required parameters of the model and the number of physical principles presented in the model. Another way to classify models involves spatial representation of the basin or modelling the basin as a whole in which spatial diversity is neglected or divided into specific spatial sub-areas. Whether randomness is included in the model and whether the time is shown is another criterion for classification.

Classification classes generally meet the following criteria:

- **Process** (extending physical principles which are used in the model structure)
  Empirical, Conceptual, Physical
- **Spatial representation** (modification of model inputs and parameters as a function of time and space)
  Lumped, Semi-distributed, Distributed
- **Time**
  Dynamic (model includes time), Static (model excludes time)

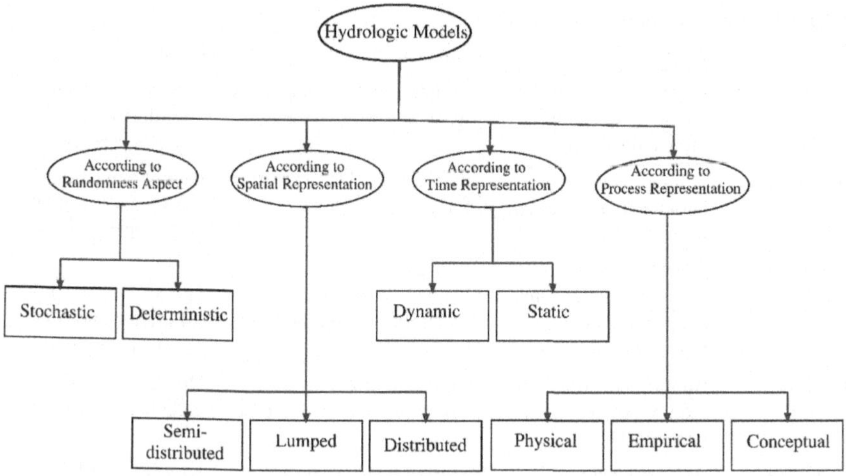

Fig. 8.2 Classification of hydrologic models

- **Aspect of randomness**

  Stochastic (outputs of these models are at least partially random), Deterministic.

Figure 8.2 represents the classifications of different hydrologic models. In the following part, the most important classification of existing models is discussed.

8.5.5 Model Structure

The structure of a model determines how runoff is computed. Several models are easy to use with a small number of variables, while others need an enormous number of interconnected variables. According to the governing equations, the structure of models fluctuates from easy to complex. These models are shown in Table 8.1 based on expanding complexity, from the simplest experimental models to the most complex mechanical-physical models (Sitterson et al. 2018). Physical and conceptual models require an exhaustive comprehension of the physics of surface water in the hydrological cycle (Sitterson et al. 2018; Srinivasulu and Jain 2008). Several models overlap in this type of structural classification (Jackson et al. 2011; Sitterson et al. 2018). Combined models merge the strengths of several models, but mainly they are categorized as three structures. The structural classification of runoff models has its own strengths and weaknesses (Sitterson et al. 2018).

Table 8.1 The main rainfall-runoff models

	Empirical	Conceptual	Physical
Method	Nonlinear relationship among outputs and inputs, black box	Simple equations that show water storage in the basin	Equations and laws according to natural hydrologic reactions
Strengths	A Small number of parameters is required, fast run time, high accuracy	Simple model structure, Easy to calibrate	Incorporate temporal and spatial variability, great scale
Weaknesses	Input data distortion, no connection among physical catchment	Do not consider spatial variability within the river basin	Site-specific, a lot of calibration and parameters required
Best use	Runoff is the only output required in ungauged watersheds	There are limits in computational data or time	On a small scale, have great data availability
Examples	Artificial Neural Networks, Curve Number	TOPMODEL, Stanford, HSPF, HBV	KINEROS, VIC, PRMS, MIKE-SHE

8.5.5.1 Empirical Models

These models are based on input data and the accuracy of inputs (Kokkonen et al. 2001; Sitterson et al. 2018). In rainfall-runoff regression models, historical rainfall-runoff inputs and runoff outputs are used at a certain location. The overall governing equation is a function of model inputs (Sitterson et al. 2018).

$$Q = F(X, Y)$$

where X and Y are the input rainfall-runoff history data set, and Q is the runoff output. Almost all experimental models are black boxes. This implies that little data is accessible on the internal procedures that control how runoff impacts (Beven 2012a; Granata et al. 2016; Sitterson et al. 2018). This method applied to convert rainfall to runoff is also without physical processes (such as the curve number (CN) procedure) or an unknown method (such as machine learning). Figure 8.3 depicts the relationship between rainfall-runoff, S is the storage parameter, I_a is the primary abstraction, Q is the runoff, and P is the rainfall (Sitterson et al. 2018; Sitterson et al. 2018).

When no other outputs are required, experimental runoff models are the best choice. For instance, these models do not have the ability to calculate the distribution of runoff values upstream and downstream (Jackson et al. 2011; Sitterson et al. 2018). Fixed catchments are also better modelled experimentally on account of the lack of particular information (about the basin) (Jackson et al. 2011; Sitterson et al. 2018). Experimental simulations are able to have acceptable performance in modelling several cases of regenerating past runoff values and long stages (Vaze

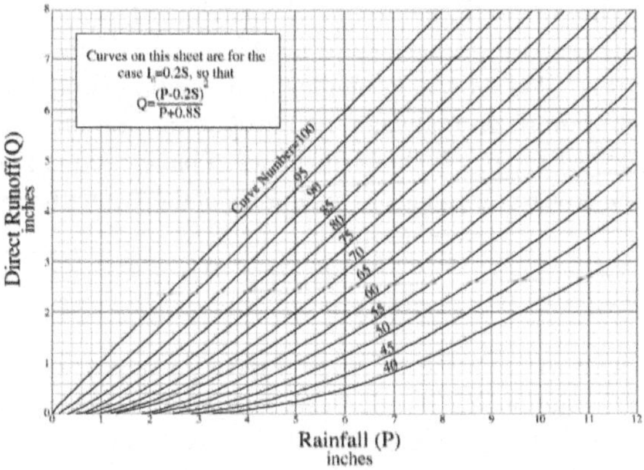

Fig. 8.3 The empirical curve number technique

et al. 2011; Xu et al. 2015). Quantitative parameters for creating case-based data models are needed, and these models are simple to use. Their parameters may have no physical meaning, since there is no realism in this model. The effortlessness of execution, quicker computational time, and cost-effectiveness are the motivations for choosing experimental models (Dawson and Wilby 2001). As these models are data-driven, the input is a wellspring of error, and the fluctuation of the input data causes severe fluctuations in the output of the model (Sitterson et al. 2018); (Srinivasulu and Jain 2008). The weaknesses of empirical models are that they might lead to several conclusions that a theoretical analysis may not support (Beven 2012a; Sitterson et al. 2018). This leads to the view that when there may be several ways to get the answer, using one way is wrong. Some examples of ML (Machine Learning) experimental models are regression equations, deep neural networks, artificial neural networks, and the SCS curve number (USDA[9] 1986); The Curve Number (CN) technique assumes that the actual runoff to potential runoff ratio is equivalent to the actual infiltration to potential infiltration ratio; however, there is no physical explanation for the correctness of the method (Beven 2012a; Sitterson et al. 2018). Regression equations find a relationship between output and input data (Devia et al. 2015). The ML used in ANNs makes output forecasts according to what has been learned from training. ANNs train themselves to learn the relationship between runoff and rainfall datasets. It can be taught with many inputs (cause over-training), which makes the model mislay its capability to distinguish one basin from another (Dawson and Wilby 2001; Sitterson et al. 2018).

8.5.5.2 Conceptual Models (Grey Box)

Conceptual Models (CMs) interpret the runoff process by connecting simplified sections to other hydrological processes. These models try to provide a conceptual

[9] United States Department of Agriculture.

framework of the hydrological behaviour of the catchment based on simple equations and reservoir storage of physical hydrological processes (Devia et al. 2015; Vaze et al. 2011). CMs show the water balance equation through the stages of rainfall conversion into runoff, groundwater, and evapotranspiration (Sitterson et al. 2018; Vaze et al. 2011). Each part is shown in the water balance equation with mathematical equations. The governing equations are copies of the "water balance equation" as shown below:

$$ds/dt = P - ET - Q_S \pm GW$$

where *ds*/*dt* is the reservoir reserve changes, *GW* is the groundwater, *P* is the precipitation, *ET* is the evapotranspiration, and *Qs* is the runoff (Sitterson et al. 2018). The schematic of the HSPF conceptual model shows that the land segment module (PERLND) combines several storage processes with the water parity equation (Fig. 8.4).

Storage of water and hydrological materials in groundwater or soil reservoirs in the model is ideal (Devia et al. 2015; Sitterson et al. 2018). The models simulate water exchange between hydrological components, storage tanks and the atmosphere according to the water balance equation. CMs differ in difficulty, depending on the complexity of equilibrium equations (Beven 2012a; Jackson et al. 2011). These models require a variety of meteorological input data and various parameters. These models have become popular among the public as they are easy to calibrate and apply. It is possible to apply a previously calibrated model to a different catchment (Sitterson et al. 2018; Vaze et al. 2011). When the calculation time is limited, it is better to use CMs, in which case the characteristics of the basin are not analyzed

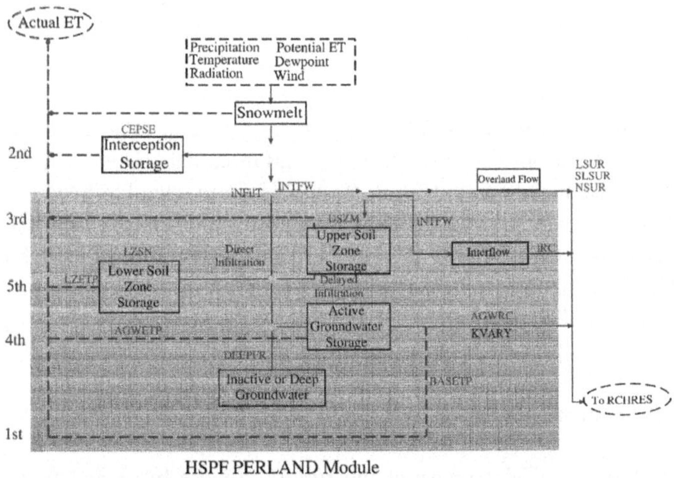

Fig. 8.4 The conceptual model HSPF schematic

in detail. TOP MODEL,[10] HBV,[11] NWSRFS[12] and HSPF are some examples of conceptual models (Sitterson et al. 2018).

8.5.5.3 Physical Models (White Box)

Physical models are known as mechanical and process-based models; these models are designed based on physical laws depending on hydrological processes (Sitterson et al. 2018; Vaze et al. 2011). These models show equations based on model physics to show several parts of the actual hydrological responses of the catchment. General physical principles and equations accepted in physical models are water balance equations, energy-saving and motion and kinematics, Darcy equations, St. Venant equations, Richards equation, and Bosinsk equation (Jackson et al. 2011; Sitterson et al. 2018).

$$\left(\frac{\partial h}{\partial t} = P_0 - I - E_0 - Q_{oc} + \sum_{j=1}^{3} Q_s^{ij}\right)_i$$

where Q_s^{ij} is the surface flow from the i element to the j element, $\partial h/\partial t$ is the water depth (at t), E_0 the evaporation, P_0 is the precipitation, Q_{OC} is the interplay between ground flow and channel routing, and I also indicates the influence (Sitterson et al. 2018).

Figure 8.5 shows the physical movement of water through any irregular triangular network with groundwater, recharge, intrusion, and evapotranspiration (Sitterson et al. 2018). Temporal and spatial–temporal changes in the catchment are included in the physical models of the catchment. The strength of this model is the relationship between the physical attributes of the catchment and the model parameters, which makes it more practical. When accurate data is available, they should be used. Physical characteristics of hydrological processes are put in an application for a small scale according to the calculation time (Sitterson et al. 2018). Several physical process parameters are required to model the calibration. A large amount of data is required to implement these models limits their employment (Uhlenbrook et al. 2004). Most physical models offer a 3D approach with respect to the exchange of water within the surface, soil and air (Sitterson et al. 2018). They are employed to

---

[10] Topography based Hydrological Model (Beven and Kirkby 1979; Beven et al. 1984) Source code for TOPMODEL written in R can be found at: https://idea.isnew.info/r.topmodel.html.

[11] Hydrologiska Byråns Vattenbalansavdelning (Bergström 1992) http://www.geo.uzh.ch/en/units/h2k/Services/HBVModel.html.

[12] National Weather Service River Forecast System (Burnash et al. 1973) http://www.nws.noaa.gov/iao/iao_hydroSoftDoc.php.

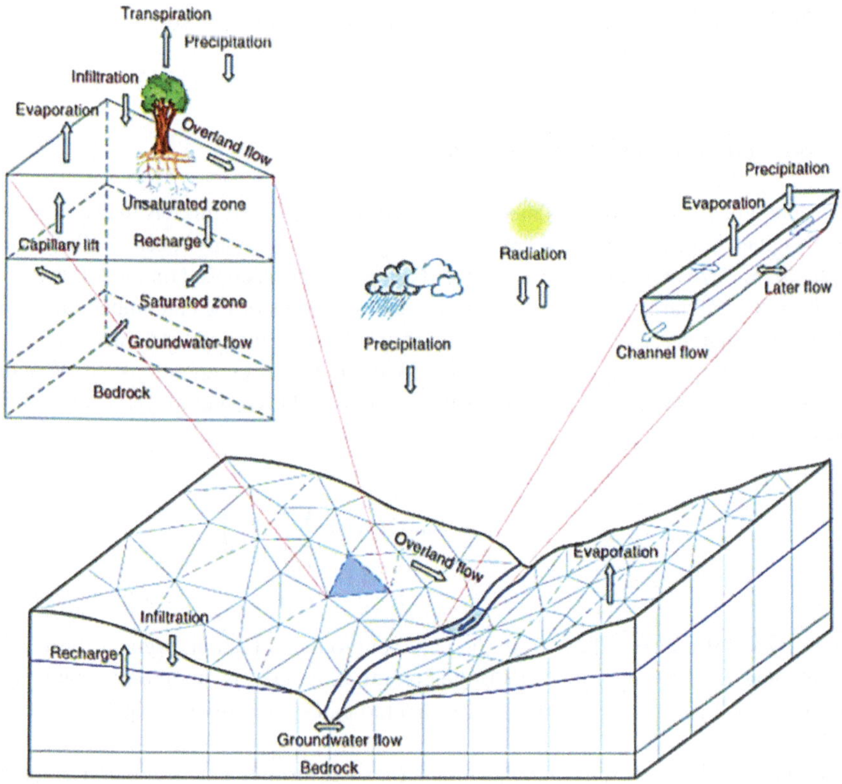

Fig. 8.5 Physically-based PIHM[13] system model

simulate groundwater movement and basin interactions with sediments. VELMA,[14] VIC,[15] MIKE SHE,[16] PIHM,[17] KINEROS[18] are examples of physical models (Singh 1995a, b). The characteristics of conceptual, experimental and physics-based models are shown in Table 8.2 (Sitterson et al. 2018).

[13] Penn State Integrated Hydrologic Modeling.

[14] Visualizing Ecosystem Land Management Assessments (Abdelnour et al. 2011; McKane et al. 2014) https://www.epa.gov/water-research/visualizing-ecosystem-land-management-assessments-velma-model-20.

[15] Variable Infiltration Capacity Model (Liang et al. 1994) http://www.hydro.washington.edu/Lettenmaier/Models/VIC/indexold.shtml.

[16] MIKE System Hydrologique European (Abbott et al. 1986) https://www.mikepoweredbydhi.com/products/mike-she.

[17] Penn State Integrated Hydrologic Modeling System (Qu et al. 2004) http://www.pihm.psu.edu/.

[18] Kinematic Runoff and Erosion Model (Woolhiser et al. 1990) http://www.tucson.ars.ag.gov/kineros/.

Table 8.2 Characteristics of models

Empirical	Conceptual	Physically-based
Metric/Data based/black box model	grey box/Parametric model (Devia et al. 2015)	white box/Mechanistic model (Devia et al. 2015)
Involve derive the value from existing time series and mathematical equations	Include physical basis semi-empirical equations and Based on modelling of reservoirs	Evaluation of parameters describing physical characteristics, Based on the spatial distribution (Devia et al. 2015)
low consideration of features and processes of the system	Parameters are derived from calibration and field data	Require data about the initial state of the model and morphology of the catchment (Devia et al. 2015)
the low explanatory High and depth predictive power	really simple, can be easily implemented in computer code	Complex model. Require human expertise and computation capability (Devia et al. 2015)
It cannot be generated in other catchments (Devia et al. 2015)	Require large hydrological and meteorological data	suffer from scale-related problems (Devia et al. 2015)
ANN, unit hydrograph	HBV model, TOPMODEL	SHE or MIKESHE model, SWAT (Devia et al. 2015)
Valid within the boundary of the given domain (Devia et al. 2015)	Calibration involves curve fitting make difficult physical interpretation (Devia et al. 2015)	Valid for several situations (Devia et al. 2015)

8.5.6 *Spatial Processes*

Spatial processes (SPs) prepare an intermediate view of the basin for modelling. Soil variability, geology, topography and vegetation affect the rainfall-runoff relationship and should be considered in modelling (Beven 2012a; Sitterson et al. 2018). In rainfall-runoff models, the spatial structure of catchment processes can be classified as semi-distributed, fully distributed, or lumped (Sitterson et al. 2018).

Lumped

Lumped models do not evaluate the spatial diversity within the catchment. Rainfall-runoff simulations cover a wide scope of SPs. Determining the type of model required is usually done according to spatial processes in rainfall-runoff models. Figure 8.6 shows the spatial structure in runoff models (Sitterson et al. 2018).

Semi-distributed

This type of model reflects some spatial variation. They consider spatial variability on a smaller scale than lumped models. But this kind of models still does not calculate runoff per cell of the network (Sitterson et al. 2018).

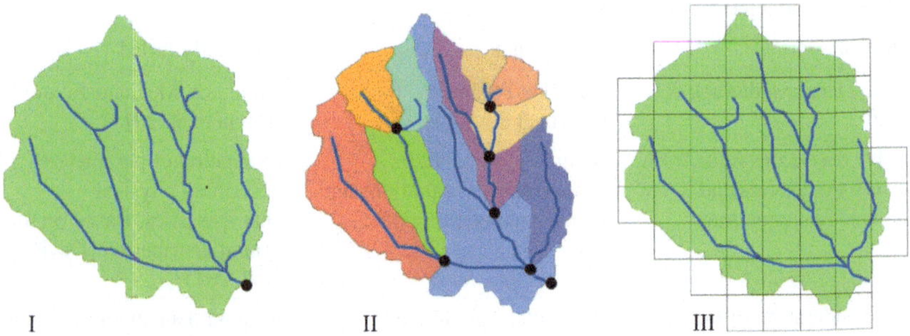

Fig. 8.6 Runoff models spatial structure. I: lumped, II: semi-distributed (sub-catchment), III: distributed (grid cell)

Fully Distributed

These models process spatial diversity with grid cells.

Runoff for each catchment is indicated by black circles in Fig. 8.6II. Distribution models compute runoff (for each grid cell), while lumped models show runoff for the whole basin at the river outlet (black circle) in Fig. 8.6I. A comparison of spatial structures is presented in Table 8.3 (Sitterson et al. 2018).

Table 8.3 Comparison of spatial structures in models

	Distributed	Semi-distributed	Lumped
Procedure	Spatial variability is considered	Series of lumped and distributed parameters (Sitterson et al. 2018)	Spatial variability is disregarded; the whole catchment is modelled as a unit
Inputs	All specific data by cell	Both averaged and specific data by sub-catchment (Sitterson et al. 2018)	All averaged data by catchment
Strengths	Physically related to hydrological processes	Represents important features in the catchment	Fast calculation time, Perform well at simulating average conditions
Weaknesses	Data intense, long computational time	Averages data into sub-catchment, spatial resolution loss	A lot of assumptions perform weakly for large areas, loss of spatial resolution (Sitterson et al. 2018)
Examples	Physically distributed models, MIKE SHE, VELMA	Conceptual and some physical models, TOPMODELSWAT (Sitterson et al. 2018)	Empirical and conceptual models, ML (Machine Learning)

8.5.6.1 Lumped Model

Lumped models look at the catchment as a homogeneous unit (Moradkhani and Sorooshian 2008; Singh 1995a, b). The mean values such as uniform values of rainfall and soil storage are applied along the catchment (Beven 2012a; Rinsema 2014; Sitterson et al. 2018). The catchment characteristics are considered to be the same for the whole area and always cause the parameters to be over-parameterized or under-regulated (Rinsema 2014; Sitterson et al. 2018). A lumped model is planned to imitate the total runoff at the outlet (not particular streams). Consequently, lumped models simulate the average runoff with a much shorter computation time than the actual value. The lumped models' average annual runoff conditions are applied for administrative purposes that gander at the existing conditions on a long-term basis (Sitterson et al. 2018). These models contain many assumptions about hydrological processes. Therefore, they tend to overestimate or underestimate runoff (UCAR 2010). They do not take into account changes within the basin (Sitterson et al. 2018; Uhlenbrook et al. 2004). Changes in land applications might alter runoff trends in specific regions, but a lumped model means examining these throughout the catchment. The inputs are completely lumped or "average" data. Obtaining or creating this type of data is relatively simple. This is achieved by averaging data across the study area (Sitterson et al. 2018). All data contain parameters, inputs and outputs that are stable over time and space in a lumped model. Considering homogeneity along the catchment, lumped models do not consider spatial separation of data (Sitterson et al. 2018). For instance, rainfall-runoff patterns vary in location and time but are fixed in lumped models (Sitterson et al. 2018; Vaze et al. 2011). Experimental and conceptual models are always implemented spatially as a lumped model. Due to moderate conditions, these perform weakly for large areas (Moradkhani and Sorooshian 2008).

8.5.6.2 Semi-distributed

These models are a modified form of lump models with the specifications of distributed models. They might contain a set of lumped parameters that are distributed in a semi-spatial manner (Sitterson et al. 2018). A semi-distributed model splits the catchment into smaller regions with several variables for each area (Rinsema 2014; Sitterson et al. 2018). Sub-regions represent the main characteristics of a basin and merge the benefits of lumped and distribution models (Pechlivanidis et al. 2011; Sitterson et al. 2018). Their input categorizes semi-distributed models. In a semi-distributed model, the inputs contain distributed and lumped variables. Most models are semi-distributed due to the range of data access between distributed models and lumped models. A semi-distributed model can separate data in the catchment but the sub-areas are homogeneous (Beven 2012a; Sitterson et al. 2018). Sub-areas could be categorized by several approaches. Based on the land use or cover type, soil type, and surface slope, a combination called hydrological reaction units (HRU) is formed from a river basin in which the area responds to rainfall in the same way (Beven 2012a; Devia et al. 2015; Sitterson et al. 2018). These models compute runoff at the starting point of each sub-catchment. Semi-distributed models consider land use characteristics and spatial diversity without a dominant model structure (Kokkonen et al. 2001). The advantage of a semi-distributed model is its quick operation time and the ability to employ parameters less than a distributed model (McMillan et al. 2011; Sitterson et al. 2018). One disadvantage of these models is the input data manipulation. For instance, distributed spatial rainfall data should be the average of rainfall data. The sub-area in particular locations should be distributed to the region applying the TPM.[19] Physical and conceptual models could be implemented in a semi-distributed mode related to input datasets. TOPMODEL is a semi-distributed concept. This model employs soil characteristics and land slope to split the basin. Some models, such as SWAT, divide catchment by using hydrological response units (Sitterson et al. 2018).

8.5.6.3 Distributed Models

These models are complicated as they take into account "spatial heterogeneity." Fully distributed models discretize the cycle by grid cells. In addition, their structure is like a physics-based model, making them more compatible with the natural hydrological cycle. Spatial distribution models affect basin management performance by preparing accurate data for small components. For more information, see the study (Bouadi et al. 2017). Each cell or small part has a distinct hydrological reaction and is computed separately but has a relationship with n bordering cells (Rinsema 2014; Sitterson et al. 2018). By computing runoff for all network cells, the model prepares accurate runoff information at several points. Distributed models direct the computed runoff from each cell to the nearest cell according to the physical equations applied to calculate the natural time delays and flow path (Sitterson et al. 2018). The distributed runoff model utilized in NASA's Global and NALDAS[20] prepares estimations on a grid cell (Sitterson et al. 2018). This extensive data assists in understanding sediment and pollutant transport in the watershed and depicts the temporal and spatial diversity of hydrological processes (Knapp et al. 1991; Sitterson et al. 2018). Distributed models examine the effects of basin change in runoff values (Singh 1995a, b; Sitterson et al. 2018). In distributed models, all input data is distributed temporally and spatially. Different types of inputs are required for a typical distributed model, such as network precipitation, soil profile, ground use images from satellites. Digital elevation models (DEM) consider and how they change over time, and watersheds and topographic features as boundaries and dimensions. One of the disadvantages of these models is their need for calibrated parameters or distributed data for each network cell. If the data is not completely scattered, estimated weight averages are applied to extract the data (Sitterson et al. 2018).

In addition, distributed models are limited by input network location or model resolution. Another drawback of distributed models is the operation time required

[19] Thiessen Polygon Method.

[20] North American Land Data Assimilation System.

to perform a simulation, depending on the size of the basin, input dataset, and calculational restraints can vary from one minute to a few hours (Vaze et al. 2011). The mentioned difficulties, when compared with lumped models, clarify why these models are not extensively employed (Rinsema 2014; Sitterson et al. 2018).

8.5.6.4 Deterministic and Stochastic Models

In Deterministic models, a given input always produces the same output. This model does not consider randomness. On the other hand, the output of stochastic models is at least somewhat random. In general, it could therefore be that if the probability of the effect of variables on the occurrence of a process is not considered and the model does not follow the laws of probability, then this model is deterministic. Otherwise, if the probability of occurrence of the desired variables and probability concepts are involved in the construction of the model then it is is stochastic.

Note that the distinction between Deterministic and Stochastic models is not clear. In many modelling studies, it is assumed that the modelled variables are not deterministic, but more advanced devices use deterministic modelling. To consider stochasticity, additional uncertainty analysis was performed assuming the probability distribution of at least some of the variables and parameters involved (Solomatine and Wagener 2011; Wilderer 2010); They are classified in terms of spatial and temporal variability, as shown in Fig. 8.7.

Some Additional Definitions for Charts

Steady and Unsteady Flow

In steady hydrological models, the flow rate does not change over time. However, in unsteady flow hydrological models, the flow rate varies with time.

Space Independence and Space Correlated

In time-independent hydrological models, the sequence of hydrological events has no effect on each other; they are time-dependent, while in the time-correlated type, the time sequence of a future event is affected by current events or happenings that are continuous and in previous time have occurred.

Data-driven models

Over the past decade, the area of experimental modelling has become important because of advancements in CI,[21] especially ML. It, therefore, enters another stage and merits a particular name, DDM[22] (Solomatine 2005; Solomatine and Wagener 2011; Wilderer 2010). A model can then be defined based on the connections between system state variables (input, internal, and output variables) with a limited number

[21] Computational Intelligence.

[22] Data Driven Modelling.

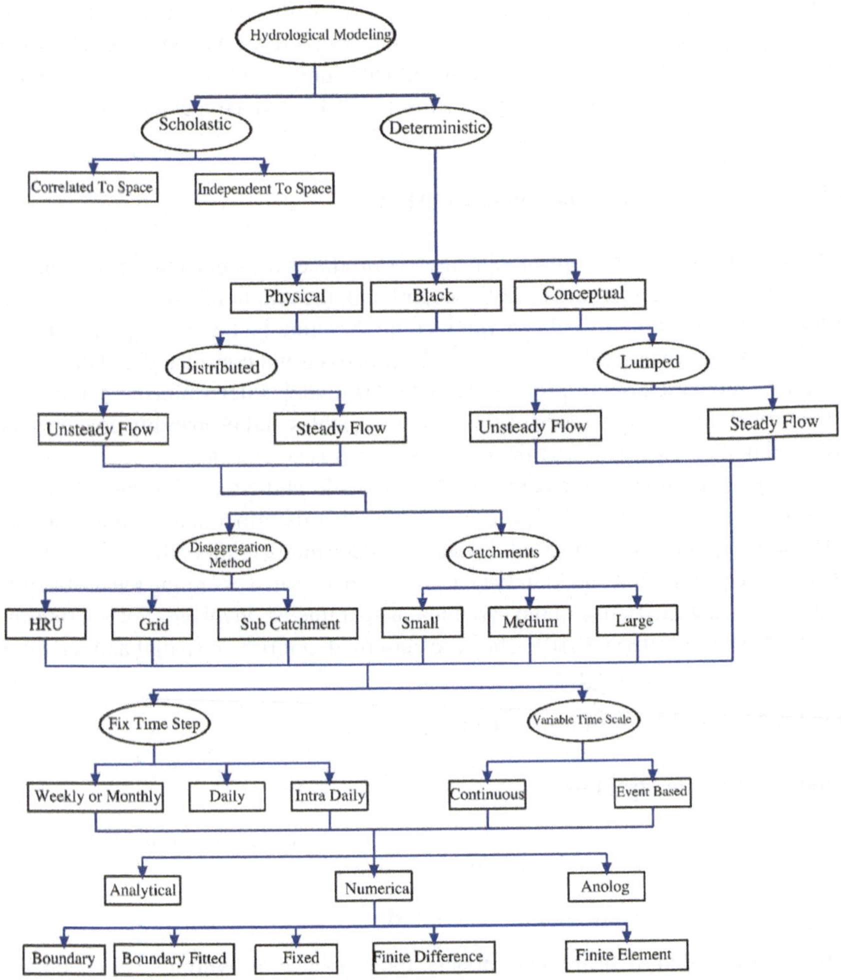

Fig. 8.7 Classification of hydrological modelling in terms of spatial and temporal variability

of assumptions about the "physical" behaviour of the system (see Fig. 8.8) (Solomatine et al. 2009). The methods used today can go far beyond the methods used in conventional experimental modelling; they are possible to solve numerical prediction problems, reconstruct nonlinear tasks, perform classification, group data, and build rule-based structures (Solomatine 2005; Solomatine and Wagener 2011; Wilderer 2010).

Contributors to DDM

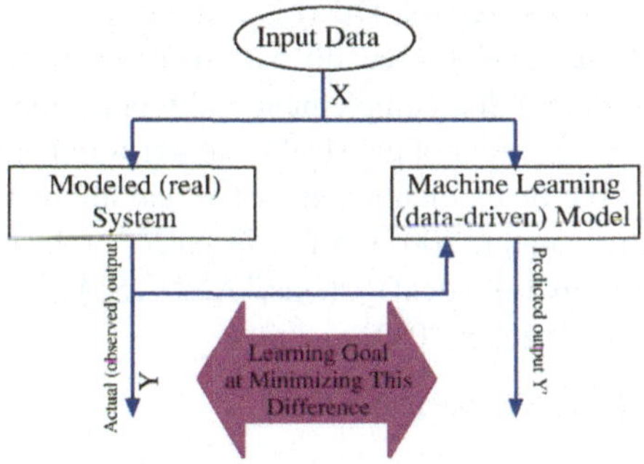

Fig. 8.8 Learning in data-driven modelling

The accompanying regions adding to DDM can be referenced as DM,[23] AI, KDD,[24] CI, ML, IDA,[25] SC,[26] and pattern recognition (Solomatine et al. 2009; Solomatine 2005; Solomatine and Wagener 2011). Terms and scopes of examination once in a while contend to name a similar interdisciplinary field. It is taught to prepare a formal definition of different regions with their proven characteristics, like NN,[27] FL,[28] PR,[29] EC,[30] BR,[31] and etc. It is hard to track down clear contrasts between them. Major areas are defined as below:

- **CI** consists of three major areas: NN, FL, and EC. CI is increasingly considering scopes that are thought of AI (except for symbolic methods) and ML (Solomatine et al. 2009; Solomatine 2005).
- **SC** is a scope that is exceptionally near to CI but with a particular emphasis on data-driven FRBS.[32]
- **ML** is a field of computer science that has for decades considered a subregion of AI and focuses on the "theoretical foundations" employed by CI and SC (Solomatine et al. 2009). Classification problems (pattern recognition) are more commonly discussed by ML than regression prediction (Solomatine 2005).
- **DM** and **KDD** are chiefly centered around large databases. In addition, they are related to applications in monetary administrations, client asset by the board and banking. DM is considered as part of the broader KDD (Solomatine 2005).
- **IDA** is a somewhat new term and appears to focus on data analysis in medicine and research (Solomatine et al. 2009; Solomatine 2005).

[23] Data Mining.

[24] Knowledge Discovery in Databases.

[25] Intelligent Data Analysis.

[26] Soft Computing.

[27] Neural Networks.

[28] Fuzzy Logic.

[29] Pattern Recognition.

[30] Evolutionary Computation.

[31] Bayesian Reasoning.

[32] Fuzzy Rule-Based Systems.

In this regard, DDM is examined as a modelling approach that concentrates on the use of CI (especially ML) methods in the construction of models (often of natural systems) that complement and replace the knowledge-driven models that describe the behaviour of physical systems (Solomatine 2005). DDM uses methods developed in the fields mentioned earlier and aligns them with specific areas of the program (Solomatine 2005). A few instances of the most famous techniques utilized in DDM are "hydrological systems," ANNs, SMs,[33] and FRBS (Solomatine 2005; Solomatine and Wagener 2011).

ML as the basis of DDM

ML can be considered the principal source of a technique for DDM (see Fig. 8.8 for more details). The ML strategy is an algorithm that estimates previously obscure mappings between system inputs and outputs from existing data (Mitchell 1997; Solomatine 2005; Solomatine and Wagener 2011). Similarly, a dependence, viz. mapping or model, is discovered (induced), which can be used to predict (or infer effectively) future system outputs from known input values. A set of examples, K, is constituted by tuples (X_k, X_k): where k = 1,...,K, $x_k = \{x_1, \ldots, x_n\}$, $y_k = \{y_1, \ldots, y_m\}$ (k & n = number of inputs, m is the number of outputs). Training is the process of creating a function (or "mapping") (Solomatine 2005; Solomatine and Wagener 2011; Wilderer 2010).

In statistics, four types of data can be considered: nominal (e.g., colour, symbolic label); ordinal (e.g., precipitation intensity is demonstrated as low, medium or high); interval (e.g., temperature); and ratio (a number with a real value, for example, water level). In ML, there is an inclination to talk only about two data categories: nominal (which includes an ordinal number and sometimes an integer) and a real value number (Solomatine 2005). The main learning styles are described as below:

- **Classification**: Based on categorized examples, find a way to assign a class label to unseen examples. For this situation, y (the output parameter) takes on nominal values (Solomatine 2005).
- **Association**: The relationship between all the parameters that characterize the system is determined (usually depends on finding the combination of values that have the most frequency).
- **Prediction** (Regression): Outcome is real numeric value or a vector ($y_k \in \acute{R}^m$).
- **Clustering**: Groups of items near in the space of input must be recognized, on which occasion there is no explicit output variable. Several numerical prediction procedures can likewise be utilized to classify, but the data must be sorted. The output number is equal to the number of class labels; the nominal output values (class labels) in the training data must be replaced by a vector of real values with zero for all outputs, except a number correlates with the main class label. While the model is trained, the class label can forecast a new item (object): the number of outputs with the maximum value is considered as the class label number (Solomatine 2005).

[33] Statistical Method.

Data-driven Modelling Process

The following list often differs from the modelling process (Pyle et al. 1999; Solomatine 2005):

- Choose a well-defined problem, identify the type of problem-solving, specify how to use the solution provided in practice, become familiar with the issue, gather domain knowledge and get it. Intelligibly specify the assumptions and debate them with domain knowledge experts, let this guide the choice of modelling methods, make the model as easy as possible, but do not make it uncomplicated; this command is occasionally set in dissimilar methods. In general, the concept is generally recognized as the Oakham Razor Principle—formulated by William Ockham in 1320 in the following form "shave all unneeded philosophy off the explanation" (Solomatine 2005), Create the model (train), Modify the model repeatedly, test and assess the outcomes (Solomatine 2005), Explore instabilities (critical regions where little changes in inputs lead to enormous shifts in outputs). Define uncertainty (important ranges in the data set that the model creates low confidence forecasts). Operate the model. If necessary, review it (Solomatine 2005).

Role of Data-Driven Approaches in Hydrological Modelling Continuum

Hydrological models are the results of an experimental process that deals with the mathematical coding of specific experimental relationships or general theories about how hydrological systems work (Gupta and Nearing 2014; Mount et al. 2016) and turn them into tools that can be understood and/or let the predictions be testable. The system is represented by an equation as below:

$$y = f(\theta, x) + \varepsilon$$

where y is the output of the hydrological model defined by the structure (f), x is the input, θ is the parameter setting, and ε is the model error (Mount et al. 2016).

Several models can be defined with respect to the permissible relative emphasis on "empirical data" or hypothetical information derived in their effectiveness (see Table 8.4 for more details) (Mount et al. 2016). Therefore, they can be placed in the range of hydrological modelling that shows the data degree or hypothetical perception for each model (Fig. 8.9). While models at different points offer a range of various types of proof to inform them, all steps involve unprovable steps or previous theory. Therefore, defining parameters is necessary to set the model, a process that accepts empiricism degree even in nearly all physical models. Accordingly, they face epistemological uncertainties, and it is important to use more empiricism to assist in describing the mentioned cases (Gong et al. 2013; Mount et al. 2016).

DDMs, such as ANNs (e.g., Abrahart et al. 2012a, b; Bowden et al. 2012; Chapuis 2012; Mount et al. 2016) have an important impact on the main role of data in the development procedure. ML methods were used to discover the model structure and enhance its variables (Mount et al. 2016). This minimizes the role of a priori

Table 8.4 Effect of data (D) or hypothetic (H) factors in estimating inputs, structure, and parameters

Technique	Inputs (x)	Structure (f)	Parameters (θ)
"Data-Driven" (e.g.. ANNs; Abrahart et al. 2012a, b)	H/D	D	D
"Flexible Conceptual" (e.g., SUPERFLEX; Fenicia et al. 2011)	H	H/D	D
"Fixed Conceptual" (e.g., TOPMODEL; Beven and Kirkby 1979)	H	H	D
Fully physical (e.g., HydroGeoSphere; Brunner and Simmons 2012)	H	H	H

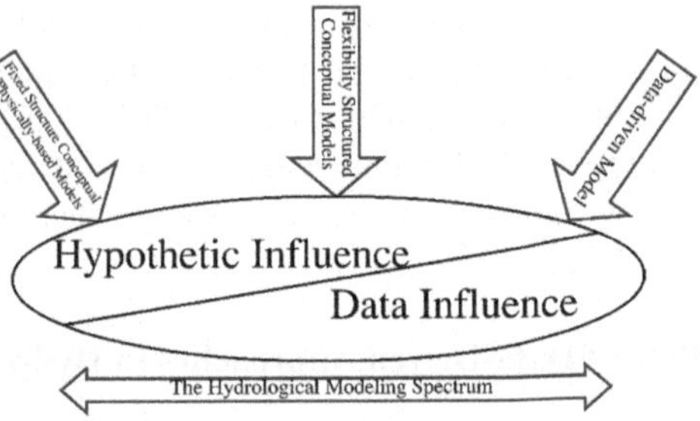

Fig. 8.9 Hypothetic impress in per model

hypotheses, despite the fact that they cannot be abolished. In DDMs, inputs are usually driven completely by hypothetical knowledge and modelling reasoning, for example, the selection of the input for a medium-discharge model with respect to basic hydrological processes (Dawson et al. 2014; Mount et al. 2016). However, this can be minimized by using more "Automated Selection Methods," for example, partial cross-information (May et al. n.d.; Mount et al. 2016). In contrast, conceptual (e.g., Beven et al. 1979; Mount et al. 2016) and physics-based models (e.g., Brunner and Simmons 2012) place a great weight on the use of hypothetical knowledge through its maturing (Mount et al. 2016; Mount et al. 2016).

The following is derived from a set of generalized hypotheses (Clark et al. 2011) about the shape and function of the hydrological system that defines the model structure and needed input (e.g., the GR4J model (Perrin et al. 2003; Mount et al. 2016). However, initial data analysis is an essential part of such models, because, in most cases, the internal parameters of each model should be calibrated for neighbourhood conditions. DDMs are commonly used for this motivation (e.g., Zhang et al. 2011; Mount et al. 2016). From this point of view, it can be argued that even several hypothesis-based models include various data-driven components (Mount et al. 2016). Lately, policies to support more flexible conceptual model structures have started to appear (Fenicia et al. 2011; Kavetski and Fenicia 2011). These recognize significant uncertainty in the choice of hypothetical structures applied in hydrological simulating, which has led to an overabundance of models (see Kampf and Burges 2007; Singh and Frevert 2002c, d). In an attempt to address this issue, flexible modelling approaches use a diagnostic method to evaluate the model that utilizes the types of data observed and the data signatures in order to accurately examine both the separate model components and the connection in model with Complete System (Clark et al. 2011; Mount et al. 2016). Thus, flexible models could be considered as intermediate positions in the hydrological modelling spectrum among hypothetical data-driven and fixed models (Mount et al. 2016).

The analytic, scientific method prescribes that the transition in the spectrum of hydrological modelling from data-influenced models to models that are increasingly hypothetically affected should be accompanied by an increase in the level of completeness, robustness and reliability in the modeller's a priori hydrologic system knowledge (Mount et al. 2016). The change is clear all through the authentic improvement of hydrological modelling (see Todini 2007; Mount et al. 2016). Starting with Horton's early experimental models (Horton 1945, 1933; Mount et al. 2016), it can be fully distributed, in terms of simple conceptual hydrological models (e.g., the Stanford watershed model, (Crawford and Linsley 1966a, b), in terms of physical-based and generalized hydrological models (e.g., Abbott et al. 1986; Mount et al. 2016). Nevertheless, despite obvious progress, the need to address the challenges related to cognitive uncertainty and recognized conflict (Beven 2013; Beven and Westerberg 2011; Mount et al. 2016). This relates to model structure, parameters, BCs, and inputs (Liu and Gupta 2007) and is a fundamental reminder that the knowledge of the system on which hydrological models are based is insufficient. In cases where systems are dynamic, fundamental questions arise about how models are structured in such a way that they can be adapted to the challenge of dealing with the knowledge of non-fixed hydrological systems. Of course, most hydrological models apply fixed structures in which external forces (i.e., occur outside the model) stimulate the dynamics of a steady-state system (Mount et al. 2016). Thus, hydrological models are conceptually limited (Mount et al. 2016). More complex, dynamic representations to illustrate the features of emerging common socio-hydrological system phenomena have not yet been achieved (Lall 2014; Mount et al. 2016).

The objectives of the PRSP show a scientific process by which such obstacles can be overcome (Mount et al. 2016). This path has three main aims: (I) Developing a new phase of hydrological models defined by a complete display of system feedback and evolutionary procedure to increase system knowledge (Mount et al. 2016); (II) increased estimation and predictability; and (III) the development of science in practice (Mount et al. 2016). This also shows a hydrological modeller with an improved level of intricacy and system uncertainty that will be modelled in the future. In terms of DDMs, which are enlightened by the experimental consequences of investigation and simulation of the "hydrological system," it has made basic progress in system knowledge, which has finally paved the way for conceptual and "physics-based models" (Mount et al. 2016). Likewise, understanding the path predicted by the PRSP requires a preliminary advance in the understanding of undetermined representations that occur at the interface between hydrology and society (Mount et al. 2016). The initial stage to achieve it certainly requires the revitalization of experimental methods, the DDM that can extract those interconnections from the novel sources of data that are accessible. These future models are probably going to bring extra difficulties (Mount et al. 2016). For instance, ways must be found to expand the execution time of more complicated models (see Abrahart et al. 2012a, b). Therefore, assignments such as "sensitivity analysis," "real-time forecasting," and "uncertainty assessment" can be executed. In addition, it would be important to discover techniques to increase the ability to predict hypothesis-based models by partially representing system procedures by modelling the salient information shown in the "model residuals" (see Corzo et al. 2009). By and large, the use of DDMs is a fundamental opportunity to assist hydrologists in providing such conditions (Mount et al. 2016).

Modelling Process

Forming a Perceptual Model

Most hydrological modelling books start with choosing a model that can be used for a specific purpose. Here, the initial stage of the modelling process will be defined by the perceptual model of rain-runoff procedures in a basin. In fact, the perceptual model is a summary of our understanding of how the basin responds to rain in various conditions (Beven 2012a, b). A perceptual model is definitely personal. The perceptual model is not related to a mathematical theory. There is no need to write, and it is not necessarily in the mind of any hydrologist. We can understand the complexities of flow processes in a small way, especially at the field sites and environments we have experienced. Therefore, it is assumed that the perceptual model of a hydrologist is different from another model. Understanding the perceptual model is essential for a special basin (Beven 2012a, b). It is vital to recognize that all mathematical descriptions made for prediction are necessarily simplifications of the perceptual model but may still be sufficiently representative. We can understand the difficulties of flow processes in a quantitative method (see, for example, (Flury et al. 1994)) which are very difficult to explain in the mathematical language (Beven 2011). However, a mathematical explanation is the first step in developing a model that makes predictions (Beven 2012a, b).

Formation of a Conceptual Model

A mathematical description is called a conceptual model. At this stage, ideas presented to clarify the explanation of procedures must be explicit (Beven 2011). Several models have been utilized to express the flow in the soil using Darcy's law, which describes that the flow corresponds to the slope of the hydraulic potential (Beven 2011). Measurements suggest that the potential hydraulic slopes in structured soils could differ considerably over short distances so that if the Darcy law is used at the soil profile scale or higher, this is essentially assumed that some moderate slopes could be applied to describe the flow and the consequences of preferential flow by macrospores on the soil. Suspicion, a simple underlying suspicion, may not be explicitly stated (Beven 2011). Normally, it is not difficult to pinpoint doubts if the background of the equations is known. This ought to be the beginning stage for estimating a special model related to the perceptual model. It is useful to make a list of all the practical model theories that we follow here to present several modelling programs. The conceptual model might be less or more complicated, from the net mass equilibrium equations for the elements constituting storage to the associated NPDE[34] (Beven 2012a, b).

Formation of a Procedural Model

A few conditions may essentially be made a straightforward interpretation into the programming code. Whether the equations are not solved analytically by presenting some BCs (Boundary Conditions) in the real system (which is regularly found in PDE[35] in hydrological models), an additional step of estimation using numerical analysis methods to determine a code is needed for the procedure model that runs on the computer (Beven 2012a). An example of this is the differential substitution of principal equations with the finite difference or finite volume equations (Beven 2012a, b).

Care must be taken at this stage: Converting concept model equations to procedural model code can cause significant errors in the actual solution of the original equations (Beven 2011). This is a special matter for solving differential equations. However, this issue is further discussed in conceptual reservoir models (Kavetski and Clark 2010). Because such models are usually nonlinear, it can be difficult to estimate the error due to just implementing a numerical solution for a conceptual model. However, it may have a fundamental impact on the conduct of a model in the adjustment interaction (e.g., Beven 2012a, b; Kavetski and Clark 2010).

Calibration and Validation (Evaluation)

The model code provides us with a software procedure that runs easily on a PC. Prior to utilizing the code to make quantitative forecasts for a specific basin, it is vital to go through a calibration phase. All models utilized in hydrology have conditions that incorporate an assortment of input and state factors (Beven 2011). Some inputs describe the geometry of the basin, which is usually considered during special modelling (Beven 2012a, b). Some variables show the BCs of the time variable during simulation, such as rain and other meteorological variables (Beven 2011). Model parameters, such as "soil water storage" or "water depth," change during modelling according to model calculations (Beven 2011). The early values of the state variables describe the status of the basin at the beginning of the modelling (Beven 2012a, b). At last, there are parameters that clarify the attributes of the basin region (Beven 2011). These variables may include the "porosity" and "hydraulic conductivity" in a spatial distribution model or the average residence time in the saturation zone for a model that applies state variables at the basin scale (Beven 2011). They are typically fixed during the simulation period (despite the fact that for some variables, such as the "storage capacity" of a canopy, there might be large time dependencies that may be necessary for some applications) (Beven 2011). Anyway, even if assumed at a fixed time, defining parameter values for a special basin is not a priori simple (Beven 2011). Of course, the most common procedure of parameter

---

[34] Nonlinear Partial Differential Equations.

[35] Partial Differential Equations.

calibration is to change the values of parameters to accomplish the best fit between model figures and any perceptions of the actual basin reaction (Beven 2012a, b, 2011).

The following step is after the evaluation or validation. The evaluation might additionally be done inside a "quantitative framework," computing at least one index of the model performance comparative with the perceptions accessible (assuming any) about the runoff response. The issue at this stage is typically hard to track down a satisfactory model. Calibration of the model variables is possible by comparing the observed discharges. Most model constructions have an adequate number of parameters that could be adjusted to fit the data. The main difficulty is that there are several amalgamations of the model structure and a set of parameter values that would fit logically with the discharge data. Therefore, the differences between the practical models and the validation of each individual model may be difficult to predict in terms of evacuation. On the other hand, discharge prediction, together with internal catchment responses, may be calculated relative to the main perceptual catchment model of interest. It is more strenuous to discover a perfectly acceptable model. Variation might prompt a correction of the parameter estimates used to reevaluate the conceptual model, or even, sometimes, to a revision of the catchment perceptual model (Beven 2012a, b).

Input data to hydrological models

Input data to hydrological models can be divided into two general categories: spatial-meteorological and hydrological data. Many hydrological models are based on physical processes that control the water flow through the soil. These models usually have two sets of input parameters, including physical and process variables (Hsu and Nguyen 1995).

- **Physical parameters**

  This data includes measurable physical characteristics of the basin, including features such as the surface of the basin, the impermeable part of the basin, the surface covered by water, and water reservoirs and surface slopes.
- **Process parameters**

  Process parameters represent the characteristics of a watershed that cannot be measured directly, such as medium depth or storage due to soil surface moisture storage, effective lateral cortex flow index, and nonlinear control coefficient of groundwater intrusion control index.

Methods for determining input parameters

There are two main methods for determining input parameters to the model: parameter measurement and estimating parameters.

- **Parameter measurement**

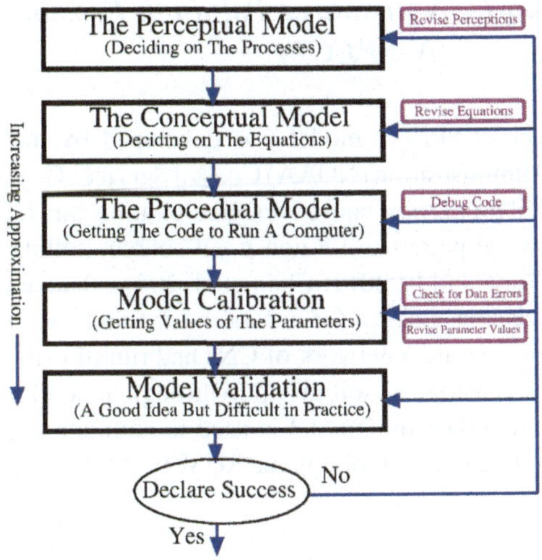

Fig. 8.10 Steps in the modelling process

In this case, we use the available data on the characteristics and behaviour of watersheds to determine the initial values of model input parameters. For physical parameters, initial values are determined utilizing measurements gotten from maps (from field data). Then, the parameters are typically adjusted using the figures obtained from measurements. Further adjustments in measurements are made when there is an error. The estimation of the amplitude (minimum and maximum data figures) of the available data for process parameters is based on hydrological judgment and watershed perception. Then, the uncertainty in the values of the parameters is reduced by the estimation process.

- **Estimation of parameters**

Various methods have been proposed to reduce the uncertainty in estimating process parameters. A common method is to initially estimate the parameters first at locations where the range has already been specified. Then, the estimated parameters similar to the model behaviour for the desired watershed are set. This setting could be done manually or by using automated methods. The steps of the modelling process are shown graphically in Fig. 8.10.

8.6 Examples of Recognized Models

Models are a simple depiction of real systems and are usually utilized to forecast the response of a modelled system affected by different administration situations. In this section, the features of some popular hydrological models are discussed (Fares and El-Kadi 2008).

8.6.1 Non-source Pollution Erosion Comparison Tool (N-SPECT)

The N-SPECT model was developed by the National Oceanic and Atmospheric Administration (NOAA) Coastal Services Center to inspect the relationships between soil properties, land cover, precipitation, and topography to evaluate the temporal and spatial patterns of a non-point source, pollution, surface water runoff, and erosion (Yang and Li 2011; Chen et al. 2021). This model has been developed as a tool for decision-making for coastal basins and uses the SCS curve number (CN) method to generate a network of CNs and runoff estimates according to the composition of a hydrological soil group hydrological conditions and land cover for each cell in a cell. When this model is used to simulate a single or annual event, soil erosion is calculated by applying the RUSLE or MUSCLE equations (Fares and El-Kadi 2008).

8.6.2 European Hydrological System (MIKE SHE)

The main MIKE SHE model was expanded in 1982 under the Système-Hydrologique Européen (SHE) system by three European organizations: the British Hydrological Institute, the French consulting firm SOGREAH and the Danish Hydraulic Institute (DHI). This distribution model is integrated, based on physics and simulation of water quality and hydrological processes at the basin scale (Fares and El-Kadi 2008). It can simulate groundwater and surface water with accuracy equal to that of models that focus separately on groundwater and surface water. The MIKE SHE modelling system affects most of the major "hydrological processes," including ground and canopy tracking after snowmelt, evapotranspiration, precipitation, saturated groundwater flow, canal flow, ground flow, and surface flow. In addition, this model simulates the main components of water quality. A network shows the model inputs, parameters, and outputs for each network. MIKE SHE uses Christensen and Jensen methods to calculate actual evapotranspiration. This includes the Muskingum-Cunge and Muskingum techniques for simple channel routing (Fares and El-Kadi 2008).

8.6.3 Soil Water Assessment Tool (SWAT)

The SWAT model is a field-based, distributed, conceptual, and continuous-scale simulation model employed as a soil and water assessment tool and field-scale model (Fares and El-Kadi 2008; Suryavanshi et al. 2017). Several SWAT versions are available, most recently SWAT2.0.3, released on April 7, 2021. For possible evapotranspiration computations, users have choices between the Hargreaves, Priestley-Taylor, and Penman–Monteith techniques. Event-based erosion due to rainfall-runoff is modelled utilizing the MUSLE.[36]

Annual Non-agricultural Source Model (AnnAGNPS)

The United States Department of Agriculture developed the AnnAGNPS (USDA ARS and NRCS), a continuously distributed simulation model employed primarily for field evaluation. This model extends the abilities of its previous AGNPS, which is a "single event" model. Runoff is computed using the CN equation but is corrected if there is a shallow soil layer (Fares and El-Kadi 2008). The number of curves is adjusted daily based on tillage operations, soil moisture, crop stage and soil moisture. Actual evapotranspiration in this model is a function of potential evapotranspiration and is computed by employing soil and water content and the Penman–Monteith equation. Soil water erosion is estimated using the modified RUSLE for field-scale runs in AnnAGNPS (Fares and El-Kadi 2008).

This model uses the GIS interface to process input and output data. However, the selection of appropriate lattice size was recognized as a major factor influencing sediment computation. The BCs before a rainfall-runoff occasion are computed by the model. In addition, long-term simulations can be performed by utilizing AnnAGNPS compared to the event-based AGNPS model (Fares and El-Kadi 2008).

8.6.4 *Chemicals, Runoff, Erosion for Agricultural Management Systems (CREAMS) and Groundwater Loading Effects of Agricultural Management Systems (GLEAMS)*

The CREAMS model was established by the USDA to help evaluate agricultural management measures to control pollution (Ahmad and Simonovic 2005; Singh and Woolhiser 2002). It is commonly used to calculate the best agricultural management methods (BMP) for pollution control. Nutrient loads and contaminants associated with daily sedimentation and erosion performance are estimated at the agricultural boundary (Fares and El-Kadi, 2008). Runoff estimation is done using the SCS CN (Curve Number) method. This model calculates daily peak flow, runoff volume, evapotranspiration, infiltration, seepage, and soil and water content. Sedimentation and erosion performance are also estimated (Fares and El-Kadi 2008).

Solute chemicals and average sediment concentrations are calculated for sediment, runoff and dense water. The GLEAMS model was established by combining an element for the vertical flux of pesticides in the root zone (Fares and El-Kadi 2008). This model is divided into three parts, namely erosion/sediment performance,

[36] Modified Global Soil Loss Equation.

pesticides, and hydrology. Surface runoff is approximated employing the SCS CN technique. Soils are divided into several layers of different thicknesses to route pesticides and water. The USDA Agricultural Research Service maintains both GLEAMS and CREAMS. The main restraint of the model is that it is a mass model. They are assuming that the whole basin is uniform in soil topography and land use is a completely unrealistic assumption (Fares and El-Kadi 2008).

8.6.5 Physical Runoff Prediction Model (TOPMODEL)

TOPMODEL is a physics-based continuously distributed simulation basin model. Boone and Kirkby (year–) advanced this model. This pattern predicts relative to soil and water saturation and drainage of the basin according to the time series information of topography, evapotranspiration and precipitation (Fares and El-Kadi 2008). This model is a set of conceptual tools that can be used to reproduce the hydrological behaviour of a watershed in a distributed or semi-distributed manner. In a model for estimating ET, the Penman–Monteith method is used. Runoff is calculated in relation to the additional infiltration mechanism. Therefore, the Green-Ampt model applies the Beven exponential function. The model presumes that the whole catchment is homogeneous, which can be unrealistic and usable for smaller watersheds. This model is subtle to parameters, such as soil transmittance at saturation, soil hydraulic conductivity, channel routing speed and root zone storage capacity in large water basins. The calibrated values are identified with the size of the grid applied in the DTA.[37] The grid and time step size are also shown to affect TOPMODEL simulation (Fares and El-Kadi 2008).

8.6.6 Hydrologically Distributed Soil Vegetation Model (DHSVM)

DHSVM is a distributed, physics-based, and continuous simulated basin model of the field. This model was established by (Wigmosta et al. 1994) at the University of Washington, Seattle. DHSVM calculates the effect of topography on groundwater, surface water, and soil moisture in complex topography. These include evaporation, melting and accumulation of snow, interception of transpiration, and runoff production through the mechanism of over-saturation. The evapotranspiration of canopy is represented by the Penman–Monteith bilayer formula that combines local pure sunlight, soil characteristics, surface meteorology, moisture status and pore resistance, and species-dependent leaf area index. "Snow accumulation" and "erosion" are modelled using an "energy balance" view that contains vegetation effects

[37] Digital terrain analysis.

and local topography. "Saturated subsurface flow" is modelled by employing a quasi-three-dimensional routing scheme (Fares and El-Kadi 2008).

8.6.7 *Water Erosion Prediction Project Model (WEPP)*

USDA-ARS has developed the WEPP erosion model, which is an incessant computer simulation that precipitates soil loss deposition, sediment shedding from dense streams in small canals, and ground flow on hill slopes and sediment in confined lands. In addition, it contains a climate element that uses a random generator to produce daily weather data, a hydrology component based on revised solutions of the wave equations and the Green-Ampt diffusion equation (Fares and El-Kadi 2008). A plant growth factor is a daily water balance. The residual decomposition component based on the EPIC model calculates the impact of erosion efficiency and an irrigation element. The WEPP model calculates the temporal and spatial distribution of sediment and soil loss and provides an explicit estimate of the time and place of erosion on the hillside and watershed to select appropriate conservation measures to better control sediment and soil loss (Fares and El-Kadi 2008). Supposedly, it could forecast how rain will interact with soil in an area through a full year or a specific rainstorm. WEPP applies soil and water balance constituents according to the corresponding simulator term for rural water resources (SWRRB). The infiltration integration of the hill slope model follows the Green and Ampt equation modified by Mein and Larson (year), and the pond time is calculated for irregular rainfall. Infiltration and water balance elements of the hill slope model are based on the SWRRB water balance component, with some modifications to improve soil evaporation parameters. WEEP uses two methods to calculate peak discharge: the wave motion model approximation and a semi-analytical solution of the wave motion model. The primary technique is employed when WEPP is run in a single event mode, whereas the second is utilized when WEPP is run in a continuous simulation model. WEPP needs a large data set that might bound the application of the model in basins where relatively little data is accessible. Several model parameters must be calibrated to avoid problems in model identification and physical interpretation of model parameters. This model does not limit the concept of ditch erosion and erosion; it limits its use to a variety of soil and field conditions. WEPP does not simulate phosphorus and nitrate losses from agricultural landscapes (Fares and El-Kadi 2008).

8.6.8 *Hydrological Simulation Program—FORTRAN (HSPF)*

The EPA-Athens Laboratory developed HSPF model is a comprehensive, conceptual and continuous simulation of the catchment that models the quantitative and

Table 8.5 Examples of recognized models

Model name	Author (year)	Remarks
Stanford Watershed Model(SWM)/Hydrologic simulation package-Forton IV (HSPF)	Crawford and Linsley (1966a, b)	steady-state, Continuous dynamic event, simulator of hydrologic and hydraulic water processes
Catchment model (cm)	Dawdy and O'Donnell (1965)	Lumped, event-based runoff model
Tennessee valley authority (TVA) model	Authority (1972)	Lumped, event-based runoff model
Utah State University (USU) Mode	Andrews et al. (1978)	Process-oriented, event/continuous stream flow-based Mode
Purdue model	Huggins and Monke (1970)	Process-oriented, physically based, event runoff model
Antidecent precipitation index (API) model	Sittner et al. (1969)	Lumped, river flow forecast model
Hydrologic engineering center hydrologic Modeling System (HES-HMS)	Feldman (1981, US and US (1981)	Physically-based, semi-distributed, event-based, runoff model
Streamflow synthesis and Reservoir regulation (SSARR) Model	U.S Army corps of Engineers (1987), Rockwood, (1982), Singh (1995a, b)	Continuous, Lumped, streamflow simulation model
National Weather Service—river forecast system (NWS-RFS)	Burnash et al. (1973), Burnash (1975)	Continuous, Lumped, river forecast system
University of British Columbia (UBC) Model	Quick (1995), Quick and Pipes (1977)	Process-oriented lumped parameter, continuous simulation model
Tank model	Sugawara (1995, 1974)	Process-oriented, semi-distributed or lumped continuous simulation model
Runoff routing model (RORB)	Laurenson et al. (n.d.), Laurenson (1964), Laurenson and Mein (1990)	Lumped, event-based runoff simulation mode
Agriculture runoff model	Donigian (1977)	Process-oriented lumped runoff simulation model
Strom water Management Model (SWMM)	Huber (1995), Huber and Dickinson (1988), Metcalf (1971)	Dynamic, Continous, Event or a steady-state simulator of Hydrologic and hydraulic, and water quality processes
Areal Non-point source Watershed Environment response simulation (ANSWERS)	Beasley et al. (1977), Bouraoui et al. (2002)	Continuous, Event-Based, lumped parameter runoff and sediment yield simulation model

(continued)

Table 8.5 (continued)

Model name	Author (year)	Remarks
National Hydrology research institute (NHRI) Model	Vandenberg (1989)	Lumped parameter, Physically based, continuous hydrologic simulation model
Technical report-20 (TR-20) Model	Soil Conservation Service (SCS), Computer model for project formulation hydrology, Tech (1965)	Event-based process-oriented lumped hydrograph Model
U.S geological survey (USGS) model	Dawdy et al. (1972)	Lumped parameter, event-based runoff simulation model
Physically-based runoff Production Model (OPMODEL)	Beven et al. (1995), Beven and Kirkby (1979)	Physically-based, distributed, continuous hydrologic simulation model
Generalized river modelling package-system hydroloque European (MIKE_SHE)	Refsgaard et al. (1995)	distributed, Physically-based, continuous hydrologic and hydraulic simulation model
Arno (Arno river model)	Todini (1996, 1988)	Continuous, Semi distributed rainfall-runoff simulation model
Waterloo flood system (WATERFLOOD)	Kouwen et al. (1993), Kouwen and Fathi-Moghadam (2000)	semi-distributed, Process-oriented, continuous flow simulation model
Topographic Kinematic Approximation and Integration (TOPIKAP) Mode	Todini (1995)	physically based, Distributed, continuous rainfall-runoff simulation model
Soil-vegetation-Atmosphere transfer (SVAT Model)	Ma et al. (1999), MA and CHENG (1996)	Macroscale, lumped parameter, streamflow simulation model
Systeme hydrologique euroepan transport (SHETRAN)	Ewen et al. (2000)	Distributed, Physically-based water quantity and quality simulation model
Daily conceptual rainfall-runoff model (hydrology) Monash Model	Chiew et al. (n.d.)	Conceptual, Lumped, continuous simulation model
Soil Water Assessment Tool (SWAT)	Arnold et al. (1998)	Distributed, conceptual, continuous simulation model
Distributed hydrological model (HYDROTEL)	Fortin et al. (2001)	Physically-based, distributed, continuous hydrologic simulation model

qualitative water processes that occur in a catchment, containing the movement of pollutants and sediment transport. HSPF is an analytical tool used in the planning of water resources systems. The program can apply probabilistic analysis in the fields of water quality management and hydrology through continuous simulation ability (Fares and El-Kadi 2008). HSPF is classified as a mass model but could produce spatial diversity by dividing the catchment into parts of the land hydrologically. Also, using several meteorological input data and basin parameters, this model can simulate runoff for each basement independently. Sediment loads, runoff flow rates, toxic chemicals, pesticides, nutrients and other concentrations of water quality can be predicted. HSPF can simulate steady-state behaviour or dynamic and continuous hydraulic/hydrological water quality processes in a basin area. In addition, this model may be used for urban watersheds through its impenetrable ground module. Several parameter requirements increase the physical significance of the model and the issue associated with parameter selection. HSPF depends on the calibration for parameterization. It is important to note that the HSPF is not able to explicitly describe the effects of agricultural management on water or runoff quality (Fares and El-Kadi 2008).

Some examples of other known models are named in Table 8.5 (Singh and Woolhiser 2002).

References

Abbott MB, Bathurst JC, Cunge JA, O'Connell PE, Rasmussen J (1986) An introduction to the European hydrological system—systeme hydrologique Europeen, "SHE", 1: history and philosophy of a physically-based, distributed modelling system. J Hydrol 87:45–59. https://doi.org/10.1016/0022-1694(86)90114-9

Abdelnour A, Stieglitz M, Pan F, Mckane R (2011) Catchment hydrological responses to forest harvest amount and spatial pattern. Water Resour Res 47:9521. https://doi.org/10.1029/2010WR010165

Abrahart RJ, Anctil F, Coulibaly P, Dawson CW, Mount NJ, See LM, Shamseldin AY, Solomatine DP, Toth E, Wilby RL (2012a) Two decades of anarchy? Emerging themes and outstanding challenges for neural network river forecasting. Prog Phys Geogr 36:480–513. https://doi.org/10.1177/0309133312444943

Abrahart RJ, Mount NJ, Shamseldin AY (2012b) Neuroemulation: definition and key benefits for water resources research. Hydrol Sci J-J Des Sci Hydrol 57:407–423. https://doi.org/10.1080/02626667.2012.658401

Ahmad S, Simonovic SP (2005) An artificial neural network model for generating hydrograph from hydro-meteorological parameters. J Hydrol 315:236–251. https://doi.org/10.1016/j.jhydrol.2005.03.032

Andrews WH, Riley JP, Masteller MB (1978) Mathematical modeling of a sociological and hydrologic decision system

Arnold JG, Srinivasan R, Muttiah RS, Williams JR (1998) Large area hydrologic modeling and assessment part I: model development. J Am Water Resour Assoc 34:73–89. https://doi.org/10.1111/j.1752-1688.1998.tb05961.x

Authority TV (1972) A continuous daily streamflow model: upper bear creek. Exp Proj Res Pap No 8

Bathurst JC, O'connell P (1992) Future of distributed modelling: the systeme hydrologique Europeen (WWW document). Hydrol Process. https://scholar.google.com/scholar_lookup?title=Futureofdistributedparametermodeling%3ATheSystemeHydrologiqueEuropeen&journal=HydrologicalProcesses&volume=6&pages=265-277&publication_year=1992&author=Bathurst%2CJ.C.&author=O%27Connell%2CP.E. Last Accessed 14 May 21

Bear J (2012) Hydraulics of groundwater. Courier Corporation

Beasley DB, Monke EJ, Huggins LF (1977) ANSWERS: a model for watershed planning, purdue agricultural experiment station. J Pap

Becker A, Nemec J (1987) Macroscale hydrologic models in support to climate research. Influ Clim Chang Clim Var Hydrol Regime Water Resour 431–445

Becker A, Serban P (1990) World meteorological organization hydrological models for water-resources system design and operation

Bergström S, Harlin J, Lindström G (1992) Spillway design floods in Sweden: I. New guidelines. Hydrol Sci J 37:505–519

Beven K (2013) So how much of your error is epistemic? Lessons from Japan and Italy. Hydrol Process 27:1677–1680. https://doi.org/10.1002/hyp.9648

Beven K (2011) Distributed models and uncertainty in flood risk management

Beven K (2012a) Rainfall-runoff modelling the primer, 2nd ed

Beven K (2012b) Down to basics: runoff processes and the modelling process. In: Rainfall-runoff modelling, pp. 1–23. Wiley & Sons, Ltd. https://doi.org/10.1002/9781119951001.ch1

Beven K, Germann P (1984) A distribution function model of channelling flow in soils based on kinematic wave theory. In: *Proceedings of the ISSS Symposium on Water and Solute Movement in Heavy Clay Soils*

Beven K, Lamb R, Quinn P, Romanowicz R, Freer J (1995) TOPMODEL. Comput Model Watershed Hydrol 627–668

Beven K, Westerberg I (2011) On red herrings and real herrings: disinformation and information in hydrological inference. Hydrol Process. https://doi.org/10.1002/hyp.7963

Beven KJ, Kirkby MJ (1979) A physically based, variable contributing area model of basin hydrology. Hydrol Sci Bull 24:43–69. https://doi.org/10.1080/02626667909491834

Beven KJ, Kirkby MJ, Kirkby AJ (1979) A physically based, variable contributing area model of basin hydrology/Un modèle à base physique de zone d'appel variable de l'hydrologie du bassin versant) A physically based, variable contributing area model of basin hydrology/Un modèle à base physique de zone d'appel variable de l'hydrologie du bassin versant A physically based, variable contributing area model of basin hydrology. Hydrol Sci J 24:43–69. https://doi.org/10.1080/02626667909491834

Biswas AK (1970) History of hydrology. North-Holland Publishing Company

Bouadi T, Cordier MO, Moreau P, Quiniou R, Salmon-Monviola J, Gascuel-Odoux C (2017) A data warehouse to explore multidimensional simulated data from a spatially distributed agro-hydrological model to improve catchment nitrogen management. Environ Model Softw 97:229–242. https://doi.org/10.1016/j.envsoft.2017.07.019

Bouraoui F, Braud I, Dillaha TA (2002) ANSWERS: a nonpoint source pollution model for water, sediment and nutrient losses. Math Model Small Watershed Hydrol Appl 833–882

Bowden GJ, Maier HR, Dandy GC (2012) Real-time deployment of artificial neural network forecasting models: understanding the range of applicability. Water Resour Res 48:10549. https://doi.org/10.1029/2012WR011984

Bras RL, Rodriguez-Iturbe I (1993) Random functions and hydrology. Courier Corporation

Brunner P, Simmons CT (2012) HydroGeoSphere: a fully integrated, physically based hydrological model. Ground Water. https://doi.org/10.1111/j.1745-6584.2011.00882.x

Burnash R, Ferral R, McGuire R (1973) A generalized streamflow simulation system: conceptual modeling for digital computers

Burnash RJC (1975) Chapter 10: the NWS river forecast system catchment modeling. Comput Model Watershed Hydrol

Carter T, Parry M, Harasawa H, Nishioka S (1994) IPCC technical guidelines for assessing climate change impacts and adaptations, in: part of the IPCC special report to the first session of the conference of the parties to the UN framework convention on climate change, intergovernmental panel on climate change. Department of Geography, University College London, UK and Center for Global

Chapuis RP (2012) Predicting the saturated hydraulic conductivity of soils: a review. Bull Eng Geol Environ 71:401–434. https://doi.org/10.1007/s10064-012-0418-7

Chen G, Hua W, Fang X, Wang C, Li X (2021) Distributed-framework basin modeling system: II. Hydrol Model Syst. Water 13:744

Chiew F, Hydrology TM-J of (1994) undefined, n.d. Application of the daily rainfall-runoff model MODHYDROLOG to 28 Australian catchments. Elsevier

Chow VT, Maidment DR, Mays LW (1988) Applied hydrology. MacGraw-Hill. Inc., New York

Clark MP, Kavetski D, Fenicia F (2011a) Pursuing the method of multiple working hypotheses for hydrological modeling. Water Resour Res 47:9301. https://doi.org/10.1029/2010WR009827

Clarke RT (1988) Stochastic processes for water scientists: development and applications. Wiley & Sons Ltd.

Clarke RT (1973) A review of some mathematical models used in hydrology, with observations on their calibration and use. J Hydrol 19:1–20. https://doi.org/10.1016/0022-1694(73)90089-9

Corzo GA, Solomatine DP, Hidayat H, De Wit M, Werner M, Uhlenbrook S, Price RK (2009) Combining semi-distributed process-based and data-driven models in flow simulation: a case study of the Meuse river basin. Hydrol Earth Syst Sci 13:1619–1634. https://doi.org/10.5194/hess-13-1619-2009

Crawford N, Linsley R (1966a) Digital simulation in hydrology'stanford watershed model 4

Crawford NH, Linsley RK (1966b) Digital simulation in hydrology' stanford watershed model 4

Darcy H (1856) Les fontaines publiques de la ville de Dijon: exposition et application. Victor Dalmont

Dawdy D, Lichty R, Bergmann J (1972) A rainfall-runoff simulation model for estimation of flood peaks for small drainage basins

Dawdy DR, O'Donnell T (1965) Mathematical models of catchment behavior. J Hydraul Div 91:123–137

Dawson CW, Mount NJ, Abrahart RJ, Louis J (2014) Sensitivity analysis for comparison, validation and physical legitimacy of neural network-based hydrological models. J. Hydroinformatics 16:1–18. https://doi.org/10.2166/hydro.2013.222

Dawson CW, Wilby RL (2001) Hydrological modelling using artificial neural networks. Prog Phys Geogr 25:80–108. https://doi.org/10.1177/030913330102500104

Devia GK, Ganasri BP, Dwarakish GS (2015) A review on hydrological models. Aquat Procedia 4:1001–1007. https://doi.org/10.1016/j.aqpro.2015.02.126

Donigian A (1977) Agricultural runoff management (ARM) model version II: refinement and testing

Duan Q, Gupta HV, Sorooshian S, Rousseau AN, Turcotte R (2004) Calibration of watershed models. American Geophysical Union

Ewen J, Parkin G, O'Connell PE (2000) SHETRAN: distributed river basin flow and transport modeling system. J Hydrol Eng 5:250–258. https://doi.org/10.1061/(asce)1084-0699(2000)5:3(250)

Fares A, El-Kadi AI (2008) Coastal watershed management. WIT Press

Feldman AD (1981) HEC models for water resources system simulation: theory and experience. In: Advances in hydroscience, pp. 297–423. Elsevier.

Fenicia F, Kavetski D, Savenije HHG (2011) Elements of a flexible approach for conceptual hydrological modeling: 1. Motivation and theoretical development. Water Resour Res 47. https://doi.org/10.1029/2010WR010174

Flury M, Flühler H, Jury WA, Leuenberger J (1994) Susceptibility of soils to preferential flow of water: a field study. Water Resour Res 30:1945–1954. https://doi.org/10.1029/94WR00871

Fortin J-P, Turcotte R, Massicotte S, Moussa R, Fitzback J, Villeneuve J-P (2001) Distributed watershed model compatible with remote sensing and GIS data. I: description of model. J Hydrol Eng 6:91–99. https://doi.org/10.1061/(asce)1084-0699(2001)6:2(91)

Gelhar LW (1986) Stochastic subsurface hydrology from theory to applications. Water Resour Res 22:135S-145S

Gleick PH (1986) Methods for evaluating the regional hydrologic impacts of global climatic changes. J Hydrol 88:97–116. https://doi.org/10.1016/0022-1694(86)90199-X

Gong W, Gupta HV, Yang D, Sricharan K, Hero AO (2013) Estimating epistemic and aleatory uncertainties during hydrologic modeling: an information theoretic approach. Water Resour Res 49:2253–2273. https://doi.org/10.1002/wrcr.20161

Granata F, Gargano R, de Marinis G (2016) Support vector regression for rainfall-runoffmodeling in urban drainage: a comparison with the EPA's storm water management model. Water (Switz) 8:1–13. https://doi.org/10.3390/w8030069

Gupta HV, Nearing GS (2014) Debates—the future of hydrological sciences: a (common) path forward? Using models and data to learn: a systems theoretic perspective on the future of hydrological science. Water Resour Res. https://doi.org/10.1002/2013WR015096

Holton JR, Staley DO (1973) An introduction to dynamic meteorology. Am J Phys 41:752–754. https://doi.org/10.1119/1.1987371

Horton RE (1945) Erosional development of streams and their drainage basins; hydrophysical approach to quantitative morphology. Geol Soc Am Bull

Horton RE (1933) The rôle of infiltration in the hydrologic cycle. Eos Trans Am Geophys Union 14:446–460. https://doi.org/10.1029/TR014i001p00446

Hsu P-C, Nguyen C (1995) Theoretical investigation of a class of new planar transmission lines from microwave and millimeter-wave integrated circuits. In: Millimeter and submillimeter waves vol. II, pp. 159–161. SPIE. https://doi.org/10.1117/12.224222

Huber WC (1995) Chapter 22: EPA storm water management model SWMM, Computer models of watershed hydrology. In: Singh VP (ed)

Huber, W.C., Dickinson, R.E., 1988. Storm water management model user's manual, version 4. Rep. No. EPA/600/3–88/001a, US Environmental Protection Agency, Athens, Ga.

Huggins LF, Monke EJ (1970) Mathematical simulation of hydrologic events of ungaged watersheds

Jackson B, Mcintyre N, Pechlivanidis IG, Jackson BM, Mcintyre NR, Wheater HS (2011) Catchment scale hydrological modelling: a review of model types, calibration approaches and uncertainty analysis methods in the context of recent developments in technology and app catchment scale hydrological modelling: a review of model types, calibration approaches and uncertainty analysis methods in the context of recent developments in technology and applications. Artic GlobalNEST Int J

Johnson F, Sharma A, Singh V (2017) Handbook of applied hydrology

Kampf SK, Burges SJ (2007) A framework for classifying and comparing distributed hillslope and catchment hydrologic models. Wiley Online Libr. 43:5423. https://doi.org/10.1029/2006WR005370

Kavetski D, Clark MP (2010) Ancient numerical daemons of conceptual hydrological modeling: 2. Impact of time stepping schemes on model analysis and prediction. Water Resour Res 46. https://doi.org/10.1029/2009WR008896

Kavetski D, Fenicia F (2011) Elements of a flexible approach for conceptual hydrological modeling: 2. Application and experimental insights of a flexible approach for conceptual hydrological modeling: 2. Application and experimental insights. Elem Water Resour Res 47:11511. https://doi.org/10.1029/2011WR010748

Kite GW (1995) The SLURP model. Comput Model Watershed Hydrol 521–562

Knapp HV, Durgunoglu A, Ortel TW (1991) Illinois state water survey a review of rainfall-runoff modeling for stormwater management, ideals.illinois.edu

Knudsen J, Thomsen A, Refsgaard JC (1986) WATBAL a semi-distributed, physically based hydrological modelling system. Nord Hydrol 17:347–362. https://doi.org/10.2166/nh.1986.0026

Kokkonen T, Koivusalo H, Karvonen T (2001) A semi-distributed approach to rainfall-runoff modelling—a case study in a snow affected catchment. Environ Model Softw 16:481–493. https://doi.org/10.1016/S1364-8152(01)00028-7

Kouwen N, Fathi-Moghadam M (2000) Friction factors for coniferous trees along rivers. J Hydraul Eng 126:732–740. https://doi.org/10.1061/(asce)0733-9429(2000)126:10(732)

Kouwen N, Soulis ED, Pietroniro A, Donald J, Harrington RA (1993) Grouped response units for distributed hydrologic modeling. J Water Resour Plan Manag 119:289–305. https://doi.org/10.1061/(asce)0733-9496(1993)119:3(289)

Kumar P, Folk M, Markus M, Alameda J (2005) Hydroinformatics: data integrative approaches in computation, analysis, and modeling

Lall U (2014) Debates-The future of hydrological sciences: a (common) path forward? One water. One world. Many climes. Many Souls Wiley Online Libr 50:5335–5341. https://doi.org/10.1002/2014WR015402

Laurenson E, Watershed RM-C models of (1995) undefined, n.d. RORB: hydrograph synthesis by runoff routing. cabdirect.org

Laurenson EM (1964) A catchment storage model for runoff routing. J Hydrol 2:141–163

Laurenson EM, Mein RG (1990) RORB-version 4, runoff routing program: user manual. Monash University Department of Civil Engineering

Liang X (1994) A two-layer variable infiltration capacity land surface representation for general circulation models. Ph. D. Thesis.

Liang S, Li X, Xie X (2013) Land surface observation, modeling and data assimilation

Ma X, Cheng W (1996) A modeling of hydrological processes in a large low plain area including lakes and ponds. J Jpn Soc Hydrol Water Resour 9:320–329

Ma X, Fukushima Y, Hashimoto E, Hiyama E (1999) Application of a simple SVAT model in a mountain catchment under temperate humid climate. J Jpn Soc Hydrol Water Resour

May RJ, Dandy GC, Maier HR, Nixon JB (n.d.) Application of partial mutual information variable selection to ANN forecasting of water quality in water distribution systems. Elsevier. https://doi.org/10.1016/j.envsoft.2008.03.008

Mays LW, Tung Y-K (2002) Hydrosystems engineering and management. Water Resour Publ

McKane R, Brookes A, Djang K, Stieglitz M, Abdelnour A, Pan F, Halama J, Pettus P, Phillips D (2014) Visualizing ecosystem land management assessments (VELMA) v. 2.0: User manual and technical documentation

McMillan H, Jackson B, Clark M, Kavetski D, Woods R (2011) Rainfall uncertainty in hydrological modelling: an evaluation of multiplicative error models. J Hydrol 400:83–94. https://doi.org/10.1016/j.jhydrol.2011.01.026

Metcalf E (1971) University of Florida and Water Resources Engineers, Inc, Storm Water Management Model, Volume I-Final Report. EPA Report 11024 DOC 07/71 (NTIS PB-203289). Environ. Prot. Agency Washington, DC, USA 352.

Mitchell TM (1997) Machine learning

Moradkhani H, Sorooshian S (2008) General review of rainfall-runoff modeling: model calibration, data assimilation, and uncertainty analysis. In: Hydrological modelling and the water cycle, pp 1–24. Springer, Berlin. https://doi.org/10.1007/978-3-540-77843-1_1

Mount NJ, Maier HR, Toth E, Elshorbagy A, Solomatine D, Chang FJ, Abrahart RJ (2016) Data-driven modelling approaches for socio-hydrology: opportunities and challenges within the Panta Rhei science plan. Hydrol Sci J 61:1192–1208. https://doi.org/10.1080/02626667.2016.1159683

Mulvany TJ (1850) On the use of self-registering rain and flood gauges in making observations of the rainfall and flood discharges in a given catchment. Trans Minutes Proceeding Inst Civ Eng Irel Sess 1

Ng HYF, Marsalek J (1992) Sensitivity of streamflow simulation to changes in climatic inputs. Nord Hydrol 23:257–272. https://doi.org/10.2166/nh.1992.0018

Pechlivanidis IG, Jackson BM, Mcintyre NR, Wheater HS (2011) Catchment scale hydrological modelling: a review of model types, calibration approaches and uncertainty analysis methods in

the context of recent developments in technology and applications. Glob Nest J. https://doi.org/10.30955/gnj.000778
Perrin C, Michel C, Andréassian V (2003) Improvement of a parsimonious model for streamflow simulation. J Hydrol 279:275–289. https://doi.org/10.1016/S0022-1694(03)00225-7
Phillips NA (1956) The general circulation of the atmosphere: a numerical experiment. QJR Meteorol Soc 82:123–164. https://doi.org/10.1002/qj.49708235202
Pinder GF, Gray WG (2013) Finite element simulation in surface and subsurface hydrology. Elsevier
Poston T, Stewart I (2014) Catastrophe theory and its applications
Pyle D, Cerra DD, Kaufmann M (1999) Data preparation for data mining
Qu T, Kim Y, Yaremchuk M, … TT-J of 2004, undefined (n.d.) Can Luzon Strait transport play a role in conveying the impact of ENSO to the South China Sea? journals.ametsoc.org
Quick MC (1995) The UBC watershed model. Comput Model Watershed Hydrol 233–280
Quick MC, Pipes A (1977) UBC watershed model/Le modèle du bassin versant UCB. Hydrol Sci J 22:153–161
Refsgaard JC, Knudsen J (1996) Operational validation and intercomparison of different types of hydrological models. Water Resour Res 32:2189–2202. https://doi.org/10.1029/96WR00896
Refsgaard JC, Storm B, Mike SHE (1995) Computer models of watershed hydrology. Water Resour Publ 809–846
Remson I, Hornberger GM, Molz FJ (1971) Numerical methods in subsurface hydrology
Rinsema JG (2014) Comparison of rainfall runoff models for the Florentine Catchment
Rockwood D (1982) Theory and practice of the SSARR model as related to analyzing and forecasting the response of hydrologic systems. Appl Model Catchment Hydrol
Ross TJ (2010) Fuzzy logic with engineering applications, 3rd ed. Wiley & Sons. https://doi.org/10.1002/9781119994374
Schulze RE (1997) Impacts of global climate change in a hydrologically vulnerable region: challenges to South African hydrologists. Prog Phys Geogr 21:113–136
Sen Z (2009) Fuzzy logic and hydrological modeling
Singh V (1995a) Computer models of watershed hydrology
Singh VP (1995b) Computer models of watershed hydrology. Water Resources Publications
Singh V, Frevert D (2002a) Mathematical models of small watershed hydrology and applications
Singh V, Frevert D (2002b) Mathematical models of large watershed hydrology
Singh VP, Frevert DK (2002c) Mathematical models of large watershed hydrology. Water Resources Publication
Singh VP, Frevert DK (2002d) Mathematical models of small watershed hydrology. Water Resources Publication
Singh VP, Woolhiser DA (2002) Mathematical modeling of watershed hydrology. J Hydrol Eng 7:270–292. https://doi.org/10.1061/(ASCE)1084-0699(2002)7:4(270)
Singh VP, Frevert DK (2006) Watershed models. CRC Press, Fort Lauderdale, Florida
Singh VP, Jain SK, Tyagi A (2007) Risk and reliability analysis: a handbook for civil and environmental engineers. Am Soc Civ Eng
Singh V (2013) Entropy theory and its application in environmental and water engineering
Singh V (2014) Entropy theory in hydraulic engineering: an introduction
Singh VP (2015) Entropy theory in hydraulic engineering: An introduction. Am Soc Civ Eng (ASCE). https://doi.org/10.1061/9780784412725
Singh VP (2017) Handbook of applied hydrology, 2nd ed. McGraw-Hill Education, New York
Singh VP (2018) Review of fluoride in drinking water: status, issues, and solutions by AK Gupta and S. Ayoob
Singh VP, Zhang L (2018) Copula–entropy theory for multivariate stochastic modeling in water engineering. Geosci Lett. https://doi.org/10.1186/s40562-018-0105-z
Sitterson J, Knightes C, Parmar R, Wolfe K, Avant B, Muche M (2018) An overview of rainfall-runoff model types
Sittner WT, Schauss CE, Monro JC (1969) Continuous hydrograph synthesis with an API-type hydrologic model. Water Resour Res 5:1007–1022

Sivakumar B, Berndtsson R (2010) Advances in data-based approaches for hydrologic modeling and forecasting
Soil Conservation Service (SCS), Computer model for project formulation hydrology, Tech (1965). USDA, Washington
Solomatine D, See LM, Abrahart RJ (2009) Data-driven modelling: concepts, approaches and experiences. Pract Hydroinformatics 17–30
Solomatine DP (2005) Data-driven modeling and computational intelligence methods in hydrology. In: Encyclopedia of hydrological Sciences. Wiley & Sons, Ltd. https://doi.org/10.1002/0470848944.hsa021
Solomatine DP, Wagener T (2011) Hydrological modeling. In: Treatise on water science, pp 435–457. Elsevier. https://doi.org/10.1016/B978-0-444-53199-5.00044-0
Srinivasulu S, Jain A (2008) Rainfall-runoff modelling: integrating available data and modern techniques. In: Practical hydroinformatics, pp 59–70. Springer Berlin. https://doi.org/10.1007/978-3-540-79881-1_5
Sugawara M (1995) Tank model. Comput Model Watershed Hydrol
Sugawara M (1974) Tank model and its application to Bird Creek, Wollombi Brook, Bikin River, Kitsu River, Sanaga River and Nam Mune. Res Notes Natl Res Cent Disaster Prev 11:1–64
Suryavanshi S, Pandey A, Chaube UC (2017) Hydrological simulation of the Betwa River basin (India) using the SWAT model. Hydrol Sci J 62:960–978. https://doi.org/10.1080/02626667.2016.1271420
Tayfur G (2014) Soft computing in water resources engineering: artificial neural networks, fuzzy logic and genetic algorithms
Todini E (2007) Hydrological catchment modelling: past, present and future. Hydrol Earth Syst Sci
Todini E (1996) The ARNO rainfall-runoff model. J Hydrol 175:339–382. https://doi.org/10.1016/S0022-1694(96)80016-3
Todini E (1995) New trends in modelling soil processes from hillslope to GCM scales. In: The role of water and the hydrological cycle in global change, pp 317–347. Springer, Berlin. https://doi.org/10.1007/978-3-642-79830-6_11
Todini E (1988) Rainfall-runoff modelling—past, present and future. J Hydrol 100:341–352. https://doi.org/10.1016/0022-1694(88)90191-6
Tung Y-K, Yen B-C (2005) Hydrosystems engineering uncertainty analysis. Asce
Uhlenbrook S, Roser S, Tilch N (2004) Hydrological process representation at the meso-scale: the potential of a distributed, conceptual catchment model. J Hydrol 278–296. https://doi.org/10.1016/j.jhydrol.2003.12.038. (Elsevier)
USDA (1986) Urban hydrology for small watersheds, second Ed. ed. United States Department of Agriculture
US HEC, US WRSC (1981) HEC-1 flood hydrograph package: users manual. US Army Corps of Engineers, Water Resources Support Center, Hydrologic
Vandenberg A (1989) A physical model of vertical integration, drain discharge, and surface runoff for layered soils. National Hydrology Research Institute
Vaze J, Jordan P, Beecham R, Frost A, Summerell G (2011) Guidelines for rainfall-runoff modelling: towards best practice model application
Wheater H, Jakeman A, Beven K (1993) Progress and directions in rainfall-runoff modelling
Wigley TML, Jones PD, Briffa KR, Smith G (1990) Obtaining sub-grid-scale information from coarse-resolution general circulation model output. J Geophys Res 95:1943–1953. https://doi.org/10.1029/JD095iD02p01943
Wigmosta MS, Vail LW, Lettenmaier DP (1994) A distributed hydrology-vegetation model for complex terrain. Water Resour Res 30:1665–1679
Wilderer PA (2010) Treatise on water science. Newnes
Woolhiser DA, Smith RE, Goodrich DC (1990) KINEROS: a kinematic runoff and erosion model: documentation and user manual
Xu CY (1999) Climate change and hydrologic models: a review of existing gaps and recent research developments. Water Resour Manag 13:369–382. https://doi.org/10.1023/A:1008190900459

Xu J, Lv C, Zhang M, Yao L, Zeng Z (2015) Equilibrium strategy-based optimization method for the coal-water conflict: a perspective from China. J Environ Manage 160:312–323. https://doi.org/10.1016/j.jenvman.2015.06.036

Yang B, Li M-H (2011) Assessing planning approaches by watershed streamflow modeling: case study of The Woodlands; Texas. Landsc Urban Plan 99:9–22

YoosefDoost A, Asghari H, Abunuri R, Sadegh Sadeghian M (2018a) Comparison of CGCM3, CSIRO MK3 and HADCM3 models in estimating the effects of climate change on temperature and precipitation in Taleghan Basin. Am J Environ Prot 6:28–34. https://doi.org/10.12691/env-6-1-5

YoosefDoost A, YoosefDoost I, Asghari H, Sadegh Sadeghian M (2018b) Comparison of HadCM3, CSIRO Mk3 and GFDL CM2.1 in prediction the climate change in Taleghan River Basin. Am J Civ Eng Archit 6:93–100. https://doi.org/10.12691/ajcea-6-3-1

Zeeman EC (1976) Catastrophe theory. JSTOR

Zhang G, Xiang X, Tang H (2011) Time series prediction of chimney foundation settlement by neural networks. Int J Geomech 11:154–158. https://doi.org/10.1061/(ASCE)GM.1943-5622.0000029

Chapter 9
Mitigation and Adaptation Measures

Sahar Baghban, Omid Bozorg-Haddad, Ronny Berndtsson, Mike Hobbins, and Nadhir Al-Ansari

9.1 Introduction

Climate change has directly and indirectly impacted natural and human systems across the globe in recent decades. The effects of climate change indicate the sensitivity of natural and human systems (Pachauri et al. 2014). Global warming is one of the most apparent features of climate change and will have significant effects on the environment. These include increasing carbon dioxide emissions, increasing atmospheric temperature, increasing intensity and frequency of storms, occurrence of droughts and extreme floods, changing precipitation patterns, and the interactions among these factors (Pathak et al. 2012). The average global temperature change for the period 2016–2035 compared to 1986–2005 under the four Representative Concentration Pathways (RCP) scenarios and is likely to be in the range of +0.3–+0.7 °C (Pachauri et al. 2014). According to the Third Assessment Report of the Intergovernmental Panel on Climate Change (IPCC), the average global temperature in the twentieth century has risen by 0.6 ± 0.2 °C.

S. Baghban · O. Bozorg-Haddad (✉)
Department of Irrigation and Reclamation Engineering, Faculty of Agriculture Engineering and Technology, College of Agriculture and Natural Resources, University of Tehran, 3158777871 Karaj, Iran
e-mail: obhaddad@ut.ac.ir

R. Berndtsson
Division of Water Resources Engineering and Centre for Advanced Middle Eastern Studies, Lund University, Lund, Sweden

M. Hobbins
Cooperative Institute for Research in Environmental Sciences (CIRES), University of Colorado-Boulder and National Oceanic and Atmospheric Administration-Physical Sciences Laboratory, Boulder, CO, USA

N. Al-Ansari
Department of Civil, Environmental and Natural Resources Engineering, Luleå University of Technology, Luleå, Sweden

O. Bozorg-Haddad (ed.), *Climate Change in Sustainable Water Resources Management*, Springer Water, https://doi.org/10.1007/978-981-19-1898-8_9

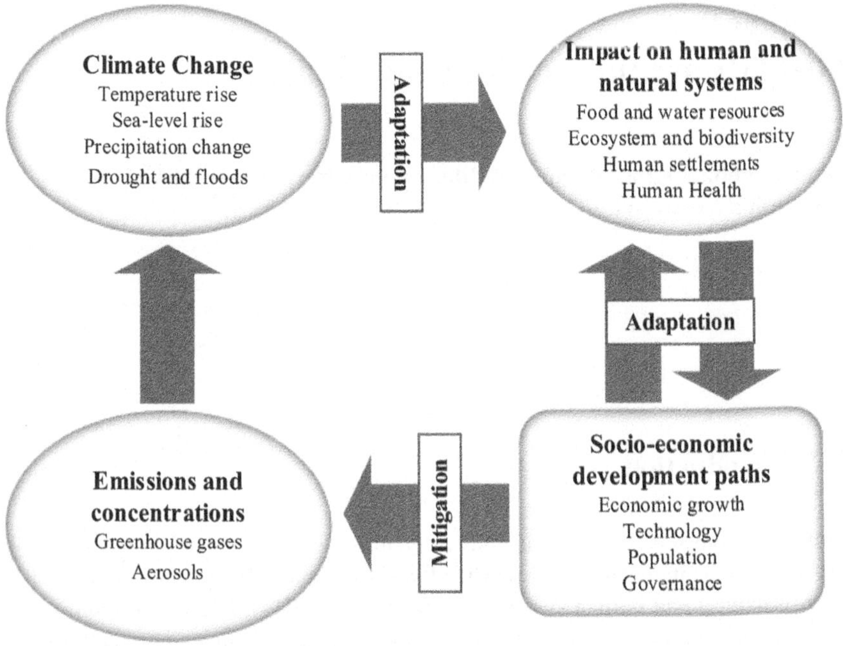

Fig. 9.1 A general schematic of the concepts of adaptation and mitigation and their interactions (IPCC 2005)

The effects of this recent climate change are evident in the planet's life systems; they threaten human health, human security, food security, economic activities, and natural resources in various aspects of the environment, and endanger the world's infrastructure, especially in arid and semi-arid regions. Scientists and policymakers have suggested the two concepts of mitigation (reducing greenhouse gas emissions and increasing carbon sequestration) and adaptation (reducing the vulnerability of humans and ecosystems and increasing resilience) to minimize the risks of climate change to human life and the planet (Duguma et al. 2014).

Governments have begun to develop adaptation plans and policies to integrate climate change considerations into broader development plans (Pachauri et al. 2014). Despite the increasing challenges of adaptation, developed countries continue to focus on mitigation, while adaptation is the most crucial factor for vulnerable and developing countries (Duguma et al. 2014) (Fig. 9.1).

9.1.1 Risks of Climate Change

Climate change poses risks to natural and human systems. These risks are unevenly distributed and are generally higher for disadvantaged communities, particularly in less-developed countries. Increasing global warming poses pervasive, irreversible risks to individuals, species, and ecosystems that threaten human well-being, food security, economic development, and ultimately human society's security (Pachauri et al. 2014). Since the extent, timing, and direction of changes in and the risks of climate change vary from region to region, prior to planning to adapt to climate change, it is essential to have sufficient information about regional changes in all meteorological drivers, including precipitation, temperature, humidity, windspeed, radiation. The risks of climate change that occur across sectors and regions can be defined as follows (Pachauri et al. 2014):

- Threats to human health and well-being and disruption of livelihoods are caused by severe storm surges, rising sea levels, coastal flooding, floods in urban areas, and periods of extreme heat.
- Systematic hazards from severe climate disasters lead to the breakdown of infrastructure networks (including water distribution network and waste water collection networks) and essential services.
- Risks of water, food, and energy insecurity, and loss of livelihood and income are elevated, especially for rural people and the poor.
- The risk of ecosystem destruction is elevated, as are threats to biodiversity, ecosystem goods, functions, and services.

It should be noted that, in most regions, water is the most influential factor in transmitting climate change risks to society. The water sector requires a greater expenditure than the agricultural, forestry, fishing, and human health sectors (Biswas and Tortajada 2016). While in developing countries, infrastructure and coastal areas require higher costs (Field 2014). By limiting the speed and extent of climate change, the overall risks of climate change impacts can be reduced in the future. Although adaptation can significantly reduce the risks of climate change effects, the greater the magnitude and severity of these effects, the more significant and severe the constraints on adaptation (Pachauri et al. 2014).

9.1.2 Vulnerability to Climate Change

Vulnerability is a measure of a system's ability to cope with the adverse effects of change—in this case climate change. In other words, vulnerability is the degree to which a system is exposed to climate change and unable to overcome it (Field 2014). In general, assessing vulnerability includes the following steps: (i) identifying the region; (ii) identifying the effects of climate change on water resources in terms of quantity and quality (considering demand and supply); (iii) determining the leading indicators of vulnerability; and (iv) prioritizing vulnerable water resources according to qualitative indicators (Ashofteh 2013, 2014). The characteristics and severity of the effects of climate change and the risks of severe disasters are related to climate change and depend on the degree of vulnerability and exposure (to hazards) of humans and natural systems (Pachauri et al. 2014). The primary effects of climate change on areas vulnerable to climate change are as follows (Biswas and Tortajada 2016):

- Africa: Water resources, food security, vector-borne diseases, and water
- Europe: Floods, water restrictions, and extreme heat events
- Asia: Floods, heat-related deaths, water, and food shortages due to drought
- Australia and Asia: Coral reefs, flood damage, coastal infrastructure, and lowland ecosystems.
- North America: Natural fires, human deaths due to heat, and floods in urban areas
- Central and South America: access to water, floods and landslides, and foodborne illnesses
- Polar regions: freshwater and land ecosystems, health and well-being, and northern communities
- Small islands: livelihoods and coastal areas
- Ocean: Fishing, coral reefs, and rising sea levels.

Vulnerabilities to climate change and adaptation and mitigation capacities are all strongly influenced by livelihood, lifestyle, behavior, and culture. Further, the social acceptability and effectiveness of climate adaptation and mitigation policies are influenced by their level of dependence on rational changes in lifestyle or behaviors (Pachauri et al. 2014). The concept of vulnerability to climate change has evolved over the years to encompass non-climatic concepts, including efforts to increase adaptation capacity and reduce projected damage (Laukkonen et al. 2009).

Rapid population growth, high levels of poverty, high reliance on rain-fed agriculture, high rates of environmental degradation, chronic food insecurity, and recurrent natural droughts all increase climate change vulnerability in communities (Kidanu et al. 2009). Areas that are already economically and geographically vulnerable will be more affected by climate change and water and food insecurity (Asian Development Bank 2016). Assessments have shown that the most vulnerable sectors are agriculture, water, and human health.

9.1.3 The Concept of Mitigation and Adaptation

Adaptation seeks to reduce vulnerability and increase coping capacity; it is defined by the IPCC as the "process of real or expected climate adjustment and its effects" on human or natural systems; In contrast, mitigation focuses on climate change drivers and is defined by the IPCC as "human intervention to reduce resources or increase greenhouse gas (GHGs) emissions" (Pachauri et al. 2014). These definitions may suggest that mitigation, and adaptations are independent concepts, but arguments have shown that the goal of both is to minimize the adverse consequences of climate change (Ayers and Huq 2009). Adaptation regulates natural and human systems in response to the effects of real and expected stimuli in future climatic conditions that can mitigate harm or take advantage of beneficial opportunities. In comparison, mitigation modifies human intervention to counteract the reduction of resources or increase greenhouse gas emissions (Klein et al. 2007).

Mitigation focuses on reducing long-term risks, while adaptation reduces the current risks due to historical dissemination or the failure to achieve the mitigation goals (Swart and Raes 2007). Adaptation's fundamental concept reduces the vulnerability of natural and human-made environments and exposes ecosystems and human habitats to the risks associated with climate change and risk minimization (Biswas and Tortajada 2016). Climate change adaptation involves a set of management actions or operations to reduce the negative consequences and reap the benefits of climate change (Pittock and Jones 2000). Adaptation leads to the well-being and health of the people, the security of resources and assets, and maintenance of the present and future ecosystems. The first step towards achieving adaptation to climate change is to reduce vulnerability and reduce exposure to climate diversity (i.e., increase reliability) (Pachauri et al. 2014). Adaptation options exist in all regions but depending upon their role in reducing vulnerability, there are various potentials and approaches. Mitigation options are also available for each region. However, adaptation is most useful when an integrated approach is used to reduce energy consumption and greenhouse gas emissions, ultimately leading to the supply of low-carbon energy, and reducing emissions (Pachauri et al. 2014). Adaptability assessment involves identifying and determining options based on accessibility, revenue, cost, efficiency, and convenience of implementing adaptation measures (Field 2014).

Mitigation and adaptation are not two independent phenomena. Because mitigation can reduce adaptation needs in the long run, more adaptation can reduce mitigation costs by increasing and improving coping capacity (Endo et al. 2017; Swart and Raes 2007). While both adaptation and mitigation can help reduce climate change risks to nature and society (Parry et al. 2005), mitigation is necessary for successful adaptation, and adaptation is necessary for successful mitigation.

9.1.4 The Relationship Between Mitigation and Adaptation

From the above definitions, it follows that mitigation will reduce the effects (both positive and negative) of climate change and thus reduce adaptation challenges. Depending on local and regional interests and priorities, both adaptation and mitigation measures applied at local and regional levels can address global dilemmas. Mitigation is recognized as a global responsibility that, in addition to providing global benefits in the long run, can also generate short-term benefits on a local scale, such as reducing air pollution. Conversely, adaptation is mainly local and regional in scale, and the benefits can be apparent in the short term (Sharifi 2020). However, it can also provide medium- and long-term global benefits (Landauer et al. 2019). Although mitigation requires enough greenhouse gas emitters to reduce their emissions, it offers many benefits globally (as well as locally and regionally). On the other hand, although adaptation is much more effective at the regional scale, adaptation measures are often implemented locally. From an economic point of view, the benefits of mitigation are more significant globally, while adaptation leads to more local benefits (Matocha et al. 2012). Emission reduction scenarios resulting from different mitigation measures can be compared and cost-effectiveness determined (Moomaw et al. 2001). In comparison, the effects and benefits of adaptation measures at the local or regional level, are assessed differently depending on the social, economic, and political context in which they occur (Klein et al., 2007). Comparisons of adaptation and mitigation vary at different scales but in order to gain the most benefits of adaptation and mitigation measures < it is generally better to do mitigation on a global scale and adaptation on a local scale (Wilbanks et al. 2003).

Analyzing the interrelationships between adaptation and mitigation is essential, because creating synergies between them can make the relevant measures cost-effective and more attractive to potential investors and other decision-makers. Synergy is an interaction of adaptation and mitigation so that their combined effect is greater than the sum of their impacts if implemented separately. However, their integration does not guarantee optimal resource use and disaster minimization (Parry et al. 2007). There are many technical and financial reasons to merge the benefits of adaptation and mitigation: Matocha et al. (2012) note that a successful adaptation is a prerequisite for successful mitigation. For example, when climate models show that, should the future climate become warmer and drier and raise the potential for wildfires, improving fire management (adaptation measures) will reduce the risks and damages and mitigate climate change through agroforestry (Schroth et al. 2009). Therefore, in climate change, adaptation measures are an essential input to the sustainable design of mitigation measures, especially in cases where climate change will affect the sustainability of future mitigation efforts (Matocha et al. 2012). As both mitigation and adaptation measures require information on climate scenarios, land use, community practices, there is an opportunity for joint planning of mitigation and adaptation projects. Therefore, in many cases, similar measures create the benefits of mitigation and adaptation, which can be integrated with little or no extra cost (Matocha et al. 2012) (Fig. 9.2).

Both adaptation and mitigation are based on reducing the adverse effects of climate change. Therefore, it is evident that they are closely related to each other from different dimensions. In general, the risks of climate change and vulnerability to it can be represented by the following simple equations (Swart and Raes 2007):

$$\text{Risk} = \text{Probability} * \text{Climate hazard} * \text{Vulnerability}$$

$$\text{Vulnerability} = \text{Exposure} * \text{Sensitivity}/\text{Adaptive capacity}$$

Adaptation can be achieved by reducing exposure to climate change, reducing sensitivity, or increasing adaptive capacity. According to these formulas, further mitigation requires less adaptation, and vice versa (Swart and Raise 2007). Adaptation and mitigation are thought to be, at worst, reciprocal, or at best, parallel strategies (Dang et al. 2003). Therefore, before examining the potential benefits of integrating adaptation and mitigation measures, it is best to consider not only their differences but also their similarities. Table 9.1 summarizes the types of responses and the 'common wisdom' as to the different dominant foci of the two approaches. Table 9.1 also shows

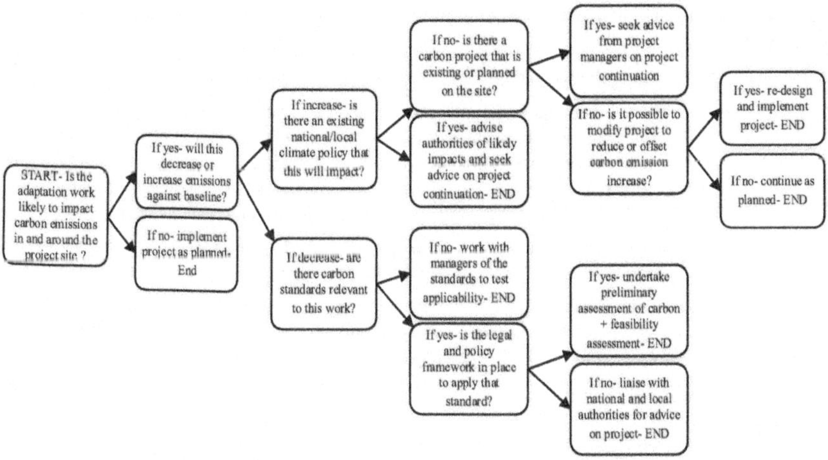

Fig. 9.2 Decision tree for the inclusion of climate change mitigation into the design of a climate change adaptation project (Matocha et al. 2012)

some exceptions to this 'common wisdom,' and notes some similarities. Pertinent discerned differences relate to spatial and temporal scales (Swart and Raise 2007).

9.2 Mitigation and Adaptation in Water Resources Management

According to the IPCC Fifth Assessment Report (2013), water scarcity has become a significant challenge for most parts around the world. Factors such as changing rainfall patterns, melting ice, and rising ocean levels will, in many areas, alter hydrologic systems and affect water resources (in both quantity and quality). As global warming increases in the twenty-first century, both water shortages and large river floods will increase. Since 2000, climate change has reduced the renewable resources of surface water and groundwater in most arid subtropical regions, resulting in intensified competition for water resources (Pachauri et al. 2014). Fundamentally, to deal with climate change, global measures need to be adopted to increase adaptation and mitigation investments to increase societies' flexibility. Countries in the Asia–Pacific region are the most vulnerable to climate change, as they face more challenges due to the multiple pressures of rapid urbanization, industrialization, and economic development (Asian Development Bank 2016).

As mentioned, variation in climatic conditions, including changes in temperature, precipitation, and evapotranspiration, will affect a region's water resources and are expected to increase the frequency of floods and droughts and reduce river flow, especially at short time scales (Asian Development Bank 2016). Interactions between

Table 9.1 Definitions, differences and similarities between mitigation and adaptation (Swart and Raes 2007)

Definition		Mitigation		Adaptation	
		Anthropogenic intervention to reduce the sources or enhance the sinks of greenhouse gasses		Adjustment in natural or human systems in response to actual or expected climatic stimuli or their effects, which moderates harm or exploit beneficial opportunities	
	Issue	Dominant focus	Examples of exception	Dominant focus	Examples of exception
Differences	Causes/effect	Primarily addresses causes	Urban design with low energy requirements and low vulnerability	Primarily addresses consequences	Drought-resistant biofuels can address both vulnerability and emissions
	Spatial scale	Primary objective is to avoid global changes	Co-benefits for short-term local air pollution, energy security, jobs	Primary objective local damage avoidance	Adaptation of temperate farmers may have global consequences
	Time scale	Long-term benefit from avoided climate change	Co-benefits for short-term, local air pollution, energy security, jobs	Often main driver short-term benefit due to reducing vulnerability to current climate	Preparing for long-term impacts
	Beneficiaries	Chiefly benefits others (altruistic)	Co-benefits for mitigating actors, local air pollution, energy security, jobs	Chiefly benefits those who implement it (egoistic)	Some adaptation may have broader benefits
	Incentives	Incentives usually needed	No-regrets options (e.g., energy efficiency)	Often incentives not needed	Anticipatory actions without immediate benefits may need incentives
	Urgency	Lower political urgency/ legitimacy	Short-term co-benefits, local air pollution, energy security, jobs enhance urgency	Higher political urgency/ legitimacy	Proactive adaptation with high costs and uncertain effect can have low urgency
Similarities	Goal	Aiming at reduction of climate change risks			
	Benefits	Having ancillary benefits that may be as significant as the climate-related benefits			
	Drivers	Driven by availability penetration of new technology and societal ability to change			

an increase in temperature, sediment, pollutants from heavy rainfall, pollutants from drought, and disruption of treatment facilities during floods endanger the quality of water resources, including drinking water (Pachari et al. 2014). A 2 °C global warming will cause a drastic reduction in water resources for approximately 15% of the world's population, consequently increasing the number of people living under absolute water scarcity to 40% compared with population growth alone (Schewe et al. 2014).

Cities worldwide are involved in climate change measures, including reducing greenhouse gas emissions (National Research Council 2009). Nevertheless, Rothausen and Conway (2011) concluded that the relationship between energy consumption and greenhouse gas emissions was not adequately considered in water resources management. Considering greenhouse gas emissions is of particular importance in water resources management for the following reasons: the high sensitivity and substantial impact of the water sector on climate change (Rothausen and Conway 2011); and the robust relationship between water and energy (Stokes and Horvath 2009), which refers to the Water-Energy Nexus (meaning the use of energy in the process of hydroelectric production, and the role of energy in water supply systems and wastewater treatment). Many climate change adaptation measures, such as desalination or pumping, require high energy consumption; if this comes from non-renewable energy sources such as fossil fuels, greenhouse gas emissions inevitably increase. There is a concern that the adaptation of water resources to climate change is likely to lead to increased greenhouse gas emissions; as greenhouse gas emissions drive climate change, climate change leads to a shortage of water resources and threatens them, water shortages are seen as a stimulus for the adaptation of water resources. As there may then be a conflict between the adaptation of water resources in response to climate change and the mitigation of its effects, these must be balanced by applying multiple goals. Therefore, reducing greenhouse gas emissions and focusing on adaptation and mitigation in response to climate change in water resources management is essential.

Rising temperatures can directly affect water resource+s access through changes in river flow regimes, especially in snow-covered or glacial basins (Biswas and Tortajada 2016); so can changes in radiation, e.g., in the Colorado River Basin, which provides water for 40 million people in the western US (Milly and Dunne 2020). The increasing exploitation of resources, including water resources, to provide food and water to an increased population leads to deeper environmental crises and the destruction of biodiversity and ecosystems. The change in land use from forest and pasture to agricultural land to harvest more limited water resources ultimately puts additional pressure on those water resources.

9.2.1 Water Resources Planning

Water resources planning presents projections of the future balance between supply and demand in each water resource zone (RZ), together with plans for ensuring an "adequate" balance between supply and demand (defined in terms of standards of service) (Charlton and Arnell 2011). Water resources planning is a broad political, economic, sociological, scientific, and technological movement (Ballentine and Stakhiv 1993). Water resource managers and planners generally make two types of decisions: firstly, with respect to new investments; and secondly, with respect to the operation and maintenance of existing systems. A third category that falls between the two—investments that improve the operational capacity of existing systems—has recently become more critical in the United States (Ballentine and Stakhiv 1993). For better informed decisions, it is essential to have sufficient information about the supply and demand of future water resources as affected by climate change. In terms of supply, water resources managers intend to estimate the variability of small-scale water resources within the basin at a specific time in the future by climatic parameters. In terms of demand, it is necessary to know the impact of climate change on water resources (Ballentine and Stakhiv 1993).

Currently, the high financial and environmental costs of water projects and the limited opportunities to build dams and reservoirs to create new resources have shifted the focus of water resource managers and planners from construction to management of existing facilities and resources (Frederick et al. 2013). In climate change studies, to support water resources planning, simulation of the global hydroclimate by General Circulation Models (GCMs) is commonly used to define scenarios of future temperature and precipitation changes and other related hydrological variables (Salathé et al. 2007).

The process of planning for adaptation to climate change, like any other planning, involves identifying potential problems, developing options to address them, ranking them by criteria, and selecting the ones that should be implemented (National Research Council 2011). Because future greenhouse gas emissions, climate sensitivity, and climate change patterns in each region are not definitively predictable, we must adapt to various scenarios in the future since we cannot predict which one happens precisely. Therefore, monitoring climate change and its effects and identifying vulnerabilities are essential; this requires decision-makers to re-evaluate and update the planning process. In fact, adaptations must meet various goals and objectives in a wide range of climatic conditions (National Research Council 2011). Government agencies and other institutions would identify their vulnerabilities and assess the risks associated with the impacts of climate change. This information should be communicated to stakeholders and relevant decision-makers to increase their awareness of current and potential problems. Using a risk-management approach, adaptation options for managing the risks associated with climate impacts can be identified, evaluated, and implemented (National Research Council 2011). In summary, the planning process consists of the steps outlined in Fig. 9.3.

In general, the development of water resources planning criteria can be considered in a range of one-purpose to multi-purpose planning criteria (Frederick et al. 2013). Much mitigation and adaptation measures can help change or improve the effects of climate change, but no single option is sufficient. Mitigation and adaptation are complementary, meaning that a lack of mitigation makes achieving adaptation more challenging, and better adaptation can reduce the need for mitigation. It

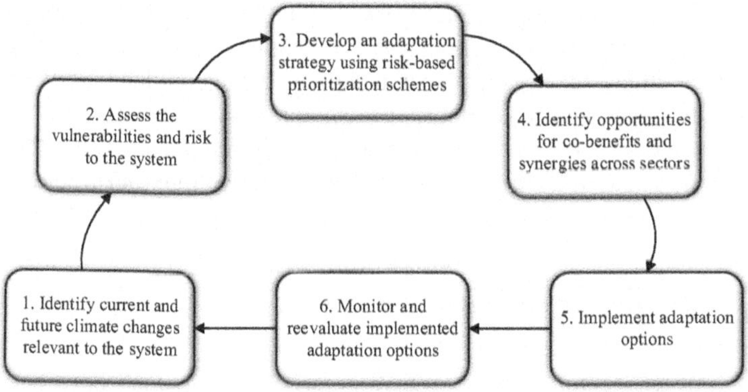

Fig. 9.3 The planning process (National Research Council 2011)

can be concluded that both mitigation adaptations at any time, and spatial scale can have many benefits. Thus, their integration in water resources planning has attracted significant attention from water resource managers (Berry et al. 2015; Ford et al. 2018).

Water resource managers should re-evaluate the performance of existing water resource systems in order to analyze changes in a wide range of climatic variables (Gohari et al. 2014). There are various measures to adapt to climate change, including technological advances, financing, innovations, or spatial measures at the institutional level (Biesbroek et al. 2009). Water resource management plans include determining additional water resources such as fog capturing and transportation technology, conservation measures, and demand reduction such as crop cultivation, release of irrigated land, and conversion to dryland agriculture (Ashofteh 2014). Possible adaptation policies in the hydroelectric sector include scaling up and scheduling planned projects and limiting downstream and irrigation flows for downstream agriculture (Gashaw et al. 2014). Other sectors may adapt to climate change by reducing demand and increasing supply reliability through transboundary water transfers to protect plant and animal habitats or converting irrigation canals and waterways into piping. These latter measures help prevent water loss, especially from evaporation due to rising temperatures (Ashofteh 2014).

Adaptation to climate change in order to minimize damage or maximize a benefit implies the involvement, coordination, and cooperation of different actors and sectors (Ledda et al. 2020). Spatial planning plays an essential role in adaptation measures, especially in water and water resources planning (White and Howe 2003; Wiering and Immink 2006). 'Spatial planning' means "actions and interventions that are based on 'critical thinking about space and place' that involves not only legislative and regulatory frameworks for the development and use of land but also the institutional and social resources through which such frameworks are implemented, challenged and transformed" (Davoudi et al. 2009). Spatial planning has been identified as a critical mechanism through which climate change adaptation can be facilitated (Hurlimann and March 2012). In this regard, Biesbroek et al. (2009) considered spatial planning a comprehensive approach to shaping spatial evolution, understanding interrelationships, and the effects of long-term spatial actions; it can coordinate different social and economic goals and aspirations, for example, development within the areas of transportation systems, economy, housing, and development within the field of natural resources and water resources management, agriculture, and the environment.

9.2.2 Water Resources Infrastructure

Over the present century, dams, reservoirs, wells, and pumps have been traditional responses to hydrological diversity and predicting increased water demands to increase and maintain water resources for use in areas with the insufficient flow. However, high costs and limited opportunities have hindered this water resource planning approach to the water supply and limited the ability to adapt to climate change through traditional infrastructure investments (Frederick et al. 2013). An essential question in this regard is whether the design of structures and infrastructure has paid enough attention to the hydrological changes caused by climate change or the increased risk of flooding of river streams (Biswas and Tortajada 2016).

To be consistent, given infrastructure suitable to a region's hydroclimate, technologies such as flood protection and drainage systems are generally useful in urban areas but at high cost. Advances in technology and the improvement of new, resilient crops are useful adaptations. When mitigation is constrained, land management, construction optimization, and integrated water resources management should be considered as adaptation options (Biswas and Tortajada 2016). The solution to improve operation and transmission efficiency includes optimizing water systems' operation to maximize productivity. It is vital to maintain and improve existing infrastructure for regional and local transmission. Improving operating efficiency can indirectly reduce greenhouse gas emissions by reducing system losses. However, many water transfers have high energy costs (Ashofteh 2014).

The economic and social effects of adaptation to climate change vary widely regarding between regions (Frederick et al. 2013). Individuals and communities are adapted to past changes and seek to moderate and reduce the side effects of future climate change and create positive changes (Goklany 1995). In the field of design and implementation in the water sector, adaptations include revision of water resources infrastructure design criteria to optimize flexibility, strength, abundance, and diversity of water resources and improve reservoir management, reduce demand through leak control, implement water conservation and processing programs, treatment and reuse water (Biswas and Tortajada 2016).

9.3 Water Consumption Management

While population growth leads to the growth of incomes, services, and amenities, water demand reduction solutions can be considered a practical approach to water efficiency measures in the agricultural, industrial, and domestic sectors. Ultimately, conserving resources can reduce greenhouse gas emissions. Changes in rainfall, temperature, and carbon dioxide emissions can affect water demand and supply. For example, demand control using pricing water and water use, economic incentives to reduce water consumption, and improved water use efficiency are reasonable climate change adaptation measures (Biswas and Tortajada 2016). In general, consumption management includes a set of policies that make better use of available water resources, protects the environment and natural ecosystems, and ensures people's well-being.

9.3.1 *Agriculture Sector*

Both water and agriculture security face significant challenges in the face of climate change as both are highly sensitive and vulnerable to climate parameter changes. It is predicted that the global average temperature will increase by 1.4–4.8 °C, which will result in a significant decrease in freshwater resources and agricultural production by the end of the twenty-first century (Misra 2014). South Asian agriculture is highly vulnerable to climate change, with rising temperatures and changing rainfall patterns posing a severe threat to the region's food security (Ahmed and Suphachalasai 2014). Agricultural production will potentially be significantly reduced due to a lack of water resources. Countries in Africa, the Middle East, and Asia have close economic ties to natural resource and climate-dependent sectors such as forestry, agriculture, water, and fisheries (Misra 2014). Farmers consider changes in crop patterns, changes in planting and harvesting times, changes in temperature and precipitation patterns, and precipitation accumulation as signs of climate change (Frederick et al. 2013).

The development of new agricultural practices and new technologies will strengthen farmers' ability to adapt to climate change, increase agricultural production, and reduce greenhouse gas emissions (Adenle et al. 2015). Increasing water efficiency and; consequently, demand management will be key factors in overcoming water scarcity. The most critical water use measures and adaptation to climate change in the agricultural sector are participatory irrigation management, fully efficient irrigation systems; crop stresses management, and improved crops (Biswas and Tortajada 2016). Unfortunately, the lack of mitigation and adaptation capacity is more pronounced in developing countries, which face low agricultural productivity levels, poor food security, and high vulnerability to climate change (Lybbert and Sumner 2012). Although agricultural technologies are directly related to the pace of climate change (Khan and Hanjra 2009), institutional and political innovations are necessary to the development of agricultural technologies and their provision to developing countries to enable them to adapt to climate change. In this regard, policies and institutions are critical for different scales (Adenle et al. 2015). It should be noted that any cost-effective policy framework for tackling climate change should accelerate beneficial Research and Development (R&D), innovation, and the dissemination of advanced technologies to reduce greenhouse gas emissions (Bosetti et al. 2009).

One effective climate change mitigation and adaptation strategy in the agricultural sector is the adoption of Integrated Soil Management Practices (ISMP) methods (Adenle et al. 2015). Prerequisites for using the ISMP strategy are conservation tillage, minimum fertilizer application, nutrient management, crop residue composition, fertilizer, mulch, compost, cover crops, and supplementary irrigation (Follett 2001; Lal 2009). In ISMP, integrated use of mineral and organic fertilizers is required, and the goal is to achieve sustainable agricultural systems (Adenle et al. 2015).

Another useful mitigation and adaptation tool is the use of biotechnology to increase the adaptability of agricultural products, increasing their productivity, and increasing food security, and thereby reducing the climate change vulnerability of human and natural systems and helping reduce greenhouse gases (Adenle 2011; Mtui 2011). Papageorgiou et al. (2009) evaluated the effects of disseminating three technologies, including mass incineration with energy recovery, biological treatment through bio-drying, and mechanical heat treatment that can be used to treat municipal solid waste for energy recovery.

9.3.2 Domestic Sector

Water consumption in the domestic sector, especially water consumption in gardens and lawn irrigation in green space, is somewhat sensitive to rainfall changes and temperature (Bardsley et al. 2013). Of course, the decrease in rainfall and temperature and its effect on the demand for water in the domestic consumption sector depends on the region, season, the amount of consumption, and how it is consumed, which is different in cities and suburbs. It is estimated that a 1% increase in temperature will increase water consumption in the home sector from 0.02 to 3.8%, and a 1% decrease in rainfall will increase water consumption from 0.02 to 0.31% (Frederick et al. 2013).

9.3.3 Industry Sector

The production, distribution, and use of most energy sources are not possible without the presence of water. Water is used for processing, washing, cooling in production facilities, producing thermal energy, producing bioenergy, and general extraction and processing of energy sources. Global warming can have significant consequences for water use in these processes, especially in cooling systems, and can reduce their efficiency and increase water demand (Frederick et al. 2013). Global warming has different effects on industrial and thermal energy consumption. For example, energy consumption in the summer for ventilation and cooling systems increases, and demand for space heating systems decreases in the winter. Therefore, changes in energy demand in terms of time and perhaps place to reduce water demand for cooling systems (Frederick et al. 2013). Reconsideration of the location of industries and agricultural products that need more water and their placement in water-rich areas and increasing water efficiency and productivity in the industrial sector will lead to adaptation to climate change (Biswas and Tortajada 2016). Replacing old fossil fuel sources with geothermal sources is not only a solution to reduce greenhouse gas emissions but can also meet some of the energy needs and be an essential factor in supporting agricultural production and reducing the risk of food security.

9.3.4 Recycled Water

Climate change, population growth, urbanization, industrialization, and rising incomes and living standards, and water and energy demand are global drivers that encourage the use of recycled water (Hanjra et al. 2012). However, the consequences of using recycled water as an climate change adaptation measure include declining productivity growth, declining investment in irrigation and agriculture worldwide, biodiversity loss, public health risks, soil salinity, land degradation, land uses problems, land covers change, freshwater scarcity (Molden 2013), and disruption of virtual water trade-offs (Wichelns 2011). Recycled water consists of 99% water, 1% suspended, colloidal, and dissolved solids; further, municipal wastewater contains organic matter and nutrients (Nitrogen, Phosphorus, Potassium); inorganic matter or dissolved minerals; toxic chemicals; and pathogens (Hanjra et al. 2012). To minimize the risks associated with the use of recycled water and maximize its benefits, the following measures should be taken (Asian Development Bank 2016):

- Improving wastewater discharge systems, both in the quantity of wastewater collected and in the proportion of wastewater collected and treated.
- ncreasing the coverage of sewerage systems to collect more wastewater and direct more flows to wastewater treatment centers.
- Increasing investment in wastewater treatment technologies to reduce organic, nutrient, and microbiological loads to suitable discharge levels.
- Increasing investment in wastewater recycling to increase the supply of water if necessary, and the management of biosolids to assess the treatment process and reduce the associated risks.

The risks of climate change in the water reuse industry can be divided into direct and indirect impacts. Direct effects are related to climatic factors on technological performance (including temperature and precipitation, rising sea levels, and extreme events). Indirect effects are mainly related to managerial and operational activities (including water consumption control and GHG emissions) (Vo et al. 2014) (Table 9.2).

Table 9.2 Typical impacts and projected adaptation strategies for wastewater treatment and reuse (NACWA, 2009)

Factors	Adaptation strategies
Changes in precipitation quantity and timing	• Reduce infiltration and inflow into sewers, flow diversion • Green infrastructure to manage site run-off • Rapid treatment
Changes in maximum temperature and other environmental variables	• Wetland treatment • Riparian restoration • Mechanical cooling • Evaporative cooling • Blending with cooler waste streams
Increased sea level Increased flood events	• Installing levees and sea walls around wastewater treatment plants (WWTP) and key infrastructure • Hardening sewer system
Collaboration between supplied water and wastewater	• A new distribution infrastructure

Today, the declining cost of seawater desalination renders it a useful climate change adaptation measure in low-water areas, even in developing countries. However, the significant energy demands of desalination do not align with climate change mitigation principles. Compared to seawater desalination, wastewater treatment—such as reusing treated wastewater for irrigation or injection into water reservoirs and aquifers—is less costly and requires less energy (Biswas and Tortajada 2016).

9.4 Extreme Events Management

Natural disasters introduce the interrelationships between extreme climatic events with human vulnerability and natural ecosystems to these events. Extreme events include changes in temperature and precipitation, including extreme heat or cold, heavy rainfall, drought, river floods; storms/storm surges/coastal floods, rising sea levels, and health and environmental hazards. Coastal cities are particularly vulnerable to these increased risks and generally, people living in cities are more affected by these events (Revi 2008). Climate change and the resulting global warming have increased the frequency of extreme events such as drought and floods (Koutsoyiannis 2011). The risk of extreme events is expected to increase sharply in the future. As global average temperatures increase, daily and seasonal maximum and minimum temperatures in most parts of the earth will increase, and heatwaves will likely occur at higher frequencies and for more extended periods. With increasing global average temperatures, severe rainfall events also become much more intense and frequent in

Table 9.3 Typology of climate extremes (Schneider et al. 2001)

Type	Description	Examples of events	Typical method
Simple extremes	Individual local weather variables exceeding critical level on a continuous scale	Heavy rainfall, high/low temperature, high wind speed	Frequency/return period, sequence and/or duration of variable exceeding a critical level
Complex extremes	Severe weather associated with particular climatic phenomena, often requiring a crucial combination of variables	Tropical cyclones, droughts, ice storms, ENSO-related events	Frequency/return period, magnitude, duration of variable(s) exceeding a critical level, the severity of impacts
Unique or singular phenomena	A plausible future climatic state with potentially extreme large-scale or global results	Collapse of major ice sheets, cessation of thermohaline circulation, significant circulation changes	Probability of occurrence and magnitude of impact

most parts of the world, particularly in the mid-latitudes and humid tropics (Pachauri et al. 2014) (Table 9.3).

A more complete assessment of the sensitivity of a water supply system to climate change would require testing under a combination of conditions of applied adaptation measures, infrastructure resilience, and the regional effects of climate change. Such an approach was demonstrated in Bardsley et al. (2013), who generated a set of scenarios for the Salt Lake City (USA) water supply system under various scenarios. These scenarios combined (i) emission‘s scenarios that led to increases in domestic consumption and municipal irrigation demands due to warming, and (ii) considered various points of failure of the water supply system in scenarios that removed components such as reservoirs. This approach requires significant expertise and data availability as it entails hydrologic modeling of contributing watersheds; simulation of runoff sensitivities to a range of changes in precipitation, temperature, and evaporative demand; applying scenarios of extreme drought and extremely low water supply; and assessing test cases for future water demand.

Managing disaster risks more successfully require systematic identification of various factors (Field et al. 2012):

- Risks are identified dynamically and play an essential role in integrating policies, strategies, environmental development measures, and management measures.
- Disaster-risk management laws are supported by specific rules applicable at various temporal and spatial scales and are integrated and complemented by other sector development and management laws.
- Disaster-risk management practices are coordinated in sectors and monitored by relevant political organizations.

- In national development plans, disaster-risk considerations can be used to protect vulnerable areas and groups by adopting climate change adaptation strategies.
- Risk is quantified and considered in national budget processes, and a range of measures are considered or implemented, including budgeting for relief costs, reserve funds, and other risk-related financing.
- Decisions are made using comprehensive information on observed changes in climate, available tools and guidelines, vulnerabilities, and methods to deal with disaster damage.
- Alert systems provide timely, relevant, and accurate hazard forecasts and are developed and implemented with public participation.
- Adaptation strategies include a combination of responses based on hard infrastructure and soft solutions (such as building individual and institutional capacity) and ecosystem-based responses, including conservation measures (such as forestry, river basins, coastal wetlands, and biodiversity).

9.4.1 Floods and Runoff

The issue of increasing flood risk due to climate change has been widely studied, with a particular focus on developing appropriate adaptation strategies, including increasing adaptation capacity, and developing advanced skills, methods, and technologies to combat climate change, assesses costs and benefits, and engages stakeholders (Zhou et al. 2012). One of the effects of global warming is on the amount and timing of runoff, in which case we can refer to premature runoff, which occurs due to the rapid melting of snow due to heating (especially for seasonal runoff). To this end, increasing reservoir capacity is an efficient flood management method, reducing flood damage and storing additional flow for periods when irrigation and other water demands are higher (Frederick et al. 2013). Whereas climate change has been broadly acknowledged as a global issue due to its foreseen impacts on urban water systems in terms of changes in water runoff and urban flooding, the use of cost-effective adaptation measures for urban drainage systems to reduce flood risk, especially in large cities with complex sewerage networks, is suggested as a solution (Zhou et al. 2012). For example, if the drainage system is designed for a specific return period only, it may sometimes have a severe capacity problem due to a climate-change-driven increase in water volume. More importantly, future drainage design must increase the frequency and intensity of precipitation to maintain an acceptable frequency of system overload (Zhou 2014).

In addition to changes in meteorological parameters, non-meteorological factors are also important drivers of extreme events, such as the change in river hydrography in some places, the conversion of agricultural lands into highly vulnerable industrial applications, reservoir exploitation policies, and management of water resources (Peterson et al. 2012). As an example, in 2011 heavy flooding occurred in northern Thailand during and after an unusual wet season (July to September), which resulted in flooding of large areas of the country, including Ayutthaya (the former capital) and neighborhoods of Bangkok (the current capital). Large-scale industrial estates were submerged for nearly two months under 2.5 m of water, resulting in significant economic damage (Peterson et al. 2012). Researchers concluded that, while rainfall in the catchment was not very unusual, other factors such as changes in hydrography and increased basin vulnerability drove the disaster's severity and scale (Peterson et al. 2012).

Flood management encompasses emergency planning, general planning (such as infrastructure improvement), and policy change. Flood management strategies can enhance a region's ability to adapt to other climate change effects, including environmental vulnerabilities and water quality (Ashofteh 2014). The most critical factors in flood management projects in adaptation measures are the area of protected habitats, the volume of natural flood reserves generated, the storm return period created for more accurate planning, and the expected damage resulting from a storm with a specific return period (Ashofteh 2014). In general, flood management can be viewed from two different perspectives: structural measures, such as embankments, dams, river canals, etc.; and non-structural measures, such as warning systems, flood preparedness programs, public awareness schemes, etc. (Biswas and Tortajada 2016).

9.4.2 Drought

Drought is caused by a lack of rainfall over a long period, usually in a season or more (Dubrovsky et al. 2009). Drought can also result from elevated evapotranspiration, or mismanagement of resources, or other factors. Drought is primarily a slow disaster with effects that often subside slowly over an extended period. Drought affects many parts of the world and is considered one of the costliest climatic hazards (Wilhite 2000). Drought can have adverse effects such as reduced crop yield, increased damage to building foundations due to ground shrinkage, reduced quantity and quality of water resources, and increased risk of forest fires (McCarthy et al. 2001). Today, there is more confidence that climate change increases the risk of drought in drought-prone areas (Alley et al. 2007). Therefore, drought has been considered by vulnerable countries, and various tools have been developed to monitor and predict it.

Definitions of drought are based on a region's specific needs and the affected sectors. Nevertheless, currently, drought is divided into four types (Heim 2002): (i) meteorological drought, resulting from significant rainfall deficits accumulating over an extended time; (ii) agricultural drought, resulting from insufficient soil moisture; (iii) hydrological drought, wherein water resources (supply) from surface water flows, and/or groundwater and reservoir water levels are depleted; (iv) socio-economic drought, which combines with the elements of the previous three types of drought but is also associated with the supply and demand of economic goods; and (v) ecological drought, wherein ecosystems are adversely affected. Definitions of drought can be categorized as conceptual or operational (Wilhite and Glantz 1985), with the former including topics such as a general description of the physical processes involved, such as lack of rainfall (meteorological drought), lack of soil moisture (agricultural drought), lack of water in lakes and streams (hydrological drought), and lack of water used in community resource management (Mishra and Singh 2010; Wilhite 2000), while the latter focuses on identifying the onset, duration, and end of drought periods, as well as their severity (Wilhite 2000).

Drought is a natural hazard that harms people and the environment and increases water demand (Mishra and Singh 2010). Human suffering, the destruction of natural resources, and lack of funding force organizations to take greater responsibility for reducing drought vulnerability and managing the impact of drought on vulnerable communities (including rural communities) (Keshavarz and Karami 2013). Doing so requires a comprehensive and adaptive system that responds to pressures and creates mechanisms to reduce negative consequences (Comfort and Kapucu 2006). Drought assessment is of particular importance for freshwater planning and management. This entails understanding a region's historical droughts and their effects (Mishra and Singh 2010). Understanding different drought concepts will help develop models to study the different properties of drought (Mishra and Singh 2010) and its forecasting (as an adaptation measure) (Fig. 9.4).

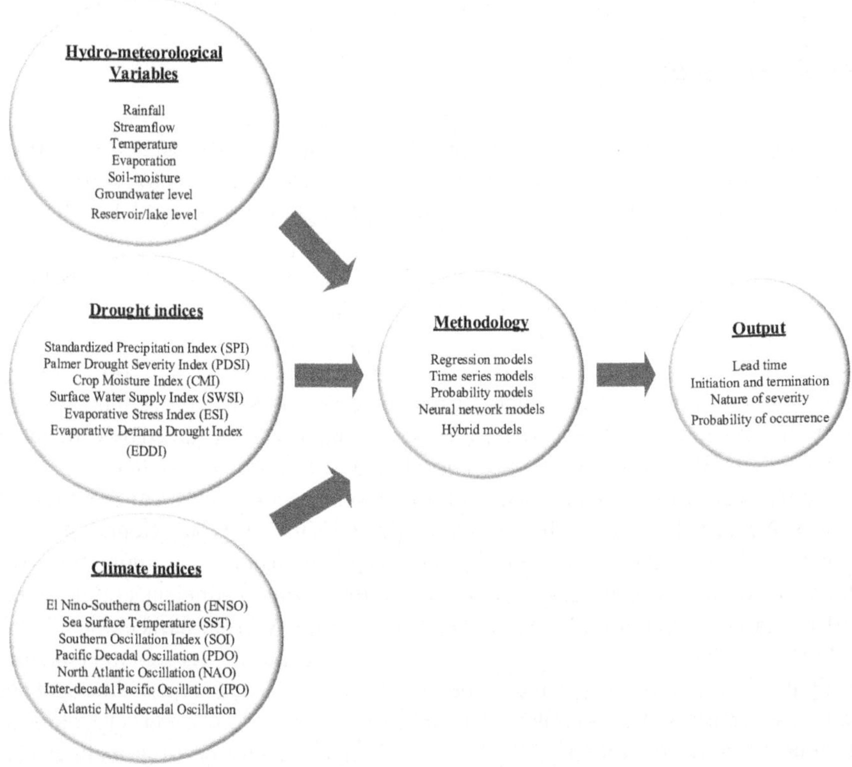

Fig. 9.4 Different components for drought forecasting (Mishra and Singh 2011)

9.5 Watershed Management

Practical approaches address basin-scale causes and effects of climate change in an integrated manner, i.e., coordination and coherence between mitigation and adaptation measures with other basin policies. In this regard, spatial planning at the basin scale to deal with climate change leads to many benefits, including (Biesbroek et al. 2009):

- Water management can contribute to both mitigation (e.g., hydropower) and adaptation (e.g., water retention);
- A holistic approach stimulates transdisciplinary research and policymaking;
- As a result, there is no dichotomy of mitigation and adaptation: planning applications will determine whether integrated responses between mitigation and adaptation are relevant;
- The river basin approach allows the assessment of possible synergies and trade-offs of mitigative and adaptive measures in a particular river basin in an integrated manner;
- Many key impacts of measures can be monitored at the river basin level, which allows evaluation of their effectiveness;
- Climate change as captured in a river basin approach forces local decision-makers to broaden the context of their planning and investment decisions;
- Adaptation and mitigation strategies can be integrated more smoothly into the (local) planning process.

In general, basin-scale adaptation strategies can be divided into three types: institutional strategies; socio-economic strategies; and natural strategies. It should be noted that adaptation strategies must first be feasible and, secondly, cost-effective (Ashofteh 2014).

9.5.1 Institutional Strategy

The goal of institutional strategies is to use programs such as optimization of water allocation among basin water users, optimization of water consumption, and optimization of cultivation pattern by optimally distributing the effects of climate change on different basin water users. This group of strategies aims to increase the understanding and cooperation of stakeholders in agriculture, industry, and other sectors towards adaptive measures against climate change and towards infrastructure adaptation strategies, including developing policies against climate change (Ashofteh 2014).

9.5.2 Socioeconomic Strategy

In socio-economic strategies, the goal is to create better water use conditions between different water users of the basin by implementing programs such as increasing water prices, insurance support, and training, and promoting the importance of savings, etc. (Ashofteh 2014). Although, these measures can reduce the adverse effects of climate change, future water demand will increase due to growth of population, agriculture and industry to such an extent that the ratio of income to cost of institutional and socio-economic strategies will be reduced. Hence, natural strategies are also considered (Ashofteh 2014).

9.5.3 Natural Strategy

The primary purpose of natural strategies is to extract and use more water resources. Natural strategies used across a basin can include cloud seeding, dam construction, feeding groundwater aquifers, water transfer between basins, wastewater treatment, and saline water desalination (Ashofteh 2014).

Adaptation and mitigation measures are divided into eight main sectors, as shown in Table 9.4. It should be noted that this table only includes measures that may involve trade-offs and maybe is not exhaustive. Based on Table, mitigation is the principal objective of 50% of the measures; 30% are mainly related to adaptation, and the rest are related to both adaptation and mitigation. As shown, most measures involve trade-offs for either mitigation or adaptation; nevertheless, some measures may lead to both adaptation and mitigation trade-offs (Sharifi 2020).

9.6 Mitigation, Adaptation, and Sustainable Development

As populations grow under global climate change, communities face challenges in achieving economic growth and sustainable development goals. Without proper management and investment, economic losses in agriculture, industry, water, energy, transportation, health, coastal and marine sectors, and tourism will increase (Asian Development Bank 2016). Communities need support to deal with climate change, extreme events, and subsequent environmental degradation. While the scarcity of freshwater resources due to global warming is a significant issue, a focus on planning for global climate change has not yet been attained. Integrating water resource planning into climate policy should be accompanied by strengthening and coordinating the relationship between improving water security, sustainable development, and reducing poverty (Asian Development Bank 2016).

Both mitigation and adaptation processes lead to the reduction and management of climate change. However, each of them has both advantages and disadvantages

Table 9.4 Adaptation and/or mitigation measures may include trade-offs, as discussed in the literature (Sharifi 2020)

Sector	Measure	Trade-off		Primary objective
		Mitigation	Adaptation	
Urban planning and land use	Compactness	✓	✓	Mitigation
	Land use mix		✓	Both
	Improved connectivity		✓	Both
	Risk zoning and relocating to avoid risk-prone areas	✓	✓	Adaptation
	Development along riverbanks to reduce exposure to heat		✓	Adaptation
	Cool roofs and pavements	✓		Both
Transport	Transit-oriented development	✓		Mitigation
	Transportation demand management		✓	Mitigation
	Congestion pricing		✓	Mitigation
	Single-tariff public-transport policy	✓		Mitigation
	Improvement of vehicle efficiency standards		✓	Mitigation
Building	Passive building design	✓		Mitigation
	Insulation		✓	Mitigation
	Building retrofit	✓		Mitigation
	Evaporative air coolers for air conditioning		✓	Both
	Conventional air conditioning	✓	✓	Adaptation
Waste	Site/neighbourhood level composting		✓	Mitigation
	Waste to energy		✓	Mitigation
	Wastewater recycling and treatment	✓		Both
Energy	Decentralization of energy supply	✓	✓	Mitigation
	Diversified energy profile based on renewable energies		✓	Adaptation

(continued)

Table 9.4 (continued)

Sector	Measure	Trade-off		Primary objective
		Mitigation	Adaptation	
Green and Blue Infrastructure	Green roofs and façades	✓	✓	Adaptation
	Network of parks, urban greenery and open spaces	✓	✓	Adaptation
	Urban nature protection	✓	✓	Adaptation
	Urban agriculture		✓	Both
Urban policy and governance	Environmental pricing and regulation		✓	Mitigation
	Low carbon investments		✓	Mitigation
	Implementing costly mitigation measures in poor areas		✓	Mitigation
Water	Rainwater harvesting	✓	✓	Adaptation
	Desalination water plants			Adaptation

because each of them shows different strategic responses to climate change and can also have different distributive effects on a local, national, and international scale (Pachauri et al. 2014). The development of coherent climate change policies is shaped by the coordination of institutions, the development of policy-making strategies, the search for practical conceptual frameworks from the international to the local level, and the consideration of climate policies in sectoral and inter-sectoral policies (Biesbroek et al. 2009). In this regard, mitigation strategies alone are not enough to deal with the severe effects of climate change in the future: adaptation strategies also play an essential role (Biesbroek et al. 2009). As a result, scientists and policymakers have concluded that both adjustment and adaptation measures are needed to reduce climate change effects, especially greenhouse gas emissions (Landauer et al. 2015).

According to Fig. 9.5, The capacity of a sector or region to adapt to climate change depends on several factors. Studies that disregard adaptive potential are likely to overestimate the costs of climatic impacts (Schneider et al. 2001). Mitigation and adaptation are both ways of tackling climate change and its effects, and since progress in each lead to other improvements, integrating both in response to climate change in the broader context of sustainable development is suggested (Wilbanks and Sathaye 2007; Bizikova et al. 2010), especially when considering policies and financial commitments in this area (Denton et al. 2014). Research has shown that, although the integration of mitigation and adaptation measures is complex, it paves the way for mitigation and adaptation towards sustainable development and is likely to be

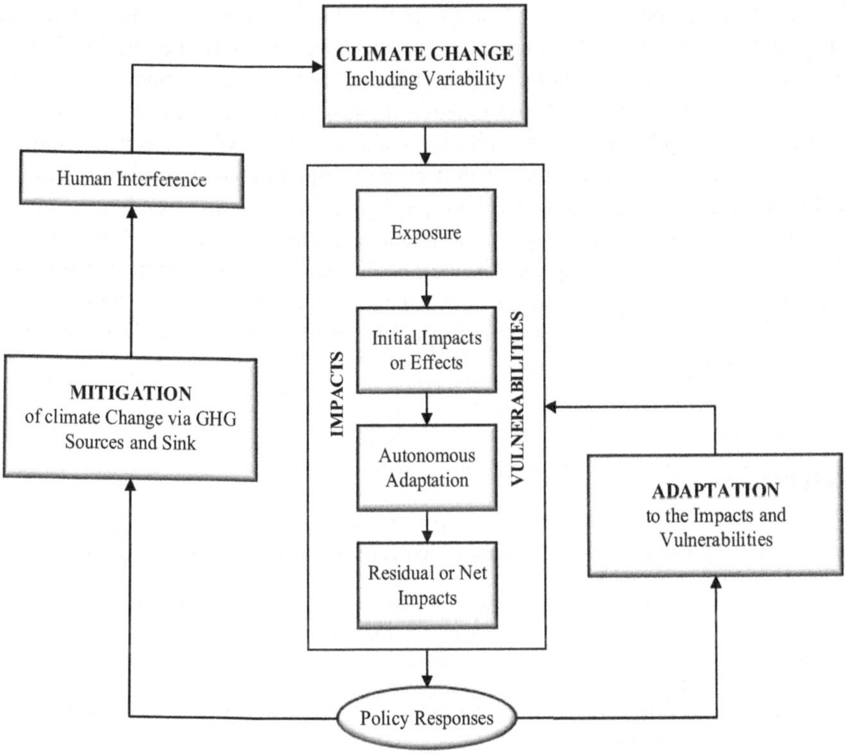

Fig. 9.5 Places of adaptation and mitigation in the climate change issue (Smit et al. 1999)

more effective. Climate change is a threat to equitable and sustainable development. Therefore, sustainable societies can be achieved by integrating mitigation, adaptation, sustainable development, and correct policies. Wilson and McDaniels (2007) cite the following motivations for integrating mitigation, adaptation, and sustainable development:

- All three decision-making areas share many important dimensions of values.
- The effects of each of the three decision-making areas may have significant consequences for others.
- Therefore, the choice among alternatives in one context is a potential means to achieving the underlying goals important in the other contexts.

9.7 Conclusion

The poorest sections of society are the most vulnerable to climate change yet have the least financial ability to deal with its effects. Therefore, climate change adaptation and mitigation at local, regional, and global scales are needed to achieve sustainable development goals. These goals can include direct assistance to support vulnerable communities' livelihoods in the face of conditions such as floods and droughts, protection of economic assets that contribute to the country's economic growth and development, and subsequent reduction in poverty. In order to adapt to climate change and mitigate its damage to society, extensive research is required to uncover monthly and annual several climatic fluctuations of different regions (with different climatic characteristics), particularly concerning trends in extreme events. Mitigating the effect of the consequences of climate change without a scientific perspective or using applied research results will not be fruitful. A review of different countries' experiences can help stakeholders develop a comprehensive plan to deal with, mitigate, and adapt to the consequences of climate change, both ongoing and in the future.

References

Adenle AA (2011) Global capture of crop biotechnology in developing world over a decade. J Genet Eng Biotechnol 9(2):83–95

Adenle AA, Azadi H, Arbiol J (2015) Global assessment of technological innovation for climate change adaptation and mitigation in developing world. J Environ Manage 161:261–275

Ahmed M, Suphachalasai S (2014) Assessing the costs of climate change and adaptation in South Asia. Asian Dev Bank

Alley RB, Berntsen T, Bindoff NL, Chen Z, Chidthaisong A, Friedlingstein P, Hoskins BJ (2007) Summary for policymakers

Ashofteh PS (2014) Climate chang and water: tools and approaches (In persian). Javdan kherad, Tehran

Ashofteh PS, Bozorg-Haddad O, Mariño AM (2013) Climate change impact on reservoir performance indexes in agricultural water supply. J Irrig Drain Eng 139(2):85–97

Asian Development Bank (2016) Asian water development outlook 2016: strengthening water security in Asia and the Pacific. Asian Development Bank

Ayers JM, Huq S (2009) The value of linking mitigation and adaptation: a case study of Bangladesh. Environ Manage 43(5):753–764

Ballentine TM, Stakhiv EZ (1993) Proceedings of the National conference on climate change and water resources management (1st) Held in Albuquerque, New Mexico on November 4–7, 1991. CORPS OF ENGINEERS, WASHINGTON DC

Bardsley T, Wood A, Hobbins M, Kirkham T, Briefer L, Niermeyer J, Burian S (2013) Planning for an uncertain future: CLIMATE change sensitivity assessment toward adaptation planning for public water supply. Earth Interact 17(23):1–26

Berry PM, Brown S, Chen M, Kontogianni A, Rowlands O, Simpson G, Skourtos M (2015) Cross-sectoral interactions of adaptation and mitigation measures. Clim Change 128(3–4):381–393

Biesbroek GR, Swart RJ, Van der Knaap WG (2009) The mitigation–adaptation dichotomy and the role of spatial planning. Habitat Int 33(3):230–237

Biswas AK, Tortajada C (eds.) (2016) Water security, climate change and sustainable development. Springer, Singapore

Bizikova L, Burch S, Cohen S, Robinson J (2010) Linking sustainable development with climate change adaptation and mitigation. na

Bosetti V, Carraro C, Duval R, Sgobbi A, Tavoni M (2009) The role of R&D and technology diffusion in climate change mitigation: new perspectives using the WITCH model

Charlton MB, Arnell NW (2011) Adapting to climate change impacts on water resources in England—an assessment of draft water resources management plans. Glob Environ Chang 21(1):238–248

Comfort LK, Kapucu N (2006) Inter-organizational coordination in extreme events: the world trade center attacks, September 11, 2001. Nat Hazards 39(2):309–327

Dang HH, Michaelowa A, Tuan DD (2003) Synergy of adaptation and mitigation strategies in the context of sustainable development: the case of Vietnam. Clim Policy 3(sup1):S81–S96

Davoudi S, Crawford J, Mehmood A (2009) Climate change and spatial planning responses. In: Davoudi S, Crawford J, Mehmood A (eds) Planning for climate change: strategies for mitigation and adaptation for spatial planners. pp 7–18

Denton F, Wilbanks T, Burton I, Chandani A, Gao Q, Lemos MC, Masui T, O'Brien K, Warner K, Dickinson T, Bhadwal S (2014) Climate-resilient pathways: adaptation, mitigation, and sustainable development

Dubrovsky M, Svoboda MD, Trnka M, Hayes MJ, Wilhite DA, Zalud Z, Hlavinka P (2009) Application of relative drought indices in assessing climate-change impacts on drought conditions in Czechia. Theoret Appl Climatol 96(1–2):155–171

Duguma LA, Wambugu SW, Minang PA, van Noordwijk M (2014) A systematic analysis of enabling conditions for synergy between climate change mitigation and adaptation measures in developing countries. Environ Sci Policy 42:138–148

Endo I, Magcale-Macandog DB, Kojima S, Johnson BA, Bragais MA, Macandog PBM, Scheyvens H (2017) Participatory land-use approach for integrating climate change adaptation and mitigation into basin-scale local planning. Sustain Cities Soc 35:47–56

Field CB (ed.) (2014) Climate change 2014–Impacts, adaptation and vulnerability: regional aspects. Cambridge University Press

Field CB, Barros V, Stocker TF, Dahe Q (eds.) (2012) Managing the risks of extreme events and disasters to advance climate change adaptation: special report of the intergovernmental panel on climate change. Cambridge University Press

Follett RF (2001) Soil management concepts and carbon sequestration in cropland soils. Soil Tillage Res 61(1–2):77–92

Ford A, Dawson R, Blythe P, Barr S (2018) Land-use transport models for climate change mitigation and adaptation planning. J Trans Land Use 11(1):83–101

Frederick KD, Major DC, Stakhiv EZ (eds.) (2013) Climate change and water resources planning criteria. Springer Science & Business Media

Gashaw T, Mebrat W, Hagos D, Nigussie A (2014) Climate change adaptation and mitigation measures in Ethiopia. J Biol Agric Healthc 148–152

Gohari A, Bozorgi A, Madani K, Elledge J, Berndtsson R (2014) Adaptation of surface water supply to climate change in Central Iran. J Water Clim Change 5(3):391–407

Goklany IM (1995) Strategies to enhance adaptability: technological change, sustainable growth and free trade. Clim Change 30(4):427–449

Hanjra MA, Blackwell J, Carr G, Zhang F, Jackson TM (2012) Wastewater irrigation and environmental health: Implications for water governance and public policy. Int J Hyg Environ Health 215(3):255–269

Heim RR Jr (2002) A review of twentieth-century drought indices used in the United States. Bull Am Meteor Soc 83(8):1149–1166

Hurlimann AC, March AP (2012) The role of spatial planning in adapting to climate change. Wiley Interdiscipl Rev Climate Change 3(5):477–488

IPCC (2005) Report of the Joint IPCC WG II & III expert meeting on the integration of adaptation, mitigation and sustainable. Île de La Réunion, France, p 236

Keshavarz M, Karami E (2013) Institutional adaptation to drought: the case of fars agricultural organization. J Environ Manage 127:61–68

Khan S, Hanjra MA (2009) Footprints of water and energy inputs in food production–global perspectives. Food Policy 34(2):130–140

Kidanu A, Hardee K, Rovin K (2009) Linking population, fertility and family planning with adaptation to climate change: views from Ethiopia
Klein RJT, Huq S, Denton F, Downing TE, Richels RG, Robinson JB, Toth FL (2007) Inter-relationships between adaptation and mitigation. Climate change 2007: impacts, adaptation and vulnerability. In: Contribution of working group ii to the fourth assessment report of the inter-governmental panel on climate change. Cambridge University Press, Cambridge, UK, pp 745, 777
Koutsoyiannis D (2011) Scale of water resources development and sustainability: small is beautiful, large is great. Hydrol Sci J 56(4):553–575
Lal R (2009) Soils and food sufficiency: A Review. In: Sustainable agriculture. Springer, Dordrecht, pp 25–49
Landauer M, Juhola S, Söderholm M (2015) Inter-relationships between adaptation and mitigation: a systematic literature review. Clim Change 131(4):505–517
Landauer M, Juhola S, Klein J (2019) The role of scale in integrating climate change adaptation and mitigation in cities. J Environ Planning Manage 62(5):741–765
Laukkonen J, Blanco PK, Lenhart J, Keiner M, Cavric B, Kinuthia-Njenga C (2009) Combining climate change adaptation and mitigation measures at the local level. Habitat Int 33(3):287–292
Ledda A, Di Cesare EA, Satta G, Cocco G, Calia G, Arras F, Congiu A, Manca E, De Montis A (2020) Adaptation to climate change and regional planning: a scrutiny of sectoral instruments. Sustainability 12(9):3804
Lybbert TJ, Sumner DA (2012) Agricultural technologies for climate change in developing countries: policy options for innovation and technology diffusion. Food Policy 37(1):114–123
Matocha J, Schroth G, Hills T, Hole D (2012) Integrating climate change adaptation and mitigation through agroforestry and ecosystem conservation. In Agroforestry-the future of global land use. Springer, Dordrecht, pp 105–126
McCarthy JJ, Canziani OF, Leary NA, Dokken DJ, White KS (eds.) (2001) Climate change 2001: impacts, adaptation, and vulnerability: contribution of working group II to the third assessment report of the intergovernmental panel on climate change, vol. 2. Cambridge University Press
Milly PC, Dunne KA (2020) Colorado River flow dwindles as warming-driven loss of reflective snow energizes evaporation. Science 367(6483):1252–1255
Misra AK (2014) Climate change and challenges of water and food security. Int J Sustain Built Environ 3(1):153–165
Mishra AK, Singh VP (2010) A review of drought concepts. J Hydrol 391(1–2):202–216
Mishra AK, Singh VP (2011) Drought modeling–a review. J Hydrol 403(1–2):157–175
Molden D (ed.) (2013). Water for food water for life: a comprehensive assessment of water management in agriculture. Routledge
Moomaw WR, Moreira JR, Blok K, Greene D, Gregory K, Jaszay T, Kashiwagi T, Levine M, MacFarland M, Prasad NS, Price L (2001) Technological and economic potential of greenhouse gas emissions reduction
Mtui GY (2011) Involvement of biotechnology in climate change adaptation and mitigation: improving agricultural yield and food security. Int J Biotechnol Mol Biol Res 2(13):222–231
National Research Council (2009) Informing decisions in a changing climate. National Academies Press
National Research Council (2011) Adapting to the impacts of climate change. National Academies Press
Pachauri RK, Allen MR, Barros VR, Broome J, Cramer W, Christ R, Church JA, Clarke L, Dahe Q, Dasgupta P, Dubash NK (2014) Climate change 2014: synthesis report. In: Contribution of Working Groups I, II and III to the fifth assessment report of the Intergovernmental Panel on Climate Change. Ipcc, p 151
Papageorgiou A, Barton JR, Karagiannidis A (2009) Assessment of the greenhouse effect impact of technologies used for energy recovery from municipal waste: a case for England. J Environ Manage 90(10):2999–3012

Parry JE, Hammill A, Drexhage J (2005) Climate change and adaptation. International Institute for Sustainable Development, Canada

Parry M, Parry ML, Canziani O, Palutikof J, Van der Linden P, Hanson C (eds.) (2007) Climate change 2007-impacts, adaptation and vulnerability: Working group II contribution to the fourth assessment report of the IPCC (Vol. 4). Cambridge University Press

Pathak H, Aggarwal PK, Singh SD (2012) Climate change impact, adaptation and mitigation in agriculture: methodology for assessment and applications. Indian Agricultural Research Institute, New Delhi, pp 1–302

Peterson TC, Stott PA, Herring S (2012) Explaining extreme events of 2011 from a climate perspective. Bull Am Meteor Soc 93(7):1041–1067

Pittock AB, Jones RN (2000) Adaptation to what and why? Environ Monit Assess 61(1):9–35

Revi A (2008) Climate change risk: an adaptation and mitigation agenda for Indian cities. Environ Urban 20(1):207–229

Rothausen SG, Conway D (2011) Greenhouse-gas emissions from energy use in the water sector. Nat Clim Chang 1(4):210–219

Salathé EP Jr, Mote PW, Wiley MW (2007) Review of scenario selection and downscaling methods for the assessment of climate change impacts on hydrology in the United States Pacific Northwest. Int J Climatol J R Meteorol Soc 27(12):1611–1621

Schewe J, Heinke J, Gerten D, Haddeland I, Arnell NW, Clark DB, Gosling SN (2014) Multimodel assessment of water scarcity under climate change. Proc Natl Acad Sci 111(9):3245–3250

Schneider S, Sarukhan J, Adejuwon J, Azar C, Baethgen W, Hope C, Moss R, Leary N, Richels R, Van Ypersele JP (2001) Overview of impacts, adaptation, and vulnerability to climate change. Climate change 75–103

Schroth G, Laderach P, Dempewolf J, Philpott S, Haggar J, Eakin H, Castillejos T, Moreno JG, Pinto LS, Hernandez R, Eitzinger A (2009) Towards a climate change adaptation strategy for coffee communities and ecosystems in the Sierra Madre de Chiapas, Mexico. Mitig Adapt Strat Glob Change 14(7):605–625

Sharifi A (2020) Trade-offs and conflicts between urban climate change mitigation and adaptation measures: a literature review. J Clean Prod 122813

Smit B, Burton I, Klein RJ, Street R (1999) The science of adaptation: a framework for assessment. Mitig Adapt Strat Glob Change 4(3–4):199–213

Stokes JR, Horvath A (2009) Energy and air emission effects of water supply

Swart ROB, Raes F (2007) Making integration of adaptation and mitigation work: mainstreaming into sustainable development policies? Climate Policy 7(4):288–303

Vo PT, Ngo HH, Guo W, Zhou JL, Nguyen PD, Listowski A, Wang XC (2014) A mini-review on the impacts of climate change on wastewater reclamation and reuse. Sci Total Environ 494:9–17

White I, Howe J (2003) POLICY AND PRACTICE: Planning and the European union water framework directive. J Environ Planning Manage 46(4):621–631

Wichelns D (2011) Assessing water footprints will not be helpful in improving water management or ensuring food security. Int J Water Resour Dev 27(3):607–619

Wiering M, Immink I (2006) When water management meets spatial planning: a policy-arrangements perspective. Eviron Plann C Gov Policy 24(3):423–438

Wilbanks TJ, Kane SM, Leiby PN, Perlack RD, Settle C, Shogren JF, Smith JB (2003) Integrating mitigation and adaptation-possible responses to global climate change. Environment 45(5):28–38

Wilbanks TJ, Sathaye J (2007) Integrating mitigation and adaptation as responses to climate change: a synthesis. Mitig Adapt Strat Glob Change 12(5):957–962

Wilhite DA (2000) Drought as a natural hazard: concepts and definitions

Wilhite DA, Glantz MH (1985) Understanding: the drought phenomenon: the role of definitions. Water Int 10(3):111–120

Wilson C, McDaniels T (2007) Structured decision-making to link climate change and sustainable development. Climate Policy 7(4):353–370

Zhou Q (2014) A review of sustainable urban drainage systems considering the climate change and urbanization impacts. Water 6(4):976–992
Zhou Q, Mikkelsen PS, Halsnæs K, Arnbjerg-Nielsen K (2012) Framework for economic pluvial flood risk assessment considering climate change effects and adaptation benefits. J Hydrol 414:539–549

Chapter 10
Case Studies Around the World

Bahareh Hossein-Panahi, Omid Bozorg-Haddad, Hugo Loáiciga, Sujo Mal Meghwar, and Martina Zeleňáková

Based on the available evidence and studies, freshwater resources can be intensely affected by climate change, with extensive consequences for human communities and ecosystems. Climate change has immediate and long-term effects on water resources such as floods, droughts, rising sea levels in estuaries, drying up of rivers, poor water quality in surface and groundwater systems, distortions of water vapor and precipitation patterns, improper distribution Ice is snow and earth and the amount of access and demand for water resources. Vulnerability due to climate change varies according to different countries, geographical location and capacity to reduce or

B. Hossein-Panahi
Faculty of Agricultural Engineering and Technology, Department of Irrigation and Reclamation Engineering, College of Agriculture and Natural Resources, University of Tehran, Karaj, Tehran, Iran
e-mail: HosseinPanahi.Ba@ut.ac.ir

O. Bozorg-Haddad (✉)
Department of Irrigation and Reclamation Engineering, Faculty of Agricultural Engineering and Technology, College of Agriculture and Natural Resources, University of Tehran, Karaj, Tehran, Iran
e-mail: OBHaddad@ut.ac.ir

H. Loáiciga
Department of Geography, University of California, Santa Barbara, CA 93016-4060, USA
e-mail: hloaiciga@ucsb.edu

S. M. Meghwar
Department of Geography, University of Sindh, Jamshoro, Sindh, Pakistan
e-mail: sujo@usindh.edu.pk

M. Zeleňáková
Department of Environmental Engineering, Technical University of Kosice, 042 00 Kosice, Slovakia
e-mail: Martina.Zelenakova@tuke.sk

O. Bozorg-Haddad (ed.), *Climate Change in Sustainable Water Resources Management*, Springer Water, https://doi.org/10.1007/978-981-19-1898-8_10

adapt to change. This chapter examines climate change in six continents: Africa, Asia, Europe, North America, South America, and Antarctica, its effects on water resources, and the policies and strategies used.

10.1 Africa

10.1.1 Geography and Climate

The total land area of the continent of Africa is 30 million km^2, which encompasses 22% of the world's deserts and a variety of climate types. Countries such as Sudan, Algeria, Democratic Republic of the Congo, Libya, and Chad are large and account for 34% Africa's surface area. The smallest five countries (all islands: Cape Verde, Comoros, Mauritius, Sao Tome, Principe, and Seychelles) constitute 3% of Africa's area. Approximately 53% of Africa is thought to be water-rich, with 61% of the continent's population living in the water-rich region, which also houses 95% of the continent's renewable water resources (Frenken 2005; Hassan and Tularam 2018). More than 160 lakes greater than 27 km^2 and more than 22 major River basins can be found in this continent (Rowledge 1999). There are 60 international river basins within Africa, which cover about 62% of the continent, and five of these (Congo, Niger, Nile, Zambezi, and Lake Chad) are shared by more than eight countries (Goulden et al. 2009). The Congo, Niagara, Ogadugne, Zambezi, Nile, Sangha, Chari-Logone, and the Volta river basins account for 75% of the land surface area in Africa.

The largest streamflow occurs in the Congo basin, while the smallest occurs in the Orange basin in South Africa. Streamflow variability is high, with coefficients variation exceeding 20% of river flowing toward the west and south of Africa (Rowledge 1999). About 90% of the surface freshwater resources of Africa are found in river basins and lakes, which are usually shared by two or more countries. Three of the river basins such as the Juba-Shibeli, Orange, and Limpopo experience water stress (less than 1700 m^3 per person per year), with the Orange and the Limpopo basins situated in southern Africa experience water scarcity (defined as less than 1000 m^3 of water per person per year). The Nile and Volta basins have the largest population densities and both are on the verge of water stress conditions. In Ethiopia's Abay/Upper Blue Nile Basin there are major changes in rainfall, temperature, and streamflow patterns, as well as possible change points. The population of the Volta basin was 4.5 million people in 2000, and it is expected to reach 8 million by 2025. There are warming trends the Black Volta Basin. Both temperature extremes, i.e., highest and lowest (annual, seasonal, and monthly scale) at upstream and downstream regions revealed an increasing trend (Abungba et al. 2020; Goulden et al. 2009).

The freshwater supplies in Africa are essential for maintaining livelihoods (particularly agriculture and fish-culture-based livelihoods), food security, and power generation, and for fulfilling increasing water demands by the domestic and industrial sectors. The freshwater sources are under pressure due to increasing competition by users. Climate change may exert heavy pressure on water availability. However, some areas may become wetter while others become drier. Natural hazards may be exacerbated by the increase in rainfall variability. River flows are changing in volume and timing, causing changes in the geographical distribution of water resources. Severe flooding was recorded in Mozambique, Namibia, South Africa Uganda in 2011; in Brazil, Columbia, and Mexico in the Americas, and in Cambodia, China, India, Korea, Pakistan, the Philippines, and Thailand in Asia, with each flood resulting in more than 50 deaths (over 1000 in the Philippines and Colombia) and significant material destruction (Goulden and Conway 2009; Kundzewicz et al. 2014).

The renewable freshwater resources in Africa are estimated to be between 4050 and 4590 km^3 $year^{-1}$ (Döll and Fiedler 2008). Groundwater contributes between 15 to 51% of the freshwater across regions of Africa. North Africa is more likely to use fossil groundwater, while renewable groundwater is more common in Sub-Saharan Africa (Taylor et al. 2009; Xu et al. 2019).

Non-renewable freshwater in North Africa, supplies are used for domestic and industrial endeavors, for instance, in Libya's Great Man-Made River Project (Kundzewicz and Doell 2009).

The subsurface water reside in 40 aquifer systems, which are shared by several countries. Seventy-two transboundary aquifers have been mapped in Africa, which cover about 40% of the continent, and 33% of the African population lives within the geographical boundaries these transboundary aquifers (Rivera and Candela 2018). There is insufficient data on Africa's groundwater availability, resources and usages. According to the BGS (2010) analysis of recharge estimation and groundwater resources in Africa it is difficult to estimate groundwater recharge in most parts of Africa due to the scarce temporal and spatial data. It is observed that nearly 80% of Africa's land area is underlain by aquifers with low permeability and limited storage capacity. Macdonald et al. (2009) defined three large rainfall recharge land areas in Africa, namely (1) scarce groundwater recharge in zones with less than 200 mm $year^{-1}$, (2) regions with about 50 mm $year^{-1}$ recharge in zones with rainfall ranging from 200 to 500 mm $year^{-1}$; and (3) more than 50 mm $year^{-1}$ recharge in zones with rainfall ranging from 500 to 1000 ^{mm} $year^{1}$.

The spatial, temporal distribution and availability of groundwater are governed by the quantity and recharge mechanism, which includes the depth of water, aquifer extent, and climatic uncertainty.

The equatorial region, the intertropical belt (between the tropics of Cancer and Capricorn) and two large deserts (The Sahara in the Northern Hemisphere, and the Kalahari in the Southern Hemisphere) encompass a wide variety of climate-driven systems, encompassing humid equatorial, seasonally-arid tropical, sub-tropical Mediterranean-type, and desert climates. This uneven climatic profile produces on average annual 20,360 km^3 of precipitation in Africa, which is equivalent to an average annual 678 mm of rainfall across the continent (ACPC 2011).

There is to about 20% the entire continent's land area in the central region of Africa that receives 37% the continent's total precipitation. In comparison, the northern African region covers a similar area and it receives about 3% of total precipitation

(Frenken 2005). 41% of the continent feature a dry climate with an average annual rainfall of less than 400 mm. 25% of Africa receives between 400 and 1000 mm yearly^{-1} of precipitation.

The average annual rainfall in the Sahara region varies between less than 1 mm year^{-1} to over $^{10,000\ mm}$ yearly^{1} in some tropical rain forest areas. The climatic regime of Africa is characterized by high variation affecting densely populated areas such as southern Africa (including the Orange and Limpopo basins), most of East Africa inclusive of the Upper Nile basin, and the East–West band, which stretches from Senegal to Sudan and covers a number of major river basins (broadly analogous to the Sahel), containing Lake Chad, the Niger, the Upper Volta, and Senegal basins. The average annual rainfall at a station in Cameroon's mountains reaches 10,000 mm year^{-1}. Much of Africa area is located within tropical or subtropical latitudes. In some warm regions the largest temperature variations occur between day and night, rather than seasonally. In desert areas the diurnal temperature range is from 10 to 15 °C (ACPC 2011).

10.1.2 Climate Change in Africa

Africa has a poor adaptation capacity and it is projected climate change and its surrounding impacts will harm mega deltas in the region. Its population has high vulnerability to see level rise, storm surges and river flooding.

In southern Africa there is moderate confidence in a rise in the frequency of warm days and a reduction in the frequency of cold days and nights. There is a moderate chance that the frequency of warm evenings in northern and southern Africa will rise (IPCC 2014).

The IPCC reports that temperatures in Africa since the 1960s have been shown a warming trend. Africa warmed at a rate of 0.5 °C in the twentieth century and the warming rate has risen in the last three decades of the same century. All of Africa, Northern Hemisphere Africa, Southern Hemisphere Africa, tropical Africa, and subtropical Africa exhibit rising temperature patterns (Collins 2011). These patterns seem to be constant across the continent. Large freshwater lakes in eastern Africa appear to be in decline. Increased precipitation in the Northern Hemisphere tropics, reduced trade winds and thereby a warmed tropical Atlantic are recognized effects of African tropical vegetation expanding northward (Pontes et al. 2020).

The warming in the region is expected to be greater than the global annual mean across Africa in all seasons within drier subtropical regions, which experience more warming than the moister tropics. In most of Mediterranean Africa region and the northern Sahara region there is a reduction in annual rainfall. On the other hand, southern Africa's winter rainfall and its western margins are expected to decrease while East Africa is expected to see a rise in mean annual rainfall. It is uncertain how the Sahel, the Guinean Coast, and the southern Sahara would do in terms of rainfall variation (IPCC 2008).

According to studies by Unganai (1996), Hughes and Bally (1996), Kruger and Shongwe (2004), Warburton et al. (2005), New et al. (2006) remote sensing-derived and observational temperature records in southern Africa indicate warming in recent decades (Kusangaya et al. 2014).

An increase in temperature may cause a higher rate of evaporation with associated heavier precipitation, and runoff (Kusangaya et al. 2014). In the Southern Hemisphere region in Africa there is an increasing temperature trend (Collins 2011). Morishima and Akasaka (2010) demonstrated the annual mean surface temperature is rising across southern Africa with especially high rates of increase of temperature in Namibia and Angola.

The African climate involves complex processes that are not well understood. These involve tropical convection and monsoon alternation, which are influenced by the shifting of temperature and rainfall patterns in regional and seasonal scales. The El Niño-Southern Oscillation (ENSO) of the Pacific Ocean has changed its origin in the southern Pacific, and it has caused a significant impact on Africa's rainfall and temperature with an irregular frequency of several years (Conway 2009).

The annual rainfall is decreasing in most areas of Africa (Niang et al. 2008). According to Conway (2011), since the early 1970s, it has been a noticeable a drying pattern across most of the Sahel, and a decadal variability in southern African rainfall related to the ENSO. However, during 1901 to 1995 areas of East Africa experienced a 10–20% risen in annual rainfall, which constituted a wetting pattern over equatorial Africa. The most significant climatic change in Africa has been formulated as a long-term decrease in rainfall in West African semiarid regions and part of the Sahel. The global Hadley cell and the regional monsoon circulation are dynamically related in the Sahel. Warming of the oceans improves the tropical atmosphere's equilibrium thus weakening the deep ascent in the Hadley circulation. The warming of the Sahara and the surrounding waters alters the composition and direction of the regional shallow circulation, allowing more extreme convective systems to form, which influence seasonal rain (Biasutti 2019).

A drop in precipitation of $2.4 \pm 1.3\%$ per decade has been reported in Africa's tropical rainforest regions, West Africa ($-4.2 \pm 1.2\%$ per decade) and north Congo ($-3.2 \pm 2.2\%$ per decade) had the higher reductions (Sonwa et al. 2020). Precipitation is the most critical meteorologic phenomenon on earth for the natural environment and human life. The high rainfall variability and river flow over a range of spatial and temporal scales in Africa has significant implications for water supply management (Twisa and Buchroithner 2019).

The Fourth Assessment of the Intergovernmental Panel on Climate Change (IPCC) Report presented an overview of climate model predictions for different African regions. These are based on a set of 21 projections from their Multi-Model Data (MMD) set, with the A1B emissions scenario focusing on climate change between 1980 and 1999 (to reflect the current climate) and 2080 to 2099 (to represent the future climate) (Christensen et al. 2007).

10.1.2.1 Impact on Water Resources

Glacier mass and snow cover reduction has occurred in the high mountains (e.g. in the Himalayas, East Africa, and the tropical Andes), according to IPCC 2019 (Medium confidence). Between 2015 and 2100, projected glacier mass decreases (excluding ice sheets) range from 18 ± 7% (likely range) for RCP2.6 to 36 ± 11% (likely range) under the greenhouse gases emissions pathway RCP 8.5, equating to a sea-level rise contribution of 94 ± 25 mm (likely range) for RCP2.6 and 200 ± 44 mm (likely range) for RCP 8.5 (medium confidence).

Assessing climate impacts caused by 1.5 and 2 °C warming is especially important for Central Africa, a vulnerable area where multiple biophysical, political, and socioeconomic pressures interact to limit the region's adaptive potential. Overall, the warming in Central Africa is predicted to be higher than the global average (Mba et al. 2018).

The effects of climate change on Africa's water resources that have been observed include (IPCC 2014): East Africa's tropical highland glaciers are retreating (high confidence, major contribution from climate change); rivers in West Africa are discharged at a lower rate (low confidence, major contribution from climate change); increases in lake surface warming and water column stratification in the Great Lakes and Lake Kariba (high confidence, substantial contribution from climate change); increases in lake surface warming and water column stratification in the Great Lakes and Lake Kariba (high confidence, significant contribution from climate change); drought in the Sahel since 1970, with slightly wetter conditions after 1990 (medium confidence, considerable influence from climate change); Beyond the loss caused by human effects one most notice that coral reefs in tropical African seas are deteriorating (high confidence, major contribution from climate change); South African farmers have developed adaptive responses to fluctuating rainfall (very low confidence, major contribution from climate change); reduced Great Lakes and Lake Kariba fisheries production, in addition to changes in fisheries management and land use (low confidence, minimal influence from climate change).Droughts and floods in Africa, as well as other extreme weather and climatic phenomena, have serious consequences for the economy, natural resources, ecosystems, livelihoods, and human health. Floods on the Zambezi River in Mozambique in 2008, for example, displaced 90,000 people, and with over 1 million people living in flood-affected regions in the Zambezi River Valley, temporary displacement is becoming a permanent characteristic (IPCC 2014).

Water supply in central Africa is complicated by population growth, changing water-use patterns, and the concentration of population and economic activities in urban areas that would bring increased pressure on African freshwater supplies in the next century (Arnell 2006). African water resources have been impacted by climate change and rising demand across the continent. Africa has been affected by climate change at different times over the last century, with the frequency of these effects increasing over the last 40 years. Many of Africa's vital water resources, such as lakes, rivers, and snow in high mountains, are demonstrating steadily dwindling according to observational evidence (ACPC 2011). Drought stress is hitting Africa's drought-prone regions particularly hard (high confidence) (IPCC 2014).

Rising temperatures may cause changing rainfall patterns, runoff distribution (both spatially and temporally), it affects soil moisture and water sources, and increases the frequency of droughts and floods (Schulze et al. 2011).

Between 75 and 250 million people have badly endured increased water stress due to climate change by 2020. Rain-fed agriculture output in some areas might be cut by up to 50%. Many African countries' agricultural productivity is expected to be seriously harmed. Food security would be harmed much more, and starvation would be exacerbated. Towards the end of the twenty-first century projected sea-level risen rise will affect low-lying coastal areas with large populations. The cost of adaptation could amount to at least 5 to 10% of gross domestic product (GDP). The land areas of arid and semi-arid are projected to increase between 5 and 8% in Africa by 2080 (TS) (IPCC 2008).

Surface drying in southern Africa is expected by the end of the century under RCP8.5 (high confidence). All parts of Africa are likely to see an increase in warm days and nights and a decrease in cold days and nights (high confidence). Summer and fall are the seasons with the warmest days (medium confidence). Heatwaves and warm spells in Africa are likely to be more common and/or last longer (high confidence) (IPCC 2014).

The water accessibility is expected to be less than 1,000 m^3 per person per year in nine countries in the eastern and southern part of Africa by 2025. Twelve countries may have per capita annual use in the range of 1,000 to 1,700 m^3 affecting 460 million people at high water risk and water shortage stress mostly in the western part of Africa (Bates et al. 2008). Projections indicate the proportion of the African population that could be pronounced at risk of water stress and shortage would rise from 47% in 2000 to 65% by 2025 (Ashton 2002). Climate change has the potential to raise the stress on Africa's water supply and demand (Bates et al. 2008). The effect of predicted climate change on water resources is not uniform across the continent. Arnell (2004) predicts that 75–250 million people and 350–600 million people in Africa may be at risk of enduring high water stress by the 2020s and 2050s, respectively.

Modeling of the severity of climate change's impact on future undernutrition in five regions of South Asia and Sub-Saharan Africa in 2050 (using the Special Report on Emissions Scenarios (SRES) A2 emissions scenario) reveals a rise in mild nutritional stunting, a condition associated with increased risk of mortality and ill health (Black et al. 2008). Between 1 and 29% of regions studied would be affected by severe nutritional stunting, such as 23% of the central Sub-Saharan Africa and 62% in South Asia (Lloyd et al. 2011).

The frequency and magnitude of floods and droughts across Africa may change due to changes in the pattern of precipitation and river flow regimes. It is observed that the increase of flood risk in coastal areas will be intensified by sea-level risen that could predominately be associated with climate change. Flooding and drought would have far-reaching indirect effects on food production, hydroelectric power generation, and domestic water sources (Kundzewicz et al. 2008). Increased occurrence of extreme rainfall incidents, water flow disruption, and interruption of hydroelectric generation have the ability to affect economic development in Sub-Saharan Africa (Gannon et al. 2018).

Several climate models project a consistent response in mean annual and seasonal temperature change in subareas Africa, who is projected to be warmer by 2080 with an increase of temperatures in a range from +3.2 °C (East Africa) to +3.6 °C (the Saharan Desert). Almost all models project wetter conditions in West and East Africa (+2% and +7% rises in precipitation, respectively), while drier conditions are projected in southern Africa and the Sahara (−4% and −6%, respectively). It's worth noting that climate projections in the Sahel are varied and are not consistent (Christensen et al. 2007). Detecting climate change is essential for a greater understanding of the climate and the development of regional and local adaptation and mitigation strategies (Gebrechorkos et al. 2019).

Strzepek and McCluskey (2006) reported projections from five climate models (CSIRO2, hadcm3, CGCM2, ECHAM, and PCM) in conjunction with two various emissions scenarios that indicated changes in all countries of South Africa, including a reduction in streamflow. The streamflow increases in South Africa under the high emission scenarios are relatively few at 10%. De Wit et al. (2006) applied six GCMs to identify a sensitive "unstable" region between Senegal and Sudan, separating the dry Sahara from wet Central Africa and projected a decrease in runoff in Southern Africa. Soliman et al. (2009) applied the ECHAM5 A1B scenario downscaled by RegCM3 to estimate a 1.5% annual rise in Blue Nile flow at El Diem in the future.

In this context, the water stress may increase over 62.0–75.8% in the total river basin land area and may decrease over 19.7–29.0% of the river's volumes (Alcamo et al. 2007) by 2050. De Wit and Stankiewicz (2006) concluded that by 2100 reduction in perennial drainage may significantly affect present access to surface water in 25% of Africa. Cavé et al. (2003) reported that expected decreases in annual rainfall in southern Africa would reduce groundwater recharge.

African water scarcity poses significant challenges in managing water sources, including high variability and regional shortage that could be compounded by a lack of institutional capacity. The transboundary nature of certain river basins complicates water management. Despite the fact that the details of climate change's effects on water supplies are unknown in detail, they are expected to worsen. River flows are especially vulnerable to climate change according to modeling studies. Climate and hydrological models project more runoff in certain parts of Africa during the twenty-first century, such as the Niger and Volta river basins in West Africa. While case studies in Central and East Africa differ on the direction and the magnitude of the change. Model projections illustrate diminishing surface runoff in the future in southern Africa, which may rise vulnerability to water shortage. However, these water demand will rise driven by economic and demographic growth and outweigh climate-related changes in the majority of river basins (Goulden and Conway 2009). In hot and dry climates runoff farming has been shown to increase land fertility and crop yields. Farmers in drought-prone agro-ecological areas could resort to rainwater harvesting to supplement crops during dry periods (Uwizeyimana et al. 2019).

Other effects of climate change on Africa's water resources that have been observed or predicted are: (1) the flow of the Upper Blue Nile is very sensitive to changes in precipitation and evapotranspiration. Its flow would be magnified by 2.9 times. Potential (reference crop) evapotranspiration increases by 2–14% when the temperature rises between 2 °C and 5 °C (Elshamy et al. 2009); (2) climate change is expected to impact Lake Tanganyika, which provides 25–40% of the surrounding human population's animal protein, and fish catches are expected to decline by 30% (O'Reilly et al. 2003); (3) Lake Chad shrank in surface area from 23,000 km^2 in 1963 to 304 km^2 in 2001 (Food and Organization 2009); (4) reduction of the snow cover in Mount Kilimanjaro (Buytaert et al. 2011); (5) the Nile Delta and the Lagos area are experiencing rising water levels, which would have an effect on the lives of millions of people (Dasgupta et al. 2009; Nicholls 1995); (6) decreasing trend of the level of Lake Malawi (Kumambala 2010) and; (7) decreasing of streamflow and evidence of flow variability are documented in Southern Africa (FANTA et al. 2001) and Western Africa (Ojo et al. 2003).

The results of the numerous studies indicate warming weather in all seasons, and higher annual mean temperature throughout the continent. Higher temperatures are expected to increase evaporative demand across Africa. Annual rainfall is expected to decline in most of Mediterranean Africa and the northern Sahara. Rainfall in southern Africa is likely to decrease in much of the Southern Hemisphere winter rainfall region and western margins. Annual rainfall in East Africa is predicted to increase, but it is unclear how rainfall might change in the Sahel, the Guinean Coast, and the southern Sahara. Vörösmarty et al. (2005) reported that water stress affects 25% of Africa's population, with another 13% experiencing water stress caused by drought once in a generation. Climate change is raising the stress on already over-tapped water sources.

10.1.2.2 Water Sector Policies and Strategies

According to IPCC 2019 some issues and prospects of adaptation include: strengthening institutional capacity for demand management, groundwater assessment, integrated water-wastewater planning, and integrated land and water governance; reducing non-climate stresses on water resources; sustainable urban development.

The majority of national governments in Africa are implementing adaptive governance structures (high confidence). The mainstreaming of adaptation into sectoral planning has begun as a result of progress on national and subnational policies and strategies, but growing institutional frameworks are unable to effectively coordinate the wide range of adaptation projects now underway. Disaster risk management, technological and infrastructural changes, ecosystem-based methods, fundamental public health interventions, and livelihood diversification are all helping to reduce vulnerability, though in isolated ways (IPCC 2014).

Climate change's effect on water resources is not directly considered in most African water sector policies (ACPC 2011). This has caused confusion in climate-change adaptations, which must be integrated into water sector policies and planning to cope with the effects of climate change on water in Africa.

Climate change has had a variety of effects on Africa's river basins. River-basin level planning can provide the capacity to address climate change impacts effectively.

One of the most significant adaptation requirements is building on traditional expertise related to water harvesting and usage (Osman-Elasha et al. 2006), signaling the need for its integration into climate change policies to ensure the implementation of successful adaptation strategies that are cost-effective, participatory, and sustainable.

The field of climate change adaptation has advanced significantly. Adapting to climate change poses significant challenges to Africa's long-term sustainability and development. In order to advance the sustainability perspective in climate adaptation research and study in Africa it is necessary to extend the knowledge base on the (relatively new) climate-smart agricultural idea (Bhatasara and Nyamwanza 2018).

Cooperative and antagonistic interactions concerning the management of water resources take place with a variety of intensities and geographical scales (Kistin 2006; Zeitoun and Mirumachi 2008; Zeitoun and Warner 2006). The nature of interactions between states and the competition for water sources are strongly affected by the power relationships between states sharing a river basin. The understanding of the essence of water conflict and cooperation has advanced over time, yet, water conflicts will become more common as populations rise and climate change manifest themselves (Petersen-Perlman et al. 2017).

Early warning systems have been utilized in Africa and other parts of the world to combat famine and food shortages, floods and other weather-related hazards, fire-related air pollution, and vector-borne and food-borne disease epidemics (IPCC 2014).

The adaptation of water resources to climate change involve policy initiatives such as the adoption of probability planning for drought, improving infrastructure construction, developing environmentally and economically sound inter-basin transfers, building new dam sites considering their impacts to achieve environmental and economic benefits, water conservation, managed aquifer recharge and sewage reuse, apply market-based mechanisms to allocate water sources, and preventing pollution (Loáiciga 2015; Smith and Lenhart 1996).

Goulden et al. (2009) proposed an agenda for more climate-change adaptation research in African international river basins. Current adaptations arising at both national and international scale, and research are required to determine what agents are causing these adaptations. Various variations of adaptation solutions must be considered to cope with the wide range of water scarcity conditions and levels of adaptive capability, including conservation, supply/demand control, and the possibility for intra-basin virtual water transfers. Physical, economic, and political conditions in African international basins demand careful assessment, especially in terms of how they promote adaptation and cooperation among stakeholders.

Basins with a wide range of GDP[1] and HDI[2] values within them show high inequalities of adaptive capacity. Some counties, such as those in the Lake Chad, Congo, and Zambezi basins could be better equipped to adapt than others. In the absence of basin-wide structures and agreements one country's adaptation may have detrimental consequences in another country within the same basin. This may also

[1] Gross domestic product.

[2] Human Development Index.

happen with respect to internationally shared groundwater deposits, which are located in an estimated 40 transboundary aquifers (Scheumann and Alker 2009). It was noted at the first African Climate Policy Center (ACPC) meeting in April 2011 that African climate change programs provide limited political guidance to RBOs,[3] RECs,[4] and AUC[5] negotiators. Climate change study findings on Africa's water supplies should be turned into strategies that can be applied by various water institutes. The uncertainties in climate change projections and water supply modeling must be taken into account by the planning agents and policymakers in the private and governmental sectors. Initiatives such as Africa Adapted demonstrates the keen knowledge sharing of climate change projections, which must be reinforced to reach a wide audience (ACPC 2011).

Some of the problems with climate change's effects on water management coordination and knowledge exchange (ACPC 2011) include:

1. Few information-sharing programs on the effect of climate change on Africa's water resources; 2. Language, infrastructure, and technological obstacles to accessing knowledge about climate change and Africa's water resources; 3. Cooperation and communication between climate scientists and impact researchers are inadequate.

10.2 Asia

10.2.1 Geography and Climate

Asia is the world's largest continent, covering 30% of the Earth's land area and 8.66% of its surface area. It is bordered on the east by the Ural Mountains, on the north by the Arctic Ocean, on the west by the Pacific Ocean, and the south by the Indian Ocean. The Himalaya Mountains is well known for its influence on climate in the region. The Yangtze (6,211 km), which flows through China, is Asia's longest and the third-longest river worldwide (Wheeler 2016).

Asian countries include Afghanistan, Bangladesh, Bhutan, India, Nepal, Pakistan, and Sri Lanka, which are seven countries in South Asia that account for 23.7% of the worlds' population. However, they account for about 4.6% of the annual renewable water supplies, which are spread unevenly across countries and river basins. These numbers are based on the population of South Asia (1.744 billion) and the world (7.349 billion) as recorded by the United Nations (2015), and estimates by the FAO's AQUASTAT database (FAO 2016): South Asia has internal annual renewable water supplies worth 1,982 km^3 and global annual renewable water resources of 42,810 km^3.

---

[3] River Basin Organizations.

[4] Regional Economic Communities.

[5] Africa Union Commission.

Water is distributed unequally throughout Asia, where large areas are experiencing water scarcity. Twenty of Asia's forty-three countries have renewable annual per capita water resources in excess of 3,000 m^3, eleven have renewable annual per capita water resources between 1,000 and 3,000 m^3, and six have renewable annual per capita water resources below 1,000 m^3 (FAO 2004a, b, c). Except for Bhutan and Nepal the per capita water availability is reported to below the world average and continues to decline as the population increases. The water impacts in many places are dramatic, with rapidly declining groundwater levels and degradation in water quality harming public health.

Agriculture uses more than 90% of all water abstracted in South Asia, compared to 70% globally, with 60% of agricultural water uses relying on surface water and 40% relying on groundwater. The agricultural sector's contribution to GDP continues to decline across Asia despite a large proportion of the workforce and total population being directly dependent on agriculture for their livelihoods (FAO 2016).

South Asia is confronted with a slow of interconnected climate risks. Floods and droughts can become more frequent and severe (associated with an increased rate of water scarcity). Changes in rainfall patterns and land cover can exacerbate erosion (possibly increasing or decreasing groundwater recharge).

The monsoonal rains from the Arabian Sea and the Bay of Bengal typically are the first to arrive in southwestern South Asia. Climate conditions vary from year to year. Some of the inter-annual climate variability is caused by anomalies in the tropical Pacific Ocean's SST[6] caused by the El Niño Southern Oscillation (ENSO) (Kumar et al. 2006). Kumar et al. (2011) used data from the NCEP/NCAR reanalysis and outgoing long-wave radiation for over 60 years (1948–2009) to investigate inter-annual variability in monsoon onset over Kerala (MOK). The study revealed that El Niño, La Niña, positive IOD, negative IOD, and concurrent years play a significant role in altering the MOK. The unusual persistence of westerlies (easterlies) many days before to MOK and increased (suppressed) deep convection across the southeastern Arabian Sea and the southern Bay of Bengal are the most notable characteristics during early (delayed) MOK years. During La Niña, negative IOD, and concurrent La Niña and negative IOD years, moisture builds up over peninsular India many pentads before to MOK, showing that it has a substantial effect on MOK, as opposed to El Niño, positive IOD, and concurrent El Niño and positive IOD years. During a delayed (early) MOK, the monsoon Hadley cell and Walker circulations become weaker (stronger). Furthermore, during delayed (early) MOK, sea surface temperature anomalies in the western Pacific are negative (positive).

Natural seasonal variations can be seen across most of South Asia, with a monsoonal rainy season beginning in April and lasting into May, followed by a dry winter season beginning in October and ending in December. This kind of climate variability is also characterized by spatial variability.

There are six main climatic zones in South Asia: tropical wet, tropical dry, semiarid, arid, humid subtropical, and highlands. Southern Bangladesh, southwest India, and Sri Lanka are among the tropical wet zones that receive the most rainfall.

[6] Sea Surface Temperature.

The region's annual rainfall averages 970 mm/year, with large differences between countries. During June and September, the monsoon season receives 70–90% of the annual precipitation. Due to orographic impact, the Himalayan range produces some of the world's highest annual rainfall. The presence of a high mountain range in the south restricts the landward advance of the monsoon, whose impact is weakened as it moves northwestwardly, resulting in low rainfall in Afghanistan and Pakistan. The majority of precipitation falls as snow in high-elevation areas during the late winter and early spring months (Lacombe et al. 2019).

10.2.2 Climate Change in Asia

People and infrastructure have been exposed to more environmental risks in recent decades due to population growth, tourism, and economic development (high confidence). The Andes, high Asian mountains, the Caucasus, and the European Alps have all been connected to changes in the cryosphere (medium confidence) (IPCC 2019).

Asian mega deltas may be especially threatened by climate change due to large populations and high exposure to sea-level rise, storm surges, and river flooding. The warming climate is projected to be well above the global mean in central Asia, the Tibetan Plateau, and northern Asia; is projected to be above the global mean in eastern Asia and South Asia; is projected to be close to the global mean in Southeast Asia. Precipitation; is expected to rise in northern Asia and the Tibetan Plateau, as well as in eastern Asia and the southern parts of Southeast Asia during the boreal winter. Summer precipitation is expected to rise in northern Asia, East Asia, South Asia, and most of Southeast Asia, but to fall in central Asia. Summer heatwaves/hot spells are likely to last longer and they could be more intense, and occur more frequently in East Asia. East Asia and South Asia are likely to have less extremely cold days. The frequency of extreme precipitation events is very likely to increase in parts of South Asia and East Asia. Tropical cyclone-related extreme rainfall and winds are expected to increase in East Asia, Southeast Asia, and South Asia (IPCC 2008).

Wide swaths of arid and semi-arid land stretch from west China and Mongolia to west Asia. Water scarcity/stress is a constraint for sustainable development even in Asia's humid and sub-humid regions. Asia has a large population that is rapidly expanding, it endures low development, and features the limited coping ability to climate change, which combined with various socioeconomic pressures is projected to intensify Asia's water-shortage situation (IPCC 2008).

AOGCM[7] simulations indicate consistency in broad aspects of Asian climate change, but there are sources of uncertainty. Model evaluation is limited in some areas due to a lack of observational data. The predicted changes in regional climatic averages and extremes have been received a limited evaluation. There are significant differences between inter-model in how they represent monsoon processes and a lack

[7] Atmosphere–Ocean General Circulation Models.

of clarity about ENSO changes contributes to the uncertainty about future regional monsoon and tropical cyclone activity. As a result, obtaining quantitative estimates of expected precipitation shifts is difficult. Because of the region's diverse topography and maritime influences, some local climate changes are likely to differ significantly from regional trends. The factors that affect monsoonal flow and precipitation are critical for understanding climate change in this area because monsoons are the dominant phenomenon across much of Asia. Both the strength of monsoonal flows and the volume of water vapor transported affect precipitation. AOGCM simulations project the weakening of monsoonal flows and the tropical large-scale circulation. The strength of the summer monsoons is linked to ENSO (Boschat et al. 2011). Therefore, the changes in ENSO would affect the monsoons. The rising strength of the studied teleconnection in the latter decades of the twentieth century In the MPI-ESM has a complicated relationship with the consensus that it was diminishing in strength in the late twentieth century (Herein et al. 2018). Furthermore, there is a connection between Eurasian snow cover and monsoon intensity, with the monsoon strengthening if the snow cover recedes. Monsoonal precipitation is further influenced by aerosols, especially absorbing aerosols in South Asia. The deposition of LAAs on snow resulted in a drop in surface albedo throughout April–May–June, triggering a series of feedback reactions, beginning with increased net surface solar radiation rapid snowmelt in the HTP and warming of the surface and upper troposphere, followed by enhanced low-level south-westerlies and increased dust loading over the Himalayas–Indo-Gangetic Plain. Higher dust aerosol heating exacerbated the warming, which was then amplified by latent heating from increased precipitation across the Himalayan foothills and northern India, via the heightened heat pump effect during June–July–August. The atmosphere-land heating induced by LAAs, particularly desert dust, plays a fundamental role in physical processes underpinning the snow–monsoon relationship proposed by Blanford more than a century ago (Lau and Kim 2018). However, most emission scenarios indicate that, at least in the South Asian region, potential changes in the regional environment would still be driven by increased greenhouse gas forcing rather than changes in sulfate and absorbing aerosols (AR4).

10.2.2.1 Impacts on the Water Resources

The combined influence of global climate change and rapid urbanization have exacerbated extreme weather events, magnified urban storm water, which has heightened stormwater flooding (IPCC 2018).

Regardless of the emissions scenario river flow in snow-dominated or glacier-fed high mountain basins is expected to shift (very high confidence), with increases in average winter flow (high confidence) and earlier spring peaks (very high confidence). Average annual and summer runoff from glaciers are anticipated to peak at or before the end of the twenty-first century (high confidence) in all emissions scenarios, for example, around mid-century in High Mountain Asia, followed by a drop in glacial runoff. Glacier flow losses projected by 2100 (RCP8.5) might lower basin flow by

10% or more in at least one month of the melt season in numerous significant river basins, notably during the dry season in High Mountain Asia (low confidence) (IPCC 2019).

Based on the pathway AR5 flooding is expected to become more frequent in parts of South, Southeast, and Northeast Asia (limited evidence, medium agreement). The current drying and warming in South and East Asia are consistent with declining trends in discharge in low and mid-latitudes (Dai 2013). In a warmer world, global means precipitation increases (almost certainly), but there are significant differences, including some losses, from place to area. In subtropical latitudes, notably in the Mediterranean, Mexico, Central America, and portions of Australia precipitation would decline; while it increases elsewhere, particularly in the high north and in India and Central Asia. At high latitudes and in the wet tropics, average annual runoff is expected to rise, but in most dry tropical regions, it is expected to fall. However, there is a great deal of uncertainty about the scale and direction of change in some places, particularly in China, South Asia, and in areas of South America. Flood hazard is expected to increase in areas of South and Southeast Asia, while it is expected to decrease in sections of Central Asia.

According to AR4[8] some of the effects of climate change include (1) freshwater availability in the Central, South, East and Southeast Asia is expected to decline by the 2050s, especially in large river basins; (2) increased flooding from the Sea, in some mega deltas and flooding from rivers would put coastal areas, especially densely populated mega delta regions in South, East and South-East Asia, at risk; (3) climate change is expected to exacerbate the strains on natural resources and the environment as a result of rapid urbanization, industrialization, and economic growth. Endemic morbidity and mortality due to diarrheal disease mainly associated with floods and droughts are expected to increase in East, South, and South-East Asia because of predicted changes in the hydrological cycle.

The AMOC is expected to diminish in the twenty-first century under all RCPs (quite likely), while a collapse is improbable (medium confidence). According to CMIP5 forecasts the collapse or lack of collapse of AMOC by 2300 are equally likely for high emissions scenarios and the AMOC collapse is extremely improbable for lower emissions scenarios (medium confidence). Lower marine productivity in the North Atlantic (medium confidence), more storms in Northern Europe (medium confidence), less Sahelian summer rainfall (high confidence), and South Asian summer rainfall (medium confidence), fewer tropical cyclones in the Atlantic (medium confidence), and a rise in regional sea-level along North America's northeastern coast (medium confidence) are all likely to come from any substantial weakening of the AMOC (IPCC 2019).

Lacombe et al. (2019) stated that all climate-related risks are becoming more severe as a result of climate change. More extreme and intense rainfall is causing sharper and more damaging flash floods in mountainous areas such as Afghanistan, northern Bangladesh, Bhutan, northern India, Nepal, and northern Pakistan. Altered

[8] Fourth Assessment Report (of the IPCC).

rainfall patterns and siltation of water bodies may increase erosion and sediment deposition, especially in Pakistan and Nepal, which would reduce water storage capacities for dry-season irrigation, hydropower generation, and groundwater recharge. The number of GLOFs[9] is increasing as the temperature rises in Bhutan, Nepal, India, and Pakistan. It also causes earlier flood peaks in the snow and ice-fed rivers, such as the Indus Basin, in the springtime. These peaks do not correspond to peak irrigation demand during the summer, putting food security and hydropower generation at risk. Increased demand for crop water, lower yields in already hot regions, increased evaporation losses from surface reservoirs and decreased groundwater recharge are all consequences of rising temperature. Sea-level has risen in combination with more severe and frequent cyclones. Which is causing coastal flooding particularly in low-lying and densely populated areas such as Bangladesh, as well as salt pollution of coastal aquifers, as in Sri Lanka. Vulnerability to meteorological and hydrological droughts in semi-arid and arid regions, including Afghanistan, northwestern India, and Pakistan is higher. Reduced snowmelt and ice melt river flow, reduced groundwater recharge, and reservoir siltation affect agricultural, industrial, and domestic water usage.

Annual mean rainfall has been decreasing in Russia, north-east and north China, Pakistan's coastal belts and arid plains, parts of north-east India, Indonesia, Philippines, and some parts of Japan. Annual mean rainfall is rising in western China, the Changjiang (River Yangtze) Basin, the south-eastern coast of China, the Arabian Peninsula, Bangladesh, and the western coasts of the Philippines. Extreme weather events linked to El Niño have been confirmed to be more frequent and severe in Southeast Asia in the last 20 years (Ahmad et al. 2020).

Asian glaciers have been melting since the 1960s. Glacial runoff and the frequency of glacial lake outbursts, which trigger mudflows and avalanches have increased due to ongoing glacier melting (Bhadra 2002; Rai and Gurung 2005). On the other hand, individual glaciers may be advancing and/or thickening owing to increased precipitation– for instance, in the central Karakoram. Given its high topography, climate-system coupling, and advancing and surge-type glaciers with complicated flow patterns, the Karakoram Himalaya is likely the least studied area. Because of several influencing elements, such as the westerlies, the Indian summer monsoon, multiple teleconnections, topographic impacts, glacier debris-cover characteristics, glacier dynamics, and geological conditions, glacier fluctuations in the Karakoram Himalaya vary in space and time (Dobreva et al. 2017).

Climate-related threats are being exacerbated by a number of anthropogenic factors (Lacombe et al. 2019), including:

1. Unsustainable pumping rates (primarily in Bangladesh, India, and Pakistan) are putting water supply and efficiency in jeopardy.
2. Changes in land use, especially in mountainous areas, are hastening erosion (Afghanistan, Bhutan, India, Nepal, and Pakistan)
3. Reduction of groundwater recharge and infiltration

[9] Glacial lake outburst flood.

4. Rising flooding risk and siltation in downstream regions.

Changes in river flow could affect the availability of surface water from major rivers like the Euphrates and Tigris. A GCM[10]estimated average rise in temperature of 1.2 °C in Lebanon under a doubled-CO_2 climate would result in a 15% reduction in annual net available water supply, while river flows would increase in winter and decrease in spring (Bou-Zeid and El-Fadel 2002).

These reports illustrated the frequency of heavy rainfall is decreasing in some countries of Asia; yet, the frequency of more extreme rainfall events have been increased in many parts of Asia, resulting in large floods, landslides, debris flows, and mudflows, while the number of rainy days and average annual precipitation have decreased. South Asian countries experience frequent drought incidents recently. The lack of drought quantification research in Bhutan and the Maldives is a major source of worry (Chandrasekara et al. 2021).

Droughts, floods, extreme temperatures (heat waves), rainfall, and sea-level risen are some of the climatic threats, which were assessed quantitatively in research in climate by Amarnath et al. (2017) in South Asia (excluding Afghanistan). Geographic information systems were applied to model these climatic threats and population densities, allowing for the calculation of human exposure and agricultural losses. India has 72% of the affected population, followed by Bangladesh and Pakistan with 12% each. An overview of individual risks reveals that floods and droughts are the most significant threats affecting agricultural areas, after that severe rainfall, extreme temperature, and sea-level rise. These are the most vulnerable regions in South Asia. This vulnerability assessment indicates the entire country of Bangladesh, as well as the Indian states of Andhra Pradesh, Bihar, Maharashtra, Karnataka, and Odisha; Sri Lanka's Ampara, Puttalam, Trincomalee, Mannar, and Batticaloa; Pakistan's Sindh and Balochistan; central and eastern Nepal; and the Indus transboundary river basins, are all severely impacted by climate change.

According to Qin (2002) by 2050 a predicted rise in surface air temperature in northwestern China is expected to result in a 27% decline in the area of glaciers, a 10–15% decline in the frozen-soil area, an increase in flood and debris flow and more extreme water shortages compared to 1961–1990 based on linear extrapolation of observed changes.

India's gross per capita water supply is expected to decline from about 1,820 m^3/yr in 2001 to as low as 1,140 m^3/yr in 2050 because of population growth (Gupta and Deshpande 2004). A report predicts that in 2050, the India's per capita availability will be 1219 cum, compared to 1434 cum in 2025. If the per capita availability reaches below 1000 cum, it is called water scarcity condition. Some of the basins that fall into this category are the Indus (up to the border), Krishna, Cauvery, Subernarekha, Pennar, Mahi, Sabarmati, East Flowing Rivers and West Flowing Rivers of Kutch and Saurashtra including Luni. Cauvery, Pennar, Sabarmati, East Flowing Rivers and West Flowing Rivers of Kutch and Saurashtra including Luni face more acute water scarcity (CWC 2019).

[10] General Circulation Model.

Hoanh et al. (2004) suggested that the maximum monthly flow of the Mekong would increase by 35–41% in the basin and by 16–19% in the delta, with a lower estimate for 2010–2038 and a higher estimate for 2070–2099, compared to 1961–1990 levels. Alternatively, the minimum monthly flow is expected to decrease by 17–24% in the basin and 26–29% in the delta.

10.2.2.2 Water Sector Policies and Strategies

Climate adaptation is being facilitated in some parts of Asia by including it into subnational development planning, early warning systems, integrated water resource management, agroforestry, and coastal mangrove regeneration (AR5).

Adaptive approaches to avail the water resource and water budget management are extensively recommended due to the variety and complexity of issues and challenges in South Asia. A better understanding of climate-related threats, the identification of information gaps, and the creation of water sector adaptation frameworks serve as a foundation for such a response. Adequate water management and the prevention of climate-related risks necessitate reliable data and an understanding of how water quality and quantity vary spatially and temporally. Floods, droughts, pollution of surface water and groundwater, flooding, and downstream siltation could all be less likely as a result of improved awareness. Climate change adaptation and risk reduction do not always necessitate diversification of strategies; instead, they may be tailored to current adaptation choices. Following that is a discussion of guidelines for dealing with the two major types of climate-related risks: droughts and floods (Lacombe et al. 2019).

Under GCM projections for the A2 scenario the water sector accounts for about 50% of total global adaptation cost, which was distributed regionally in the following percentages:: East Asia/Pacific, 20%; Europe/Central Asia, 10%; Latin America/Caribbean, 20%; the Middle East/North Africa, 5%; South Asia, 20%; sub-Saharan Africa, 20% (AR5).

Recent studies in Asia demonstrated the marginal livelihood groups that are dependent on primary sources. These are especially vulnerable to climate change impacts if their natural resource base is depleted due to overuse, or if their governance structures are incapable of responding effectively (Leary et al. 2013). Climate change adaptation includes infrastructure projects such as coastal protection in the Maldives, and the prevention of glacial lake outburst flooding in Nepal. Shrestha and Shrestha (2004) Lacombe et al. (2019) provided suggestions for dealing with droughts:

1. Improve water-use efficiency and productivity information transfer;
2. Carefully design and built reservoirs to increase their water storage capacity
3. Improve understanding of the surface water/groundwater continuum and the management of combined surface and groundwater uses;
4. Improve broad data analysis, drought monitoring, and early warning systems.

From a data and information management standpoint there are many approaches to coping with floods and landslides, including mapping flood-prone and landslide-prone areas to restrict vulnerable settlements in these exposed areas, building flood refuges and flood protection systems, and predicting floods and landslides to boost preparedness of the exposed population and give them more time to avoid hazards.

Here are some examples of adaptation that could prevent the majority of the predicted losses include (IPCC 2014): exposure reduction via structural and non-structural measures, effective land-use planning, and elective relocation, discount in the vulnerability of lifeline substruction and services (such as, water, energy, waste management, food, biomass, mobility, local ecosystems, telecommunications), and constructing monitoring and early warning systems; measures to identify exposed areas, assist vulnerable areas and households, and diversify livelihoods, economic diversification, Heat health warning systems, Urban planning to reduce heat islands; Improvement of the built environment; Development of sustainable cities, disaster preparedness including early-warning systems and local coping strategies, adaptive/integrated water resource management, water infrastructure and reservoir development, water infrastructure and reservoir development, water sources diversification, including re-use, and more efficient water usage (e.g., improved agricultural practices, irrigation management, and resilient agriculture).

Peatland drainage contributes 1.3 to 3.1% of current global CO_2 emissions from the burning of fossil fuels, according to study by Hooijer et al. (2010) and Couwenberg et al. (2010) in Southeast Asia, and peatland rewetting might significantly reduce net GHG emissions (AR5).

Education systems can assist in the education and training among the citizens to meet the ongoing and evolving demands of climate change. Local governments have a significant role in providing autonomous adaptation (Resurreccion et al. 2008).

The IPCC warns against focusing too much on short-term results or failing to anticipate outcomes. Resilience in other sectors or ecosystems may be reduced by poorly designed development programs or sector-specific adaptation strategies due to that climate change cuts through sector borders. Some development paths could make certain groups more vulnerable to future climate change. This is a challenging task in South Asia where economies are rapidly expanding populations (AR5, IPCC 2014).

The various ecological, social, economic, technological, institutional, and political constraints will continue to limit effective adaptation and adaptive ability, especially in developing Asian countries. Wastewater recycling (or reuse) is a long-term cost-effective and environmentally sustainable solution to climate change adaptation, but it is difficult to make in practical lever due to the lack of the technology facility in under-developing and developing countries in Asia. However, when compared to alternative water sources such as the use of imported water or groundwater, the treatment of wastewater for reuse, and the construction of distribution systems, as is now practiced in Singapore, that can be initially costly but it could not possible across the continent. Nevertheless, they are potentially valuable adaptation options in many Asian countries. Water wastage and leakage should be reduced to compensate for reductions in water supply caused by decreased precipitation and rising temperatures. Market-based approaches to minimizing excessive water usage may also help mitigate the negative effects of climate change on water supplies. Planned water management interventions such as dams and reservoirs could slightly reduce wet-season flows while significantly increasing dry-season flows in rivers like the Mekong, where wet-season discharge is expected to increase while dry-season flows are expected to decrease (IPCC 2008).

10.3 Europe

10.3.1 Geography and Climate

Europe's continent features are dominated by numerous permanent rivers, and many of them flow outward from the continent center. It also has a lot of low-relief areas with maritime, transitional, continental, polar, and Mediterranean climates, which are the primary types of climate in Europe (IPCC Technical Paper VI).

Water availability is relatively high in Europe compared with other continents (EEA 2009). The freshwater resources in Europe are about 2270 km^3/year. Furthermore, just 13% of this resource is withdrawn indicating that enough water is available to satisfy demand. Overexploitation by a variety of economic sectors, on the other hand, presents a threat to Europe's water supplies in many places, and demand frequently exceeds supply. The 44% of total water abstraction in the EU countries are used for energy production, 24% for agriculture 21% for public water supply, and 11% for industrial purposes. Agricultural water accounted for more than half of total national abstraction in southern Europe, which reaches more than 80% in some areas, while more than half of water abstracted in Western Europe is used as cooling water for energy production. The 'consumptive' use of water differs greatly among these industries.

The proportion of total water abstraction devoted to the public water supply section varies by Member State (MS) and can be very small. Most EU Member States have annual freshwater withdrawal rates between 50 and 100 m^3 per capita, according to Eurostat. This water use implies special circumstances. For example, public water is delivered without a use fee to customers in Ireland (141 m^3 per capita), and there are especially high losses in the public water distribution network in Bulgaria (129 m^3 per capita). In some Nordic and Alpine non-member countries, especially Iceland, Norway, and Switzerland, water supplies are plentiful and supply is low. Low withdrawal rates were reported by Estonia and Lithuania owing in part to below-average connection rates to the public supply, while Malta and Cyprus have partially replaced groundwater with desalinated seawater. Austria, Denmark, and Latvia depend solely on groundwater withdrawal for freshwater supply, while Bulgaria, Spain, and Ireland rely primarily on surface water. Furthermore, Cyprus and Malta depend heavily on desalination to meet their drinking water needs (BIOIntelligenceService 2012).

The average river flow in Europe is about 450 mm/year, but it varies greatly, varying from less than 50 mm/year in southern Spain to more than 1500 mm/year in parts of the Atlantic coast and the Alps. The amount of seasonal change in river flow varies across Europe. River flow in the south, for example, maybe minimal during the summer months, accompanied by sporadic and extreme rainfall events that cause drastic but brief increases in river flow. This flow regime makes maintaining a reliable supply of water from rivers without storing it in reservoirs extremely difficult. The Atlantic maritime climate has much less inflow variation during the year in Western Europe. Most of the winter precipitation falls as snow in the north and east, and spring snowmelt account for a significant portion of river flow. The seasonality of the flow regime is often determined by hydrogeological characteristics; rivers fed primarily by groundwater have a higher dry season flow than those fed primarily by surface runoff (EEA 2009).

The combined effects of latitude, topography and distance from the sea are physical features that produce a widely varying distribution of precipitation across Europe, which ranges from less than 400 mm/year in parts of the Mediterranean region and the central plains of Europe to more than 1000 mm/year along the Atlantic coasts from Spain to Norway, the Alps and their eastern extension. However, much of this precipitation is lost to evapotranspiration and the remaining 'effective rainfall' in much of Europe is no more than 250 mm/year. Efficient rainfall in some parts of Southern Europe is less than 50 mm per year due to the physical features (JRC 2006).

In general, precipitation in Europe has been increased during the twentieth century with the range of 6–8% on average between 1901 and 2005. However, there are significant regional variations with a decrease in the Mediterranean and Eastern Europe. In addition, there have been seasonal changes, including an increase in winter precipitation for most of western and northern Europe and a decline in summer precipitation (EEA 2008).

10.3.2 Climate Change in Europe

Europe's annual mean temperature is expected to increase more than the global average. The greatest warming is expected to occur in northern Europe during the winter and in the Mediterranean region during the summer. Northern Europe's minimum winter temperatures are expected to rise further than the rest of the continent. Maximum summer temperatures are expected to rise higher than the average in southern and central Europe. Annual precipitation is projected to increase across most of northern Europe while falling across most of the Mediterranean. Precipitation in central Europe is likely to rise in the winter but fall in the summer. That leads to the extremes of daily rainfall. Precipitation is predicted to increase throughout Northern Europe. The annual number of precipitation days in the Mediterranean is very likely to decrease. The risk of summer drought in Central Europe and the Mediterranean is likely to rise because of global climate change. In most of Europe, the period of the snow season is likely to shorten, and snow depth is likely to decrease (AR4).

Climate change has rendered Europe more vulnerable to adverse impacts. One such effect is the north–south gradient. Southern Europe would be the most negatively affected (EEA 2004). Southern Europe features a warm and semi-arid climate which is predicted to become hotter and drier, posing a danger to its rivers, hydropower, agricultural production and timber harvests. Summer precipitation is expected to decrease in Central and Eastern Europe, resulting in increased water stress. Northern countries are also vulnerable to climate change, but there may be some advantages in the early stages of warming, such as increased crop yields and forest growth (IPCC 2008).

Several aspects of simulated climate change in Europe and the Mediterranean region are qualitatively consistent across models and qualitatively well known in physical terms. But there remain significant uncertainties. Simulated seasonal mean temperature variations differ by a factor of two to three among the current generation of AOGCMs[11] that could also occur at the sub-continental scale. Similarly, climatic models agree on a large-scale increase in winter half-year precipitation in northern areas and a decrease in summer half-year precipitation in southern areas; yet, they disagree on the extent and geographical specifics of precipitation change. These uncertainties indicate the European climate's susceptibility to the magnitude of warming and to shifts in atmospheric circulation and the Atlantic MOC.[12] Modeling flaws in the processes that govern local water and energy cycles in Europe often introduce uncertainty, both in terms of average conditions and extremes. The pronounced natural variability of the European climate is a main source of uncertainty, particularly for short-term climate predictions in the region (Hulme et al. 1999). Several other factors, in addition to global warming and its direct thermodynamic effects, such as increased water vapor transport from low to high latitudes, may influence future climate changes in Europe and the Mediterranean region. Using the CSIRO conformal-cubic atmospheric model (CCAM), McGregor et al (2016) simulated the regional climate of Nusa Tenggara Barat (NTB) Province in eastern Indonesia under the SRES A2 Delayed Development or "Business as Usual" emissions scenario (1971–2100). The findings reveal that there are steep climatic gradients due to orographic influences, resulting in significant local disparities in climate forecasts.

According to the AR4 the European climate is influenced by variations in atmospheric circulation on both inter-annual and longer time scales. Central European faced heatwave in summer 2003, which was characterized by a long period of anticyclonic weather, the extreme cyclone-induced flooding in central Europe in August 2002 and the heavy warming of northern European winters from the 1960s to 1990s, which was influenced by a trend toward a more positive phase of the NAO,[13] are all recent examples (Second Assessment of Climate Change for the Baltic Sea Basin 2015). Topography, especially in areas of complex terrain, modify

[11] Atmosphere–Ocean General Circulation Model.

[12] Meridional Overturning Circulation.

[13] North Atlantic Oscillation.

the effects of atmospheric circulation at fine geographical scales (Bojariu and Giorgi 2005; Fernández et al. 2003). Part of Europe's relatively mild climate, especially in its northwestern regions, can be attributed to the Atlantic MOC's northward heat transport (Stouffer et al. 2006). On the other hand, the greenhouse gas concentrations have been earnestly. According to most climatic models the weak MOC would limit the warming in Europe. However, based on current understanding, reversing the warming to a cooling trend is very unlikely in the northern part of the continent. It is observed through local thermodynamic factors, those headed to have an effect on the European climate, which might have caused the significant change in the climate. A reduction in snow cover in parts of Europe is already occurring in winter, which is likely to cause positive feedback and amplify the warming. In the current climate feedbacks associated with summer soil drying are significant in the Mediterranean region and at times in central Europe. For example, they contributed to the 2003 heatwave (Black et al. 2004; Fink et al. 2004).

10.3.2.1 Impact on Water Resources

In most regions of Europe, North America, and Japan, current snowmaking systems are expected to be less successful in reducing ski tourism in a warmer climate, particularly at 2 °C global warming and beyond (high confidence) (IPCC 2019).

Some of the impacts of climate change and variability on Europe's water resources include (IPCC):

1. Climate change is expected to magnify regional differences in Europe's natural resources and assets. Increased risk of inland flash floods and more frequent coastal flooding and increased erosion would be the negative effects of storms and the rise in sea-level.
2. Under high emissions scenarios mountainous areas would see glacier retreat, decreased snow cover and winter tourism, and extensive species declines in some areas of up to 60% by 2080.
3. Heatwaves and the frequency of wildfires are also predicted to become more extreme due to climate change.
4. Climate change is expected to exacerbate conditions such as high temperatures and drought in southern Europe. It has been observed that the vulnerable to climate change and reduce water accessibility, hydropower generation, summer tourism, and crop productivity in general.

In snow-dominated and glacier-fed river basins, changes in snow and glaciers have impacted the volume and seasonality of runoff and water supplies (very high confidence). Hydropower plants in central Europe, Iceland, Western USA/Canada, and the tropical Andes, for example, have undergone seasonal fluctuations as well as increases and reductions in water inflow from high mountain ranges (medium confidence) (IPCC 2019).

Klein Tank et al. (2002) studied average winter rainfall in the Atlantic Ocean and northern Europe from 1946 to 1999, and the results indicate that average winter rainfall has risen in most of the Atlantic Ocean and northern Europe during the study period. Norrant and Dogudroit (2006) concluded that annual rainfall in the Mediterranean region was negative in the eastern part between 1950 and 2000. Another study found an increase in the average rainfall in wet days in most parts of the continent, even in some areas that are drier (Alexander et al. 2006). The average annual rainfall has been to rise in northern Europe and decline in southern Europe (Giorgi et al. 2004; Räisänen and Alexandersson 2003).

According to Good et al. (2006) the length of the longest yearly dry spell could increase by up to 50% particularly in France and central Europe.

It is projected that climate change will have a range of impacts on water resources (IPCC 2008):

1. Annual runoff is expected to rise in Atlantic and northern Europe, while it will fall in southern, Mediterranean, and Eastern Europe (Falloon and Betts 2010).
2. Under the A2 and B2 scenarios and climate scenarios from two separate climate models annual average runoff in northern Europe (north of 47°N) is expected to increase by approximately 5–15% in the 2020s and by 9–22% by the 2070s (Alcamo et al. 2007). However, runoff in southern Europe (south of 47°N) is expected to decline by 0–23% through the 2020s and by 6–36% through the 2070s.
3. In Central and Eastern Europe groundwater recharge is likely to be limited, with a greater reduction in valleys and lowlands, such as the Hungarian steppes (Falloon and Betts 2010).
4. Seasonality of flow increases, with higher flows during the peak flow season and either lower flows or longer dry periods during the low-flow season (Arnell 2003, 2004).
5. By 2020s the risk of winter floods increases in northern Europe and flash floods occur throughout Europe, and the risk of snowmelt floods changes from spring to winter. Summer flow would be reduced (Falloon and Betts 2010).
6. By 2050s annual runoff would decrease by up to 20–30% in south-eastern Europe (Arnell 2004).
7. By 2070s the summer low flow would decrease by up to 80% (Arnell 2004). Drought risk in Northern Europe is decreasing, while drought risk in Western and Southern Europe is rising. In parts of Spain and Portugal, western France, Poland's Vistula Basin, and western Turkey today's 100-year droughts are expected to return every 10 (or fewer) years by the 2070s (Lehner et al. 2006).

Lehner et al. (2006) project today's 100-year floods may occur more frequently in northern and north-eastern Europe (Sweden, Finland, N. Russia), Ireland, central and eastern Europe (Poland, Alpine rivers), and Atlantic areas of South Europe (Spain, Portugal); and less frequently in large parts of Southern Europe.

10.3.2.2 Water Sector Policies and Strategies

Adaptation policies have been created at all levels of government in Europe, with some adaptation planning being integrated into coastal and water management, environmental protection and land planning, and disaster risk management (AR5).

According to the Fifth Assessment Report of the Intergovernmental Panel on Climate Change there is ample experience in climate adaptation in Europe, exemplified by deployment of flood-protection technologies, growing knowledge in wetlands restoration, investments in of updating flood protection, implementation of best practices and governance instruments in river basin management plans, and integrated water management strategies (e.g., for irrigation, crop species, land cover, industries, and household use), proven adaptation potential from the adoption of more water-efficient technologies and water-saving techniques (e.g., for irrigation, crop species, land cover, industries, and domestic use), implementation of warning systems, adaptation of homes and workplaces, as well as transportation and energy infrastructure, emissions reductions to enhance air quality, improved wildfire management, development of insurance products against weather-related yield variability. This experience may prevent the hazards posed by climate change (high confidence).

Climate change would present Europe with two major water management challenges: increased water stress, primarily in southeastern Europe, and increased flood risk across the continent. Adaptation for dealing with these difficulties are well known (IPCC 2001b). Biodiversity, landscape, soil and land erosion, forest degradation, natural hazards, water conservation, and recreational environments are all examples of key environmental stresses. The majority of habitats in Europe are controlled or semi-managed, and they are often fragmented and stressed by pollution and other human impacts (IPCC 2008).

Ecosystem-based adaptation, such as protected areas, conservation agreements, and community management of natural areas, is taking place across Central and South America. In some parts of the agricultural industry, resilient crop varieties, climate projections, and integrated water resource management are being embraced (AR5).

Reservoirs and dykes are likely to remain the primary structural flood-prevention measures in both highland and lowland areas (Hooijer et al. 2004). Other planned adaptation options, such as expanded floodplain areas (Helms et al. 2002), emergency flood reservoirs (Somlyódy 2002), floodwater preservation areas (Silander et al. 2006), and flood forecasting and warning systems, especially for flash floods, are becoming more common. Flood and drought adaptation measures are provided by multi-purpose reservoirs.

The most common and the planned solutions needed to adapt to the arising water stress remain supply-side measures such as damming rivers to create instream reservoirs (Iglesias et al. 2005). However, environmental restrictions and high investment costs (Schröter et al. 2005) are limiting new reservoir construction in Europe. Other supply-side approaches, such as wastewater reuse and desalination, are becoming extremely taken into account, although their popularity is diminished by health issues associated with wastewater usage (Gereš 2004).

Some plans demanded pivotal side strategies, such as reducing domestic, manufacturing, and agricultural water use, reducing urban water leakage systems and irrigation, and water pricing. These are also pragmatic plans. Regional and watershed-level climate change adaptation strategies are being incorporated into proposals for integrated water management, while national strategies are being developed to fit into current governance frameworks (IPCC Technical Paper VI).

Some countries and regions (e.g., the Netherlands, the United Kingdom, and Germany) are developing water sector adaptation procedures and risk management practices that recognize the volatility of expected hydrologic changes (IPCC 2008).

Water pricing, effective water use (water conservation), raising awareness and combating are illicit water withdrawals. Those are all comprehensive policies and activities that must be implemented in order to achieve sustainable water resource management. In these policy areas, the EU and its Member States will play a critical role by using public spending and grants to build and sustain appropriate infrastructure, encourage technical innovation and incentivize behavioral change.

Parts of the 'Green New Deal' public investment programs being considered by some governments in reaction to the ongoing global economic downturn. This is especially important for the compacts plan to solve the major issues of climate change (EEA 2009):

- promoting effective water pricing across all industries, including the installation of metering to enable volume-based charging, yet making water affordable by communities to meet basic needs (drinking, cooking, bathing, cleaning);
- assuring that agricultural subsidies are related to improved water efficiency;
- Investing in emerging technologies to improve water quality and improve water infrastructure networks;
- concentrating investment in the long-term viability of renewable water sources where demand management strategies have already been exhausted.

Water resources management is a matter of public policy. The policy should be so that all national and international activities for assessing water resources are coordinated and financed over the long term. Individual countries' approaches to accomplishing this aim will vary, but they will almost always require the imposition of rules and administrative choices, particularly in terms of distributing financial resources (Negm et al. 2020).

10.4 North America

10.4.1 Geography and Climate

The population of North America was about 520,000,000 people in 2005, being the fourth most populous continent in the world the accounting for around 5.5% of the global population. North America is composed of 38 countries and islands. The most heavily populated areas are the eastern United States, Mexico, Central America, and the Caribbean, while the least densely populated areas are Canada, Alaska, and Greenland (Wheeler 2016). The climate of North America varies on a large scale from the frost-free tropical climate of southernmost Florida to the perennial ice and snow of the Canadian Archipelago's northernmost islands, from the rain-drenched mountains of the northwest coast to the drought-stricken deserts of the southwest. Because of its vast size (about 25 million km^2) this continent encompasses a wide range of latitudes (about 5° to the north pole in the northern hemisphere). It has a diverse range of climates that are very similar to those of the much larger continent of Eurasia at similar latitudes (Corcoran and Johnson 2005).

The vegetation of North America may be classified into eight distinct groups, namely (1) deciduous forest (with four distinct seasons with warm summers and cold, wet winters), (2) coniferous forest (also known as taiga, cold and dry with snowy winters and warmer summers), (3) Mediterranean (warm to high temperatures with rainfall in the autumn and winter months), (4) Grassland (hot summers and cold winters with above average rainfall), (5) tundra (this area is characterized by a layer of permafrost (soil that has remained below freezing for at least two years) whose Winters are very cold, summers are warm and there is little rainfall), (6) alpine/mountain (cold, windy and snowy) where winter is from October to May with temperatures below freezing, while summer is from June to September where the temperature can reach 15 °C, (7) rainforest (high temperatures and high rainfall throughout the year), and (8) desert (Warm to high temperatures with very little rainfall).

North America is bordered on the east by the Atlantic Ocean, on the north by the Arctic Ocean, and on the west by the Pacific Ocean. Mexico, Central America, the Caribbean, and the southern United States have more constant temperatures during the year, and feature rainy and dry seasons from May to October and November to April, respectively (Wheeler 2016).

10.4.2 Climate Change in North America

In these regions the warming is projected to exceed the global mean warming. Warming is most likely to be largest in the winter in northern regions and in the summer in the southwestern United States. Minimum winter temperatures are expected to rise faster than in other seasons in Northern North America's. In the southwest maximum summer temperatures are expected to rise faster than the national average. Annual mean precipitation is expected to rise in Canada and the northeastern United States, but it is projected to fall in the southwestern United States. Precipitation in southern Canada is expected to increase in the winter and spring, but it is to projected decrease in the summer. In most of North America the snow season length and snow depth are expected to decrease with the exception of the northernmost part of Canada where maximum snow depth is expected to increase. Annual mean warming is projected to exceed global mean warming in most areas.

Warming is most likely to occur in northern regions during the winter and in the southwestern United States during the summer. The minimum winter temperatures would occur in the northern portion. The North American continent is to projected to experience faster warming than the global average. The ability of AOGCMs to capture the dynamical features affecting the area is closely linked to the uncertainties in regional climate changes across North America. ENSO and the NAO/AO[14] responses to climate change show significant model-to-model variations in Atmosphere–Ocean General Circulation Models. The extent of reduced warming in the extreme north-eastern part of North America as a result of changes in the Atlantic MOC is uncertain; cooling here cannot be completely ruled out. AOGCMs' resolution of the Hudson Bay and Canadian Archipelago is low, which is leading to uncertainty in ocean circulation and sea ice shifts as well as their effect on northern regions' climate. The MMD models do not predict tropical cyclones accurately. Shifts in the frequency, intensity, and tracks of disturbances making landfall in southeast regions are unclear. High-altitude terrain is poorly resolved in MMD models with a coarse horizontal resolution, which is resulting in an underestimation of warming associated with snow-albedo feedback at high elevations in western regions of the United States and Canada. There is scant information about the dynamic effects of warming over land than over the ocean, which could impact the subtropical anticyclone's northward displacement and intensification off the West Coast. The North Pacific eastern boundary current is useful for trade that could be helpful to the offshore Ekman shipping, upwelling and its cooling effect on SST, persistent marine stratus clouds, and thus precipitation in the southwest United States (Fourth Assessment of the IPCC).

According to the IPCC mid-latitude cyclones are affecting the central and northern areas of North America. This caused a small poleward change in storm tracks. These atmospheric has anomalies have increased the number of powerful cyclones, and have decreased medium-strength cyclones. Atmospheric moisture transport and convergence in the northern part of the continent is expected to increase the temperature as a result of the predicted warming over the surface. This is resulting in an increase in annual precipitation across most of the continent, with the exception rate of the south and south-western parts of the United States and Mexico.

A subtropical ridge of high pressure is associated with the thermal contrast between land and adjacent ocean dominates the southwest area, which is making it extremely arid. But, in early in July the North American Monsoon System evolves (Higgins and Mo 1997). The prevailing winds that circulate over the Gulf of California shift from northerly in the winter to southerly in the summer. This circulation would bring a significant increase in rainfall to the southwestern United States and bring to an end the late spring wet period in the Great Plains (Bordoni et al. 2004). The subtropical anticyclone's amplification and northward relocation, as well as smaller warming over the Pacific Ocean is expected to result in a decrease in annual precipitation in the southwestern United States and northern Mexico.

[14] Arctic Oscillation.

10.4.2.1 Impact on Water Resources

The impacts of climate change and variability in North American water resources can be noted to warming in the western mountains of the United States and Canada. This would likely result in less snowpack, more winter flooding, and lower summer flows, as well as intense competition for over-allocated water resources. Cities are currently subjected to heatwaves, those are expected to face increased numbers, intensities, and duration of heatwaves over the course of the century. Heat waves have had adverse health consequences. The climate change impacts could be interacting with development and pollution that will exert greater stress on coastal communities and habitats. The moderate climate change is expected to increase aggregate rain-fed agriculture yields by 5 to 20% in the early decades of the twenty-first century, with significant regional variability. The crops that are nearing the warm end of their suitable range or depend on heavily depleted water supplies would face significant challenges (AR4).

Climate change would increase competition among agricultural, municipal, industrial, and ecological uses in North America. That has already overburdened water supplies. It brought changes in surface and groundwater hydrology, which are expected to have some of the most significant societal and ecological consequences of climate change in this area. Most of the regions in North America could expect changes in the timing, volume, and frequency of streamflow, which impact temporal and spatial distribution of freshwater that is available for human settlements, agriculture, and industrial users. Warming is projected to accelerate in the coming decades. Climate change affects runoff, streamflow, and groundwater recharge in North America. Canada and United States are assessing the differences in income and geography that may lead to an unequal distribution of possible impacts, weaknesses, and adaptability (IPCC 2008).

The annual runoff in North American catchments varies from area to the annual mean precipitation is expected to decrease in the southwestern United States but increase across most of North America. The increasing rates in precipitation in Canada are expected to be in the range of 20% for the annual mean and 30% for the winter under the A1B greenhouse gas emissions scenario. Several studies project widespread increases in severe precipitation, and droughts linked to increased temporal variability in precipitation. The atmospheric changes in precipitation extremes are expected to be larger than changes in mean precipitation (IPCC Technical Paper VI).

The warming climate has modified the depth, intensity, timing, and duration of precipitation in the western mountains of the United States and Canada. There would early melting and substantial reductions in the snowpack. This means snowmelt runoff would rise, winter and early spring flow would increase (raising flooding potential). The summer flows would decrease subtly in mountain snowmelt-dominated watersheds. The increasing winter precipitation, more intense spring floods on the coast demonstrated the summer droughts along the south coast and southern interior are all expected to occur in British Columbia. This is reducing streamflow and affecting fish survival and water sources in the summer when water demand is at the highest level during the summer season. Lower water levels are expected to intensify problems related to water quality, navigation, tourism, hydropower generation, water transfers, and binational relationships in the Great Lakes. Three main factors are likely to affect groundwater availability: withdrawals (which are growth, water demand, and the availability of other sources), evapotranspiration (which rises with temperature), and recharge (determined by temperature, timing and amount of precipitation, and surface water/groundwater interactions). Annual groundwater base flows and aquifer levels are stimulated to respond to temperature, precipitation, and pumping, with higher pumping in drier scenarios and larger flows under wetter scenarios. Streamflow in projected to increase during the winter and decrease in the spring and early summer. Salinization of shallow aquifers can occur as a result of increased evapotranspiration or groundwater pumping in semi-arid and arid regions of North America. Furthermore, sea level has risen more rapidly since the 1950s. This is likely to induce saltwater intrusion into coastal aquifers (IPCC 2008).

Hodgkins and Dudley (2011) studied the changes in the timing and magnitude of winter-spring streamflow at gaging stations in eastern North America north of the 41° north parellel over periods of 50, 60, 70, and 80 years. Their results show that over these periods approximately 32% of the stations north of the 44° parallel may have earlier flows; 64% may have significantly earlier flows over the 80-year period, and there are no stations with significantly later flows over any time period examined. The flow timing findings are supported by changes in monthly mean runoff over time. Summer air temperature in New England increased by 1.1 °C from 1950 to 2006, which may have contributed to decreasing base flows in northern and coastal Maine by increasing evapotranspiration. Many variables rose at a slower rate between 1930 and 2006 than they did between 1950 and 2006.

Several studies indicate the key effects of climate change, which would be the depletion of mountain snow accumulation and declines in the snowmelt-derived water supply. These studies project a trend toward earlier spring snowmelt-derived streamflow, which has been observed in many regions across western North America since the late 1940s (Cayan et al. 2001; Regonda et al. 2005; Stewart et al. 2005).

10.4.2.2 Water Sector Policies and Strategies

Governments across North America are assessing and preparing for gradual adaptation, particularly at the local level. Proactive adaptation is taking place (AR5) to safeguard longer-term investments in energy and public infrastructure. North America has an ample capacity to adapt to climate change's water-related aspects. But this capacity has not always shielded people and property from floods, droughts, forest fires, landslides, hurricanes, and other severe weather-related events. Indigenous communities and those who are socially or economically marginalized are among the most vulnerable groups. North American traditions and organizations have favored a decentralized response system in which adaptation is reactive, unevenly distributed, and aimed at coping rather than preventing problems. There are few examples of adaptive behavior affected solely or primarily by climate change predictions and their impacts on water supplies (IPCC Technical Paper VI).

North America's water vulnerability is determined by the efficacy of adaptation and the distribution of coping capability, which are currently inequitable and have not always shielded vulnerable communities from the adverse effects of climate variability and severe weather events. The United States and Canada have developed economies with advanced infrastructure and established institutions, and significant zonal and socioeconomic diversity (Waldhoff et al. 2015). Over the last decade agriculture in North America has been subjected to several extreme weather events. More variable weather, combined with emigration from rural areas and economic pressures have heightened the agricultural sector's vulnerability, raising questions about the sector's potential ability to cope with a more variable environment (Oliver and Wiebe 2003; Wheaton et al. 2005). Agriculture in North America is a dynamic resource. It has been adapting to a variety of pressures and opportunities, such as market and weather shifts. It is a common occurrence in the industry. Crop, sector diversification soil, and water conservation are often used to mitigate weather risks (Wall and Smit 2005).

The private sectors in Canada and United States are already investigating water supply adaptations. The following are some examples of these forms of adaptations (IPCC 2008):

- The City of Peterborough, Canada, suffered two 100-year flood events within three years; in response, it flushed drainage systems and replaced trunk sewage systems to meet more severe 5-year flood requirements (Burrell 2011).
- Droughts in six major US cities, including New York and Los Angeles, prompted adaptive measures, including investments in water management, systems, and new water supply/distribution facilities as a result of recent droughts (Changnon and Changnon 2000).
- Burlington and Ottawa, Ontario, used both structural and non-structural measures to cope with a 15% rise in heavy precipitation, including guiding downspouts to lawns in order to promote infiltration and raising depression and street detention storage (Waters et al. 2003).
- Water usage in Los Angeles, California, has increased by just 7% since 1970, despite population growth of over 35% (nearly one million people) (Wilkinson et al. 2002). This is primarily due to conservation practices.
- The Regional District of Central Okanagan in British Columbia created in 2004 a water management plan for the Trepanier Landscape Unit planning area, which specifically addresses climate scenarios, projected changes in water supply and demand, and adaptation options (Cohen et al. 2004).
- Insurance firms are spending money to avoid potential hazard harm to insured property and to adjust pricing models (Mills and Lecomte 2006).
- Ski resort operators are investing in lifts to reach higher altitudes and in equipment to compensate for decreased snow cover (Scott and Jones 2006).
- New York has reduced overall water consumption by 27% and per capita consumption by 34% since the early 1980s (City of New York 2005).

- In the Los Angeles area local water districts have reward and information programs to promote water conservation (MWD 2005).
- Managed aquifer recharge (MAR) has expanded thus providing underground storage for winter runoff and treated sewage.
- Farmers are changing crop and variety selection, irrigation techniques, and pesticide applications based on highly detailed weather details (Wall and Smit 2006).

10.5 South America

10.5.1 Geography and Climate

South America is the world's fourth-largest continent, with a land area of 17,840,000 km^2 and a surface land area of 3.5% of the Earth's surface. The Atlantic Ocean, the Pacific Ocean, the Southern Ocean, and North America border it to the east, to the west, to the south, and to the north respectively. The continent's eastern side is generally lower than the western side, which includes the Andes mountain range. The Amazon River (6,400 km) is the world's second-longest river (After the Nile), which is flows across the continent eastwardly. Argentina is home to the world's largest waterfall Iguazu (979 m wide), and the world's highest peak outside Asia, Aconcagua (6960 m). There are 16 countries and Islands in South America (Wheeler 2016). In terms of water resources South America is the most diverse continent. The continent is home to the world's largest river in terms of volume of runoff discharged (the Amazon). The driest area on the planet (the Atacama Desert), and the world's longest mountain chain (the Andes). It features large gradients in water supply and quality arising from geographical diversity. The Amazon and Orinoco river basins, as well as the Pacific coasts of Colombia and Ecuador, which are located in the North and Northeast parts of the continent, receive plenty of rain (Marengo 2009; Sanso and Guenni 1999; Urrutia and Vuille 2009). Because of its geographical location, South America has a wide range of climates, including arid and semi-arid areas. From snowy, icy, high elevations to temperate and tropical climates the climatic spectrum is vast. In recent decades glaciers have generally receded and many small glaciers have vanished (IPCC Technical Paper VI). The Amazon, Parana, and Orinoco rivers together bring more than 30% of the world's clean freshwater into the Atlantic Ocean. These water supplies are inequitably distributed to its surroundings and its vast areas found limited water supply (Mata et al. 2001). Many areas in Latin America face severe constraints on their water supply due to droughts that are statistically related to ENSO events.

There are six climatic provinces in South America. The first ("Tropical Cordilleran") covers the far northwest (Colombia and Ecuador's coasts), with "perpetual spring" climates at high altitudes, high temperatures near sea-level, and a tropical regime of precipitation. The second ("South American Tropical") encompasses the vast northern and northeastern territory east of the Andes and extends to the south of the tropical belt. This region is influenced by trades and features equatorial rains, with mean annual temperatures exceeding 25 °C. The third ("Peruvian") extends all the way from the Pacific coast to 30° degrees south, including northern Chile. The weather in this province is unusually cool and dry. Its southern neighbor, the "North Chilian" province, has a sub-tropical climate with winter rains. Further south, the "South Chilian" province encompasses the continent's most southern reaches, is extremely rainy, and has mild year-round temperatures with cool summers. The sixth (" Pampa ") province, which includes the area east of the Andes and south of the "Tropical" province, has a wide range of temperatures, especially in the north, although rain is scarce (DEC. WARD 2021).

Large-scale agriculture and livestock raising, as well as growing urbanization and intensification of economic activities, characterize the southeast of the continent, which includes the south of Brazil, Uruguay, Paraguay, and Argentina. This area is largely semi-arid, with the South American Monsoon system influencing climate variability and thus water availability (Marengo et al. 2012).

10.5.2 Climate Change in South America

According to the Fourth Assessment of the IPCC the annual mean warming in southern South America is likely to be close to the global mean warming. The changes in atmospheric circulation may cause significant local variability in precipitation response in mountainous areas, while annual precipitation is likely to decrease in most of Central America and the southern Andes. Winter precipitation in Tierra del Fuego (Argentina) is expected to rise by summer precipitation in south-eastern South America. It is unknown how annual and seasonal mean rainfall in northern South America, including the Amazon, would change. In some areas the simulations, however, are qualitatively consistent (rainfall increasing in Ecuador and northern Peru, and decreasing at the northern tip of the continent and in southern northeast Brazil). The wide inter-model variations in predicted future changes in El Niño amplitude and the systemic errors in simulating the current mean tropical climate and its variability prevent a definitive evaluation of regional changes over large areas of Central and South America. Most MMD models, especially over most of Amazonia (AMZ), are poor in reproducing regional precipitation patterns in their control experiments and have a low signal-to-noise ratio. Lower resolution models do not perform well in the high and sharp Andes Mountains, thus affecting the assessment of the future climate in South America. The feedbacks from changes in land use and land cover, as with all landmasses, are poorly accommodated. This produces ambiguity in model projections. The possibility of sudden biogeochemical system changes in AMZ remains a source of uncertainty. Large variations occur in predicted climate sensitivities by climate models (Friedlingstein et al. 2003).

Tropical cyclones over Central America may become an additional source of uncertainty for regional climate change scenarios, as systemic changes in hurricane tracks and intensity may influence summer precipitation in parts of northern South America. Changes in the strength and position of tropical convection are the primary concern over most of Central and South America, extratropical disturbances, on the other hand, influence the winter climate of Mexico and the climate of southern South America throughout the year. Specific geographical features that influence the climate in the region include a continental barrier over Central America and along the Pacific coast of South America, and in the world's largest rainforest (Amazonia). Most of Mexico and Central America feature a dry season from November through April, and a rainy season from May to October (Magaña et al. 1999). The rainy season's seasonal evolution is primarily determined by air-sea interactions over the Americas' warm oceans, topographic effects over a dominant easterly flow, and the ITCZ's[15] temporal evolution. The seasonal fluctuation of the Subtropical North Atlantic Anticyclone dominates atmospheric circulation over the Gulf of Mexico and the Caribbean Sea during the boreal winter with invasions of extratropical systems affecting mostly Mexico and the western portion of the Great Antilles. The mean seasonal period of precipitation in tropical and subtropical latitudes over South America is dominated by a warm-season precipitation maximum associated with the South American Monsoon System (Vera et al. 2006). Chen et al. (2001) reported that, despite deforestation, Amazonia has experienced an increase in rainfall over the last 40 years due to global-scale water vapor convergence (Fourth Assessment of the IPCC). The rainforest's future is critical for its ecological value, and the global carbon cycle's future evolution, and the driver of regional climate change. ENSO has a strong influence on the monsoon system (Lau and Zhou 2003). Therefore, the future changes in ENSO would result in complementary changes in the South American region. Liebmann et al. (2004) reported that the displacement of the South Atlantic Convergence Zone has substantial regional ramifications, such as the recent strong positive precipitation trend across southern Brazil.

10.5.2.1 Impact on the Water Resources

The rapid rise in water demand was underpinned by rapid economic and demographic growth rates. That is the main driver of water-related issues in South America. The continent's population grew by 1.1% on average in 2011 (Cepal 2011).The South American economic growth has focused strongly on industries like agriculture and mining that are at the same time highly water-intensive and polluting. Agriculture uses about 70% of available water supplies in South America, which is close to the global average. An area of rising precipitation covers the Ecuadorian and Peruvian coasts, the Andes region, and south Amazonia on a continental scale. The Caribbean coast and northeast Amazonia, as well as the Chilean and Bolivian highlands, are flanked by an area with declining precipitation. Industrial activities such as agriculture (The Peruvian coast), mining (the tropical Andes, northern Chile, Bolivia, and the Brazilian Amazon), and agriculture (southern and eastern Brazil, Uruguay,

[15] Inter-Tropical Convergence Zone.

Paraguay, and Argentina) are hotspots of pressure on gross water supplies (Buytaert and Breuer 2013). Mining in the Brazilian Amazon resorts to hydropower production for aluminum processing.

Climate change has many impacts on water resources (AR4):

1. Increases in temperature and associated decreases in soil water are expected to gradually replace tropical forests in eastern Amazonia with savanna by the mid-twenty-first century. Arid-land vegetation would gradually replace semiarid vegetation.
2. Species extinction poses a major threat to biodiversity in several parts of the tropical American continent.
3. The loss of glaciers and changes in rainfall patterns are expected to have a major impact on water supply for human use, agriculture, and energy production.
4. Some essential crops' productivity is expected to fall, livestock productivity is expected to fall posing a threat to food security. Soybean yields are expected to rise in temperate zones. The number of people at risk of hunger is expected to rise overall.

Droughts linked to La Niña have severely limited water supply and irrigation water supply in central-western Argentina and central Chile. El Niño -related droughts in Colombia reduced the flow of the Cauca River (Pabón 2003).

The South American climate's inter-annual and inter-decadal variability is the result of the superposition of several large-scale phenomena. The El Niño Southern Oscillation (ENSO) has a powerful, direct impact on coastal Ecuador, Peru, and northern Chile, as well as an indirect impact (via atmospheric teleconnections) on most subtropical South America, even at high altitudes. Likewise, the climate and weather of eastern South America are affected by the sea surface temperature (SST) meridional gradient over the tropical Atlantic. Droughts in Amazonia and North-eastern Brazil have been linked to unusually warm tropical North Atlantic surface waters. Climate variability over South America appears to be influenced by northern hemisphere high-latitude forcings, such as the Antarctic Oscillation (AAO) and the North Atlantic Oscillation (NAO) (Garreaud et al. 2009).

Over the last three decades, climate disasters such as floods, droughts, and land-slides have increased in Latin America (e.g., heavy precipitation in Venezuela (1999 and 2005); the flooding in the Argentinean Pampas (2000 and 2002), the Amazon drought (2005), catastrophic hail storms in Bolivia (2002) and Buenos Aires (2006), Cyclone Catarina in the South Atlantic (2004), and the record hurricane season of 2005 in the Caribbean region). Between 1970 and 1999 and 2000–2005 the number of climate-related disasters increased by 2.4 times, continuing a pattern that began in the 1990s (Nagy et al. 2006).

Precipitation is projected to increase in southern Brazil, Paraguay, Uruguay, north-east Argentina (Pampas), and parts of Bolivia, north-west Peru, Ecuador, and north-west Mexico. The higher precipitation caused a 10% rise in flood frequency in the Amazon River at Obidos, a 50% increase in streamflow in the Uruguayan, Parana, and Paraguayan rivers, and further floods in the Bolivian Amazon's Mamore Basin. There has also been a rise in extreme rainfall events and consecutive dry days in the later basins. In Chile, south-western Argentina, north-eastern Brazil, southern Peru, and western Central America (e.g. Nicaragua), on the other hand, precipitation has been decreasing (Camilloni 2005; Lagomarsino et al. 2005; Marengo 2004; Penalba and Vargas 2004; Ronchail et al. 2005).

22.2 million people lived in water-stressed watersheds (i.e., with reserves of less than 1,000 m^3/capita/yr) in 1995, a number expected to rise by 12 to 81 million in the 2020s, and by 79 to 178 million in the 2050s, according to the SRES scenarios (Arnell 2004). The joint negative impact of rising demands for water supply and irrigation due to an increasing population coupled with drier conditions in many basins would exacerbate the need for water.

The Chacaltaya Glacier in Bolivia (16°S) is a good example of a small glacier that is disintegrating and most likely disappearing. Studies indicate that its area had shrunk from 0.22 km^2 in 1940 to less than 0.01 km^2 in 2005 (Berger et al. 2005; Francou et al. 2003; Ramirez et al. 2001). The glacier lost 90% of its surface area and 97% of its ice volume between 1992 and 2005 (Berger et al. 2005). A linear extrapolation of these observed data predicts it may vanish entirely before 2010 (Coudrain et al. 2005). Glacier mass balance has occurred in the tropics is sensitive to changes in precipitation and humidity; the Chacaltaya's shrinkage since the 1980s is consistent with a 50 m/decade ascent of the 0 °C isotherms in the tropical Andes (Vuille et al. 2003).

10.5.2.2 Water Sector Policies and Strategies

Implementing urban drainage management is costly and disruptive to metropolitan areas. Less impervious surfaces enhance groundwater recharge. Green infrastructure, and rooftop gardens are examples of low-regret techniques with co-benefits. As sea-levels rise water levels in coastal outfalls also rise, obstructing drainage. Many earlier rainfall design standards are still in use, and they must be modified to suit current climatic conditions; Wetland protection, particularly mangroves, and land-use planning policies might assist to minimize the severity of flood occurrences that are examples of how adaptation might assist to minimize the bulk of the expected losses (AR5).

There are examples of adaptation that could prevent the majority of the predicted losses: integrated water resource management; urban and rural flood management (including infrastructure), early warning systems, better weather and runoff forecasts, and infectious disease control; development of early warning systems for disease control and mitigation based on climatic and other relevant inputs. Many factors augment vulnerability. It is imperative to establish programs to extend basic public health services (IPCC 2014).

The absence of sufficient adaptation strategies in Latin American countries to deal with the hazards and risks of floods and droughts is owing to low GNP, population increase in flood-prone, landslide-prone, or drought-prone areas, as well as a lack of sufficient political, institutional, and technological frameworks (Solanes and Jouravlev 2006).

The IRDB (2000) reports that many poor inhabitants have been encouraged to relocate from flood-prone areas to safer places. They erected new dwellings with the help of IRDB and IDFB financing, such as resettlements in Argentina's Paraná River Basin following the 1992 disaster (IPCC 2008).

Self-organization programs for strengthening water delivery is practiced in very disadvantaged communities contribute to people's adaptive capacity to water stresses. The Water and Sanitation Clusters of Business Partners for Development have been working in Latin America on four "focus" plans: Cartagena (Colombia), La Paz (Bolivia), and El Alto (Bolivia), as well as several unidentified initiatives. Between 2000 and 2002 the safe water programs e constructed ten rainwater catchment and storage systems for local communities in arid regions of Santiago del Estero Province, Argentina (Basán Nickisch 2002).

Rainwater cropping and storage systems in the semi-arid tropics are critical components of long-term development. The NGO Network Articulaço no Semi-rido (ASA) Project has formed a collaborative project called the P1MC Project in Brazil, which calls for one million cisterns to be built by civilians in a decentralized manner. The aim is to provide drinking water to one million rural households in Brazil's semi-arid tropics (BSATs), where drought is a constant. By the end of 2004 ASA and the Brazilian Ministry of the Environment had installed 12,400 cisterns, with another 21,000 expected (Gnadlinger 2003).

Water management policies in Latin America must be achieved to the level of adaptation to climate change. This will strengthen the region's ability to better handle the water supply. Adapting to drier conditions in roughly 60% of Latin America would necessitate significant investments in water supply systems. Managing trans-basin diversions has been successful in some instance (e.g., Yacambu Basin in Venezuela, Alto Piura, and Mantaro Basin in Peru). Water recycling and water use optimization have been suggested to cope with water management during water-stressed times (COHIFE 2003).

10.6 Polar Region (Antarctica)

10.6.1 Geography and Climate

Antarctica is the world's coldest continent. Throughout the year the average interior temperature is −57 °C with a minimum temperature of −90 °C during the winter season. During the summer temperatures can reach a maximum between −2 °C and 8 °C along the coast. On average, it is the coldest, windiest, and driest continent on the planet. Antarctica and the Southern Ocean represent 20% of the planet's surface.

The low average incidence of direct solar radiation at high southern latitudes makes Antarctica a cold land. The eccentricity of the earth's orbit around the earth causes the Antarctic region to receive slightly more solar radiation than the Arctic. During their summers Polar Regions absorb about 30% of the energy obtained at the equator, but during their winters neither Pole receives any solar radiation. The high latitudes of the southern hemisphere have a different distribution of land and sea than those of the northern hemisphere, which makes the Antarctic area much colder than the Arctic regions. The Antarctic Circle, which is surrounded by the Southern Ocean is almost fully filled by Antarctica and its associated ice shelves. The region of floating ice covering Antarctica (17–20 million km^2) in the winter, that is larger than the continent itself (14 million km^2). The ice sheet that covers the majority of the continent is 2000 m high on average (Campbell and Claridge 2009).

The Polar Regions are composed of large cryosphere components, which also dominate their hydrologic processes and they are the water source. That is likely to undergo some of the earliest and most profound climate-induced changes. The emphasis in the Antarctic has been on the mass balance of the major ice sheets and their effect on sea level. An ice sheet, from a high central ice plateau with a slight average surface slope, flows outwards. The margins typically have a steeper slope and the majority of the ice is discharged into fast-flowing ice streams or outlet glaciers. It was recorded sometimes into the Sea or onto floating ice shelves. In the modern world there are only three massive ice sheets: one on Greenland and two on Antarctica (the East and West Antarctic ice sheets, divided by the Transantarctic Mountains). Others existed during glacial times. The ice shelf is a large floating sheet of ice extending from the coast (usually with a large horizontal extent and a flat or gently sloping surface), often filling embayment in the ice sheets' coastline. Antarctica is the home to almost all of the world's ice shelves.

Mean annual air temperatures vary from −18 °C near the coast to −24 °C on Mt Fleming at 1700 m altitude, with −18 °C at Marble Point near the coast, −20 °C on the Wright Valley floor, and −24 °C on Mt Fleming at the head of the Wright Valley. Snowfall and infrequent rains are common in the MDVs throughout the summer. Precipitation is lowest in the mid-valley floors, with a mean annual precipitation of 45 mm recorded at Vanda Station during two years. Snowfall is heavier near the valley's eastern and western ends, as well as at higher elevation (Balks and O'Neill 2016).

10.6.2 Climate Change in Antarctica

Based on studies by Nicolas and Bromwich (2014); Jones et al. (2016); Turner et al. (2016) in contrast to the Arctic, temperature changes across the Antarctic continent have been less consistent during the last 30–50 years, with warming in regions of West Antarctica and no major change in East Antarctica.

Based on what has been reported in Special Report on the Ocean and Cryosphere in a Changing Climate (IPCC 2019) the Antarctic Circumpolar Current (ACC), the world's biggest ocean current, circumnavigates Antarctica, moving $173.3 \pm 10.7 \times 10^6$ $m^{3\ s-1}$ (Donohue et al. 2016) of water eastward in a geostrophic balance built up by the contrasting characteristics of waters surrounding Antarctica and those inside the subtropical gyres to the north of ACC.

Antarctica will be warmer and precipitation is likely to increase over the continent. In the Polar Regions it is uncertain how much the frequency of severe temperature and precipitation occurrences in the Polar Regions will shift. On inter-annual, decadal, and longer time scales, it has been observed that the polar climate has many natural variabilities, which is a major source of uncertainty. The patterns in the underlying teleconnections, such as the NAM[16] or ENSO, are subject to considerable uncertainty in projections. Furthermore, due to the dynamic atmosphere-land-cryosphere-ocean ecosystem interactions involving a number of distinct feedbacks, our understanding of the polar climate system remains incomplete. Clouds, planetary boundary layer processes, and sea ice are examples of processes that are poorly represented in models. Furthermore, global models' structure and resolution are insufficient to address critical processes in the polar seas. All of this adds up to a wide variety of current and future simulations, thus lowering confidence in future predictions. The lack of observations against which to evaluate models and build process knowledge, especially over Antarctica, constitutes an evidence defficiency (Fourth Assessment of the IPCC).

During the austral summer the SAM has shown a positive trend in recent decades, indicating a strengthening of the surface westerly winds near Antarctica. According to palaeoclimate reconstructions (Abram et al. 2014; Dätwyler et al. 2018) this protracted positive phase of the SAM is unusual in at least 600 years and is related to colder conditions throughout the continent (Special Report on the Ocean and Cryosphere in a Changing Climate).

The Southern Ocean mediates the atmospheric circulations of Antarctica and lower latitudes across almost all longitudes; therefore, the inter-annual variability and longer-term shifts in the climatic climate of the southern extra-tropics have significant global consequences for human society (Simmonds et al. 1998).

Changes in snow accumulation indicated the result of global climate change, as well as the trend of temperature change, which are of particular interest over Antarctica, especially any variations in warming between the peninsula and the interior of the ice sheet. Warming of the troposphere, as in the Arctic, is expected to increase precipitation. Changes in ocean and atmosphere circulation, on the other hand, can change the pattern of air masses, affecting precipitation and temperature patterns significantly across the area. Many studies have found signatures of these patterns in the Antarctic (Carleton 2003).

Cold anomalies over most Antarctica and warm anomalies over the Antarctic Peninsula are correlated with the positive phase of the SAM. The SAM has been drifting in recent decades. Observational studies have shown that the Antarctic Peninsula has warmed significantly, but the majority of the continent has remained relatively unchanged over the last half of the twentieth century. The SAM exhibits a robust positive trend in transient warming simulations, and ozone hole in the late twentieth century is also a positive perturbation to the SAM. That made an ordinary extrapolation of current trends into the future uncertain. The SO has a poorer statistical association with surface temperature over Antarctica than the SAM, but it has

[16] Northern Annular Mode.

a significant connection with SST and sea ice fluctuations in the Pacific sector of the Southern Ocean. The positive SOI trend might explain two-thirds of the winter ice-edge trend in this sector, which is connected to ice drift and surface winds, exposing the ice edge to substantial decadal SO variability. If this association persists, the negative SOI trend previous to the satellite period predicts ice edge trends on a comparable time scale that is contrary to the recent record. Significant low-frequency ice edge changes, associated with natural SO variability, are overlaid over any anthropogenic forcing-related changes (Kwok et al. 2016).

The Southern Oscillation Index has been correlated to Antarctic precipitation and accumulation, but the signal's persistence is unknown. Recent research indicates that the nonlinear interactions between ENSO and SAM differ on decadal time scales. Strong negative correlations are associated with strong positive (negative) anomalies in the Inter-decadal Pacific Oscillation and the Amundsen Sea Low during periods when SAM and ENSO indices are of opposite (equal) sign (Dätwyler et al. 2020).

Over the last few decades, the SO index has been trending downward (corresponding to a trend toward more El Niño-like conditions in the equatorial Pacific. Which is correlated with Sea of ice cover anomalies in the Pacific sector, namely negative (positive) anomalies in the Ross and Amundsen Seas). However, it is impossible to make a conclusive evaluation of ENSO amplitude and frequency shifts in the twenty-first century (Kwok and Comiso 2002).

Warming projections for the twenty-first century indicate scenario-independent geographical trends close to those seen in recent decades. Warming would be greater over land and at the highest northern latitudes with the least warming predicted over the Southern Ocean (near Antarctica) and northern North Atlantic, following recent trends.

10.6.2.1 Impact on Water Resources

According to IPCC (2019) under RCP2.6 the rise in global mean sea-level (GMSL) is expected to be 0.39 m (0.26–0.53 m, probable range) between 2081 and 2100, and 0.43 m (0.29–0.59 m, probable range) between 1986–2005 and2100. The GMSL rise for RCP8.5 is 0.71 m (0.51–0.92 m, probable range) from 2081 to 2100, and 0.84 m (0.61–1.10 m, probable range) in 2100. Under RCP8.5 the mean sea-level rise estimates are 0.1 m greater than AR5 in 2100, and the probable range expands beyond 1 m in 2100 owing to a larger predicted ice loss from the Antarctic ice Sheet (medium confidence). The ice sheets, particularly in Antarctica, are a major source of uncertainty near the end of the century. According to studies by Fogwill et al. (2016) and Shadwick et al. (2017) due to the recent calving of Mertz Glacier Tongue in East Antarctica sea ice and ocean stratification have been changed, and polynyas there are now twice as productive.

Climate change has many impacts on water resources, such as reduced thickness and extent of glaciers, ice sheets, and sea ice, as well as changes in natural ecosystems are the key predicted biophysical impacts, with negative consequences for a variety of species. It is including migratory birds, mammals, and superior predators. The destructive effects include the infrastructure causes and traditional native ways of life. Unique ecosystems and habitats are vulnerable in both Polar Regions as climatic barriers to species invasions are lowered (IPCC).

On century and longer time scales processes dominating the timing of future ice-shelf loss and the amount of ice sheet instabilities might boost Antarctica's contribution to sea-level rise to levels far greater than the expected range (low confidence). This high-impact risk deserves consideration because of the repercussions of sea-level rise that a collapse of portions of the Antarctic Ice Sheet would entail (IPCC 2019).

The Antarctic's freshwater aquatic ecosystems are highly vulnerable to climate change, especially changes in air temperature, though such trends differ across the continent. In the Dry Valleys, for example, lake productivity has been observed to decrease as air temperature drops (Doran et al. 2002).

According to the IPCC (2019), Antarctic Ice Sheet lost mass at a mean pace of 155 ± 19 Gt yr^{-1} (0.43 ± 0.05 mm yr^{-1}) between 2006 and 2015, owing mostly to accelerated thinning and retreat of key outlet glaciers draining the West Antarctic Ice Sheet (very high confidence). Outside of Greenland and Antarctica, glaciers lost mass at a pace of 220 ± 30 Gt yr^{-1} (equal to 0.61 ± 0.08 mm yr^{-1} sea-level rise) from 2006 to 2015. In the Amundsen Sea Embayment of West Antarctica and Wilkes Land, East Antarctica, ice flow and retreat have accelerated, potentially leading to sea-level rise of several meters within a few centuries (very high confidence) (IPCC 2019).

The Southern Ocean is one of the world's most significant CO_2 sinks. However, in recent decades it has become less successful at absorbing global CO_2 due to increased westerly winds, which have induced an upwelling of CO_2 -rich water from other regions, reducing its ability to absorb additional CO_2. The stronger winds of the Southern Ocean have driven more deep and warm subsurface water to well up to the surface in the Amundsen Sea region of West Antarctica. Warm water has seeped under the ice shelves of West Antarctica, which was causing them to erode. These ice shelves serve as "brakes" for glaciers nearby. As the shelves thin, the glaciers accelerate and drain ice from the main ice sheet at a faster rate than previously recorded. The massive Pine Island Glacier (50 km wide) and its neighboring glaciers are speeding up as a result of this operation placing the West Antarctic ice sheet in jeopardy. The collapse of the ice sheet could release as much ice into the sea as Greenland melting, but this is unlikely in this century (ACCE). Increasing air temperatures on the maritime sub-Antarctic island of Signy have resulted in some of the quickest and most severe changes in lake temperature ever recorded in the Southern Hemisphere (Quayle et al. 2002) (AR4).

Warming and changes in sea ice are expected to boost marine net primary production in the Arctic (medium confidence) and around Antarctica (low confidence), with nutrient availability varying owing to upwelling and stratification shifts (IPCC 2019).

According to Special Report on the Ocean and Cryosphere in a Changing Climate on polar regions (IPCC 2019) the significant ice loss from the Greenland and Antarctic ice sheets during the early twenty-first century is very likely to continue into the near future, contributing to global sea-level rise. Recent Antarctic Ice Sheet

(AIS) mass losses may be irreversible over decades to millennia according to limited data and widespread consensus.

10.6.2.2 Water Sector Policies and Strategies

The Antarctic Treaty Consultative Parties continue to work closely with the Scientific Committee on Antarctic Research, the Council of Managers of National Antarctic Programs, and the Subsidiary Group of the Committee, the International Association of Antarctica Tour Operators, and other NGOs to understand, mitigate, and adapt to impacts associated with changes to the Southern Ocean and Antarctic cryosphere for Environmental Protection on Climate Change Response, through the Committee for Environmental Protection (CEP) and its Subsidiary Group of the Committee for Environmental Protection on Climate Change Response. Understanding, mitigating, and adapting to climate change are among the primary research goals recognized in the area (Kennicutt et al. 2014a, b), and bilateral and multilateral programs are now underway (IPCC 2019).

Adaptation might mitigate the majority of the expected losses in polar regions in the following ways: A better understanding of the problem via scientific and indigenous knowledge, resulting in more effective remedies and/or technical breakthroughs; improved monitoring, regulation, and warning systems to ensure that ecosystem resources are used safely and sustainably; if possible, hunting or fishing for a variety of species and diversifying revenue sources; Enhanced observation, monitoring, and warning systems; co-production of more resilient solutions that integrate science and technology with indigenous knowledge; improved communications, education, and training; shifting resource bases, land use, and/or settlement areas; co-production of more robust solutions that integrate science and technology with indigenous knowledge; enhanced observation, monitoring, and warning systems; Improved communications, education, and training (IPCC 2014).

The Antarctic Treaty System (ATS) refers to the Antarctic Treaty and its associated agreements as a whole. The Consultative Parties to the Antarctic Treaty decided that a Climate Change Response work plan would address these issues (XXXIX 2016). As a result, the Committee for Environmental Protection's Subsidiary Group on Climate Change Response was formed (Secretariat 2017). Consensus, on the other hand, is now constraining CCAMLR's (2017a) work program-level responses to climate change, despite chances to incorporate climate concerns into mechanisms for implementation and monitoring targeted at conserving ecosystems and the environment (Brooks et al. 2018).

Based on Special Report on the Ocean and Cryosphere in a Changing Climate on Polar Regions the growth of polar glaciers and ice sheets, as well as their effects on global sea-level, must be better understood. Longer and more accurate quantifications of their changes are needed, particularly where mass losses are largest, as well as clearer attribution of natural vs human sources. It is necessary to have a better knowledge of Antarctica's vulnerability to marine ice sheet instability, as well as if recent developments in West Antarctica signal the start of irreversible change (IPCC 2019).

The International Panel on Climate Change reports that future adaptation steps range from measures to facilitate the use of water supplies (for example, improvements in ice-road construction techniques has increased open-water transportation, which caused flow control for hydroelectric production, harvesting strategies, and drinking-water access methods) to adaptation strategies to cope with increased/decreased freshwater hazards (for example, ice jams are frequent in cold regions, resulting in catastrophic floods and major damage due to a quick rise in water levels upstream of ice jams. Citizens, authorities, insurance companies, and government organizations are all concerned about these floods since they can be crucial hydrological and hydraulic catastrophes, (Madaeni et al. 2020)).

Antarctica is a frozen desert with low rainfall, with the majority of falls as snow. The Antarctic Strategy is being developed as part of the Director General's Organizational Development and Change process, and the IUCN Program 2009–2012. An IUCN Antarctic Strategy will strengthen IUCN's ability to control, promote, and assist countries, organizations, and the private sector to protect the Antarctic ecosystem's integrity and diversity, and the non-material wilderness and scientific values. Whenever the use of natural resources is justified it must be done in a fair and environmentally friendly manner (International Union for Conservation of Nature).

Bibliography

2015, The 2015 Revision of the UN's world population projections: population and development review 41:557–561

Abram NJ, Mulvaney R, Vimeux F, Phipps SJ, Turner J, England MH (2014) Evolution of the Southern Annular Mode during the past millennium. Nature Climate Change 4:564–569

Abungba JA, Khare D, Pingale SM, Adjei KA, Gyamfi C, Odai SN (2020) Assessment of hydro-climatic trends and variability over the black volta basin in Ghana. Earth Syst Environ 4:739–755

ACCE, IAATO fact sheet summarizing the antarctic climate change and the environment report. Climate Change in Antarctica - Understanding the Facts

ACPC, 2011, Climate change and water in Africa: analysis of knowledge gaps and needs, United nations Economic Commission for Africa. African Climate Policy Centre; United Nations. Economic Commission for Africa, p 18

Ahmad F, Uddin MM, Goparaju L, Rizvi J, Biradar C (2020) Quantification of the land potential for scaling agroforestry in South Asia. KN-J Cartogr Geograph Inform 70:71–89

Alcamo J, Flörke M, Märker M (2007) Future long-term changes in global water resources driven by socio-economic and climatic changes. Hydrol Sci J 52:247–275

Albert JS, Destouni G, Duke-Sylvester SM, Magurran AE, Oberdorff T, Reis RE, ... Ripple WJ (2020) Scientists' warning to humanity on the freshwater biodiversity crisis. Ambio, 1–10

Alexander LV, Zhang X, Peterson TC, Caesar J, Gleason B, Klein Tank AMG, Haylock M, Collins D, Trewin B, Rahimzadeh F, Tagipour A, Rupa Kumar K, Revadekar J, Griffiths G, Vincent L, Stephenson DB, Burn J, Aguilar E, Brunet M, Taylor M, New M, Zhai P, Rusticucci M, Vazquez-Aguirre JL (2006) Global observed changes in daily climate extremes of temperature and precipitation. J. Geophys. Res. Atmosph. 111

Amarnath G, Alahacoon N, Smakhtin V, Aggarwal P (2017) Mapping multiple climate-related hazards in South Asia. Colombo, Sri Lanka. Int Water Manag Inst (IWMI). 41 p. (IWMI Research Report 170). http://www.iwmi.cgiar.org/publications/iwmi-research-reports/iwmi-research-report-170/. Accessed May 17, 2018

AQUASTAT (2011) African dams database. Food and Agriculture Organization of the United Nations, Rome

Arnell NW (2003) Relative effects of multi-decadal climatic variability and changes in the mean and variability of climate due to global warming: future streamflows in Britain. J Hydrol 270:195–213

Arnell NW (2004) Climate change and global water resources: SRES emissions and socio-economic scenarios. Global Environ Change 14:31–52

Arnell NW (2006) Climate change and water resources: a global perspective. Cambridge University Press, Cambridge, pp 167–175

Ashton PJ (2002) Avoiding conflicts over Africa's water resources: AMBIO. J Human Environ 31:236–242

Balks MR, O'Neill TA (2016) Soil and permafrost in the Ross Sea region of Antarctica: stable or dynamic?

Basán Nickisch M (2002) Sistemas de captación y manejo de agua. Estación Experimental Santiago del Estero. Instituto Nacional de Tecnología Agropecuaria. http://www.inta.gov.ar/santiago/info/documentos/agua/0001res_sistemas.htm

Bates B, Kundzewicz Z, Wu S, Palutikof J (2008) Climate change and water technical paper of the intergovernmental panel on climate change (Geneva: IPCC Secretariat). Climate Change 95:96

Berger T, Mendoza J, Francou B, Rojas F, Fuertes R, Flores M, Noriega L, Ramallo C, a. H. Baldivieso ER (2005), Glaciares Zongo – Chacaltaya – Charquini Sur – Bolivia 16°S. Mediciones Glaciológicas, Hidrológicas y Meteorológicas, Año Hidrológico 2004–2005: Informe Great Ice Bolivia, IRD-IHHSENMAHI-COBEE, 171

Bhadra B (2002) Regional cooperation for sustainable development of Hindu Kush Himalaya region: opportunities and challenges: Keynote paper presented at the alpine experience–an approach for other mountain regions, Berchtesgaden, Germany, June, pp 26–29

Bhatasara S, Nyamwanza A (2018) Sustainability: a missing dimension in climate change adaptation discourse in Africa? J Integr Environ Sci 15:83–97

Biasutti M (2019) Rainfall trends in the african sahel: characteristics, processes, and causes: wiley interdisciplinary reviews. Climate Change 10:e591

BIOIntelligenceService B (2012) Literature review on the potential climate change effects on drinking water resources across the EU and the identification of priorities among different types of drinking water supplies, Final report -ADWICE projectprepared for, European Commission DG Environmentunder contract number 070326/SER/2011/610284/D1

BGS (2010) A literature review of recharge estimation and ground water resource assessment in Africa. Keyworth, Nottingham: Ground water resources prlogramme, British Geological Survey (BGS)

Black E, Blackburn M, Harrison G, Hoskins B, Methven J (2003) 2004, Factors contributing to the summer. Europ Heatwave Weather 59:217–223

Black RE., Allen LH, Bhutta ZA, Caulfield LE, De Onis M, Ezzati M, Mathers C, Rivera J, Maternal, C. U. S. Group (2008) Maternal and child undernutrition: global and regional exposures and health consequences. Lancet 371:243–260

Bojariu R, Giorgi F (2005) The North Atlantic Oscillation signal in a regional climate simulation for the European region. Tellus A: Dyn Meteorol Oceanogr 57:641–653

Bordoni S, Ciesielski PE, Johnson RH, McNoldy BD, Stevens B (2004) The low-level circulation of the North American Monsoon as revealed by QuikSCAT. Geophys. Res. Lett. 31

Boschat G, Terray P, Masson S (2011) Interannual relationships between Indian summer monsoon and Indo-Pacific coupled modes of variability during recent decades. Climate Dyn 37:1019–1043

Bou-Zeid E, El-Fadel M (2002) Climate change and water resources in Lebanon and the Middle East. J Water Resour Planning Manag 128:343–355

Brooks CM, Ainley DG, Abrams PA, Dayton PK, Hofman RJ, Jacquet J, Siniff DB (2018) Antarctic fisheries: factor climate change into their management. Nature Publishing Group

Burrell BC (2011) Inland flooding in Atlantic Canada, Fredericton, New Brunswick : Atlantic climate adaptation solutions association, [2011], ©2011, Ottawa, Ontario : Canadian Electronic Library, 2015

Buytaert W, Breuer L (2013) Water resources in South America: sources and supply, pollutants and perspectives: proceedings of the IAHS-IAPSO-IASPEI Assembly, Gothenburg, Sweden, pp 106–113

Buytaert W, Cuesta-Camacho F, Tobón C (2011) Potential impacts of climate change on the environmental services of humid tropical alpine regions. Global Ecol Biogeogr 20:19–33

Camilloni I (2005) endencias climáticas. El Cambio Climático en el Río de la Plata, V. Barros, A. Menéndez and G.J. Nagy,: T Eds., CIMA/CONICET-UBA, Buenos Aires, pp 13–19

Campbell IB, Claridge GG (2009) Antarctic permafrost soils. Springer, Permafrost soils, pp 17–31

Carleton AM (2003) Atmospheric teleconnections involving the Southern Ocean. J Geophys Res Oceans 108

Cavé L, Beekman HE, Weaver J (2003) Impact of climate change on groundwater resources. UNESCO IHP Series 64

Cayan DR, Kammerdiener SA, Dettinger MD, Caprio JM, Peterson DH (2001) Changes in the onset of spring in the western United States. Bull Amer Meteorol Soc 82:399–416

CEPAL N (2011) Economic survey of Latin America and the Caribbean 2010–2011: Internacional integration and macroeconomic policy challenges amid global economic turmoil

Chandrasekara SS, Kwon H-H, Vithanage M, Obeysekera J, Kim T-W (2021) Drought in South Asia: Aa review of drought assessment and prediction in South Asian countries. Atmosphere 12:369

Changnon SA, Changnon D (2000) Long-term fluctuations in hail incidences in the United States. J Climate 13:658–664

Chen T-C, Yoon J-H, St KJ, Croix, and E.S. Takle, (2001) Suppressing impacts of the Amazonian deforestation by the global circulation change. Bull Am Meteorol Soc 82:2209–2216

Christensen JH, Hewitson B, Busuioc A (2007) Regional climate projections. In: Solomon S, Qin D, Manning M, Chen Z, Marquis M, Averyt KB, Tignor M, Miller HL (eds) Climate change 2007: the physical science basis. Contribution of Working Group I to the fourth assessment report of the intergovernmental panel on climate change, pp. 847–940, Cambridge University Press: Cambridge, UK

City of New York (2005), Water Supply - DEP - NYC.gov. The City of New York Department of Environmental Protection, New York, New York. http://www.ci.nyc.ny.us/html/dep/html/watersup.html

Cohen S, Neilsen D, Welbourn R (2004) Expanding the dialogue on climate change and water management in the Okanagan Basin: British Columbia: Final Report January 1, 2002 to June 30, 2004

COHIFE (2003) Principios rectores de Política Hídrica de la República Argentina. Acuerdo Federal Del Agua, Consejo Hídrico Federal, COHIFE 8, August 2003, Argentina

Collins JM (2011) Temperature variability over Africa. J Climate 24:3649–3666

Conway G (2009) The science of climate change in Africa: impacts and adaptation. Grantham Inst Climate Change Discussion Paper 1:24

Conway D (2011) Adapting climate research for development in Africa. Climate Change, 428–450

Corcoran WT, Johnson E (2005) North America, Climate of. In: Oliver JE (ed) Encyclopedia of World Climatology: Dordrecht. Springer, Netherlands, pp 525–535

Coudrain A, Francou B, Kundzewicz ZW (2005) Glacier shrinkage in the Andes and consequences for water resources. Hydrol Sci J

Couwenberg J, Dommain R, Joosten H (2010) Greenhouse gas fluxes from tropical peatlands in south-eastAsia.Global Change Biol 16(6):1715–1732

CWC (2019) Water and Realated Statistics, New Delhi, Water related statistics directorate information system organisation water planning & project wing central water commission

Dai A (2013) Increasing drought under global warming in observations and models. Nature Climate Change 3:52–58

Dasgupta S, Laplante B, Meisner C, Wheeler D, Yan J (2009) The impact of sea level rise on developing countries: a comparative analysis. Climatic Change 93:379–388

Dätwyler C, Grosjean M, Steiger NJ, Neukom R (2020) Teleconnections and relationship between the El Niño–Southern Oscillation (ENSO) and the Southern Annular Mode (SAM) in reconstructions and models over the past millennium. Climate Past 16:743–756
Dätwyler C, Neukom R, Abram NJ, Gallant AJ, Grosjean M, Jacques-Coper M, Karoly DJ, Villalba R (2018) Teleconnection stationarity, variability and trends of the Southern Annular Mode (SAM) during the last millennium. Climate Dyn 51:2321–2339
DEC. WARD, R. (2021) The climate of south America. Taylor & Francis Ltd 35(4):353–360
De Sales F, Okin GS, Xue Y, Dintwe K (2019) On the effects of wildfires on precipitation in Southern Africa. Clim Dyn 52(1):951–967
De Wit M, Jacek S (2006) Changes in surface water supply accross Africa with predicted climate change. Africa Earth Observatory Network, university of Capt Town, Rondebosch
Dobreva ID, Bishop MP, Bush AB (2017) Climate–glacier dynamics and topographic forcing in the Karakoram Himalaya: concepts, issues and research directions. Water 9:405
Döll P, Fiedler K (2008) Global-scale modeling of groundwater recharge. Hydrol Earth Syst Sci 12:863–885
Donohue K, Tracey K, Watts D, Chidichimo MP, Chereskin T (2016) Mean antarctic circumpolar current transport measured in drake passage. Geophys Res Lett 43:11,760–11,767
Doran PT, Priscu JC, Lyons WB, Walsh JE, Fountain AG, McKnight DM, Moorhead DL, Virginia RA, Wall DH, Clow GD (2002) Antarctic climate cooling and terrestrial ecosystem response. Nature 415:517–520
EEA (2004) Impacts of Europe's changing climate: an indicatorbased assessment. EEA Report No 2/2004, European Environment Agency, Copenhagen, Denmark (or: Luxembourg, Office for Official Publications of the EC)
EEA (2008) Impacts of Europe's changing climate — 2008 indicator-based assessment, EEA Report No 4/2008. European Environment Agency, Copenhagen (or: Luxembourg, Office for Official Publications of the EC)
EEA (2009) Water resources across Europe — confronting water scarcity and drought, European Environment Agency, Copenhagen, Denmark (or: Luxembourg, Office for Official Publications of the EC)
Elshamy ME, Seierstad IA, Sorteberg A (2009) Impacts of climate change on Blue Nile flows using bias-corrected GCM scenarios. Hydrol Earth Syst Sci 13:551–565
Falloon P, Betts R (2010) Climate impacts on European agriculture and water management in the context of adaptation and mitigation—The importance of an integrated approach. Sci Total Environ 408:5667–5687
Fanta B, Zaake B, Kachroo R (2001) A study of variability of annual river flow of the southern African region. Hydrol Sci J 46:513–524
FAO (2004a) Yearbook of fishery statistics 2002. Capture Production 94(1):654
FAO (2004b) Yearbook of fishery statistics 2002. Aquaculture Production 94(2):206
FAO (2004c) Data base. Food and Agriculture Organization of the United Nations, Rome
FAO (2016) AQUASTAT database, Food and Agriculture Organization of the United Nations, p http://www.fao.org/aquastat
Fernández J, Sáenz J, Zorita E (2003) Analysis of wintertime atmospheric moisture transport and its variability over southern Europe in the NCEP Reanalyses. Climate Res 23:195–215
Fink AH, Brücher T, Krüger A, Leckebusch GC, Pinto JG, Ulbrich U (2004) The 2003 European summer heatwaves and drought-synoptic diagnosis and impacts. Weather 59:209–216
Fogwill CJ, van Sebille E, Cougnon EA, Turney CSM, Rintoul SR, Galton-Fenzi BK, Clark GF, Marzinelli EM, Rainsley EB, Carter L (2016) Brief communication: impacts of a developing polynya off Commonwealth Bay, East Antarctica, triggered by grounding of iceberg B09B. Cryosphere 10:2603–2609. https://doi.org/10.5194/tc-10-2603-2016
Food, Organization A (2009) Adaptive water management in the Lake Chad Basin. Addressing current challenges and adapting to future needs. FAO Water Seminar Proc World Water Week

Francou B, Vuille M, Wagnon P, Mendoza J, Sicart JE (2003) Tropical climate change recorded by a glacier in the central Andes during the last decades of the twentieth century: Chacaltaya, Bolivia, 16 S. J Geophys Res Atmosph 108

Frenken K (2005) Irrigation in Africa in figures: AQUASTAT Surv 29, Food & Agriculture Org

Friedlingstein P, Dufresne J-L, Cox P, Rayner P (2003) How positive is the feedback between climate change and the carbon cycle?. Tellus B: Chem Phys Meteorol 55:692–700

Gannon KE, Conway D, Pardoe J, Ndiyoi M, Batisani N, Odada E, Olago D, Opere A, Kgosietsile S, Nyambe M (2018) Business experience of floods and drought-related water and electricity supply disruption in three cities in sub-Saharan Africa during the 2015/2016 El Niño. Global Sustain 1

Garreaud RD, Vuille M, Compagnucci R, Marengo J (2009) Present-day south american climate: palaeogeography. Palaeoclimatol Palaeoecol 281:180–195

Gebrechorkos SH, Hülsmann S, Bernhofer C (2019) Long-term trends in rainfall and temperature using high-resolution climate datasets in East Africa. Scient Reports 9:1–9

Gereš D (2004) Analysis of the water demand management. Proc. XXII conference of the Danubian countries on the hydrological forecasting and hydrological Bases of Water Management, Brno

Giorgi F, Bi X, Pal JS (2004) Mean, interannual variability and trends in a regional climate change experiment over Europe. I. Present-day climate (1961–1990). Climate Dyn 22:733–756

Gnadlinger J (2003) Rainwater catchment and sustainable development in the Brazilian semi-arid tropics (BSATs): an integrated approach. 11th IRCS Conference, Mexico City, pp 25–29

Good P, Bärring L, Giannakopoulos C, Holt T, Palutikof J (2006) Non-linear regional relationships between climate extremes and annual mean temperatures in model projections for 1961–2099 over Europe. Clim Res 31:19–34

Goulden M, Conway D (2009) Cooperation and adaptation to climate change in transboundary river basins in Africa: evidence from the Nile Basin: IOP conference series. Earth Environ Sci

Goulden M, Conway D, Persechino A (2009) Adaptation to climate change in international river basins in Africa: a review/Adaptation au changement climatique dans les bassins fluviaux internationaux en Afrique: une revue. Hydrol Sci J 54:805–828

Gupta S, Deshpande R (2004) Water for India in 2050: first-order assessment of available options. Current Sci, 1216–1224

Hamududu BH, Ngoma H (2020) Impacts of climate change on water resources availability in Zambia: implications for irrigation development. Environ Dev Sustain 22(4):2817–2838

Hassan OM, Tularam GA (2018) The Effects of climate change on rural-urban migration in sub-saharan Africa (SSA)—the cases of democratic republic of Congo, Kenya and niger. Appl Water Syst Manag Model, 2

Helms M, Büchele B, Merkel U, Ihringer J (2002) Statistical analysis of the flood situation and assessment of the impact of diking measures along the Elbe (Labe) river. J Hydrol 267:94–114

Herein M, Drótos G, Bódai T, Lunkeit F, Lucarini V (2018) Reconsidering the relationship of the El Ni\~ no--Southern Oscillation and the Indian monsoon using ensembles in Earth system models: arXiv preprint arXiv: 1803.08909

Higgins RW, Mo KC (1997) Persistent north pacific circulation anomalies and the tropical intraseasonal oscillation. J Climate 10:223–244

Hoanh CT, Guttman H, Droogers P, Aerts J (2004) Will we produce sufficient food under climate change? Mekong Basin (South-east Asia). In: Aerts JCJH, Droogers P (eds) Climate Change in Contrasting River Basins: Adaptation Strategies for Water, Food, and Environment, Aerts. CABI Publishing, Wallingford, pp 157–180

Hodgkins GA, Dudley RW (2011) Historical summer base flow and stormflow trends for New Englandrivers. Water Resour Res 47:W07528. https://doi.org/10.1029/2010WR009109

Hofste RW, Reig P, Schleifer L (2019) 17 countries, home to one-quarter of the world's population, face extremely high water stress

Hooijer A, Klijn F, Pedroli GBM, Van Os AG (2004) Towards sustainable flood risk management in the Rhine and Meuse river basins: synopsis of the findings of IRMA-SPONGE. River Res Appl 20:343–357

Hooijer A, Page S, Canadell JG, Silvius M, Kwadijk J, Wösten H, Jauhiainen J (2010) Current and future CO_2 emissions from drained peatlands in Southeast Asia. Biogeosciences 7(5):1505–1514
Hughes WS, Balling RC (1996) Urban influences on South African temperature trends. Int J Climatol 16:935–940
Hulme M, Barrow EM, Arnell NW, Harrison PA, Johns TC, Downing TE (1999) Relative impacts of human-induced climate change and natural climate variability. Nature 397:688–691
Iglesias A, Estrela T, Gallart F (2005) Impactos sobre los recursos hídricos: evaluación Preliminar de los Impactos en España for Efecto del Cambio Climático, pp 303–353
International union for conservation of nature, strategy for IUCN's Programme and Policy on Antarctic issues, as approved by the 72nd meeting of council on 2–4 February 2009
IPCC (Intergovernmental Panel on Climate Change), 2001b: Climate Change 2001: impacts, adaptation, and vulnerability. contribution of working group ii to the third assessment report of the intergovernmental panel on climate change, McCarthy JJ, Canziani OF, Leary NA, Dokken DJ, White KS (Eds) Cambridge University Press, Cambridge, 1032 p
IPCC (Intergovernmental Panel on Climate Change. Climate Change 2007: The physical science basis contribution of working group I to the fourth assessment report of the intergovernmental panel on climate change .Cambridge United Kingdom and New York, NY, USA: 2007
IPCC (Intergovernmental Panel on Climate Change. 2008. Climate Change and Water. Contribution of working group II technical support unit to the fourth assessment report of the technical paper of the intergovernmental panel on climate change, Bates BC, Kundzewicz ZW, Wu S, Palutikof JP (Eds), Cambridge University Press, Cambridge, IPCC Secretariat, Geneva, 210 p
IPCC, 2014: climate change 2014: impacts, adaptation, and vulnerability. part a: global and sectoral aspects. contribution of working group II to the fifth assessment report of the intergovernmental panel on climate change [Field CB, Barros VR, Dokken DJ, Mach KJ, Mastrandrea MD, Bilir TE, Chatterjee M, Ebi KL, Estrada YO, Genova RC, Girma B, Kissel ES, Levy AN, MacCracken S, Mastrandrea PR, White LL (eds)]. Cambridge University Press, Cambridge, United Kingdom and New York, NY, USA, 1132 p
IPCC, 2014: Climate Change 2014: mitigation of climate change. Contribution of working group III to the fifth assessment report of the intergovernmental panel on climate change [Edenhofer O, Pichs-Madruga R, Sokona Y, Farahani E, Kadner S, Seyboth K, Adler A, Baum I, Brunner S, Eickemeier P, Kriemann B, Savolainen J, Schlömer S, von Stechow C, Zwickel T, Minx JC (eds)]. Cambridge University Press, Cambridge, United Kingdom and New York, NY, USA
IPCC, 2018. SR1.5. Global Warming of 1.5°C: An IPCC special report on the impacts of global warming of 1.5 °C above pre-industrial levels and related global greenhouse gas emission pathways, in the context of strengthening the global response to the threat of climate change, sustainable Development, and Efforts to Eradicate Poverty; Cambridge University Press: Cambridge, UK
IPCC, 2019: Summary for Policymakers. In: IPCC special report on the ocean and cryosphere in a changing climate [Pörtner H-O, Roberts DC, Masson-Delmotte V, Zhai P, Tignor M, Poloczanska E, Mintenbeck K, Alegría A, Nicolai M, Okem A, Petzold J, Rama B, Weyer NM (eds.)]. In press
IPCC, 2019: Meredith M, Sommerkorn M, Cassotta S, Derksen C, Ekaykin A, Hollowed A, Kofinas G, Mackintosh A, Melbourne-Thomas J, Muelbert MMC, Ottersen G, Pritchard H, Schuur EAG (2019) Polar Regions. In: IPCC special report on the ocean and cryosphere in a changing climate [Pörtner H-O, Roberts DC, Masson-Delmotte V, Zhai P, Tignor M, Poloczanska E, Mintenbeck K, Alegría A, Nicolai M, Okem A, Petzold J, Rama B, Weyer NM (eds.)]. In press
IRDB, 2000, Gestión de los Recursos Hídricos de Argentina. Elementos de Política para su Desarrollo Sustentable en el siglo XXI. Oficina Regional de América Latina y Caribe. Unidad Departamental de Argentina y los Grupos de Finanzas, Sector Privado y Infraestructura, y Medio Ambiente y Desarrollo Social Sustentable. Informe No. 20.729-AR. August 2000
Jones JM et al (2016) Assessing recent trends in high-latitude Southern Hemisphere surface climate. Nat Clim Chang 6:917. https://doi.org/10.1038/nclimate3103
JRC (2006) An atlas of pan-European data for investigating the fate of agrochemicals in terrestrial ecosystems, Joint Research Centre report EUR 22334 EN

Karamage F, Liu Y, Fan X, Francis Justine M, Wu G, Liu Y, Wang R (2018) Spatial relationship between precipitation and runoff in Africa. Hydrol Earth Syst Sci Discus, 1–27

Kennicutt MC et al (2014a) Polar research: Six priorities for Antarctic science. Nature 512:23–25. https://doi.org/10.1038/512023a

Kennicutt MC et al (2014b) A roadmap for Antarctic and Southern Ocean science for the next two decades and beyond. Antarct Sci 27(1):3–18. https://doi.org/10.1017/S0954102014000674

Kistin EJ (2006) Qualifying cooperation over transboundary waters: water Governance for Africa-The challenge of uncertainty and change. University of Bradford

Klein Tank AMG, Wijngaard JB, Konnen GP, Bohm R, Demaree G, Gocheva A, Mileta M, Pashiardis S, Hejkrlik L, Kern-Hansen C, Heino R, Bessemoulin P, Muller-Westermeier G, Tzanakou M, Szalai S, Palsdottir T, Fitzgerald D, Rubin S, Capaldo M, Maugeri M, Leitass A, Bukantis A, Aberfeld R, VanEngelen AFV, Forland E, Mietus M, Coelho F, Mares C, Razuvaev V, Nieplova E, Cegnar T, López JA, Dahlstrom B, Moberg A, Kirchhofer W, Ceylan A, Pachaliuk O, Alexander LV, Petrovic P (2002) Daily dataset of 20th-century surface air temperature and precipitation series for the European Climate Assessment. Int J Climatol 22:1441–1453

Kruger AC, Shongwe S (2004) Temperature trends in South Africa: 1960–2003. Int J Climatol 24:1929–1945

Kumambala PG (2010), Sustainability of water resources development for Malawi with particular emphasis on North and Central Malawi, University of Glasgow

Kumar, K. K., B. Rajagopalan, M. Hoerling, G. Bates, and M. Cane, 2006, Unraveling the mystery of Indian monsoon failure during El Niño: Science, v. 314, p. 115–119.

Kumar MR, Sankar S, Reason Ch (2011) On the relative roles of el nino and indian ocean dipole events on the monsoon onset over Kerala. Climatol 103(3-–4):359–374

Kundzewicz ZW, Doell P (2009) Will groundwater ease freshwater stress under climate change?. Hydrol Sci J 54:665–675

Kundzewicz ZW, Kanae S, Seneviratne SI, Handmer J, Nicholls N, Peduzzi P, Mechler R, Bouwer LM, Arnell N, Mach K (2014) Flood risk and climate change: global and regional perspectives. Hydrol Sci J 59:1–28

Kundzewicz ZW, Mata L, Arnell NW, Döll P, Jimenez B, Miller K, Oki T, Şen Z, Shiklomanov I (2008) The implications of projected climate change for freshwater resources and their management. Hydrol Sci J 53:3–10.

Kusangaya S, Warburton ML, Van Garderen EA, Jewitt GP (2014) Impacts of climate change on water resources in southern Africa: a review: Physics and Chemistry of the Earth. Parts a/b/c 67:47–54

Kwok R, Comiso J, Lee T, Holland P (2016) Linked trends in the South Pacific sea ice edge and Southern oscillation index. Geophys Res Lett 43:10,295–10,302

Kwok R, Comiso JC (2002) Spatial patterns of variability in Antarctic surface temperature: connections to the Southern Hemisphere annular mode and the Southern oscillation. Geophys Res Lett 29:50–1–50–4

Lacombe G, Chinnasamy P, Nicol A (2019), Review of climate change science, knowledge and impacts on water resources in South Asia. Int Water Manag Ins (IWMI)

Lagomarsino, J., E. Forbes, and G. Nagy, 2005, CAPÍTULO 14 TENDENCIAS CLIMÁTICAS, HIDROLÓGICAS Y OCEANOGRÁFICAS EN EL RIO DE LA PLATA Y COSTA URUGUAYA: EL CAMBIO CLIMÁTICO EN EL RÍO DE LA PLATA, p. 137.

Lau KM, Zhou J (2003) Anomalies of the South American summer monsoon associated with the 1997–99 El Nino–Southern oscillation. Int J Climatol J Royal Meteorol So 23:529–539

Lau WK, Kim K-M (2018) Impact of snow darkening by deposition of light-absorbing aerosols on snow cover in the Himalayas–Tibetan Plateau and influence on the Asian summer monsoon: a possible mechanism for the Blanford hypothesis. Atmosphere 9:438

Leary N, Adejuwon J, Bailey W, Barros V, Batima P, Caffera RM, Chinvanno S, Conde C, De Comarmond A, De Sherbinin A (2013) For whom the bell tolls: Vulnerabilities in a changing climate. Routledge, Climate Change and Vulnerability and Adaptation, pp 19–46

Lehner B, Döll P, Alcamo J, Henrichs T, Kaspar F (2006) Estimating the impact of global change on flood and drought risks in Europe: a continental. Integr Anal Climatic Change 75:273–299

Liebmann B et al (2004) An observed trend in Central South American precipitation. J Clim 17:4357–4367

Lloyd SJ, Kovats RS, Chalabi Z (2011) Climate change, crop yields, and undernutrition: development of a model to quantify the impact of climate scenarios on child undernutrition. Environ Health Perspect 119:1817–1823

Loáiciga HA (2015) Managing municipal water supply and use in water-starved regions: looking ahead. Amer Soc Civil Eng

MacDonald AM, Calow RC, Macdonald DM, Darling WG, Dochartaigh O, B. E, (2009) What impact will climate change have on rural groundwater supplies in Africa? Hydrol Sci 544:690–703

Madaeni F, Lhissou R, Chokmani K, Raymond S, Gauthier Y (2020) Ice jam formation, breakup and prediction methods based on hydroclimatic data using artificial intelligence: a review. Cold Regions Sci Technol 174:103032

Magaña V, Amador JA, Medina S (1999) The midsummer drought over Mexico and Central America. J Climate 12:1577–1588

Marengo J, Liebmann B, Grimm A, Misra V, Silva Dias PD, Cavalcanti I, Carvalho L, Berbery E, Ambrizzi T, Vera CS (2012) Recent developments on the South American monsoon system. Int J Climatol 32:1–21

Marengo JA (2004) Interdecadal variability and trends of rainfall across the Amazon basin. Theoret Appl Climato 78:79–96

Marengo JA (1920) 2009, Long-term trends and cycles in the hydrometeorology of the Amazon basin since the late. Hydrol Proces Int J 23:3236–3244

Mata LJ, Campos M, Basso E, Compagnucci R, Fearnside P, Magri G, Marengo J, Moreno AR, Suaez A, Solman S, Villamizar A, Villers L (2001) Latin America. Climate Change 2001, impacts, adaptation, and vulnerability. Contribution of working group II to the third assessment report of the intergovernmental panel on climate change, McCarthy JJ, Canziani O, Leary N, Dokken D, White K (Eds), Cambridge University Press, Cambridge, pp 691–734

Mba WP, Longandjo G-NT, Moufouma-Okia W, Bell J-P, James R, Vondou DA, Haensler A, Fotso-Nguemo TC, Guenang GM, Tchotchou ALD (2018) Consequences of 1.5 C and 2 C global warming levels for temperature and precipitation changes over Central Africa. Environ Res Lett 13:055011

McGregor L, Nguyen C, Kirono GC, Katzfey J (2016) High-resolution climate projections for the islands of Lombok and Sumbawa, Nusa Tenggara Barat Province, Indonesia: Challenges and implications. ELSEVIER Climate Risk Manag 12(2016):32–44

Mills E, Lecomte E (2006) From risk to opportunity how insurers can proactively and profitably manage climate change, United States

MWD (2005) The family of southern california water agencies. metropolitan water district of Southern California. http://www.bewaterwise.com/index.html

Nagy GJ, Caffera RM, Aparicio M, Barrenechea P, Bidegain M, Jiménez JC, Lentini E, GM a. Co-authors (2006) Understanding the potential impact of climate change and variability in Latin America and the Caribbean. Report prepared for the Stern Review on the Economics of Climate Change, p 34

Negm AM, Romanescu G, Zelenakova M (2020) Water resources management in Balkan countries (Springer Water). Springer International Publishing Ag, Gewerbestrasse 11, Cham, Ch-6330, Switzerland

Nevitt MP (2020) On environmental law, climate change, & national security law. Harv. Envtl. l. Rev. 44:321

New M, Hewitson B, Stephenson DB, Tsiga A, Kruger A, Manhique A, Gomez B, Coelho CAS, Masisi DN, Kululanga ME, Adesina F, Saleh H, Kanyanga J, Adosi J, Bulane L, Fortunata L, Mdoka ML, Lajoie R (2006) Evidence of trends in daily climate extremes over Southern and West Africa. J Geophys Res 111:1–11

Niang I, Osman-Elasha B, Githeko A, Yanda PZ, Medany M, Vogel A, Boko M, Tabo R, Nyong A (2008) Africa climate change 2007: Impacts, adaptation and vulnerability: Contribution of working group II to the fourth assessment report of the intergovernmental panel on climate change. Cambridge University Press

Nicholls RJ (1995) Coastal megacities and climate change. GeoJournal 37:369–379

Nicolas JP, Bromwich DH (2014) New reconstruction of antarctic near-surface temperatures: Multi-decadal trends and reliability of global reanalyses. J Clim 27(21):8070–8093. https://doi.org/10.1175/JCLI-D-13-00733.1

Norrant C, Douguédroit A (2006) Monthly and daily precipitation trends in the Mediterranean. Theor Appl Climatol 83:89–106

O'Reilly CM, Alin SR, Plisnier P-D, Cohen AS, McKee BA (2003) Climate change decreases aquatic ecosystem productivity of Lake Tanganyika, Africa. Nature 424:766–768

Ojo O, Oni F, Ogunkunle O (2003) Implications of climatic variability and climate change on water resources availability and water resources management in West Africa. Int Assoc Hydrol Sci Public, 37–47

Oliver D, Wiebe J (2003) Climate change: we are at risk: interim report

Osman-Elasha B, Goutbi N, Spanger-Siegfried E, Dougherty B, Hanafi A, Zakieldeen S, Sanjak A, Atti HA, Elhassan HM (2006) Adaptation strategies to increase human resilience against climate variability and change: lessons from the arid regions of Sudan: assessments of impacts and adaptations to climate change (AIACC) working paper, v 42

Pabón JD (2003) El cambio climático global y su manifestación en Colombia: Cuadernos de Geografía: Revista Colombiana de Geografía, pp 111–119

Penalba OC, Vargas WM (2004) Interdecadal and interannual variations of annual and extreme precipitation over central-northeastern Argentina. Int J Climatol J Royal Meteorol Soc 24:1565–1580

Petersen-Perlman JD, Veilleux JC, Wolf AT (2017) International water conflict and cooperation: challenges and opportunities. Water Int 42:105–120

Pontes GM, Wainer I, Taschetto AS, Gupta AS, Abe-Ouchi A, Brady EC, Chan W-L, Chandan D, Contoux C, Feng R (2020) Drier tropical and subtropical Southern Hemisphere in the mid-Pliocene warm period. Scient Reports 10:1–11

Qin DH (2002) Assessment of environment change in western China, 2nd Volume, Prediction of environment change in western China. Science Press, Beijing, 64, 73, 115, 132, 145–154, 160–161

Quayle WC, Convey P, Peck LS, Ellis-Evans CJ, Butler HG, Peat HJ (2003) Ecological responses of maritime Antarctic lakes to regional climate change. Antarctic Res Series 79:159–170

Quayle WC, Peck LS, Peat H, Ellis-Evans J, Harrigan PR (2002) Extreme responses to climate change in Antarctic lakes.(Climate Change). Science 295:645–646

Rai SC, Gurung T (2005) An overview of glaciers, glacier retreat, and subsequent impacts in Nepal, India and China, p 80

Räisänen J, Alexandersson H (2003) A probabilistic view on recent and near future climate change in Sweden: Tellus A. Dyn Meteorol Oceanogr 55:113–125

Ramirez E, Francou B, Ribstein P, Descloitres M, Guerin R, Mendoza J, Gallaire R, Pouyaud B, Jordan E (2001) Small glaciers disappearing in the tropical Andes: a case-study in Bolivia: Glaciar Chacaltaya (16o S). J Glaciol 47:187–194

Regonda SK, Rajagopalan B, Clark M, Pitlick J (2005) Seasonal cycle shifts in hydroclimatology over the western United States. J Climate 18:372–384

Resurreccion BP, Sajor EE, Fajber E (2008) Climate adaptation in Asia: Knowledge gaps and research issues in South East Asia; full report of the South East Asia team, Institute for Social and Environmental Transition (ISET), Kathmandu, NP

Rivera A, Candela L (2018) Fifteen-year experiences of the internationally shared aquifer resources management initiative (ISARM) of UNESCO at the global scale. J Hydrol Regional Stud 20:5–14

Ronchail J, Bourrel L, Cochonneau G, Vauchel P, Phillips L, Castro A, Guyot J-L, De Oliveira E (2005) Inundations in the Mamoré basin (south-western Amazon—Bolivia) and sea-surface temperature in the Pacific and Atlantic Oceans. J Hydrol 302:223–238

Rowledge L (1999) Global environment outlook 2000: UNEP's Millenium report on the environment. Earthscan Publishers, London, UK, United Nations Environment Programme

Sanso B, Guenni L (1999) Venezuelan Rainfall data analysed by using a Bayesian space-time model. J Royal Stat Soc. Series C (applied Statistics) 48:345–362

Santos FD, Forbes K, Moita R (Eds) (2002) Climate Change in Portugal: scenarios, impacts and adaptation measures. SIAM Project Report, Gradiva, Lisbon, 456 p

Scheumann W, Alker M (2009) Cooperation on Africa's transboundary aquifers—conceptual ideas. Hydrol Sci J 54:793–802

Schröter D, Cramer W, Leemans R, Prentice IC, Araújo MB, Arnell NW, Bondeau A, Bugmann H, Carter TR, Gracia CA, de la Vega-Leinert AC, Erhard M, Ewert F, Glendining M, House JI, Kankaanpää S, Klein RJT, Lavorel S, Lindner M, Metzger MJ, Meyer J, Mitchell TD, Reginster I, Rounsevell M, Sabaté S, Sitch S, Smith B, Smith J, Smith P, Sykes MT, Thonicke K, Thuiller W, Tuck G, Zaehle S, Zierl B (2005) Ecosystem service supply and vulnerability to global change in Europe. Science 310:1333–1337

Schulze R, Hewitson B, Barichievy K, Tadross M, Kunz R, Horan M (2011) Methodological approaches to assessing eco-hydrological responses to climate change in South Africa: WRC Report

Scott D, Jones B (2006) Climate change and seasonality in Canadian outdoor recreation and tourism–executive summary: Waterloo. University of Waterloo, ON

Second assessment of climate change for the Baltic Sea Basin (Ebook), 2015, Springer International Publishing. The BACC II Author Team. 501

Secretariat AT (2017) Final Report of the fortieth Antarctic treaty consultative meeting, Beijing

Shadwick EH, Tilbrook B, Currie KI (2017) Late-summer biogeochemistry in the Mertz Polynya: East Antarctica. J Geophys Res Oceans 122(9):7380–7394. https://doi.org/10.1002/2017jc013015

Shrestha ML, Shrestha AB, 2004, Recent Trends and Potential Climate Change Impacts on Glacier Retreat, Glacier Lakes in Nepal and Potential Adaptation Measures. ENV, EPOC, GF, SD, RD, (2004) 6/FINAL. OECD, Paris, p 23

Silander J, Vehviläinen B, Niemi J, Arosilta A, Dubrovin T, Jormola J, Keskisarja V, Keto A, Lepistö A, Ollila M, Pajula H, Pitkänen H, Sammalkorpi I, Merja S, Veijalainen N (2006), Climate change adaptation for hydrology and water resources. FINADAPT Working Paper 6

Simmonds I, Jones DA, Walland DJ (1998) Multi-decadal climate variability in the Antarctic region and global change. Ann Glaciol 27:617–622

Smith JB, Lenhart SS (1996) Climate change adaptation policy options. Climate Res 6:193–201

Solanes M, Jouravlev A (2006) Water governance for development and sustainability, ECLAC

Soliman SA, Sayedz MA, Jeulands M (2009) Impact assessment of future climate change for the Blue Nile basin using a RCM nested in a GCM. Nile Basin Water Eng Scient Mag

Somlyódy L (2002) Strategic issues of the Hungarian water resources management: Budapest. Acad Sci Hungary 164:8074–79

Sonwa D, Oumarou Farikou M, Martial G, Félix FL (2020) Living under a Fluctuating Climate and a Drying Congo Basin. Sustainability 12:2936

Stewart IT, Cayan DR, Dettinger MD (2005) Changes toward earlier streamflow timing across western North America. J Climate 18:1136–1155

Stouffer RJ, Yin J, Gregory JM, Dixon KW, Spelman MJ, Hurlin W, Weaver AJ, Eby M, Flato GM, Hasumi H, Hu A, Jungclaus JH, Kamenkovich IV, Levermann A, Montoya M, Murakami S, Nawrath S, Oka A, Peltier WR, Robitaille DY, Sokolov A, Vettoretti G, Weber SL (2006) Investigating the causes of the response of the thermohaline circulation to past and future climate changes. J Climate 19:1365–1387

Strzepek K, McCluskey A (2006) District level hydro-climatic time series and scenario analysis to assess the impacts of climate change on regional water resources and agriculture in Africa. Pretoria: Centre for environmental economics and policy in Africa (CEEPA)

Taylor RG, Koussis AD, Tindimugaya C (2009) Groundwater and climate in Africa—a review. Hydrol Sci J 54:655–664
Turner J et al (2016) Absence of 21st century warming on Antarctic Peninsula consistent with natural variability. Nature 535(7612):411–415. https://doi.org/10.1038/nature18645
Twisa S, Buchroithner MF (2019) Seasonal and annual rainfall variability and their impact on rural water supply services in the Wami river basin, Tanzania. Water 11:2055
UNDP (2006) Human development report 2006. Beyond scarcity: power, poverty, and the global water crisis. New York: Palgrave Macmillan, United Nations Development Program (UNDP)
Unganai LS (1996) Historic and future climatic change in Zimbabwe.Climate Research. 6: 137 145. United Nations, 2011. Population Distribution, Urbanization, Internal Migration and Development: An International Perspective. United Nations Department of Economic and Social Affairs Population Division
Urrutia R, Vuille M (2009) Climate change projections for the tropical Andes using a regional climate model: temperature and precipitation simulations for the end of the 21st century. J Geophys Res Atmosph 114
Uwizeyimana D, Mureithi SM, Mvuyekure SM, Karuku G, Kironchi G (2019) Modelling surface runoff using the soil conservation service-curve number method in a drought prone agro-ecological zone in Rwanda. Int Soil Water Conserv Res 7:9–17
Vera C, Higgins W, Amador J, Ambrizzi T, Garreaud R, Gochis D, Gutzler D, Lettenmaier D, Marengo J, Mechoso C (2006) Toward a unified view of the American monsoon systems. J Climate 19:4977–5000
Vörösmarty CJ, Douglas EM, Green PA, Revenga C (2005) Geospatial indicators of emerging water stress: an application to Africa. Ambio 34(3):230–236
Vuille M, Bradley RS, Werner M, Keimig F (2003) 20th century climate change in the tropical Andes: observations and model results, Climate variability and change in high elevation regions: Past, present & future, Springer, pp 75–99
Waldhoff ST, Martinich J, Sarofim M, DeAngelo B, McFarland J, Jantarasami L, Shouse K, Crimmins A, Ohrel S, Li J (2015) Overview of the special issue: a multi-model framework to achieve consistent evaluation of climate change impacts in the United States. Climatic Change 131:1–20
Wall E, Smit B (2005) Climate change adaptation in light of sustainable agriculture. J Sustain Agric 27:113–123
Wall E, Smit B (2006) Agricultural adaptation to climate change in the news. Int J Sustain Dev 9:355–369
Warburton ML, Schulze RE (2005) Chapter 15: detection of climate change: a review of literature on changes in temperature, rainfall and streamflow, on detection methods and data problems, pp257 274. In: Schulze RE, (ed) climate change and water resources in southern Africa: studies on scenarios, impacts, vulnerabilities and adaptation. Water Research Commission, Pretoria, RSA, WRC Report 1430/1/05
Waters D, Watt WE, Marsalek J, Anderson BC (2003) Adaptation of a storm drainage system to accommodate increased rainfall resulting from climate change. J Environ Planning Manag 46:755–770
Wheaton E, Wittrock V, Kulshreshtha S, Koshida G, Grant C, Chipanshi A, Bonsal B, Adkins P, Bell G, Brown G (2005) Lessons learned from the Canadian drought years of 2001 and 2002: synthesis report, Saskatchewan Research Council Publication No
Wheeler HY (2016) Natural history on the net, https://www.naturalhistoryonthenet.com/Continents/. Last accessed Thursday, July 21, 2016
Wilkinson R, Clarke K, Goodchild M, Reichman J, Dozier J (2002) The potential consequences of climate variability and change for California: the California regional assessment. US Global Change Research Program, Washington, DC
Wu F, Pennings SC, Tong C, Xu Y (2020) Variation in microplastics composition at small spatial and temporal scales in a tidal flat of the Yangtze Estuary, China. Sci Total Environ 699:134252
Xu Y, Seward P, Gaye C, Lin L, Olago DO (2019) Preface: groundwater in sub-Saharan Africa. Hydrogeol J 27:815–822

XXXIX A (2016) Final report of the thirty-ninth antarctic treaty consultative meeting: Santiago
Zeitoun M, Mirumachi N (2008) Transboundary water interaction I: reconsidering conflict and cooperation: international environmental agreements: politics. Law Econ 8:297–316
Zeitoun M, Warner J (2006) Hydro-hegemony–a framework for analysis of trans-boundary water conflicts. Water Policy 8:435–460

Summary

Climate change is one of the main environmental challenges facing the world today. The water resources are increasingly being stressed due to the climate change variability and increased demand over the continents. According to existing reports and studies, climate change has had a profound effect on the planet, such as reduced snow cover and glacier mass in the high mountains of Africa, increased frequency of heavy rainfall in many parts of Asia that leading to major floods, In Central and Eastern Europe, groundwater recharge is limited and the risk of winter floods increases in northern Europe. The effects of climate change on North American water resources include warming in the western mountains of the United States and Canada, leading to reduced snowfall, winter flooding, and reduced summer currents, as well as severe competition for more water resources. In South America, the disappearance of glaciers and changes in rainfall patterns have a major impact on water supply for human use, agriculture and energy production. The Antarctic ice sheet lost its mass as a result of the rapid thinning and retreat of the main outlet glaciers reducing the West Antarctic Ice Sheet. The results of researches and studies also predict worrying prospects for the future. A useful solution in terms of the impact of climate change on water resources and its management is to develop methods related to water resources and provide a conceptual model on the effects of climate change on water resources, vulnerability and adaptation with it. Hydrological modeling can be defined as the description of real hydrological features, which uses small-scale physical models, mathematical analogs, and computer simulations to solve a hydrology issue. Hydrological models are helpful to simulate the process of rain-runoff and hydraulic models to analyze the runoff flow in the river and study how it spreads. Climate change adaptation and mitigation at local, regional, and global scales at short, medium, and long-term levels are needed to achieve sustainable development goals and minimizing the effect of climate change on water resources. A review of different countries' experiences can help stakeholders in developing a comprehensive plan to deal with, mitigate, and adapt to the current and future consequences of climate change.

Correction to: Introduction to Key Features of Climate Models

Mahsa Jahandideh Tehrani, Omid Bozorg-Haddad, Santosh Murlidhar Pingale, Mohammed Achite, and Vijay P. Singh

Correction to:
Chapter 6 in: O. Bozorg-Haddad (ed.), *Climate Change in Sustainable Water Resources Management*, Springer Water, https://doi.org/10.1007/978-981-19-1898-8_6

In the original version of the book, the author name has been updated from "Santosh Murlidhar Pingal" to "Santosh Murlidhar Pingale" in the Chapter "Introduction to Key Features of Climate Models". The chapter and book have been updated with the changes.

The updated version of this chapter can be found at
https://doi.org/10.1007/978-981-19-1898-8_6

O. Bozorg-Haddad (ed.), *Climate Change in Sustainable Water Resources Management*,
Springer Water, https://doi.org/10.1007/978-981-19-1898-8_11

The manufacturer's authorised representative in the EU is Springer Nature Customer Service Centre GmbH, Europaplatz 3, 69115 Heidelberg, Germany. If you have any concerns regarding our products, please ontact ProductSafety@springernature.com

'ed and bound by CPI Group (UK) Ltd, Croydon, CR0 4YY
30/09/2026
02225250-0015